Tectonic Geomorphology

Tectonic Geomorphology

Editor:

Navodita Bhatnagar

Delve Publishing

www.delvepublishing.com

Tectonic Geomorphology

Editor: Navodita Bhatnagar

708 3rd Avenue, 6th Floor
New York, NY 10017
United States of America
www.delvepublishing.com

ISBN: 978-1-68095-834-8

Library of Congress Control Number: 2016951576

About the Editor

Navodita Bhatnagar

Navodita Bhatnagar finished her masters in technology and currently pursuing her PhD in environmental sciences from Institute of Technology, Carlow, Ireland. She's a keen enthusiast of environment protection and management. An alumunus of Indo-German centre for sustainibility which is essentially a collaborative interface to gather scientists in the field of environment sciences to come together and work for the common cause which is environment protection.

Contents

List of Contributors

S. Hergarten
Institut für Geo- und Umweltnaturwissenschaften, Albert-Ludwigs-Universität Freiburg, Freiburg im Breisgau, Germany

J. Robl
Institut für Geographie und Geologie, Universität Salzburg, Salzburg, Austria

K. Stüwe
Institut für Erdwissenschaften, NAWI Graz, Karl-Franzens-Universität Graz, Graz, Austria

T. J. Coulthard
Department of Geography, Environment and Earth Sciences, University of Hull, UK

M. J. Van de Wiel
Department of Geography, University of Western Ontario, London, Ontario, Canada

Mehran Arian
Department of Geology, Science and Research Branch, Islamic Azad University, Tehran, Iran

Hamideh Noroozpour
Department of Geology, Faculty of Science, Payame Noor University, Tehran, Iran

Engielle M. R. Paguican
Department of Geology, University at Buffalo, The State University of New York, Buffalo, NY, USA
Department of Geography, Earth System Sciences, Vrije Universiteit Brussel, Brussels, Belgium

Marcus I. Bursik
Department of Geology, University at Buffalo, The State University of New York, Buffalo, NY, USA

Elahe Javadi Mosavi
Department of Geology, Science and Research Branch, Islamic Azad University, Tehran, Iran

Mehran Arian*
Department of Geology, Science and Research Branch, Islamic Azad University, Tehran, Iran

Jiren Xu and Zhixin Zhao
Institute of Geology, Chinese Academy of Geological Sciences, Beijing 100037, China

Lev Eppelbaum
Department of Geophysics and Planetary Sciences, Raymond and Beverly Sackler Faculty of Exact Sciences, Tel Aviv University, Ramat Aviv, Israel;

Youri Katz
Paleontological Division of the Zoological Museum, Department of Zoology, Faculty of Life Sciences, Tel Aviv University, Ramat Aviv, Israel.

L. Y. Kagho
Laboratoire de Mécanique et de Modélisation des Systèmes Physiques (L2MSP), Faculté des Sciences, Université de Dschang, BP 69, Dschang, Cameroon

M. W. Dongmo
Laboratoire de Mécanique et de Modélisation des Systèmes Physiques (L2MSP), Faculté des Sciences, Université de Dschang, BP 69, Dschang, Cameroon

F. B. Pelap
Laboratoire de Mécanique et de Modélisation des Systèmes Physiques (L2MSP), Faculté des Sciences, Université de Dschang, BP 69, Dschang, Cameroon

Alicia García
Institute IGEO, CSIC-UCM, J. Gutierrez Abascal, 2, Madrid 28006, Spain

Servando De la Cruz-Reyna
Instituto de Geofísica, Universidad Nacional Autónoma de México, C. Universitaria, México D.F. 04510, México
CUEIV, Universidad de Colima, Colima 28045, México

José M. Marrero
Instituto Geofísico, Escuela Politécnica Nacional, Ladrón de Guevara, P.O. Box 1701-2759, Quito, Ecuador
Unidad de Gestión, Investigación y Desarrollo, Instituto Geográfico Militar, El Dorado, 170413, Quito, Ecuador

Ramón Ortiz
Institute IGEO, CSIC-UCM, J. Gutierrez Abascal, 2, Madrid 28006, Spain

M. K. Kachakhidze
Department of Natural Sciences, St. Andrew The First-Called Georgian University of The Patriarchy of Georgia, Tbilisi, Georgia

R. Kiladze
Department of Natural Sciences, Ilia State University, Abastumani Astrophysical Observatory, Tbilisi, Georgia

N. Kachakhidze
Department of Natural Sciences, St. Andrew The First-Called Georgian University of The Patriarchy of Georgia, Tbilisi, Georgia

V. Kukhianidze
Department of Natural Sciences, Ilia State University, Abastumani Astrophysical Observatory, Tbilisi, Georgia

G. Ramishvili
Department of Natural Sciences, Ilia State University, Abastumani Astrophysical Observatory, Tbilisi, Georgia

U. Lombardo
Universität Bern, Geographisches Institut, Hallerstrasse 12, 3012 Bern, Switzerland

Girish Ch. Kothyari and B. K. Rastogi
Institute of Seismological Research, Raisan, Gandhinagar, Gujarat 382 009, India

B. K. Rastogi
Institute of Seismological Research, Raisan, Gandhinagar, Gujarat 382 009, India

A. Margirier
Institut des Sciences de la Terre, Université Grenoble I, CNRS, 1381 rue de la Piscine, 38400 Grenoble CEDEX 09, France

L. Audin
Institut des Sciences de la Terre, Université Grenoble I, CNRS, 1381 rue de la Piscine, 38400 Grenoble CEDEX 09, France
Institut de recherche pour le développement, Institut des Sciences de la Terre, 1381 rue de la Piscine, 38400 Grenoble

J. Carcaillet
Institut des Sciences de la Terre, Université Grenoble I, CNRS, 1381 rue de la Piscine, 38400 Grenoble CEDEX 09, France

S. Schwartz
Institut des Sciences de la Terre, Université Grenoble I, CNRS, 1381 rue de la Piscine, 38400 Grenoble CEDEX 09, France

C. Benavente
Instituto Geológico Minero y Metalúrgico, A. Canadà 1470, San Borja, Lima, Peru

G. De Guidi and S. Scudero
University of Catania, Department of Biological, Geological and Environmental Sciences, Earth Science Section, Corso Italia 57, 95129, Catania, Italy

S. Scudero
University of Catania, Department of Biological, Geological and Environmental Sciences, Earth Science Section, Corso Italia 57, 95129, Catania, Italy

Hongyuan Zhang
State Key Laboratory of Geological Processes and Mineral Resources, China University of Geosciences, Beijing 100083, China
Faculties of Earth Sciences and Resources, China University of Geosciences, Beijing 100083, China

Junlai Liu
State Key Laboratory of Geological Processes and Mineral Resources, China University of Geosciences, Beijing 100083, China
Faculties of Earth Sciences and Resources, China University of Geosciences, Beijing 100083, China

Wenbin Wu
Faculties of Earth Sciences and Resources, China University of Geosciences, Beijing 100083, China

Fenghua An
National Engineering Research Center for Coal & Gas Control, Faculty of Safety Engineering, China University of Mining & Technology, Xuzhou 221116, China

Yuanping Cheng
National Engineering Research Center for Coal & Gas Control, Faculty of Safety Engineering, China University of Mining & Technology, Xuzhou 221116, China
State Key Laboratory of Coal Resources and Mine Safety, China University of Mining & Technology, Xuzhou 221116, China

H. Perea
LATTEX – IDL, GeoFCUL, Universidade de Lisboa, Ed.C6, Campo Grande 1749-016 Lisbon, Portugal
Unitat de Tecnologia Marina – CSIC, Centre Mediterrani d'Investigacions Marines i Ambientals, Psg. Mar′ıtim de la Barceloneta, 37–49, 08003, Barcelona, Spain

E. Gracia
Unitat de Tecnologia Marina – CSIC, Centre Mediterrani d'Investigacions Marines i Ambientals, Psg. Mar′ıtim de la Barceloneta, 37–49, 08003, Barcelona, Spain

P. Alfaro
Dept. de Ciencias de la Tierra y del Medio Ambiente, Universidad de Alicante, Apdo. 99, 03080 Alicante, Spain

R. Bartolome
Unitat de Tecnologia Marina – CSIC, Centre Mediterrani d'Investigacions Marines i Ambientals, Psg. Mar′ıtim de la Barceloneta, 37–49, 08003, Barcelona, Spain

C. Lo Iacono
Unitat de Tecnologia Marina – CSIC, Centre Mediterrani d'Investigacions Marines i Ambientals, Psg. Mar′ıtim de la Barceloneta, 37–49, 08003, Barcelona, Spain

X. Moreno
Unitat de Tecnologia Marina – CSIC, Centre Mediterrani d'Investigacions Marines i Ambientals, Psg. Mar′ıtim de la Barceloneta, 37–49, 08003, Barcelona, Spain

E. Masana
RISKNAT, Dept. Geodinamica i Geof` ′ısica, Universitat de Barcelona, Mart′ı Franques, s/n, 08028 Barcelona, Spain

EVENT-SHELF Team
M. Farran (ICM-CSIC), E. Andara (IGME), S. Perez and M. Rom ′ an Alpiste (UG)

Quanlin Hou
Graduate University of Chinese Academy of Sciences, Beijing 100049, China

Qing Liu,
Graduate University of Chinese Academy of Sciences, Beijing 100049, China

Hongyuan Zhang
School of Earth Sciences and Resources, China University of Geosciences, Beijing 100083, China

Xiaohui Zhang
Institute of Geology and Geophysics, Chinese Academy of Sciences, Beijing 100029, China

Jun Li
Institute of Geology and Geophysics, Chinese Academy of Sciences, Beijing 100029, China

A. Argnani
ISMAR-CNR, Istituto di Scienze Marine, Sede di Bologna, Via Gobetti 101, 40129 Bologna, Italy

A. Armigliato
University of Bologna, Department of Physics, Sector of Geophysics, Viale Carlo Berti Pichat 8, 40127 Bologna, Italy

G. Pagnoni
University of Bologna, Department of Physics, Sector of Geophysics, Viale Carlo Berti Pichat 8, 40127 Bologna, Italy

F. Zaniboni
University of Bologna, Department of Physics, Sector of Geophysics, Viale Carlo Berti Pichat 8, 40127 Bologna, Italy

S. Tinti
University of Bologna, Department of Physics, Sector of Geophysics, Viale Carlo Berti Pichat 8, 40127 Bologna, Italy

C. Bonazzi
ISMAR-CNR, Istituto di Scienze Marine, Sede di Bologna, Via Gobetti 101, 40129 Bologna, Italy

Preface

Tectonic geomorphology is the study of the interplay between tectonic and surface processes that shape the landscape in regions of active deformation and at time scales ranging from days to millions of years. Tectonics is the branch of geology relating to the structure of the Earth's crust and the large-scale processes which take place within it. The field of tectonic geomorphology is in a state of tension. The widespread availability of high-quality, high-resolution digital topographic data encourages the development of simple morphological 'tools' which can be used to comprehend recent tectonic evolution. Meanwhile, process geomorphologists recognize that current models have a noteworthy empirical basis, and lack insight into the underlying physics of erosion processes. Most tectonic geomorphology research is concerned with rivers, but glaciers, debris flows and hill slope processes also play a key role in hypotheses linking climate to tectonics, through surface processes, while submarine geomorphology has hardly been investigated in a tectonic context. Over the past decade, recent advances in the quantification of both rates and the physical basis of tectonic and surface processes have underpinned an explosion of new research in the field of tectonic geomorphology. Modern tectonic geomorphology is an exceptionally integrative field that utilizes techniques and data derived from studies of geomorphology, seismology, geochronology, structure, geodesy, stratigraphy, meteorology and Quaternary science.

This book entitled with Tectonic Geomorphology focuses on the current understanding of the dynamic interplay between surface processes and active tectonics. As it ranges from the timescales of individual earthquakes to the growth and decay of mountain belts. Syntheses of recent and earlier works, combined with new results and interpretations, are illustrated for varied tectonic settings. Most of chapters contain up-to-date materials of comprehensive geological-geophysical investigations, which can help more clearly understand the essence of mechanisms of different tectonic processes. General problems of tectonics are discussed and the approaches required for further progress have been outlined. Over the past several decades, significant advancements in digital tools and dating techniques including increased quality and accessibility of high-resolution digital topographic data, and innovation and refinement of geochronology, low-temperature thermochronolgy, and surface dating, has led to an explosion of high quality tectonic geomorphology research. Moreover, field and GIS-based studies are increasingly paired with sophisticated numerical models to explore landscape evolution, geodynamics, structural, and time-temperature histories from the scale of individual faults and folds, to mountain ranges, or entire orogens. In this book we solicit contributions that address this subject, particularly those integrating multiple methods and applying innovative approaches. Tectonic Geomorphology will be of valuable for upper-level undergraduate and graduate earth science students and for practicing geologists and engineers because it applies concepts taught in geochemistry, sedimentology, seismology, soils genesis, stratigraphy, structural geology, and tectonics courses in an interdisciplinary manner that has obvious environmental applications.

Editor
Navodita Bhatnagar

INTRODUCTION

Geomorphology is the scientific study of the surface of a planet and those processes responsible for forming it. Scientists involved in this field often study historical changes, through events such as erosion, in order to understand how a particular geographical region came into existence. They may also study current data to better predict how landforms might change in the future and to understand how people can help maintain current features. This allows scientists to anticipate changes in the general structure of the earth.

Natural Processes of Change

Landforms on any world, including earth, are not static; they are part of a dynamically changing system. There are various geomorphic processes that can alter the surface of a world, including plate tectonics, changes in climate, and human activities. Wind can shape landscapes, as can water — both liquid and ice, in the form of glaciers. Volcanic activity, including violent eruptions and the steady flow of lava from some sites, can create new islands or devastate a landscape. Plants and animals can also alter landforms, whether a beaver damming a river or a grove of trees that anchor the soil in a particular location.

Tectonic Changes

Slow movements of the earth's tectonic plates contribute to the uplift and elevation of landforms. There are two common types of tectonic uplift: orogenic and isostatic. Orogenic tectonic uplift is caused when tectonic plates crash together, which raises the land where they meet to create forms such as mountains. Isostatic uplift, on the other hand, refers to how landforms can become higher after the weight on the land is reduced; as land is eroded or glaciers melt, it is believed that the land that was being weighed down can rise.

Changes Caused by Water

The geomorphic effect of water bodies are studied in fluvial geomorphology, which examines how bodies of water alter the landscape. As waterways such as rivers flow, they often carry sediment, which reduces the land around the river itself but increases areas where this sediment is released. Water from rain and flash floods can also be responsible for erosion, which physically alters rocks and other land areas.

Changes Caused by Glaciers

Glaciers also change the landscape. As these heavy sheets of ice advanced across the landscape during the last ice age, they scoured the softer land areas in their way; they also picked up some of this material and moved it. When the ice melted, valleys and fjords — coastal valleys that are filled with water — were left behind, as were the rocks and soil, called "till," that the glacier picked up.

Volcanic Changes

On the opposite side, volcanoes can both create and destroy landforms. Often found at the edges of tectonic plates, underwater volcanoes have shaped islands like Hawaii, the Philippine Islands, and New Zealand. On land, they can form large volcanic mountains. The violent explosion of a volcano can radically change the landscape, and wipe out plants and animals in the area.

Changes Caused by Wind

Although it often works much more slowly, wind can also alter the land. Called eolian geomorphology, wind can erode landforms, breaking them down, and build others up, as material is moved from one place to another. The Nebraska Sand Hills, for example, is an area where ancient winds created huge sand dunes that have since stabilized and became a regular part of the landscape.

Bio geomorphology

Plants and animals can have a big impact on the landscape as well. Animals dig tunnels and dens, move rocks and soil, and block rivers, among other things. Plant roots can grow through the cracks in rocks, breaking them apart, or help to hold the soil in an area together, decreasing erosion caused by water and wind. Living things can also combine with other forces to cause changes; a volcanic eruption may destroy a stand of trees, for example, leaving the land in the area exposed to the weathering caused by wind and rain.

Changes Caused by People

Human interventions can also contribute to changes on the earth. With the expansion of civilization, humans began to enact direct changes to their surroundings. The most radical changes to landforms are possible due to technological and organizational advances; the building of the Panama or Suez canals, for example, were significant alterations to the earth's natural form. People have straightened rivers or prevented them from naturally changing their course, created lakes and other bodies of water, and prevented beaches from expanding or eroding in some cases. The long-term effects of many of the changes that human beings have made is not fully known, and it may take centuries for the side effects — good and bad — to become fully clear.

Geomorphology on Other Planets

Geomorphology is not restricted to questions about landforms on earth; it applies to all terrestrial planets. The field of "extraterrestrial geomorphology" is expanding due to the influx of scientific data from satellites and space expeditions. For example, volcanic eruptions on the surface of Io, one of Jupiter's moons, have created many unique features, including tall mountains and flat plains. Scientists studying Mars, Venus, and other planets examine formations like channels and valleys, and theorize about the different processes that might have created them.

Related Fields of Study

A number of different research methods and fields of inquiry are often used for this type of study. Archaeology, for example, can be invaluable in understanding how past human populations have changed and shaped the environment and geography. Study of global surveys through ground and orbital photography is also beneficial, as it allows geomorphologists to gain a better perspective of various landforms. Soil scientists investigate the composition and formation of soils, which helps explain how an area has changed over time.

Geomorphology is also used in a number of other fields. Civil engineers, for example, build and maintain structures like roads, bridges, and dams; understanding how the landscape was formed and how it may change is crucial for such projects. Environmental resource management involves finding ways to make the best use of resources, including water and land, so an understanding of how human activities can change those resources is vital.

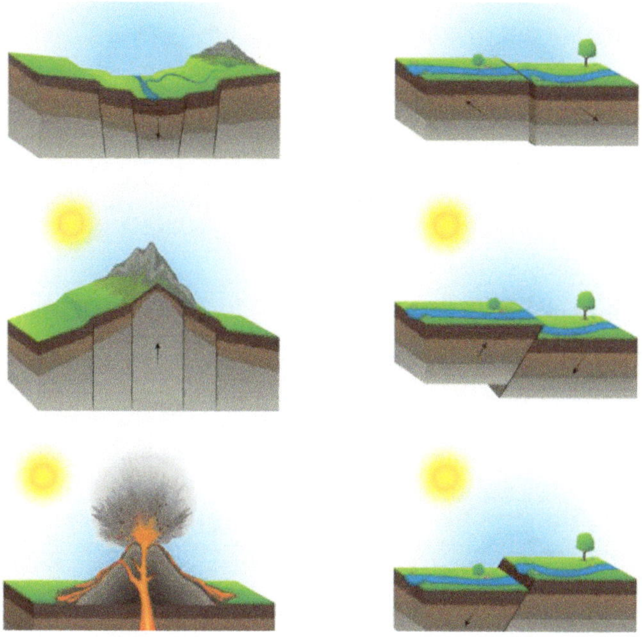

Significance of Plate Tectonics

There is a good reason why Plate Tectonics is the most important unifying theory in geology. In addition to determining organic processes, the assembly of continents, the formation and destruction of ocean basins, and the distribution of most earthquakes and volcanoes, Plate Tectonics also strongly influences:

- global climate by affecting
    1. Ocean and atmospheric circulation
    2. Degree of continentally
    3. rates of weathering related to uplift - (weathering removes CO_2 and promotes cooling)
- tidal range, and wave energy reaching the coast - influenced by the width of the continental shelf
- the size, orientation, and gradients of large drainage basins (e.g. South America)
- Large scale transgressed and regressive phases, which are determined by rates of sea-floor spreading. Regression is associated with slower rates of spreading.
- glaciation by positioning continents at high latitudes

Gravity and isostasy

Isostasy is the theory that the lithosphere floats on the denser asthenosphere, and that surface elevations are largely a function of variations in crustal density and thickness. Isostacy accounts for elevation differences between continents and ocean basins and between the lower order features they contain.

- Pratt Theory: Crustal density block model
- Airy Theory: Crustal thickness block model
- Composite model: Elevation is related to both crustal thickness and density
- Flexural isostasy: Crust behaves like a flexible sheet that warps in response to changes in stress. (e.g. peripheral bulge, foreland basin, glacial rebound, etc)

Isostatic anomalies resulting in uplift or subsidence are caused by loading or unloading (e.g. glacial expansion and retreat, or erosion and deposition), sub-lithospheric flow, and crustal thickening through compression or under plating, and thinning through extension. Rather than acting like a series of blocks that bob up and down the crust behaves more like a semi-rigid sheet that warps in response to changes in stress, this is known as flexural isostasy.

Global (Eustatic) and Local (Isostatic) Sea Level Fluctuations

Erosion and deposition are intricately linked to base level, which is ultimately controlled by sea level. Decreasing sea level initiates erosion whereas rising sea level drives deposition.

Eustatic, or world-wide changes in sea level will occur if

1. there is a fluctuation in the volume of water in the ocean, or
2. there is a change in the size of the oceanic basin

Volume changes are climate driven, whereas basin size is driven by rates of plate motion and isostasy. The most important climatic event influencing sea level is glaciation. Less water is held in the oceans when continental glaciers are at their maximum. As glaciers melt sea level rises. Eustatic rises in sea level are also linked to rapid rates of sea floor spreading and subduction, leading to more buoyant crust on the ocean floor and shallower basins. On the other hand, slower rates of crustal turnover leads to a higher percentage of older, denser oceanic crust and deeper basins.

Isostatic sea level changes are local relative fluctuations related to some isotactic adjustment of the land. For example glacial rebound may cause relative sea level to drop even though ecstatic sea level is rising.

Review of plate tectonics

Classification of plate boundaries: A boundary is classified by its relative plate motion and the type of crust in contact across it.

Table 1. Basic classification of tectonic plate boundaries.

CONVERGENT	DIVERGENT	TRANSFORM
oceanic-oceanic (Philippines)	**oceanic-oceanic** (Mid-Oceanic Ridges and Rises)	(San Andreas and Levant Fault, Middle East, Anatolian Fault, Turkey)
continental-continental (Alps, Himalayan mountain belt)	**continental-continental** (Africa, Rift system, Basin and Range)	
oceanic-continental (western Cordillera of North and South America)		

- Stresses are generally not purely tensional, compressional, or shear (e.g. transtensional and transpersonal -- both are associated with the San Andreas fault).

Driving mechanisms of plate motion

- slab pull - most important
- ridge push
- gravity sliding

Plate Margins and Related Landforms

Oceanic-Oceanic Convergence: (Intra-oceanic arc) When two oceanic plates converge the older, denser plate will typically subduct.

Morphological elements of intra-oceanic arcs

- Backarc zone (backarc basin)
  - Formation of backarc basins
  - entrapment: oceanic crust behind the arc is old
  - backarc spreading: spreading is initiated by upwelling along a secondary convection cell (e.g. Sea of Japan)
- Volcanic arc zone: composed of active region and frontal anticline
  - Andesitic volcanism predominates unless arc basement is continental

    - volcanic zone is located 80 to 150 km above and 75-175 km landward of the Wadati-Benioff zone. Distance is a function of subduction angle (30-50°)
- volcanism results partial melting of overlying mantle

Forearc zone:

- frontal arc: older uplifted accretionary prism rocks
- forearc basin (flysch--turbidites)
- accretionary prism (outer-arc ridge): imbricate thrust packages
- trench: the thickness of turbidite fill (flysch) is dependent on proximity to sediment source and sediment entrapment by neighboring basins
- peripheral bulge or outer swell (linear upwarp caused by an upward flexure of the crust fronting the trench)

Oceanic-Continental Convergence

Oceanic crust invariably subducts beneath continental crust which is too buoyant to sink. Thickening along the orogenic belt is related to shortening by plastic flow, underplating of magma, and thrust faulting.

CHAPTER 1 TECTONIC GEOMORPHOLOGY AT SMALL CATCHMENT SIZES

S. Hergarten, J. Robl, and K. Stüwe

INTRODUCTION

The vast majority of the approaches used to derive information on tectonic processes from topography are based on the analysis of longitudinal river profiles. The fundamental relationship between channel slope S and upstream catchment size A,

$$S = k_S A^{-\theta} \tag{1}$$

which is used to infer such information, dates back to a seminal empirical study of Hack (1957) and is often referred to as Flint's law (Flint, 1974). The parameters k_s and θ denote steepness index and concavity index, respectively.

Understanding and quantitative interpretation of Eq. (1) hinges on the stream-power approach (e.g., Howard, 1994;

Whipple and Tucker, 1999; Whipple, 2004; Wobus et al., 2006), where it is assumed that the rate of fluvial erosion in a bedrock channel depends on the product $A^{\theta}S$. In this context, Eq. (1) reflects a constant erosion rate along the river as it occurs, e.g., in equilibrium with homogeneous uplift. In the simplest version of the stream-power approach it is assumed that the erosion rate E is linearly proportional to $A^{\theta}S$. The more general approach implements a power-law relationship

$$E = K\left(\left(\frac{A}{A_0}\right)^{\theta} S\right)^{n} \tag{2}$$

where K denotes erodibility. The arbitrary reference catchment size A_0 has been introduced as a scaling parameter in order to avoid an odd physical dimension of K. Using this scaling, K describes the erosion rate at a catchment size A0 and a (hypothetical) channel slope of 1. Although called erodibility, K not only refers to the properties of the channel bed but also contains the effect of precipitation, as the erosion rate in principle depends on the discharge instead of the catchment size.

Physically based models of bedrock incision suggest that the concavity index θ of a steady-state bedrock river under homogeneous conditions depends not only on the constitutive laws of the erosion process but also on the crosssectional geometry of the channels (e.g., Whipple, 2004; Whipple et al., 2013; Lague, 2014).

This explains some variation in θ around the value θ ≈ 0.5 originally found by Hack (1957) or around the reference value θ_{ref} = 0.45 being widely assumed for perfect bedrock channels under homogenous steady-state conditions (Whipple et al., 2013; Lague, 2014). A range of θ between about 0.4 and 0.7 has been found under relatively homogeneous conditions (e.g., Whipple, 2004; Whipple et al., 2013), while a wider range from less than 0.2 in steep headwater channels to more than 1 in some alluvial channels has been reported (Brummer and Montgomery, 2003; Montgomery, 2001; Sofia et al., 2015). Apparent variations in θ may also arise from spatial inhomogeneity or non-steady topography. Analyzing channel slopes at constant catchment sizes, Hergarten et al. (2010) found a strong positive correlation between surface elevation and slope in several orogens, suggesting a correlation between uplift rate and elevation. This correlation will lead to a higher apparent concavity index when following individual rivers, which may explain why the majority of the values of θ found in nature are greater than θ_{ref} = 0.45.

Compared to the concavity index θ, less is known about the exponent n as it cannot be determined from individual equilibrium river profiles under uniform conditions. According to Eq. (4), the exponent n can be determined by comparing

Citation: Hergarten, S., Robl, J., & Stüwe, K. (2015). Tectonic geomorphology at small catchment sizes, 3(3), 689-714. http://dx.doi.org/10.5194/esurfd-3-689-2015.

river segments being in equilibrium with different uplift rates, and the results tentatively suggest that n should not be far away from 1 (Wobus et al., 2006). Using Eq. (2), the evolution of the surface elevation H(x,t) along the stream profile through time under a given uplift rate U follows the partial differential equation

$$\frac{\partial H}{\partial t} = U - K\left(\left(\frac{A}{A_0}\right)^{\theta}\frac{\partial H}{\partial x}\right)^{n} \quad (3)$$

where the linear coordinate x follows the upstream direction of the considered river. Both U and K may vary spatially and temporally.

The simplest interpretation of Eq. (3) refers to steady-state topography where uplift and erosion are in local equilibrium. Under these conditions, the ratio of uplift rate and erodibility can be directly obtained from the steepness index (Eq. 1) according to

$$\frac{U}{K} = \left(\frac{k_s}{A_0^{\theta}}\right)^{n} \quad (4)$$

The most interesting applications of the stream-power erosion equation (Eq. 3), however, concern non-equilibrium river profiles due to temporally changing uplift rates or due to climate-induced changes in the erodibility. If such changes are discontinuous, they result in distinct knick points propagating in the upstream direction.

THE χ TRANSFORMATION AND ITS LIMITATION

Recently, the so-called χ plot (or χ transformation) has introduced the perhaps most important methodic progress in evaluating and interpreting longitudinal river profiles since the seminal work of Howard (1994). It transforms the upstream coordinate x into a new coordinate χ in such a way that the inherent curvature of equilibrium profiles due to the reduction in catchment size in the upstream direction vanishes. The catchment size A can be eliminated from Eq. (3) if the transformation satisfies the condition

$$\frac{dx}{d\chi} = \left(\frac{A}{A_0}\right)^{\theta} \quad (5)$$

which can be achieved by

$$\chi(x) = \int_{x_0}^{x}\left(\frac{A(\xi)}{A_0}\right)^{-\theta} d\xi \quad (6)$$

where x_0 is an arbitrary reference point. As the channel slope is $S = \partial H/\partial x$, the erosion rate (Eq. 2) can be written in the form

$$E = K\left(\frac{dx}{d\chi}\frac{\partial H}{\partial x}\right)^{n} \quad (7)$$

$$= K\left(\frac{\partial H}{\partial \chi}\right)^{n} \quad (8)$$

Thus, the local erosion rates is directly related to the slope of the river profile in the H vs. χ representation, and Eq. (3) simplifies to

$$\frac{\partial H}{\partial t} = U - K\left(\frac{\partial H}{\partial \chi}\right)^{n} \quad (9)$$

The solutions of this equation and their potential for unraveling the uplift and erosion history have been discussed by Royden and Perron (2013), and a formal inversion procedure for the linear case (n = 1) has been presented by Goren et al. (2014).

The most striking property of the χ transformation is immediately recognized in Eq. (9): if U and K are spatially homogeneous, all upstream paths starting from x0 are described by the same differential equation, so that the H vs. χ curves of all tributaries must collapse with the H vs. χ curve of the main stream. Conversely, spatial inhomogeneity results in a deviation of the curves belonging to different branches that increases in the upstream direction. Thus, a narrow bunch of H vs. χ curves with a nonlinear overall shape is the fingerprint of temporal variations under spatially homogeneous conditions, while a wide but overall straight bunch points towards spatial heterogeneity under steady-state conditions. This simple interpretation, however, only holds as long as the drainage pattern has not changed in the past, since changes in catchment sizes also result in deviations between different branches (Willett et al., 2014; Yang et al., 2015). Since a clear distinction requires the consideration of a large number of tributaries, the inherent limitation of the stream-power approach to the fluvial regime also limits the χ method.

As addressed in several studies, Flint's law (Eq. 1), and thus the stream-power erosion equation (Eq. 2) with a constant concavity index θ, breaks down at small catchment sizes where lower limits between about 0.1 and 5 km2 have been reported (Montgomery and Foufoula-Georgiou, 1993; Stock and Dietrich, 2003; Wobus et al., 2006). The transition from a fluvial regime at large catchment sizes to a regime dominated by hillslope processes is explored by an example from Taiwan in Fig. 1. Based on the recently released SRTM1 digital elevation model (DEM) with a mesh width of 1 arcsecond, flow directions (D8 algorithm; O'Callaghan and Mark, 1984), catchment sizes, and channel slopes were computed for the entire island after filling all local depressions. For comparison, the same analysis was performed on the older SRTM3 DEM with a mesh width of 3 arcseconds.

The mean slope (black markers) follows Eq. (1) well above a catchment size of a few square kilometers with a steepness index $\theta_{ref} = 0.45$. Clear deviations from this behavior are visible at catchment sizes below about 2 km2 in the Taiwan data set. These deviations are even more pronounced in the finer SRTM1 data set than in the SRTM3 data set, suggesting that they indeed arise from a limitation of the stream-power approach and not from the potentially inadequate representation of the drainage pattern on coarse DEMs (Stock and Dietrich, 2003). On the other hand, the number of nodes with a catchment size of A or larger roughly decreases like A−0.5 (Maritan et al., 1996). For a DEM with a mesh width of 1 arcsecond, this means that only some 2 % of all DEM nodes have a catchment size $A \geq 2$ km^2 thus, about 98 % of all nodes cannot be used in the χ method here. Therefore, an extension of the χ method towards smaller catchment sizes is desirable for deriving maps of χ in order to separate temporal variations from effects of spatial heterogeneity. Beyond increasing the pure data density, it is also helpful with regard to the contest of catchments brought into discussion by Willett et al. (2014). Growing or shrinking catchments are characterized by the curvature in the H vs. χ plot becoming more and more significant close to the migrating drainage divide. Therefore,

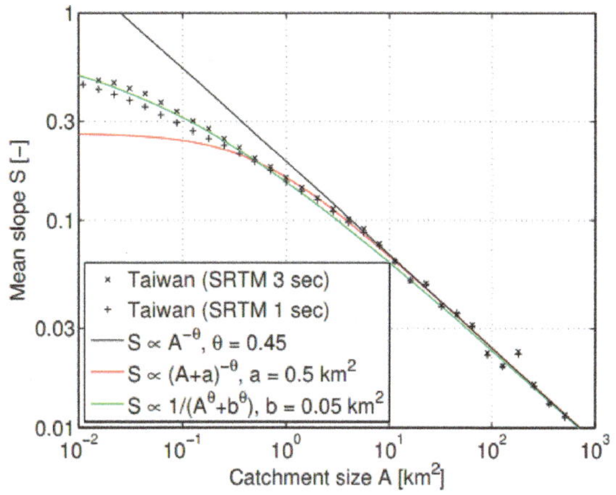

Figure 1. Relationship between mean channel slope and catchment size for the topography of Taiwan. Channel slopes and catchment sizes were derived from the SRTM1 and SRTM3 DEMs, and mean slopes were obtained from logarithmic bins with a factor of √2 (black markers).

analyzing the migration of drainage divides quantitatively requires a χ transform free of any bias at small catchment sizes induced by the limited applicability of the stream-power law.

EXTENDING THE χ METHOD TO SMALL CATCHMENT SIZES

In the following we present two extensions of the basic relationship between channel slope and catchment size (Eq. 1) towards small catchment sizes and their implementation in the χ method. In most applications of the χ method, the concavity index θ is considered an adjustable parameter and used to improve either the straightness of the H vs. χ plot or the collinearity with tributaries. In the following, the approach with adjustable concavity index θ is denoted χ_θ, while χ represents the version with the reference value $\theta_{ref} = 0.45$.

However, the curvature of the data in Fig. 1 already suggests that the adjustment of θ may only introduce a limited improvement at small catchment sizes compared to the reference value θ_{ref}. The approaches presented in the following are intended to be as simple as possible. First, we aim at a representation by a uniform equation without distinguishing different regimes, although the domain below (concerning catchment size) the region where Flint's law holds is sometimes described as the debris flow regime (Stock and Dietrich, 2003). Second, it should involve as few parameters as possible in order to limit the numerical effort of parameter estimation. The data shown in Fig. 1 suggest that the erosion rate still depends on the catchment size at least for $A \geq 0.01$ km^2 , but this dependence is weaker than predicted by the streampower law. A simple modification of Eq. (1) consists in adding a constant value a to the catchment size, i.e., to assume

$$S = k_s(A+a)^{-\theta} \tag{10}$$

The respective modification of the χ transformation is obtained by replacing A in Eq. (6) with A + a:

$$\chi_a(x) = \int_{x_0}^{x} \left(\frac{A(\xi)+a}{A_0}\right)^{-\theta} d\xi \tag{11}$$

This extension can be either considered a one-parameter approach where a is an adjustable parameter, while $\theta = \theta_{ref}$ is pre-defined, or a two-parameter approach with both a and θ being free parameters. For consistency, the latter is denoted $\chi_{\theta a}$ in the following.

As an alternative approach, a constant value can be added to the term A^θ. In order to avoid odd physical dimensions, this term is written in the form b^θ , where b has the dimension of an area. With this extension, Eq. (1) turns into

$$S = \frac{k_s}{A^\theta + b^\theta} \tag{12}$$

and the erosion rate (Eq. 2) becomes

$$E = K\left(\frac{A^\theta + b^\theta}{A_0^\theta} S\right)^n \tag{13}$$

In the linear case (n = 1), this extension can be interpreted as an erosion rate consisting of two additive components being both proportional to the channel slope. One of them depends on the catchment size according to the stream-power law, while the second one is independent of the catchment size and may correspond, for example, to hillslope erosion. Equation (12) is essentially the same as the empirical relationship

$$S = \frac{s_0}{1 + a_1 A^{a_2}} \tag{14}$$

suggested by Stock and Dietrich (2003) for debris flow valleys. Here, s0 is the hypothetic slope at the valley head (A = 0), and a_2 is the counterpart of the concavity index θ. The parameters a_1 and s_0 are related to those from Eq. (12) by $a_1 = b^{-\theta}$ and $s_0 = ksb^{-\theta}$. In this sense the difference between the approaches only concerns the considered regime and the definition of the parameters. We use a parameter b characterizing the catchment size where fluvial erosion and the sum of surface processes independent of the catchment size contribute equally to total erosion, while Stock and Dietrich (2003) used a more abstract parameter a_1. The respective modified χ transformation reads

$$\chi_b(x) = \int_{x_0}^{x} \frac{A_0^\theta}{A(\xi)^\theta + b^\theta} d\xi \quad (15)$$

Similar to the first approach, χb(x) refers to the oneparameter version with $\theta = \theta_{ref}$ in the following, while χθ b denotes the two-parameter version with adjustable parameters b and θ.

As shown by the red and green lines in Fig. 1, both extensions with $\theta = \theta_{ref}$ do not capture the behavior of the mean slope at small catchment sizes perfectly. While the first version (χ_a) should be better at catchment sizes moderately below the range where the original stream-power approach is valid, the second version (χ_b) should be preferable if the entire range shown in Fig. 1 is considered. Each of the approaches contains one or two adjustable parameters (a, b and/or θ) where the optimum value differs from catchment to catchment. As already pointed out by Perron and Royden (2013) for the one-parameter version χ_θ with variable θ, determining the respective optimum parameter value is nontrivial. In the simplest situation, a steady-state topography under homogeneous uplift and erodibility, the H vs. χ plot should be a straight line.

Here, the R^2 value (coefficient of determination) of a linear fit and Pearson's correlation coeffi- cient provide equivalent objective functions to be minimized. However, this may lead to systematic bias for transient topographies. An extension based on fitting piecewise linear functions was recently suggested by Mudd et al. (2014), but this algorithm may become numerically expensive, in particular if applied to a large number of catchments or if two adjustable parameters are involved. Including small catchment sizes in the analysis even facilitates the determination of the adjustable parameters since the collinearity of a large bunch of lines in the H vs. χ plot can be tested. Therefore, a criterion that measures how well the data follow a monotonic relationship between H and χ without being too sensitive to the shape of this relationship (such as R^2 and Pearson's correlation coefficient preferring linear relations) should be used. Spearman's rank correlation coefficient is the most widely used criterion in this context. Here, both the H values and the χ values are sorted independently.

Then, an H rank and a χ rank are assigned to each data point, and the correlation coefficient of the two ranks is computed. However, this approach suffers from the χ rank being a discontinuous function of the χ values and thus of the adjustable parameters. As a consequence, the rank correlation coefficient is a piecewise constant function of the parameters with a huge number of discontinuities, which makes its numerical maximization at least theoretically problematic. This problem could be avoided by considering the correlation between the χ values themselves and the H rank as the elevations are fixed values. However, this would introduce a bias towards a certain overall relationship depending on the hypsographic curve of the catchment, so that there is no advantage to the R^2 value or Pearson's correlation coefficient (preferring a linear relationship). Due to the problems with the rank correlation coefficients discussed above, we suggest an alternative criterion for assessing the collinearity of all rivers in the H vs. χ plot. In a first step, all pairs of χ_i and H_i are sorted in order of increasing H, and the sum

$$S = \sum_i |\chi_{i+1} - \chi_i| \quad (16)$$

is computed. This sum becomes minimal if H and χ are related monotonically and increases with each pair of subsequent points where χ decreases. However, S linearly scales with the absolute χ values; therefore, minimizing S would introduce a bias towards parameters leading to small overall χ values. Thus, S must be rescaled appropriately. As the χ values start from zero, the lowest possible value of S is the maximum χ value, χ_{max}, occurring for the given parameter set. Thus, the straightforward rescaled objective function to be minimized is

$$D = \frac{S - \chi_{max}}{\chi_{max}} \quad (17)$$

$$= \frac{\sum_i |\chi_{i+1} - \chi_i| - \chi_{max}}{\chi_{max}} \quad (18)$$

denoted χ disorder in the following. A perfect monotonic relationship between H and χ is characterized by D = 0. Some attention should be paid to pairs of identical elevation values occurring frequently in integer-valued DEMs. Here we suggest to assume that all χ values belonging to the same elevation are always in ascending order, so that they do not increase D artificially.

RESULTS AND DISCUSSION

In the following we compare the different approaches χ_a, χ_b, χ_θ , $\chi_{\theta a}$, and $\chi_{\theta b}$ using the recently released SRTM1 DEM with a mesh width of 1 arcsecond. Taiwan was selected as a region with high tectonic activity where glaciation only affected rather small regions around the highest mountains (Ono et al., 2005). Therefore, Taiwan should be an almost perfect example of a fluvial landscape. In order to get a sufficient number of catchments of similar sizes where each catchment contains a significant portion in the fluvial regime, a procedure to automatically delineate catchments with a size $A \approx 100$ km^2 was implemented. In a first step, all sites with catchment sizes $A < 100$ km^2 where the catchment size of the respective flow target is greater than 100 km^2 , and where the site itself makes the largest contribution to its flow target, are determined. These points or, more precisely, their flow targets are considered the base points (x0) of the respective catchments. The drainage pattern is then followed in the upstream direction down to a catchment size of 0.01 km^2 , and the different methods are applied to each of the catchments. All DEM nodes without valid elevation data or where the surface elevation had to be increased when filling local depressions were disregarded. The topography of Taiwan yields 89 catchments meeting these criteria, with each of them containing between 6464 and 27 732 valid SRTM1 DEM nodes. These catchments are shown in Fig. 2, while Fig. 3 displays the resulting cumulative distribution of the χ disorder of the five approaches – χ_a, χ_b, χ_θ , $\chi_{\theta a}$, and $\chi_{\theta b}$ – compared to the reference χ. As expected, the three approaches involving an adjustable parameter are much better than χ, and χ_b is the best among these

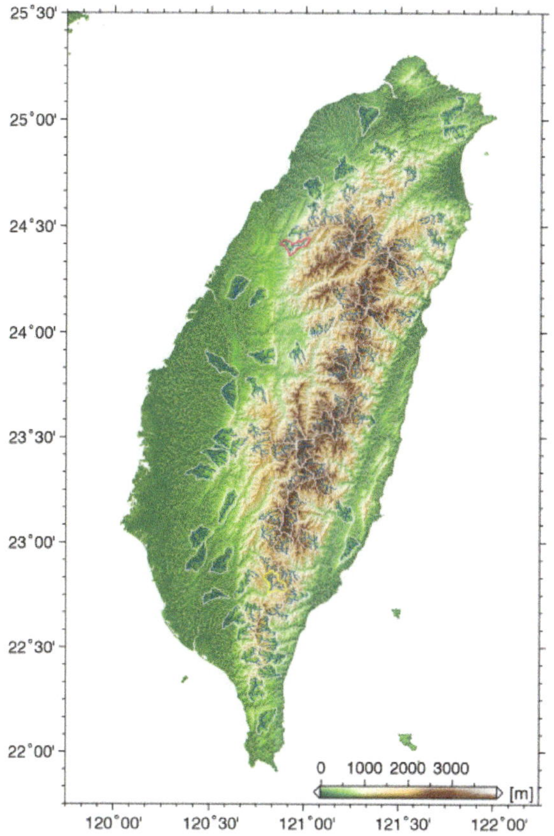

Figure 2. Map of the 89 considered catchments in Taiwan with catchment sizes $A \approx 100$ km^2 . The two catchments bordered in magenta and yellow are considered in detail in Figs. 4–6.

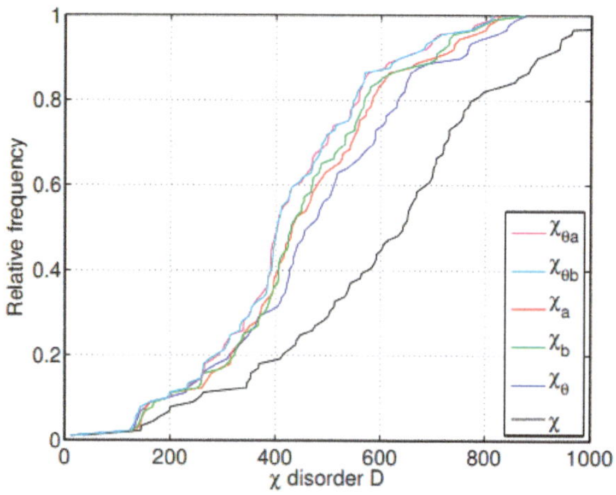

Figure 3. Cumulative distribution of the χ disorder for the 89 considered catchments in Taiwan for 0.01 km$^2 \leq A \leq 100$ km^2 . Each curve describes the relative number of the catchments with a χ disorder lower than or equal to the value D on the x axis.

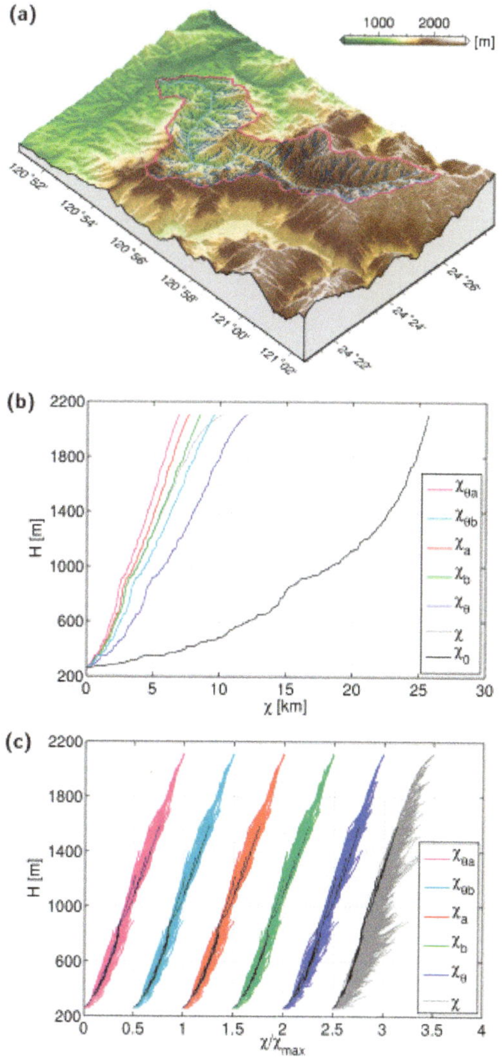

Figure 4. The mountainous catchment in Taiwan with the lowest χ disorder. (a) Topography and drainage pattern for $A \geq 0.01$ km^2 . The largest river is drawn in light blue. (b) H vs. χ plots of the main river. χ_0 refers to $\theta = 0$, so that $\chi_0 = x$, and the plot describes the original river profile. (c) H vs. χ plots for the entire drainage network. The plots are shifted horizontally in order to avoid overlapping curves. The black lines show the part of the drainage network with $A \geq 1$ km^2 .

approaches if the entire range 0.01 km$^2 \leq A \leq 100$ km^2 is considered. For 46 out of the 89 catchments, χ_b yields the best approximation (lowest χ disorder) among the three oneparameter approaches, while χ_a was found to be best for 17 catchments, and χ_θ for 26 catchments. As an immediate consequence of the additional parameters, both two-parameter approaches $\chi_{\theta a}$ and $\chi_{\theta b}$ yield a further improvement. The benefit is, however, weaker than that of the one-parameter approaches towards the version χ involving no adjustable

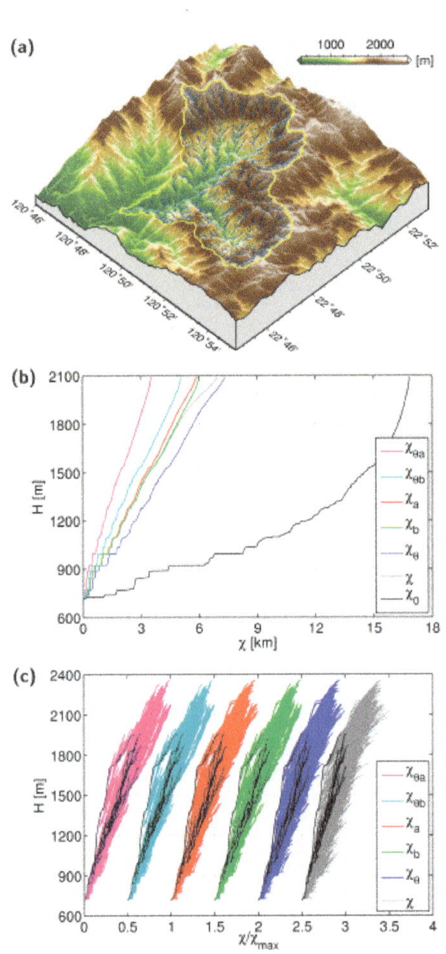

Figure 5. A catchment in Taiwan with a rather high χ disorder. (a) Topography and drainage pattern for $A \geq 0.01km^2$. The largest river is drawn in light blue. (b) H vs. χ plots of the main river. χ_0 refers to $\theta = 0$, so that $\chi_0 = x$, and the plot describes the original river profile. (c) H vs. χ plots for the entire drainage network. The plots are shifted horizontally in order to avoid overlapping curves. The black lines show the part of the drainage network with $A \geq 1\ km^2$.

parameters. The difference between $\chi_{\theta\,a}$ and $\chi_{\theta\,b}$ seems to be negligible. Figures 4 and 5 show two examples of catchments and their representation in the H vs. χ plot. The first one (Fig. 4) is the catchment with the lowest χ disorder in all approaches except χ (ranging from D = 199 for $\chi_{\theta\,b}$ to D = 230 for χ_θ) located in the mountain range. The χ disorder of the second example (Fig. 5) is more than 2 times higher than in the first catchment (ranging from D = 474 for $\chi_{\theta\,a}$ to D = 502 for χ_θ). These values are close to the two-thirds quantile of the respective distribution, which means that about two-thirds of the 89 considered catchments have a lower χ disorder than this example. Taking into account the width of the H vs. χ bunches, none of the two catchments shows a significant overall curvature in the H vs. χ plot except for the lowermost region of the first example (Fig. 4). However, this small part of the catchment is located at the edge of the mountain belt and even in an anthropogenically disturbed region. This finding suggests that spatial heterogeneity has a stronger effect on the H vs. χ plot than potential temporal changes in uplift rate in the past. This heterogeneity may be due to spatial variations in uplift rate, precipitation or, perhaps most likely at the catchment scale, the resistance of the rocks to erosion. The relevance of spatial heterogeneity to the H vs. χ plot is obviously related to the topology of the drainage network.

The example shown in Fig. 4 is an elongated catchment consisting of one main river and several smaller tributaries. The drainage network of the other example (Fig. 5) is characterized by confluences of rivers of similar sizes, and thus subcatchments with comparable χ values may occur at quite large spatial distances. For such a topology, spatial heterogeneity will likely generate diverging segments in the overall H vs. χ plot. In this example, it is readily recognized that a large part of the heterogeneity even arises from a small region when plotting χ (here $\chi_{\theta\,b}$) and elevation in a map (Fig. 6). In a region east of the center of the map, the contour lines of χ are at significantly higher elevations than elsewhere in the domain. This behavior corresponds to the single very steep tributary visible in Fig. 5. Therefore, a limited region east of the center of the map seems to be characterized by a lower erodibility than the rest of the domain. If only a smaller range of catchment sizes is considered, the differences between the methods partly disappear.

Figure 7 shows the same analysis applied down to catchment sizes of 1 km^2 instead of 0.01 km^2 . Here the approach χ_θ outcompetes the other one-parameter concepts by providing a better approximation in 78 out of 89 catchments and even comes close to the two-parameter approaches. As the extensions involving the parameters a and b were designed to capture the behavior at small catchment sizes, this result is not surprising. It is also immediately recognized that restricting the lower limit of catchment size reduces the absolute values of the χ disorder. The reduction mainly arises from the number of valid DEM nodes decreasing by more than 1 order of magnitude to a range from 670 to 2174. However, the χ disorder is diminished by less than 1 order of magnitude here. This nonlinear scaling is presumably related to the scale dependence of spatial heterogeneity. Nonlinear scaling properties can also be expected with respect to the total catchment size and to the resolution of the DEM, but investigating this in detail would go beyond the scope of this study. Thus the χ disorder is found to be a good criterion for comparing different extensions of the χ method and for determining the

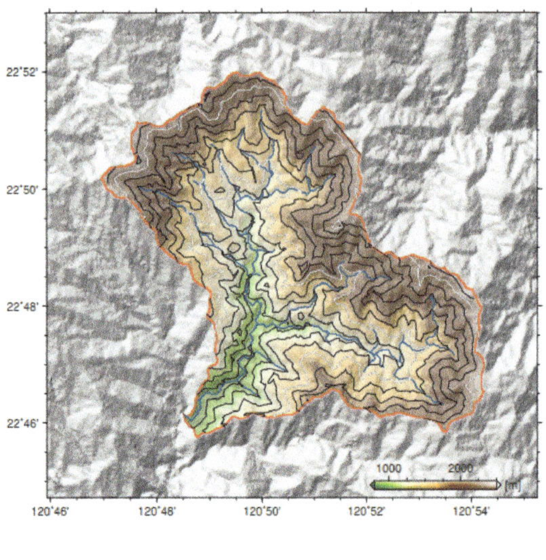

Figure 6. Map of elevation (coded by color) and $\chi_{\theta\,b}$ values (contour lines) of the catchment considered in Fig. 5. The con-

tour line interval is 0.5 km, and the lines of $\chi_{\theta\,b}$ = 2.5 km and $\chi_{\theta\,b}$ = 5 km are emphasized in white.

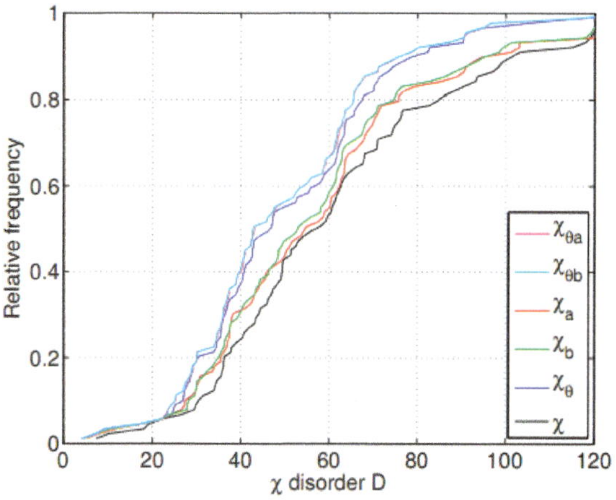

Figure 7. Cumulative distribution of the χ disorder for the 89 considered catchments in Taiwan for 1 km^2 $\leq A \leq$ 100 km^2 . Each curve describes the relative number of the catchments with a χ disorder lower than or equal to the value D on the x axis.

respective parameter values, but it cannot be used for comparing catchments of different sizes and data obtained from different DEMs. As a second example we consider the European Alps as an orogen that was heavily affected by glacial erosion in the past. For simplicity, we define the region as the domain inside the 600 m elevation contour line as previously done by Hergarten et al. (2010). Although the properties having an influence on the absolute values of the χ disorder (upper and lower limit of catchment size and DEM resolution) are the same as for the Taiwan example, the values of the χ disorder

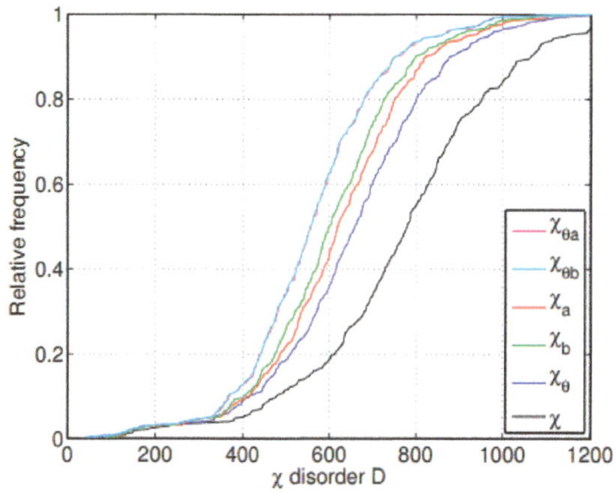

Figure 8. Cumulative distribution of the χ disorder for the 371 considered catchments in the European Alps for 0.01 km^2 $\leq A \leq$ 100 km^2 . Each curve describes the relative number of the catchments with a χ disorder lower than or equal to the value D on the x axis.

shown in Fig. 8 are significantly higher than those obtained for Taiwan. This increase is presumably related to glaciation causing strong local deviations from fluvial topography. The observed differences between the different approaches, however, persist or become even more pronounced. Here, the method χ_b yields the best results among the one-parameter approaches for more than 75 % of all catchments (280 out of 371). Beyond the goodness of the fit expressed by the χ disorder, the best-fit parameter values may also be taken into account when comparing the different approaches. In this context, the concavity index θ is the most important parameter as it has already been addressed in numerous studies on larger catchment sizes.

Figure 9 compares the statistical distributions of the best-fit θ values for the three methods involving θ as an adjustable parameter in the Taiwan example. If θ is the only parameter (χ_θ), the best-fit θ values tend to be below the widely used reference value θ_{ref} = 0.45. This effect becomes more pronounced if points with small catchments size are included. The median for 0.01 km^2 $\leq A \leq$ 100 km^2 is θ = 0.33, and 82 out of 89 catchments have $\theta < 0.45$.

As the deviations from Flint's law at small catchment sizes can only be compensated for by smaller θ values here, the significant bias towards smaller θ values found for $\chi\theta$ is not surprising. In return, the two-parameter approach $\chi_{\theta\,a}$ exhibits a tendency towards values $\theta > \theta_{ref}$, reflected in median of θ = 0.56. The other two-parameter approach, $\chi_{\theta\,b}$, yields best-fit θ values with a median of θ = 0.47 close to the reference value θ_{ref} = 0.45. While $\chi_{\theta\,a}$ and $\chi_{\theta\,b}$ are evenly matched with respect to the χ disorder on average, $\chi_{\theta\,a}$ obviously needs artificially increased θ values for achieving the best fit. The approach $\chi_{\theta\,b}$ turns out to be more robust against this bias, although some tendency towards larger θ values occurs if the catchment sizes are restricted to a narrower

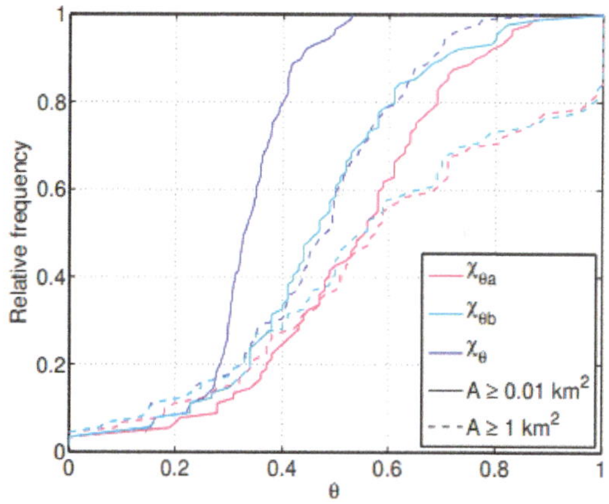

Figure 9. Cumulative distribution of the concavity index θ for the 89 considered catchments in Taiwan. Each curve describes the relative number of the catchments with an estimated concavity index lower than or equal to the value θ on the x axis. Solid lines refer to fits over the entire range 0.01 km^2 $\leq A \leq$ 100 km^2 , while dashed lines correspond to fits for $A \geq$ 1 km^2 only.

range (here, 1 km^2 $\leq A \leq$ 100 km^2). Under this aspect, the approach $\chi_{\theta\,b}$ should be superior to $\chi_{\theta\,a}$ if a wide range of catchment sizes is taken into account.

CONCLUSIONS

We have presented and investigated several concepts of extending Flint's law and the χ method towards small catchment sizes. Including points with small catchment sizes into the analysis of stream profiles strongly increases the data density and thus allows for a better distinction between effects of temporal changes in uplift rate or climate and spatial heterogeneity. Among the approaches considered in this study, an extension of Flints's law similar to an equation originally suggested for debris flow channels (Stock and Dietrich, 2003) turned out to be the most suitable concept if a wide range of catchment sizes is included. The respective definition of the extended χ transform (Eq. 15) can be implemented either as a two-parameter approach where both θ and b are adjustable parameters or as a one-parameter approach where b is variable and $\theta = \theta_{ref}$ with $\theta_{ref} = 0.45$ or any other fixed reference value. Minimizing the χ disorder defined in Eq. (18) provides a simple way to determine the best values of the adjustable parameters. The χ disorder refers to the collinearity of tributaries and does not require any further assumptions such as spatial homogeneity or an uplift rate being constant over distinct time intervals and should thus be applicable in a wide context.

ACKNOWLEDGEMENTS

The authors would like to thank W. Schwanghart and J. Jasiewicz for their constructive reviews and G. Sofia for the editorial handling.

REFERENCES

1. Brummer, C. J. and Montgomery, D. R.: Downstream coarsening in headwater channels, Water Resour. Res., 39, 1294, doi:10.1029/2003WR001981, 2003.
2. Flint, J. J.: Stream gradient as a function of order, magnitude, and discharge, Water Resour. Res., 10, 969–973, doi:10.1029/WR010i005p00969, 1974.
3. Goren, L., Fox, M., and Willett, S. D.: Tectonics from fluvial topography using formal linear inversion: Theory and applications to the Inyo Mountains, California, J. Geophys. Res.-Earth, 119, 1651–1681, doi:10.1002/2014JF003079, 2014.
4. Hack, J. T.: Studies of longitudinal profiles in Virginia and Maryland, no. 294-B in US Geol. Survey Prof. Papers, US Government Printing Office, Washington D.C., USA, 1957.
5. Hergarten, S., Wagner, T., and Stüwe, K.: Age and prematurity of the Alps derived from topography, Earth. Planet. Sc. Lett., 297, 453–460, 2010.
6. Howard, A. D.: A detachment-limited model for drainage basin evolution, Water Resour. Res., 30, 2261–2285, doi:10.1029/94WR00757, 1994.
7. Lague, D.: The stream power river incision model: evidence, theory and beyond, Earth Surf. Proc. Land., 39, 38–61, doi:10.1002/esp.3462, 2014.
8. Maritan, A., Rinaldo, A., Rigon, R., Giacometti, A., and RodriguezIturbe, I.: Scaling laws for river networks, Phys. Rev. E, 53, 1510–1515, 1996.
9. Montgomery, D. R.: Slope distributions, hillslope thresholds, and steady-state topography, Am. J. Sci., 301, 432–454, doi:10.2475/ajs.301.4-5.432, 2001.
10. Montgomery, D. R. and Foufoula-Georgiou, E.: Channel network source representation using digital elevation models, Water Resour. Res., 29, 3925–3934, doi:10.1029/93WR02463, 1993.
11. Mudd, S. M., Attal, M., Milodowski, D. T., Grieve, S. W. D., and Valters, D. A.: A statistical framework to quantify spatial variation in channel gradients using the integral method of channel profile analysis, J. Geophys. Res.-Earth, 119, 138–152, doi:10.1002/2013JF002981, 2014.
12. O'Callaghan, J. F. and Mark, D. M.: The extraction of drainage networks from digital elevation data, Computer Vision, Graphics, and Image Processing, 28, 323–344, 1984.
13. Ono, Y., Aoki, T., Hasegawa, H., and Dali, L.: Mountain glaciation in Japan and Taiwan at the global Last Glacial Maximum, Quatern. Int., 138–139, 79–92, doi:10.1016/j.quaint.2005.02.007, 2005.
14. Perron, J. T. and Royden, L.: An integral approach to bedrock river profile analysis, Earth Surf. Proc. Land., 38, 570–576, doi:10.1002/esp.3302, 2013.
15. Royden, L. and Perron, J. T.: Solutions of the stream power equation and application to the evolution of river longitudinal profiles, J. Geophys. Res.-Earth, 118, 497–518, doi:10.1002/jgrf.20031, 2013.
16. Sofia, G., Tarolli, P., Cazorzi, F., and Dalla Fontana, G.: Downstream hydraulic geometry relationships: Gathering reference reach-scale width values from LiDAR, Geomorphology, 250, 236–248, doi:10.1016/j.geomorph.2015.09.002, 2015.
17. Stock, J. and Dietrich, W. E.: Valley incision by debris flows: Evidence of a topographic signature, Water Resour. Res., 39, 1089, doi:10.1029/2001WR001057, 2003. Whipple, K. X.: Bedrock rivers and the geomorphology of active orogens, Annu. Rev. Earth Pl. Sc., 32, 151–185, doi:10.1146/annurev.earth.32.101802.120356, 2004.
18. Whipple, K. X. and Tucker, G. E.: Dynamics of the stream power river incision model: Implications for height limits of mountain ranges, landscape response time scales and research needs, J. Geophys. Res., 104, 17661–17674, doi:10.1029/1999JB900120, 1999.
19. Whipple, K. X., DiBiase, R. A., and Crosby, B. T.: Bedrock rivers, in: Fluvial Geomorphology, edited by: Shroder, J. and Wohl, E., vol. 9 of Treatise on Geomorphology, Academic Press, San Diego, CA, USA, 550–573, 2013.
20. Willett, S. D., McCoy, S. W., Perron, J. T., Goren, L., and Chen, C.-Y.: Dynamic reorganization of river basins, Science, 343, 1248765, doi:10.1126/science.1248765, 2014.
21. Wobus, C., Whipple, K. X., Kirby, E., Snyder, N., Johnson, J., Spyropolou, K., Crosby, B., and Sheehan, D.: Tectonics from topography: Procedures, promise, and pitfalls, in: Tectonics, Climate, and Landscape Evolution, edited by: Willett, S. D., Hovius, N., Brandon, M. T., and Fisher, D. M., vol. 398 of GSA Special Papers, Geological Society of America, Boulder, Washington, D.C., USA, 55–74, doi:10.1130/2006.2398(04), 2006.
22. Yang, R., Willett, S. D., and Goren, L.: In situ low-relief landscape formation as a result of river network disruption, Nature, 520, 526–529, doi:10.1038/nature14354, 2015.

CHAPTER 2

CLIMATE, TECTONICS OR MORPHOLOGY: WHAT SIGNALS CAN WE SEE IN DRAINAGE BASIN SEDIMENT YIELDS?

T. J. Coulthard and M. J. Van de Wiel

INTRODUCTION

Sediment yields from upland basins are driven by tectonics and climate. The magnitude of sediment delivered by this "erosional engine" (Whittaker et al., 2009) is a major control on the size and location of sedimentary units found in depositional basins, and thus creates an opportunity to invert sedimentary records to establish climate and tectonic histories of the source basin. However, the presence of two major external forcings (climate and tectonics) as well as the internal autogenic processing of these signals (e.g. Jerolmack and Paola, 2010; Van De Wiel and Coulthard, 2010) leads to a plurality of possible interpretations for each sedimentary record. The difficulty of inverting this plurality to establish whether individual or identifiable combinations of forcings can be determined is a fundamental limitation to our present capability to determine past climates and landscape histories from sedimentary records. At its simplest, the sedimentary system can be split into three components, an upland production "erosional engine" (Whittaker et al., 2009), transfer of sediment, and deposition in a basin or store.

Historically, research has focused on controls in the depositional setting; only more recently have researchers examined how tectonic and climatic changes can alter sediment production. For example, Willgoose et al. (1991), Whipple and Tucker (2002), Tucker and Whipple (2002), and others have demonstrated with numerical models how following uplift there is an increase in basin sediment discharge associated with a "wave" of incision that migrates upwards through a basin. Densmore et al. (2007) showed that for small fans the timing and amplitude of sediment flux to basins is controlled by changes in fault slip rates.

Most recently, Armitage et al. (2011) modelled a coupled fan basin system and showed how fault slip rates altered grain size trends – with grain size changing spatially and temporally away from the fan apex. Other researchers have shown how climate changes can also change or increase sediment delivery (Tucker and Slingerland, 1997; Coulthard et al., 2002) through mechanisms such as the extension of the stream network and increased stream powers. Allen and Densmore (2000) in their simulations from uplifting catchments suggest that sedimentary records more faithfully represent climate forcings than those from fault variability. Furthermore, physical experiments (Bonnet and Crave, 2003) also showed how erosional landscapes responded sharply to climate change. However, Armitage et al. (2013) found that sediment response was relatively insensitive to short-term climate variability. One common theme from the above studies and others (Allen, 2008; Humphrey and Heller, 1995; Métivier and Gaudemer, 1999) is that basin response to external forcings is complex and strongly contingent upon the internal basin processing of these forcings. Therefore, there is clearly a need to disentangle the relative impacts of climate, tectonics and autogenics on sediment delivery and how these may manifest themselves in the sedimentary record. In addition, comparatively little research has examined the transport of sediment from uplands to basin, with many theoretical and modelling studies assuming a direct link between the two.

Yet in numerous natural settings there is a considerable length and width of alluvial "conduit" between areas of sediment erosion and deposition and therefore this conduit must play a key role in how signals are transmitted from uplands to basin (Castelltort and Van Den Driessche, 2003). Geomorphological research indicates that this conduit operates in a non-linear way as part of a complex response (e.g. Schumm, 1979) acting to buffer sediment supply signals

Citation: Coulthard, T. & Van de Wiel, M. (2013). Climate, tectonics or morphology: what signals can we see in drainage basin sediment yields?. Earth Surface Dynamics, 1(1), 13-27. http://dx.doi.org/10.5194/esurf-1-13-2013.

(Castelltort and Van Den Driessche, 2003; Métivier, 1999), and more recent research has suggested that (geologically) short-term storage in floodplains may "shred" any upstream signals of forcings (Jerolmack and Paola, 2010). Simpson and Castelltort (2012) used a reach-based model to simulate how sediment pulses and water discharge perturbations may be transmitted to depositional settings.

Their simulations showed that changes in discharge input and sediment storage (reflected in increased valley floor gradients) led to spikes in sediment associated with increased water inputs. Coulthard et al. (2005) calculated individual sediment budgets for reaches of a medium sized drainage basin from basin simulations carried out over a 9000 yr period. Their research showed that during wetter periods (increased rainfall magnitudes) there was a dramatic increase in sediment yields and this was largely sourced from the lower, valley floor sections of the catchment. In short, wetter climates mined out the valley floor of sediment. Clearly, the behaviour of the conduit between upland erosional areas and the ultimate depositional setting can have a major impact on sediment delivery. In this paper we apply a numerical landscape evolution model to demonstrate that there are unexpected and major differences between the relative impacts of climate and tectonics on sediment delivery, and illustrate how the shape of the basin and length of conduit imparts an important control on this relationship.

METHODS

The CAESAR Model

The simulations were carried out with the CAESAR landscape evolution model (Coulthard et al., 2002; Van de Wiel et al., 2007). The main features of CAESAR are a combined hydrological and hydraulic flow model that operates on a subevent time step, with multi-grain size erosion and deposition as well as slope processes (diffusive creep and landslides). CAESAR was initially developed to examine the relative roles of climate and land cover change on geomorphology and sediment yield and has been applied to a range of real drainage basins with outputs successfully compared to independent field data. These examples include patterns of sedimentation in Alpine environment (Welsh et al., 2009) sediment yields and longer term lowering rates from Northern Australia (Hancock et al., 2010), comparisons to field plot experiments (Coulthard et al., 2012a), predicting patterns of contaminated sediment dispersal (Coulthard and Macklin, 2003) and simulating 9000 yr of drainage basin evolution in the UK (Coulthard and Macklin, 2001).

CAESAR models landscape evolution by routing water over a grid of regular sized cells and changing elevations according to erosion and deposition from fluvial and slope processes. Caesar can be run in two ways: firstly in catchment mode (as used here), where there are no external in-fluxes other than rainfall; and secondly in reach mode, with one or more points at which sediment and water are input into the system. For both modes of operation CAESAR requires several parameters or initial conditions including surface elevation, grain sizes and rainfall (catchment mode) or a flow input (reach mode). In concept, the operation of Caesar is simple, where precipitation falling on the modelled surface drives fluvial and hillslope processes that determine the erosion and deposition for the modelled time step. This changes the topography, which then becomes the starting point for the following time step. Outputs of the model are the elevation changes across the whole modelled topography as well as water discharges and sediment fluxes at the outlet(s) over time.

There are four main components in Caesar, a hydrological model, a surface flow model, fluvial erosion and deposition and slope processes. When running in catchment mode, Caesar uses rainfall precipitation input to generate runoff over the catchment using an adaptation of TOPMODEL (Beven and Kirkby, 1979). TOPMODEL contains a lumped soil moisture store that creates surface runoff when it exceeds a threshold value. The movement of the surface runoff is then simulated using a surface flow model. CAESAR's flow model uses a "flow-sweeping" algorithm, which calculates a steady-state, uniform flow approximation of the 2-D flow field. In this water discharge is distributed to all cells within a 2–5 cell range in front of a cell according to the differences between the water surface elevation of the donor cell and bed elevations of the receiving cells. Flow depths and velocity are calculated from discharges between cells using Manning's equation. These flow depths and velocities are then used to simulate the transport and deposition of sediment. Caesar calculates sediment transport over nine grain size fractions, which can be transported either as bed load or as suspended load, depending on the user specification. Caesar provides two options to calculate sediment transport, using Einstein (1950) or the Wilcock et al. (2003) equations. These equations are used as they calculate sediment transport for each size fraction that allows easy integration with the multiple grain size model components described later. For this study we have chosen to use the Einstein (1950) method where the calculation of sediment transport for each size fraction i involves first determining the balance between the forces moving and restraining a particle:

$$\psi = \frac{(\rho_s - \rho)D_i}{\rho d S} \qquad 1$$

where ρs is the sediment density, ρ water density, Di the grain size of the i fraction, d flow depth and S slope. In the above equation the term ρdS is replaced by τ/g, where τ denotes shear stress and g gravitational acceleration. A dimensionless bedload transport rate, φ, can then be estimated from ψ using the relationship established by Einstein (1950):

$$\phi = 40(1/\psi)^3 \qquad 2$$

The value of φ is then used to rearrange the following equation to estimate q_i, the rate of sediment transport (m3 s^{-1}):

$$\phi = q_i \sqrt{\frac{\rho}{(\rho_s - \rho) g D_i^3}} \qquad 3$$

Deposition of sediments differs between bed load and suspended load. At each iteration all transported bed load is deposited in the receiving cells, whereas the deposition of suspended sediments is determined from fall velocities and concentrations for each suspended sediment fraction. This procedure allows the selective erosion, transport and deposition of the different size fractions resulting in spatially variable sediment size distributions. Since this variability is expressed not only horizontally, but also vertically, CAESAR requires a method of storing sub-surface sediment data. This is enabled with a system of layers comprising an active layer representing the stream bed; multiple buried layers (strata); a base layer; and, if required, an immovable bedrock layer. These strata layers have a fixed thickness and their position is fixed relative to the bedrock layer and up to 20 strata can be stored at any cell on the grid. The active layer represents the river bed and has a variable thickness between 25 % and 150 % of the strata thickness. Erosion removes sediment and causes the active layer thickness to decrease. If the thickness becomes less than a threshold value, then the upper strata layer is incorporated in the active layer to form a new, thicker active layer. Conversely, deposition adds material to the active layer, causing it to grow. If the active layer becomes greater than a set value, a new stratum is created leaving a thinner active layer. Slope processes are also modelled, with mass movement represented as an instantaneous removal process. When the slope between adjacent cells exceeds a threshold (e.g. 0.5 or 45 degrees) material is moved from the uphill cell to the one below until the angle is lower than the threshold. As a small slide in a cell at the base of a slope may trigger more movement uphill, the model uses an iterative procedure to check the adjacent cells until there is no more movement. Soil creep is also modelled by CAESAR using the equation below, where Crate is the user-specified rate of soil creep (m yr^{-1}), T = time (years) and Dx grid-cell size. This represents diffusion-like processes whereby sediment flux is linearly proportional to surface slope (Carson and Kirkby, 1972).

$$\text{Creep} = \frac{SC_{\text{rate}}T}{Dx} \qquad 4$$

These mass movement and soil creep formulations allow material from slopes to be fed into the fluvial system as well as the input from landslides (both large scale and small – e.g. bank collapse). After the fluvial erosion/deposition and slope process amounts are calculated, the elevations and grain size properties of the cells are updated simultaneously. CAESAR required modification to simulate uplift. A simple uplift algorithm was added which raises the elevations of designated cells by a specified amount, ΔZu, at a specified time.

Model set-up and experimental procedure

Here, CAESAR was applied to the River Swale located in the North of England (Fig. 1). The section of the Swale modelled in this study lies upstream of Catterick Bridge, consisting of a total basin area of 490 km^2 . The average elevation within the basin is 357 m, the altitudinal range is 68–712 m and the average river gradient is 0.0064. At the headwaters the landscape is characterized by steep valleys and the geology is Carboniferous limestone and millstone grit (Bowes et al., 2003). Further downstream, valleys become less pronounced and the underlying geology becomes Triassic mudstone and sandstone (Bowes et al., 2003). For the numerical experiments, three different size subbasins of the Swale were used (Fig. 1), herein termed small, medium and large, and located upstream of points S1, S2 and S3, respectively. The initial DEM for the simulations was obtained from the Ordnance Survey "OpenData" DEM library. The climatic inputs into the hydrological component of the model are a 10 yr hourly rainfall data set for the area generated by the UKCP09 weather generator (see Coulthard et al., 2012b), repeated for the length of the simulations. Basin

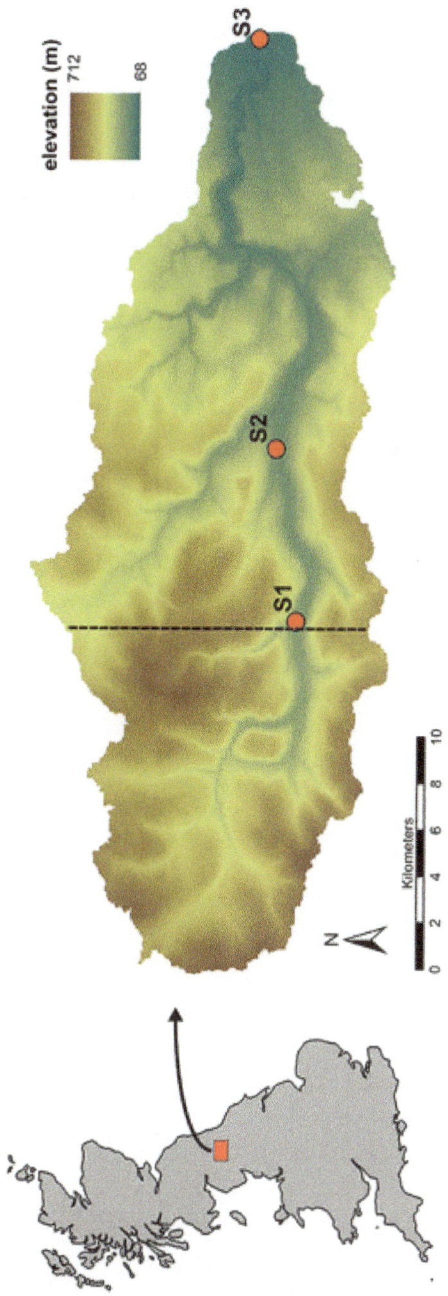

Figure 1. Location map (left) and elevation map (right) of the Swale basin. The area to the west of the fault line (dashed black line) was uplifted in the uplift simulations. Points S1, S2 and S3 indicate three locations where simulated sediment yields were recorded, respectively corresponding to a small, medium and large upstream sub-basin. Point S1 is located immediately downstream of the fault line. Points S2 and S3 are 10 and 30 km downstream from the fault line.

Table 1. Sediment yields in uplift scenarios for 100 yr simulations

scenario	S1		S2		S3	
	cum. yield (10^6 m^3)	rel. change (%)	cum. yield (10^6 m^3)	rel. change (%)	cum. yield (10^6 m^3)	rel. change (%)
base	2.18	0.0	1.63	0.0	16.16	0.0
1 m uplift	2.23	2.5	1.65	1.2	16.24	0.5
2.5 m uplift	2.46	13.0	1.68	3.1	16.23	0.4
5 m uplift	2.98	36.9	1.71	4.8	16.41	1.5
10 m uplift	3.96	81.8	2.02	24.1	16.58	2.6
25 m uplift	6.74	209	2.67	63.7	17.78	10.0
50 m uplift	11.82	442	3.75	130	19.90	23.1
100 m uplift	21.03	865	6.12	275	23.76	47.0
250 m uplift	40.67	1767	12.78	684	30.64	89.6

hydrology parameters were held constant during all simulated periods. There is no bedrock representation in these simulations, so all incision is into multi-grain sized material. Nine groups of numerical experiments were conducted (E1 to E9), with each experiment consisting of several simulations in which either climate or uplift were increased over a range of values (Fig. 2). An additional simulation with no change provides a base reference to assess the impact of the changes in external driving conditions. Different experiments were conducted to investigate comparatively short-term impacts, over a period of 100 yr (E1, E2, E6, E7), and longer term impacts, over a period of 900 yr (E3, E4, E5, E7, E8).

In these simulations, a year is defined as 365 days, that is, the extra day in leap years is ignored. Although CAESAR runs on sub-daily or sub-hourly time steps, output for the simulations is aggregated to facilitate analysis and interpretation. For the 100 yr simulations, data are aggregated in daily intervals. For the 900 yr simulations a 10 day interval was used. Additional analysis on annual and decadal intervals was done on further aggregation of these daily data. In all experiments, the changes in rainfall or uplift were initiated at 50 yr into the simulation. For all uplift experiments (E1, E3, E5, E6, E8), only the area to the west of point S1 was uplifted (Fig. 2). Uplift is instantaneous in our simulations, with the exception of experiment E5 where a continued gradual uplift is applied every year for the last 850 yr of the simulation. For the climate experiments (E2, E4, E7, E9), the changes in climate were defined as a percentage increase in the base rainfall. This increase is incurred after 50 yr into the simulation, is sustained throughout the remainder of simulations and affects the entire basin (Fig. 2).

RESULTS

Impacts of climate and tectonics on sediment totals

To investigate the role of tectonics we conducted a series of simulations (experiment E1) on the medium basin by instantaneously uplifting the upper section (i.e. the area west of S1; Figs. 2 and 3) after 50 yr of simulation, and then continuing the simulation for a further 50 yr. This was to establish whether there was any rapid/instant response as well as short-term (decadal) changes as the basin adjusts to its new topography. The simulations were repeated several times

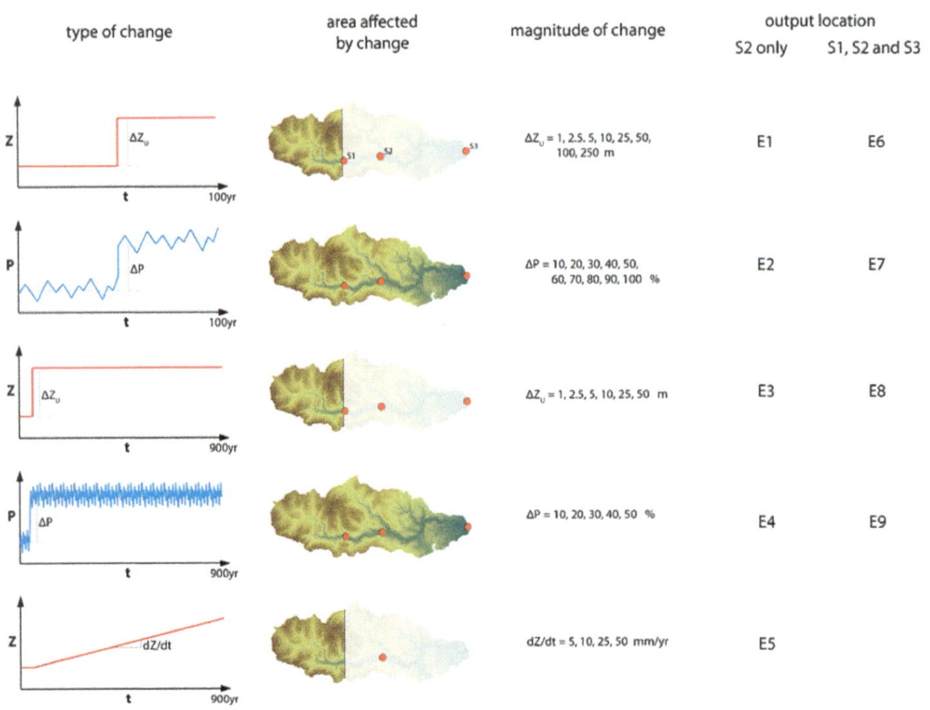

Figure 2. Schematic representation of experiment design. Nine numerical experiments were carried out (E1 to E9). Each experiment consists of a group of 4 to 10 simulations, where either rainfall (blue) or uplift (red) was altered after 50 yr into the simulation. In uplift experiments,

only the western part of the basin is uplifted. Changes in rainfall affect the entire basin. Red dots on the maps indicate position of output points S1, S2 and S3.

with differing amounts of uplift (ΔZu = 1 m, 2.5 m, 5 m, 10 m, 25 m, 50 m, 100 m, 250 m). Figure 3a shows cumulative sediment yields from the basin measured at point S2 and Table 1 contains cumulative yields and percentage increases in sediment yield. The red lines are from simulations with different instantaneous uplift amounts and generally show an increase in sediment yield following the uplift events. However, uplift of 10 m or more is required to create a noticeable shift in the cumulative line.

Using the same medium basin but disabling the uplift, we then investigated the role of climate by increasing the magnitude of the rainfall input to the hydrological model after 50 yr (E2). The simulations were run with increased rainfall rates (ΔP = 10 to 100 %, in 10 % intervals). The cumulative sediment yields from these simulations are plotted as blue lines on Fig. 3a. Increasing rainfall magnitude creates a very similar effect as tectonic uplift, though with a small but identifiable instantaneous response (a vertical increase in the cumulative). Overall, comparatively small increases in rainfall magnitude result in relatively large increases in sediment yield (e.g. 20 % increase in rainfall increases sediment yield by 53 %; Table 1, Fig. 3a). To achieve a similar increase of sediment yield, a large amount of instantaneous uplift is needed (e.g. 25 m of uplift results in 64 % increase in yield). In Fig. 4a, we compare the sediment yield totals after 100 yr from all the tectonic and climate simulations that comprised E1 and E2. Relatively modest rises in rainfall magnitudes (+20 %, +30 %) result in 50 to 100 % sediment yield increase, which is matched only by large levels of instantaneous uplift (i.e. 25 m or more). At larger rainfall increases, the difference becomes even starker. For example, approximately 90 m of uplift is required to achieve the cumulative sediment output from a 50 % increase in rainfall (5.6×10^6 m^3).

These simulations, however, are only over short 100 yr timescales. To investigate longer term trends we extended a

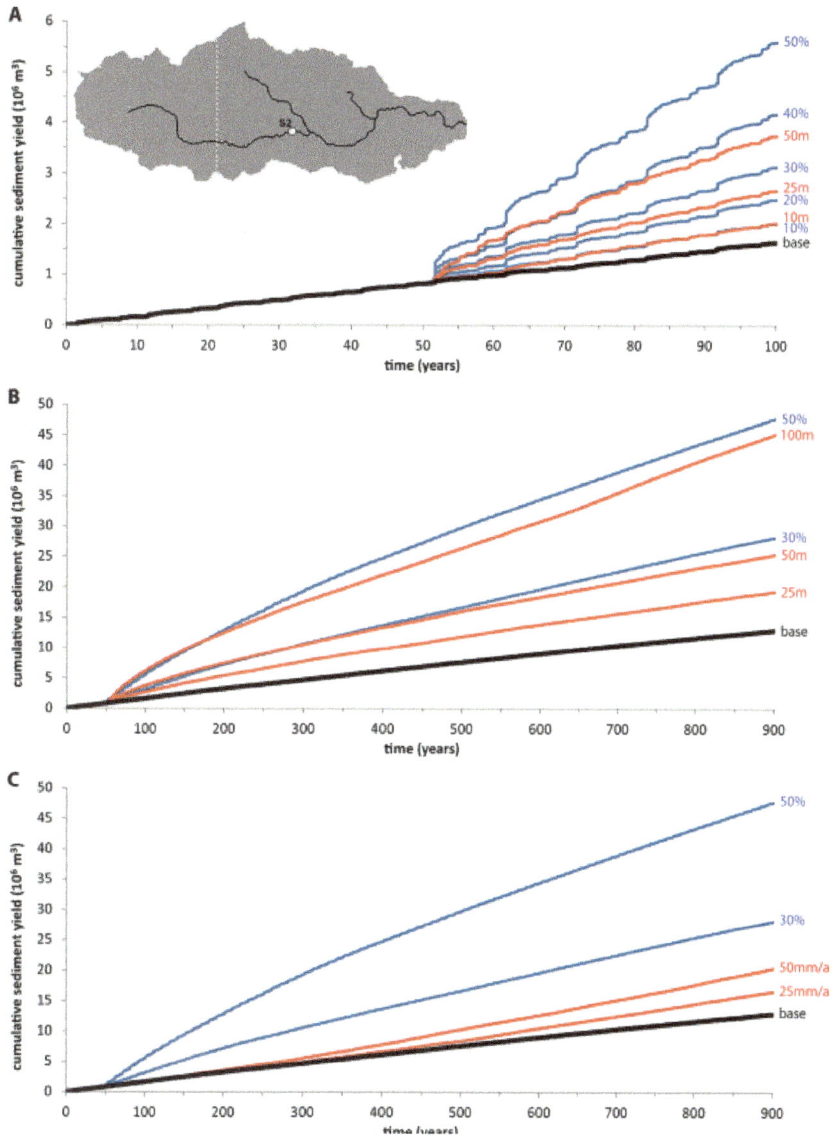

Figure 3. Simulated cumulative sediment yields at point S2 for a range of different scenarios from E1–E5. (A) 100 yr scenarios for rainfall increase and instantaneous uplift (E1, E2). (B) 900 yr scenarios for rainfall increase and instantaneous uplift (E3, E4). (C) 900 yr scenarios for rainfall increase and gradual uplift (E3, E5). The uplift or increase in rainfall occurred at 50 yr. The base scenario (no uplift, no rainfall increase) is shown as a thick black line, rainfall scenarios are shown as blue lines, and tectonic scenarios are shown as red lines. Inset shows location of fault line and measurement point S2.

select number of these runs to continue for 900 yr (E3, E4). Cumulative sediment yields show a very similar response to the shorter term simulations (Fig. 3b). There is a slight decay in the rate of increase in sediment yields after ca. 200 yr from both uplift and climate change – which is in response to locally increased gradients (tectonics) and expanding drainage areas (climate) leading to an initial surge from readily available sediment.

Clearly, the trends observed in Fig. 3a are not transient conditions and persist over longer time periods. We also investigated how gradual uplift (dZ / dt = 5, 10, 25 and 50 mm per year) compared to changes in climate (E5). As per previous results we can see that even large rates of gradual uplift are superseded by modest increases in climate (Fig. 3c).

To establish how basin shape and the length of channel between zones of uplift and deposition (the conduit) alter sediment delivery, an additional experiment (E6) was carried out uplifting the same upper section of the basin, but increasing

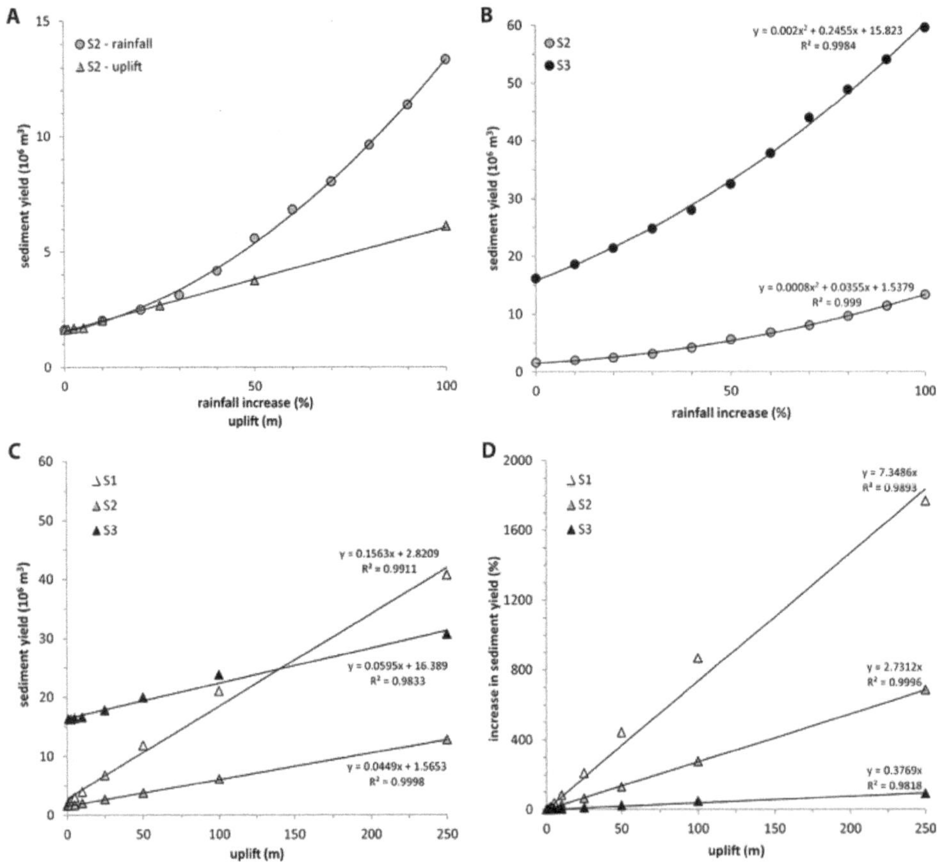

Figure 4. Impact of rainfall increase or tectonic uplift on simulated sediment yields after 100 yr, at point S1, S2 and S3 (see Fig. 1 for locations). (A) Impact of rainfall increase and tectonic uplift on sediment yield after 100 yr, at S2. (B) Total sediment yield for rainfall scenarios, at S2 and S3. (C) Total sediment yields for instantaneous uplift scenarios, at S1, S2 and S3. (D) Percentage change in sediment yield, relative to the base scenario, for instantaneous uplift scenarios, at S1, S2, S3.

the distance between uplift and basin outlet. Three size basins were used, with outlets at S1, S2 and S3, respectively 500 m, 10 km and 30 km downstream of the uplifted area (Figs. 1, 2). As per the previous experiments, uplift was instantaneous and added after 50 yr of simulation. As the temporal response was similar as previous simulations, only the impacts on total sediment yields are shown (Fig. 4c–d).

These data demonstrate that the percentage increase in sediment yield due to uplift is highest in the small basin, and progressively smaller in the medium and large basins (Fig. 4d). In absolute values, sediment yield increases linearly with uplift (Fig. 4c). In the larger basins a noticeable increase in sediment yield only occurs after 25 m of uplift, and is driven by the relatively substantial changes uplift causes in basin relief (without uplift basin relief is 500 m). That aside, however, there are two other processes in operation. Firstly, adding a downstream section of valley floor to accommodate storage of sediment, as is the case in the medium and large basins, removes or "shreds" the part of the signal from the uplift events (Fig. 4c). The physical mechanism for this is evident studying the surface morphology of the simulations, where an alluvial fan forms immediately downstream of the uplift locus (Fig. 5). However, the larger basin does not seem to be more "effective" at shredding this signal than the medium basin, as the curves rise at very similar rates, albeit offset (Fig. 4c). Yet for the larger basin, the percentage change in sediment yield is far less (Fig. 4d).

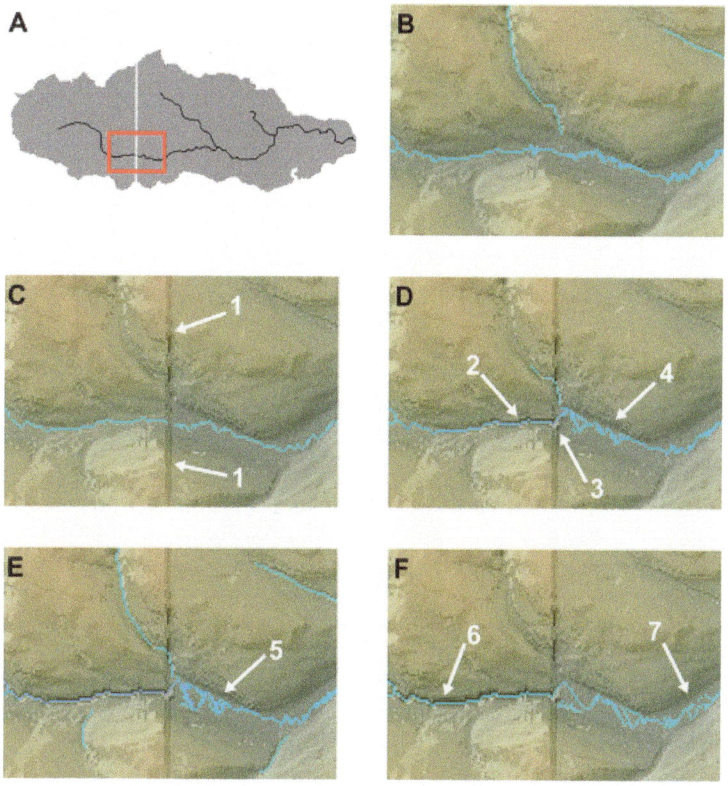

Figure 5. Geomorphic changes occurring after the 50 m instantaneous uplift scenario. (A) location of fault (solid white line) and area selected for elevation maps (red rectangle). (B–F) Elevation maps of selected area, respectively after 47.9, 50.1, 55.5, 63.5 and 99.2 yr of simulation. Arrows indicate specific features: 1. Escarpment after uplift; 2. Incision of channel upstream of fault line; 3. Formation of small alluvial fan; 4. Braided river pattern downstream of fault line; 5. Expansion of braiding; 6. Continued incision upstream of fault line; 7. Downstream propagation of braided river pattern.

This demonstrates the second process, where the uplift signal is simply diluted by sediment being added from the non-uplifted parts of the larger basin as the ratio of uplifted area to total basin area decreases. Next we compared increases in sediment yield from uplift to those from rainfall in different size basins (E7). Unlike the uplift scenarios, where only part of the basin is uplifted, increases in rainfall affect the whole basin (Fig. 2), giving 2 to 3 times larger sediment yields for the larger basin, reflecting its larger area (Fig. 4b). Figure 5 shows the impacts of 50 m of uplift on the geomorphology of the simulations around the area uplifted. This shows the before and ,at four time points, after uplift. Here there are clear changes in the nature of the channel both upstream and downstream. Upstream there is an entrenching of the main stream with a nick point migrating upstream ultimately leaving a river terrace behind. Downstream there is an immediate change in the channel pattern from single thread to braided (indicative of increased sediment yields) and the development of a small alluvial fan.

Impacts of climate and tectonics on sediment grain size

As well as providing information on sediment yields, the multiple grain sizes incorporated within CAESAR allow changes in grain size to be simulated in response to climate and tectonic forcings. Here we focus our discussion on the median grain size, D50, for experiment E6, that is, for the 100 yr simulations with uplift, for the S1, S2 and S3 basins (Fig. 6). These results show the annual mean D50 for all events within a year, and they thus increase and decrease with years that have fewer and greater large flood events. Daily data show similar patterns, but are omitted here as the yearly totals present a clearer picture. Results for D84 show similar trends and are not discussed. For S1, the 5 m uplift initially leads to a 20 yr increase in the D50 (Fig. 6a). The initial coarsening agrees with field observations (e.g. Whittaker et al., 2009) associated with higher stream powers due to increased channel gradients (Whittaker et al., 2009). However, this effect begins to attenuate after 20 yr with the D50 diminishing to slightly less than that prior to uplift. For 10 m and 25 m of uplift there is a very different reaction with both a reduction in D50 and a smoothing of the signal from different annual events. For 10 m uplift there is a partial recovery in this signal after 30 yr and possibly beginning at 50 yr for 25 m uplift. The simulated fining associated with uplift is clearly different from expected results of coarsening (Whittaker et al., 2009) and we offer two explanations for this.

Firstly, the 10 and 25 m uplift events we have simulated are very unusual in the field – and therefore it is quite possible we are simulating an effect not observed in the field. Secondly, our model has a high level of connectivity between channel, floodplain and hillslopes. Within our high uplift scenarios significant fluvial incision occurs (see Fig. 5), leading to the immediate introduction of fresh sediment from the collapse of river banks that will be sustained as a wave of incision passes up through the valley floor (as indicated in Fig. 5).

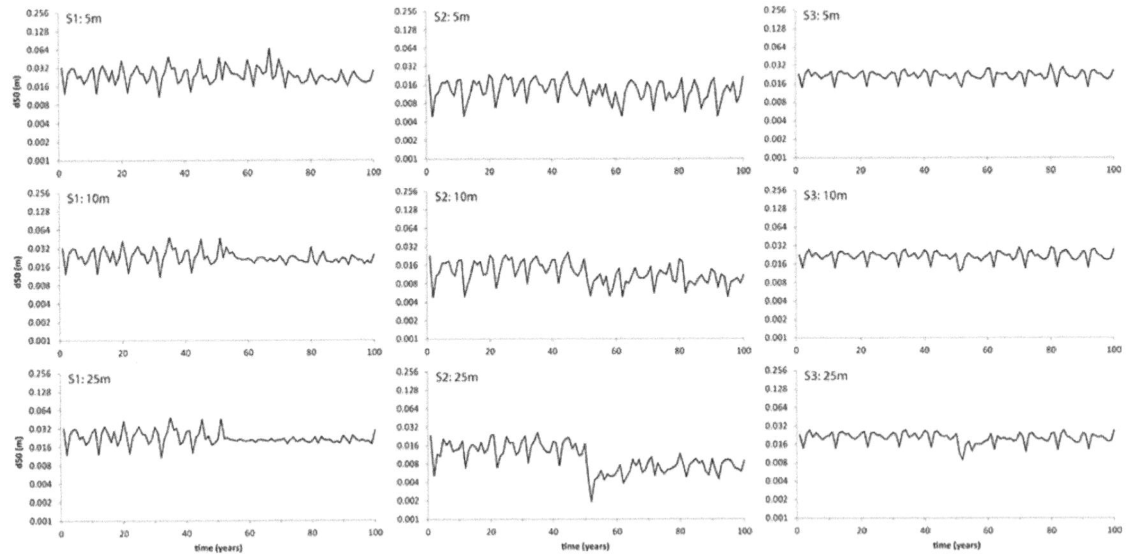

Figure 6. Median annual grain sizes for S1, S2 and S3 for 5, 10 and 25 m uplift simulations. Values are calculated by summing total daily sediment yields for each of 9 grain sizes output by CAESAR, over a simulated year. This was carried out in preference to calculating daily median grain sizes and then averaging this over a year, which would skew the median grain size by over-weighting the contribution of days with small daily sediment totals.

The bank and slope material has a relatively higher concentration of fines than sediments in the river bed. In addition, this effect will also translate into tributary streams leading to a sustained input of finer sediment. The reduction in the annual variation in D50 in the 10 m and 25 m uplift simulations is a different example of signal shredding. Here the system has been overloaded with finer sediment creating a situation where climatic variations are no longer observed in the D50 (although they are in the total sediment yield). At S2 all simulations show a fining of the sediment post uplift, with the effect becoming more amplified with greater amounts of uplift (Fig. 6).

This shows that some of the finer sediment released from S1 after uplift is being transported through the reach – though, since the grain size reduction is reduced from S1 to S2, a substantial volume and proportionally more of the coarser material is being stored between S1 and S2. The reduction in interannual variability of the D50 that was observed at S1 is less noticeable at S2, although there still is some reduction in the variability as the amount of uplift increases. For S3 there is little or no variation in D50 pattern in the 5 m and 10 m uplift scenarios. Only in the 25 m uplift scenario is there an observable impact on the D50, namely that the variability decreases.

Location S3 is thus sufficiently far downstream of the fault-line for grain size not be affected by the uplift of 5 and 10 m with only a small impact from the 25 m uplift. In other words, the valley between S1 and S3 absorbs the effects of the uplift on transported sediment sizes. Figure 7 describes the impacts of climate on grain size for basins S1, S2 and S3 (experiment E7). Unlike uplift overall there is a far weaker response in the grain size signal to increased wetness. For S1, S2 and S3 there is negligible change in grain size after the 10 % increase in wetness, with any variations indiscernible from variations prior to the increase. However, for larger increases in wetness drops or dips in grain size are amplified. These drops in grain size correspond to wetter years indicating that these correspond to th

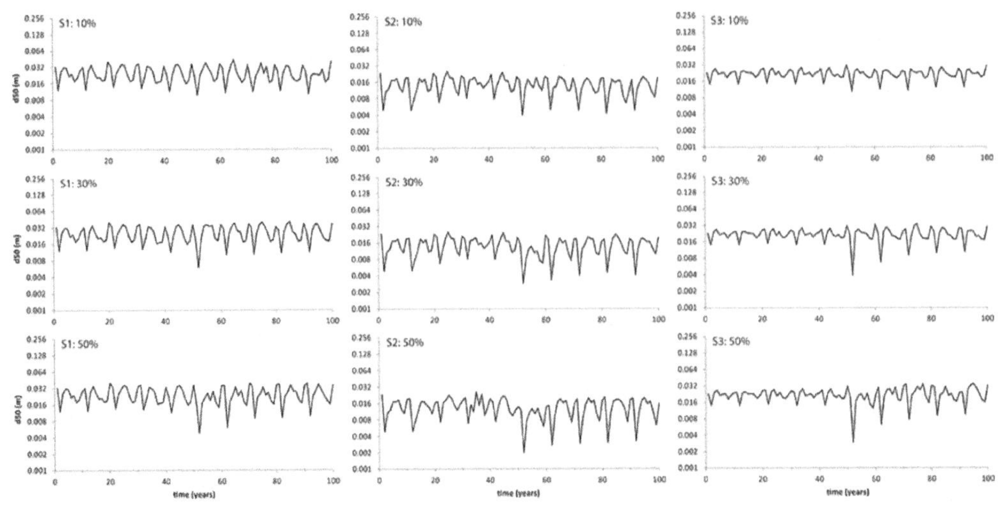

Figure 7. Grain size changes after rainfall changes for S1, 2 and 3 with 10, 30 and 50 % increases in precipitation magnitude after 50 simulated

years. Median grain sizes were calculated as per the method outlined in Fig. 6.

increased delivery of fine sediments (from slopes, banks and also incision and expansion of tributaries) during wetter periods.

The role of autogenic processes

Within our results, we believe we can see evidence of autogenic, or internal processes that are creating signals within the basin outputs. In Fig. 8, we have plotted the annual and decadal sediment outputs from the 900 yr uplift simulations (E8), with actual and relative values (relative values are normalised to the mean pre-uplift sediment output). Signals generated in post uplift sediment curve of S2 and to a lesser extent S3 are of similar (10 m) and greater (5 m) magnitude than peaks immediately following the uplift itself. Therefore, at both annual and decadal timescales the S2 and S3 basins are capable of generating sediment yield peaks equivalent to or greater than those seen immediately after uplift. There is also evidence that uplift changes the nature of sediment outputs from the system. For example, at S2, sediment yield variability increases following uplift (Table 2). This effect is more pronounced and longer lasting as the amount of uplift increases. At S3, only the 25m uplift has an impact on the sediment yield variability (Table 2). This is also visible in Fig. 8, where the relative changes (right side graphs) also show there is a greater indication of the uplift signal in S2 than S3 (supporting earlier findings).

Table 2. Variability of relative sediment yield in uplift scenarios for 900 yr simulations

time period	S2			S3		
	5 m	10 m	25 m	5 m	10 m	25 m
pre-uplift: 1–50 yr	0.709	0.709	0.709	0.558	0.558	0.558
post-uplift: 51–100 yr	0.775	0.952	1.626	0.497	0.520	0.600
post-uplift: 451–500 yr	0.570	0.669	0.810	0.333	0.346	0.341
post-uplift: 851–900 yr	0.558	0.592	0.685	0.307	0.312	0.317

Variability is shown as standard deviations of relative annual sediment yield over the specified time period. Relative annual yields are normalized to the mean yield of the 50 yr pre-uplift period.

DISCUSSION

These results provide us with considerable insight into how a drainage basin processes different external forcings. In these experiments climate changes clearly generate greater increases in sediment yield than uplift. In this study we have deliberately explored extreme values of climate and uplift, but if we restrict the results to more reasonable values (e.g. instantaneous uplift up to 10 m and rainfall/climate increases of 30 %) then in systems where the sedimentary deposits are not proximal to the sediment source, climate changes clearly have a far greater impact on sediment delivery. In our simulations elevated sediment yields from 10 m of uplift are equivalent to a 10 % increase in rainfall magnitude. Even looking at longer timescales and continual uplift rates of 25 mm yr−1 equate to a 10 % increase in rainfall. However, the converse is apparent when examining grain size. Here (especially at S1 and S2) there is a mixed reaction to uplift that generates an increase or decrease in D50 according to the magnitude of the change (increases associated with smaller amounts of uplift up to 5 m). For smaller (< 10 m) amounts of uplift our findings largely agree with those of Armitage et al. (2011) who note the deposition of larger sediment close to the uplift point (equivalent to our S1) with reductions in grain size further away (our S2 and S3). However, for larger uplift values, that is, 10 m and greater, we observed fining at S1 instead of coarsening. This would suggest a switch in sediment response above certain levels of uplift. Alterations in climate have a less apparent impact on D50 with the exception of the largest increases in rainfall magnitude (30 % and 50 %) leading to an increased drop in grain size during wetter years.

The role of the sedimentary system "shredding" input signals (Jerolmack and Paola, 2010) is also apparent – with the addition of less than 10 km of non-uplifted floodplain (the difference between S1 and S2) removing much of the uplift signal from both sediment volume and grain size. In Jerolmack and Paola's paper, they used a physical model of a rice pile – where the storage and (non-linear) release of rice in the pile led to the removal or shredding of any input signal in the output response. We would argue that a floodplain operates in a similar manner to the rice pile by storing and releasing sediment in a non-linear manner. Indeed, other workers have noted non-linear, self organising mechanisms in operation in floodplains with meandering and cut-offs (Hooke, 2003; Stølum, 1998). Thus, we would argue that for a given flood event the amount of sediment released from the end of a floodplain (bedload not suspended) may or may not bear any relation to the size of the flood event. Interestingly, our simulations show that as basin size increases the impact of "shredding" does not increase, thus indicating that only a short area of accommodation for storage and re-working of sediment is required. Furthermore, as the total area of basin relative to the area uplifted increases, the relative importance of shredding decreases as any signal from the uplift is diluted by sediment from tributaries in non-uplifted parts of the basin.

Our findings support previous work indicating long lag times from tectonic changes that are "buffered" by the fluvial system (Allen, 2008; Métivier and Gaudemer, 1999), in particular within areas of valley floor and floodplain (Castelltort and Van Den Driessche, 2003; Métivier and Gaudemer, 1999; Simpson and Castelltort, 2012) as found in our expanding catchment settings. It is worth noting that increases in sediment delivery are generated by all uplift and climate scenarios – but as catchment area and valley floor length grow, peaks in sediment were smoothed, or lost within the noise of the autogenic signals.

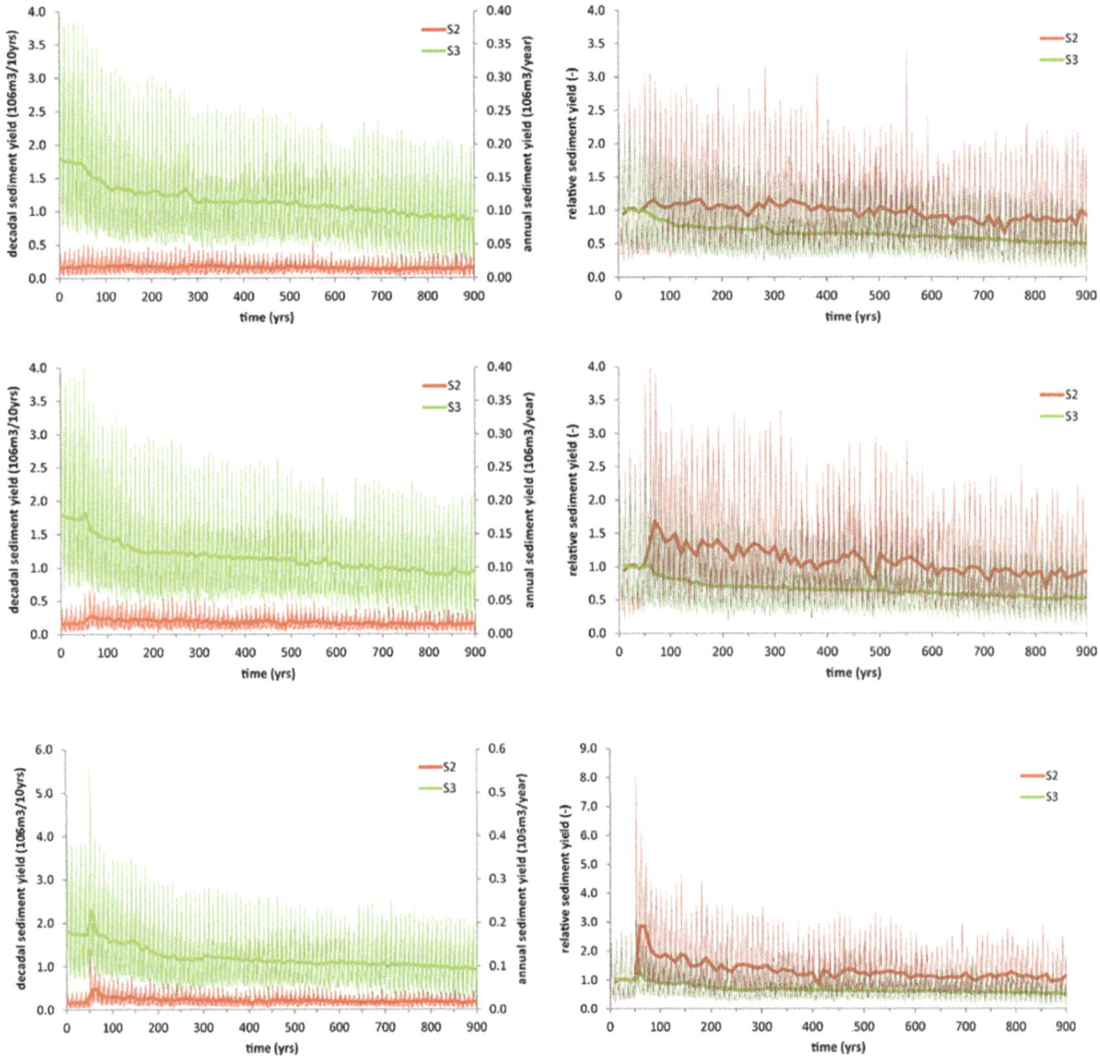

Figure 8. Annual and decadal sediment yields for S2 and S3 over 900 simulated years for 5, 10 and 25 m of uplift respectively (top to bottom). The left side shows absolute values and the right side relative values (normalised to pre-uplift average).

There may be important implications from this research for the interpretation of stratigraphy. Firstly, our results only generate sediment yields at the edge of our simulated drainage basins. We, therefore, do not account for any depositional settings or changes in accommodation space that may occur subsequently downstream. However, we suggest that the thickness of a sedimentary unit found in basins is more likely to represent changes in climate rather than any uplift history, except for where there is a direct and high level of connectivity between source and deposit. Closer to the point of uplift, coarsening in grain size is likely to be indicative of small/moderate uplift, but a fining in grain size indicates either large amounts of tectonic uplift or major increases in rainfall magnitude. In short, changes in the volume of sediment exported from a basin is more contingent on climate than tectonics, yet changes in grain size in this sediment are more likely to represent tectonic changes rather than climate.

These findings re-enforce those from Simpson and Castelltort (2012) and of Allen and Densmore (2000) who suggest that discharge/climate variability may be better represented in the sediment signal than tectonic events and comment on the difficulties of inverting the sedimentary record (Simpson and Castelltort, 2012). These results also indicate that short-term peaks in sediment volume are more likely to be generated by increases in rainfall than uplift events (as per the physical experiments of Bonnet and Crave, 2003). Therefore, marked increases in the thickness of sedimentary units – if representing increases in sediment supply – are more likely to indicate climatic changes rather than tectonic. It is important to note that this "short-term" sediment peak equates to a year or more of increased sediment delivery, rather than that from individual flood events, which may be very difficult to identify (Van de Wiel and Coulthard, 2010). Furthermore, for moderate levels of uplift, the autogenic factors generate annual and even decadal peaks in sediment of equivalent or greater magnitude than uplift. In other words, the noise generated within the system is greater than the signal from the input. We are aware that the timescales we have simulated here are considerably shorter than those used in other studies (e.g. Armitage et al., 2011, 2013).

However, for sedimentary records over the Holocene and Quaternary scales, we would argue that the changes we are simulating are highly relevant. There are several studies that link increases in Holocene and Quaternary sedimentation rates to shorter term environmental forcings (e.g. Macklin et al., 2006) and even to individual events (e.g. Hinderer and Einsele, 1997; Hinderer, 2001; Larsen et al., 2013). We believe our findings are highly relevant in this context. Conversely, for records spanning tens of millions of years our findings may be less relevant and could just be part of the "noise" of the system. It is also important to note that our simulations are not run to any form of landscape steady state where there is equilibrium between (for example) uplift and erosion, and at these shorter timescales the catchments are always in a state of transience. An important control on long-term basin development is bedrock – yet in these simulations we have not included its role.

Within CAESAR, bedrock can be represented as a layer that is difficult/impossible to erode beneath layers of regolith/alluvium. However, when designing these experiments incorporating this would then introduce another parameter (the depth to bedrock) and we wished to keep the experiments as parsimonious as possible. It is important to remember that whilst there is an unlimited depth of regolith in our simulations, this is of different grain sizes interacting through an active layer system. Therefore, the model will behave in a supply limited manner. This means that during the model operation the model will preferentially erode the finer fractions from channels and "wear down" any obstructions or high spots in the channel network (termed spinning up or conditioning, Coulthard et al., 2012b). As a result of this conditioning process the basin is left with an armoured channel bed that is relatively stable under the range of events that it formed under (i.e. the topography and rainfall events that formed it). If you perturb the basins driving forces (e.g. via increased precipitation or uplift) there follows an increase in sediment yield as the channel works through the armour layer that it had already developed.

Therefore, in many respects, this scenario is identical to a the behaviour of a bedrock channel that experiences uplift or a change in climate, except the reaction speed of the bedrock basin will be far slower. For example, Godart et al. (2013) examine how basin lithology alters the response time of basins to climate changes with different "resonance periods" for harder and softer lithologies. For example a basin with a highly resistant bedrock would not have enough time to respond to short changes in climate – yet softer lithologies could respond (and thus generate sediment flux) to such changes (Godart et al., 2013). In addition, bedrock controls will have an increasing importance as the duration of simulations increases – so for the comparatively short times simulated here (100–900 yr) the effects may be less important. Further complications of including bedrock include the position of channel bedrock sections relative to the point of uplift. For example if the uplift were located where bedrock was on the stream bed, then uplift would have a minimal impact on sediment delivery – as the controlling base level (the bedrock) would rise with the section of the catchment uplifted. However, if at the point of uplift bedrock were 2 m below the alluvial bed then we would expect a similar response to these simulations – equivalent to 2 m of uplift.

Therefore, the introduction of bedrock below a surface would probably have the impact of reducing any spikes or pulses in sediment yield associated with uplift in effect limiting them to the depth to bedrock. Bedrock at a certain depth could also reduce the sediment output from increases in climate though the reduction would be smaller than for tectonics as climate change would impact across the whole basin. Vegetation is another factor that could alter the relative result of climate and tectonic change on sediment yield. We did not model vegetation changes in these experiments, but in CAESAR, a simple vegetation model can act to bind the surface together until an erosional threshold is exceeded. This acts to reduce basin sediment output but would have a greater role on reducing the impact of climate change than tectonics – as increased sediment from climate is driven by channel network extension (and incision) into what would be vegetated slopes. Whereas tectonic increases in sediment are driven by local gradient changes that would largely be within non-vegetated (and thus unaffected) channels.

The responses of a basin to bedrock and vegetation change are certainly worthy of further research, especially contrasting over the shorter (as per here) and longer timescales. There are also limitations with the location and model parameterization. Firstly, the Swale is not a tectonically active basin, but was chosen as CAESAR has been extensively evaluated and validated on the Swale over decadal to centennial timescales. For determining how the distance from uplift affected any tectonic or climatic change signal the Swale has a relatively straightforward valley floor and is typical of many upland basins. Importantly, we chose a natural basin over an artificial landscape (e.g. Van De Wiel and Coulthard, 2010) as we wanted to include topographic heterogeneity (tributaries, floodplains, alluvial fans) as well as actual rainfall records and grain sizes. In addition, finding a field site with enough available data (in particular initial conditions e.g. topography) and with a tectonic and climatic history was difficult (as discussed later). We deliberately manipulated uplift rates and climate changes in a rather unrealistic manner, but this is to establish the outer limits of the relationships between climate, uplift and morphology on sediment yield.

Further modes of uplift were tried (fore tilt, back tilt, sideways tilt, gradual and instant) and all gave similar results. Considering the model and parameterization, compared to alternative landscape evolution models, CAESAR has a different level of process representation. Therefore, some of the phenomena we have simulated here (e.g. different grain size responses to varying uplift) may be missed by different model configurations and parameterizations. For example the recent work by Armitage et al. (2011, 2013) is based on a one-dimensional model that whilst accounting for slope processes does not explicitly include them within a twodimensional framework. It is also worth considering that in our study and that of Simpson and Castelltort (2012) hydraulics and flow processes are simulated and both models show a sensitivity of sediment output/throughput to climate. Whereas Armitage et al. (2013) do not simulate the hydraulics and show an insensitivity to climate (essentially impacting water discharge).

These observations raise the question as to what level of process detail, spatial and temporal resolution is required to simulate landscape dynamics. If high levels of detail are required, this may expose a particular problem for long-term landscape modelling, as field data to drive simulations (e.g. initial landscapes, grain sizes, climates) and validate model runs (e.g. stratigraphy, topographic data) are especially hard to come by. In addition, a circularity can develop whereby stratigraphy may be used to validate models, as well as generate driving data – yet could be highly variable and not truly reflect drivers or products of the basins dynamics (e.g. Coulthard et al., 2007; Van De Wiel and Coulthard, 2010). Finally, assessing the reliability of numerical landscape evolution models is difficult given many of the issues outlined above (initial landscape, grain sizes, past climates etc.).

As stated in the model description, CAESAR has successfully simulated landscape evolution in a range of environments (Coulthard and Macklin, 2001, 2003; Coulthard et al., 2012a; Hancock et al., 2011; Welsh et al., 2009), but it is important to question whether the non-linear response of the model is simply a model by-product or a representation of actual basin dynamics? CAESAR has a long history of modelling the non-linear reaction of catchments from 1998 through to 2010 (Coulthard and Van De Wiel, 2007; Coulthard et al., 1998, 2005; Van De Wiel and Coulthard, 2010) and given that similar non-linear dynamics have been well documented in fluvial systems (Cudden and Hoey, 2003; Gomez and Phillips, 1999; Hooke, 2003; Stølum, 1998), we consider the simulation of non-linear sediment dynamics in CAESAR genuine. In addition, CAESAR is certainly not the only model to show non-linear sediment responses. Examples of this include the Lapsus model (Temme and Van De Wiel, 2012), the results of Simpson and Castelltort (2012), and the Zscape model (Allen and Densmore, 2000). The notion that different numerical and physical models, with very different underlying algorithms produce non-linear responses indicates that what we simulate is a fundamental process of sediment transport dynamics rather than a model side effect.

CONCLUSIONS

A series of numerical experiments were carried out, in which the impacts of climate change and tectonic uplift on catchment sediment yield was evaluated. Our results indicate that both have an impact on the sediment yield, but the nature of that impact is different. Climate changes are more likely to impact the total volume of the sediment yield. Tectonic uplift, on the other hand is more likely to affect the grain size distribution of the sediment yield. In our simulations, the impacts of tectonic uplift on the volume sediment yield are pronounced immediately downstream of the uplift zone, but considerably less notable the further downstream you check. In effect the tectonic signal is diluted by storage of sediment in the floodplain. In addition, our results also indicate that autogenic variability of sediment yield, due to temporary storage and release of sediment within the basin, can be of the same magnitude as the spikes in sediment yield associated with the external disturbance, that is, climate change or tectonic uplift. All of these findings have implications for the reliability and meaning of inverting sedimentary records for determining past environmental and tectonic conditions.

ACKNOWLEDGEMENTS

M. J. Van de Wiel was supported by a NSERC Discovery Grant. We are very grateful for review comments from Arnaud Temme, Kevin Norton and Greg Hancock. In addition Sebastien Castelltort provided invaluable commentary on certain parts of the manuscript. The CAESAR model and source code is freely available from https://code.google.com/p/caesar-lisflood/.

REFERENCES

1. Allen, P. A.: Time scales of tectonic landscapes and their sediment routing systems, Geological Society, London, Special Publications, 296, 7–28, doi:10.1144/SP296.2, 2008.
2. Allen, P. A. and Densmore, A. L.: Sediment flux from an uplifting fault block, Basin Res., 12, 367–380, doi:10.1046/j.1365- 2117.2000.00135.x, 2000.
3. Armitage, J. J., Duller, R. A., Whittaker, A. C., and Allen, P. A.: Transformation of tectonic and climatic signals from source to sedimentary archive, Nat. Geosci., 4, 231–235, doi:10.1038/ngeo1087, 2011.
4. Armitage, J. J., Dunkley Jones, T., Duller, R. A., Whittaker, A. C., and Allen, P. A.: Temporal buffering of climate-driven sediment flux cycles by transient catchment response, Earth Planet. Sc. Lett., 369–370, 200–210, doi:10.1016/j.epsl.2013.03.020, 2013.
5. Beven, K. J. and Kirkby, M. J.: A physically based, variable contributing area model of basin hydrology / Un modèle à base physique de zone d'appel variable de l'hydrologie du bassin versant, Hydrological Sciences Bulletin, 24, 43–69, doi:10.1080/02626667909491834, 1979.
6. Bonnet, S. and Crave, A.: Landscape response to climate change: Insights from experimental modeling and implications for tectonic versus climatic uplift of topography, Geology, 31, 123, doi:10.1130/0091-7613(2003)0312.0.CO;2, 2003.
7. Bowes, M. J., House, W. A., and Hodgkinson, R. A.: Phosphorus dynamics along a river continuum, Sci. Total Environ., 313, 199– 212, doi:10.1016/S0048-9697(03)00260-2, 2003.
8. Carson, M. A. and Kirkby, M. J.: Hillslope Form and Process, Cambridge University Press, New York, 1972. Castelltort, S. and Van Den Driessche, J.: How plausible are highfrequency sediment supply-driven cycles in the stratigraphic record?, Sediment. Geol., 157, 3–13, doi:10.1016/S0037- 0738(03)00066-6, 2003.
9. Coulthard, T. J. and Macklin, M. G.: How sensitive are river systems to climate and land-use changes? A model-based evaluation, J. Quaternary Sci., 16, 347–351, doi:10.1002/jqs.604, 2001.
10. Coulthard, T. J. and Macklin, M. G.: Modeling longterm contamination in river systems from historical metal mining, Geology, 31, 451, doi:10.1130/0091-7613(2003)0312.0.CO;2, 2003.
11. Coulthard, T. J. and Van De Wiel, M. J.: Quantifying fluvial non linearity and finding self organized criticality? Insights from simulations of river basin evolution, Geomorphology, 91, 216–235, doi:10.1016/j.geomorph.2007.04.011, 2007.
12. Coulthard, T. J., Kirkby, M. J., and Macklin, M. G.: Nonlinearity and spatial resolution in a cellular automaton model of a small upland basin, Hydrol. Earth Syst. Sci., 2, 257–264, doi:10.5194/hess-2-257-1998, 1998.
13. Coulthard, T. J., Macklin, M. G., and Kirkby, M. J.: A cellular model of Holocene upland river basin and

alluvial fan evolution, Earth Surf. Proc. Land., 27, 269–288, doi:10.1002/esp.318, 2002.

14. Coulthard, T. J., Lewin, J., and Macklin, M. G.: Modelling differential catchment response to environmental change, Geomorphology, 69, 222–241, doi:10.1016/j.geomorph.2005.01.008, 2005.
15. Coulthard, T. J., Hicks, D. M., and Van De Wiel, M. J.: Cellular modelling of river catchments and reaches: Advantages, limitations and prospects, Geomorphology, 90, 192–207, doi:10.1016/j.geomorph.2006.10.030, 2007.
16. Coulthard, T. J., Hancock, G. R., and Lowry, J. B. C.: Modelling soil erosion with a downscaled landscape evolution model, Earth Surf. Proc. Land., 37, 1046–1055, doi:10.1002/esp.3226, 2012a.
17. Coulthard, T. J., Ramirez, J., Fowler, H. J., and Glenis, V.: Using the UKCP09 probabilistic scenarios to model the amplified impact of climate change on drainage basin sediment yield, Hydrol. Earth Syst. Sci., 16, 4401–4416, doi:10.5194/hess-16-4401- 2012, 2012b.
18. Cudden, J. R. and Hoey, T. B.: The causes of bedload pulses in a gravel channel: the implications of bedload grainsize distributions, Earth Surf. Proc. Land., 28, 1411–1428, doi:10.1002/esp.521, 2003.
19. Densmore, A. L., Allen, P. A., and Simpson, G.: Development and response of a coupled catchment fan system under changing tectonic and climatic forcing, J. Geophys. Res., 112, F01002, doi:10.1029/2006JF000474, 2007.
20. Einstein, H. A.: The Bed-load Function for Sediment Transportation in Open Channel Flows, in: Technical Bulletin No. 1026, USDA Soil Conservation Service, p. 71, US Department of Agriculture, 1950.
21. Godard, V., Tucker, G. E., Burch Fisher, G., Burbank, D. W., and Bookhagen, B.: Frequency-dependent landscape response to climatic forcing, Geophys. Res. Lett., 40, 859–863, doi:10.1002/grl.50253, 2013.
22. Gomez, B. and Phillips, J. D.: Deterministic Uncertainty in Bed Load Transport, J. Hydraul. Eng., 125, 305–308, 1999.
23. Hancock, G., Lowry, J., Coulthard, T., Evans, K. and Moliere, D.: A catchment scale evaluation of the SIBERIA and CAESAR landscape evolution models, Earth Surf. Proc. Land., 35, 863–875, doi:10.1002/esp.1863, 2010.
24. Hancock, G. R., Coulthard, T. J., Martinez, C., and Kalma, J. D.: An evaluation of landscape evolution models to simulate decadal and centennial scale soil erosion in grassland catchments, J. Hydrol., 398, 171–183, doi:10.1016/j.jhydrol.2010.12.002, 2011.
25. Hinderer, G. and Einsele, M.: Terrestrial sediment yield and the lifetimes of reservoirs , lakes , and larger basins, Geol. Rundsch., 86, 288–310, 1997.
26. Hinderer, M.: Dénudation quaternaire récente dans les Alpes, remplissage des vallées et des lacs, charge solide des rivières actuelles, Geodin. Acta, 14, 231–263, doi:10.1016/S0985- 3111(01)01070-1, 2001.
27. Hooke, J.: River meander behaviour and instability: a framework for analysis, T. I. Brit. Geogr., 28, 238–253, doi:10.1111/1475- 5661.00089, 2003.
28. Humphrey, N. F. and Heller, P. L.: Natural oscillations in coupled geomorphic systems: An alternative origin for cyclic sedimentation, Geology, 23, 499–502, doi:10.1130/0091-7613(1995)023, 1995.
29. Jerolmack, D. J. and Paola, C.: Shredding of environmental signals by sediment transport, Geophys. Res. Lett., 37, 1–5, doi:10.1029/2010GL044638, 2010.
30. Larsen, A., Bork, H.-R., Fuelling, A., Fuchs, M., and Larsen, J. R.: The processes and timing of sediment delivery from headwaters to the trunk stream of a Central European mountain gully catchment, Geomorphology, doi:10.1016/j.geomorph.2013.06.022, in press, 2013.
31. Macklin, M. G., Benito, G., Gregory, K. J., Johnstone, E., Lewin, J., Michczynska, D. J., Soja, R., Starkel, L., and ′ Thorndycraft, V. R.: Past hydrological events reflected in the Holocene fluvial record of Europe, Catena, 66, 145–154, doi:10.1016/j.catena.2005.07.015, 2006.
32. Métivier, F.: Diffusive like buffering and saturation of large rivers, Phys. Rev. E, 60, 5827–5832, doi:10.1103/PhysRevE.60.5827, 1999.
33. Métivier, F. and Gaudemer, Y.: Stability of output fluxes of large rivers in South and East Asia during the last 2 million years: implications on floodplain processes, Basin Res., 11, 293–303, doi:10.1046/j.1365-2117.1999.00101.x, 1999.
34. Schumm, S. A.: Geomorphic Thresholds: The Concept and Its Applications, T. I. Brit. Geogr., 4, 485–515, 1979.
35. Simpson, G. and Castelltort, S.: Model shows that rivers transmit high-frequency climate cycles to the sedimentary record, Geology, 40, 1131–1134, doi:10.1130/G33451.1, 2012.
36. Stølum, H.: Planform geometry and dynamics of meandering rivers, Geol. Soc. Am. Bull., 11, 1485–1498, 1998.
37. Temme, A. J. A. M. and Van De Wiel, M. J.: Filtering signals?: complexity experiments with LAPSUS, Geophysical Research Abstracts, EGU General Assembly, 14, 5299, 2012.
38. Tucker, G. E. and Slingerland, R.: Drainage basin responses to climate change, Water Resour. Res., 33, 2031, doi:10.1029/97WR00409, 1997.
39. Tucker, G. and Whipple, K.: Topographic outcomes predicted by stream erosion models: Sensitivity analysis and intermodel comparison, J. Geophys. Res., 107, 2179, doi:10.1029/2001JB000162, 2002.
40. Van De Wiel, M. J. and Coulthard, T. J.: Self-organized criticality in river basins: Challenging sedimentary records of environmental change, Geology, 38, 87–90, doi:10.1130/G30490.1, 2010.
41. Van de Wiel, M. J., Coulthard, T., Macklin, M., and Lewin, J.: Embedding reach-scale fluvial dynamics within the CAESAR cellular automaton landscape evolution model, Geomorphology, 90, 283–301, doi:10.1016/j.geomorph.2006.10.024, 2007.
42. Welsh, K. E., Dearing, J. A., Chiverrell, R. C., and Coulthard, T. J.: Testing a cellular modelling

approach to simulating lateHolocene sediment and water transfer from catchment to lake in the French Alps since 1826, The Holocene, 19, 785–798, doi:10.1177/0959683609105303, 2009.

43. Whipple, K. and Tucker, G.: Implications of sediment-fluxdependent river incision models for landscape evolution, J. Geophys. Res., 107, 1–20, 2002.

44. Whittaker, A. C., Attal, M., and Allen, P. A.: Characterising the origin, nature and fate of sediment exported from catchments perturbed by active tectonics, Basin Res., 22, 809–828, doi:10.1111/j.1365-2117.2009.00447.x, 2009.

45. Wilcock, P. R., Crowe, J. C., and Wilcock, P. Crowe, J.: Surface-based transport model for mixed-size sediment, J. Hydraul. Eng., 129, 120–128, doi:10.1061/(ASCE)0733- 9429(2003)129:2(120), 2003.

46. Willgoose, G., Bras, R. L., and Rodriguez-Iturbe, I.: A coupled channel network growth and hillslope evolution model 1. Theory, Water Resour. Res., 27, 1671–1684, doi:10.1029/91WR00935, 1991.

CHAPTER 3 TECTONIC GEOMORPHOLOGY OF SALT STRUCTURES

Mehran Arian, Hamideh Noroozpour

Keywords: Tectonic Geomorphology, Salt Structures, Diapir, Iran, Halokinesis

INTRODUCTION

Tectonic geomorphology is the study of the interaction between tectonics and surface processes that form the structure during geologic times. Salt diapir or piercement structure with a central, nearly equidimensional salt plug, generally one to two kilometers or more in diameter, which has risen through the enclosing sediments from a mother salt bed beneath the top of the plug, is called salt dome [1] . For the first time, the economic importance of salt domes was expressed at the beginning of the recent century, when discovery of oil and gas had developed, because oil and gas are trapped around and under the some salt domes. About four fifth of oil and gas, which is reserved at south of united states of America as well as some oil reserves in the middle east, are in some way affiliated to the salt structure. In addition, extracted crude oil and natural gas are mostly preserved at salt caves, for future use. Meanwhile, due to non-permeable character of salt, it has been suggested to store radioactive wastes in the salt crevasses [2] . The main aim in this paper is determination of diapiric provinces in Iran. Iran has composed from different plates. Arabian plate in south and west (Zagros and Persian Gulf), Cimmerian miniplate in north and east, Eurasian plate in northeast margin [3] and [4] . There are three physiographic-tectonic zones that they have salt diapirs: 1) Zagros Mountain; 2) Persian Gulf basin; and 3) Cental Iran basin.

Citation: Arian, M. and Noroozpour, H. (2015) Tectonic Geomorphology of Salt Structures. Open Journal of Geology, 5, 61-72. doi: 10.4236/ojg.2015.52006.

GEOLOGIC SETTING

Zagros Mountain

Dominant structural trends in Zagros Mountain (Figure 1) are NW-SE in northwestern part and E-W in south- eastern part. From tectonics point of view, it contains the overthrusted and simple folded belts of Zagros that formed on the northeastern part of Arabian plate's passive margin. Zagros is external platform (folded and thrusted belt) of north margin of Arabian Craton. Vergence of folding in this zone is toward south and southwest. Hormuz Salt had been deposited within main sub-basins of Zagros basin in Late Precambrian-Early Cambrian. The Hormuz formation is formed by salt, anhydrite, black dolomite, shale, red siltstone and sandstone and volcanic rocks. The most of 107 emergent salt structures are related to central and eastern part Zagros. The Hormuz and equivalent series were deposited in an Evaporite basin during the late Precambrian-early Cambrian [5] . Coeval salt basins crop out in a large domain including the Central Iran as Ravar series [6] .

Persian Gulf

Dominant structural trend in Persian Gulf basin (Figure 1) is NW-SE. From tectonics point of view, it contains the south margin of simple folded belts of Zagros that formed on the northeastern part of Arabian plate's passive margin. Persian Gulf is north margin of internal platform of north margin of Arabian Craton and foreland basin of Zagros Mountain.

Hormuz Salt had been deposited within in Persian Gulf basin in Late Precambrian-Early Cambrian. The name of the Hormuz Salt was taken from the Iranian Hormuz Island (Figure 2) cored by a salt diapir, which also gives its name to the opening strait of the Persian Gulf. Eight emergent salt structures or Iranian diapiric islands such as Hormuz, Larak, Tonb-Kochak, Tond-e Bozorg, Abu Musa, Faror, Bani-Faror and two emergent salt structures in Qeshm and Hengam Islands are related to northeastern part of Persian Gulf.

Central Iran

Dominant structural trends in Central Iran (Figure 1) are NW-SE in northwestern part and N-S in eastern part. From tectonics point of view, it contains Iranian plateau. The Iranian plateau is one of the two main plateaus in the Alpine-Himalayan collision system. The basement of the plateau consists of Cimmerian miniplate. Volcanics of Late Cretaceous-Early Miocene age in Central Iran represent Andean type magmatism during Neo-Tethyan subduction [8].

Volcanic rocks, evaporates and turbidites successions up to 3 km thick represent Eocene back-arc extension across Central Iran. This succession is commonly overlain in Central Iran by terrestrial clastics, evaporates and volcanics of the Lower Red Formation; of Oligocene age [9] . Marine deposition resumed across much of Central Iran with the carbonates of the largely lower Miocene Qom Formation. The Qom Formation is overlain by middle Miocene terrestrial clastics and evaporates of the Upper Red Formation.

The salt deposits belonging to Cenozoic of Central Iran basin have two separate origins: The old type related to Upper Eocene deposits and Lower Red formation; and new type related to Upper Red formation [10] . In the Central Iran basin, there are many salt domes, which as compared with the salt domes of Zagros and Persian Gulf basin, less attention have been paid to them. The reason of this issue, also lays in presence of oil and gas reserves at later cases. Extrusion of Hormuz salt domes of Zagros [11] and Persian Gulf basin are the deeper and older than Central Iran basin [12] , [13] and [14].

DIAPIRIC PROVINCES OF IRAN

Salt diapirs of Iran present a large variety of shapes and elevation from high relief to entirely eroded structures. In addition, they have different sizes at outcrop scale, ranging from craters a few km wide to diapirs and glaciers more than 5 km wide.

These different morphologies are controlled by various factors, for example, position related to other structures, the time of emergence, the rate of salt dissolution, which is controlled by the rate of erosion, the bulk rate of salt emission from the source layer, and eventually by the tectonic activity [15] . Based on the previous investigations such as [16] and this research, nine diapiric provinces have distinguished in Iran (Figure 3).

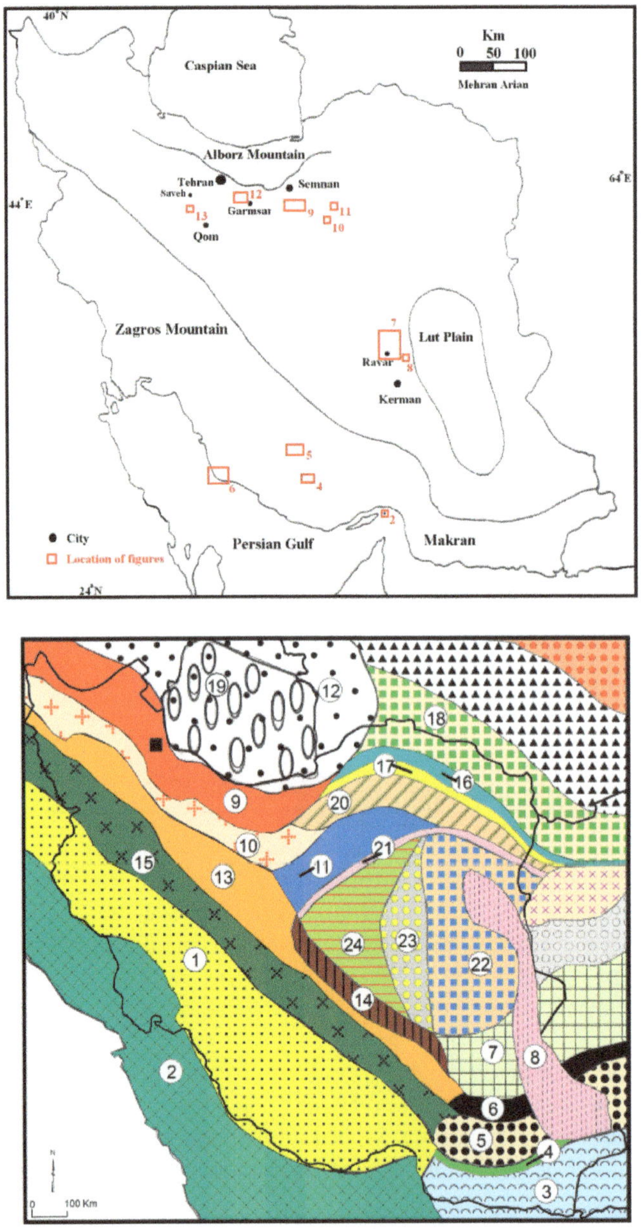

Figure 1. Geographical distribution of salt diapiric areas in Iran. Locations of figures 2, 4 to 13 are shown by red rectangles, grids marked by border of figure (up) and Physiographic-tectonic zoning map of Iran's sedimentary basins Iran (down), modified from [7] . Numbers in this figure are, 1: Zagros-East Taurus hinterland, 2: Persian Gulf-Mesopotamian foreland basin, 3: Makran accretionary prism, 4: Bashagard Mountains, 5: Jazmorian-Mashkel fore arc basin, 6: Shah-savaran-Soltan magmatic arc, 7: South Lut-South Helmand back arc basin, 8: East Iran Mountain belt, 9: West-Central Alborz and lesser Caucasus hinterland, 10: Great Kavir-Northen Urmieh lake foreland basin, 11: South Great Kavirfold and thrust belt, 12: South Caspian-Black Sea foreland basin, 13: Urmieh-Dokhtar magmatic arc, 14: Naien-Kerman retro arc foreland basin, 15: Sanandaj-Sirjanoverthrust belts, 16: East Alborz or Binalod hinterland, 17: Torbat-e am-Neysh-abour retro arc foreland basin, 18: Kopet Dagh hinterland, 19: South Caspian remnant basin, 20: Maiamay-Taibad Inverted back arc basin, 21: Khaf-Kavir Plain magmatic arc, 22: Lut Plain-Gonabad back arc basin, 23: Tabas hinterland, 24: Yazd-Khour Piggy back basin.

Figure 2. ETM+ satellite image of Hormuz Island in the east- ern part of Persian Gulf. Map located in Figure 1.

Hormuz province

Hormuz province (Figure 3) in southeastern part of Zagros, with 84 emergent diapirs can be considered the most extensive salt basin of Zagros. Most of salt domes are pre-orogenic emergent diapirs, and had a prominent role in determining the shape and location of folds during the Zagros orogeny. But, some of the salt diapirs have been located on the plunging noses of anticlines. So, they are post-orogenic emergent diapirs. Also, salt glaciers (Figure 4) are common feature in this diapiric province.

Nyriz-Jahrum province

Nyriz-Jahrum province (Figure 3) in eastern part of Zagros, with a few emergent diapirs, many blind salt domes such as Sim, Meymand, Bavush, Kharman, Kuh-e Gavbast anticline and diapiric fold such as Kuh-e Bavush and Khaftar anticlines (Figure 5) can be considered as the blind salt basin of Zagros. The Southern part of Nyriz- Jahrum province has been formed by Gavbandi High [17] . An old regional uplift was formed by the Qatar Arc and its northward continuation towards Zagros. Many researchers believe that, Gavbandi High has more height related to around area during the Late Pre Cambrian-Early Cambrian and so, there was very shallow evaporate basin as same as western part of Zagros.

Shiraz-Kazerun province

Shiraz-Kazerun province (Figure 3) in central part of Zagros, with 23 emergent diapirs can be considered the second most extensive salt basin of Zagros.

Most of the salt diapirs have been located on the main segments of transverse strike slip faults such as Kazerun fault (Figure 6) with 6 salt diapirs, Kareh Bas fault with 6 salt diapirs and Sarvestan fault with 3 salt diapirs. So, most of them are pre-orogenic emergent diapirs, and they had a prominent role in determining the shape and location of folds during the Zagros orogeny. Also, salt glaciers such as Kuh-e Namak (Figure 6) are common feature in this diapiric province.

Ravar province

Ravar province (Figure 3) with north-south trending is bound to main fault of Nayband at east, and Kalmard at west. The subsidence of this province, during upper Jurassic, can be considered as the cause of evaporate con-

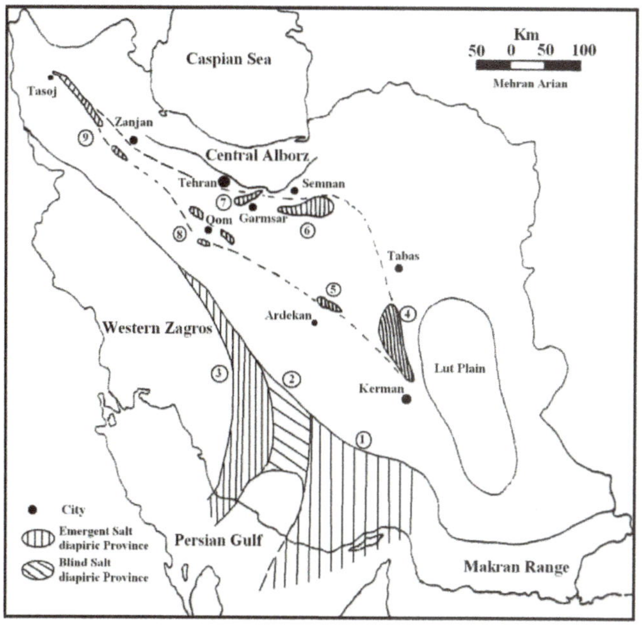

Figure 3. Geographical distribution of diapiric provinces in Iran. Locations of diapiric pro- vinces are shown by numbered circles.

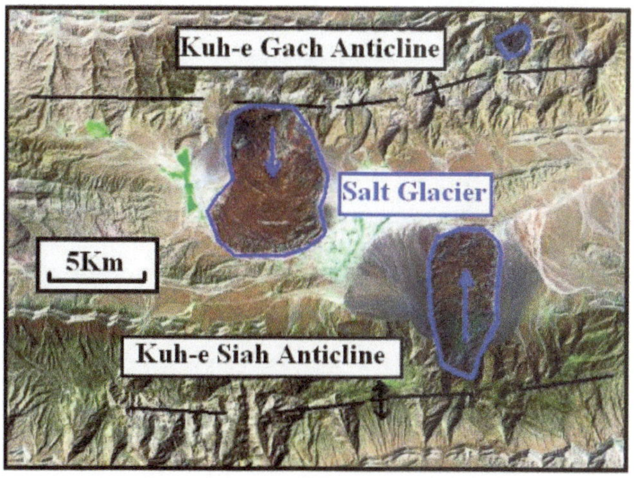

Figure 4. Interpreted ETM+ satellite image of two salt glaciers (blue color) at Hormuz dia- piric province, between two anticlines (black color). Map located in Figure 1.

centrations (Mago Group) in it. However, due to regional compression, we are now encountering its additional diapirism. Meanwhile, as the salt domes have projected in a dominant north-south direction, the external trigger impact on the region's diapirism shall be assumed effective [18] - [20] .

At this province, there are separate individual salt domes, at a north-south trending faults. The salt domes are related to upper Jurassic salt formation [21] along Ravar fault and diapiric unit of Ravar series along Lakarkuh fault (Figure 7).

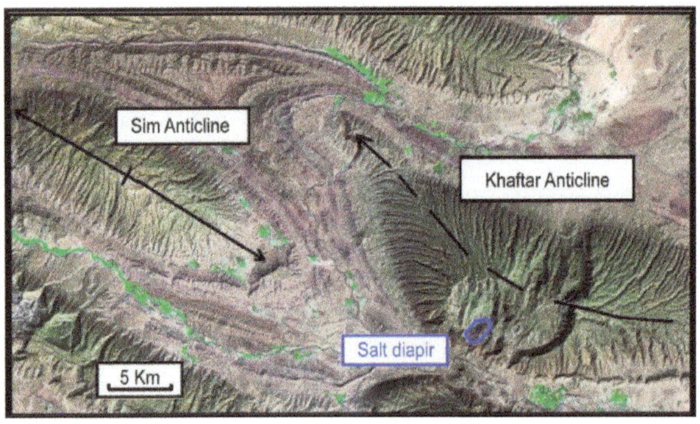

Figure 5. Interpreted ETM+ satellite image of Sim Anticline (a blind salt dome) and Khaftar anticline (a diapiric fold) with a small salt plug (blue color) at Nyriz-Jahrum province. Map located in Figure 1.

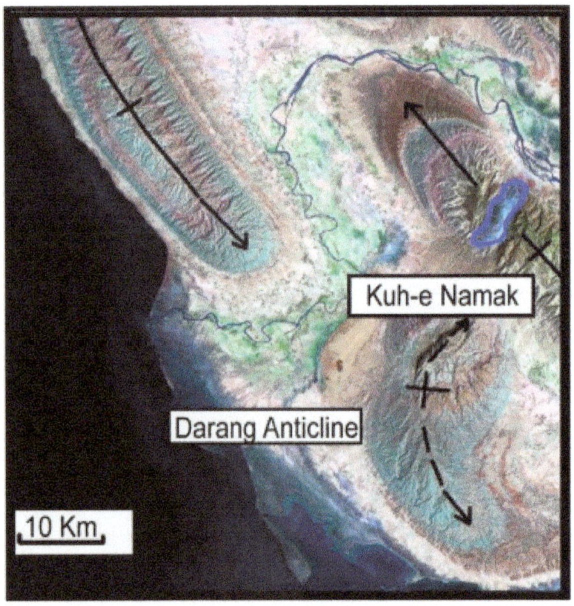

Figure 6. Interpreted ETM+ satellite image of Kuh-e Namak salt gla- cier (blue color) and Darang Anticline (a diapiric fold) at south of Shiraz-Kazerun pro- vince. Map located in Figure 1.

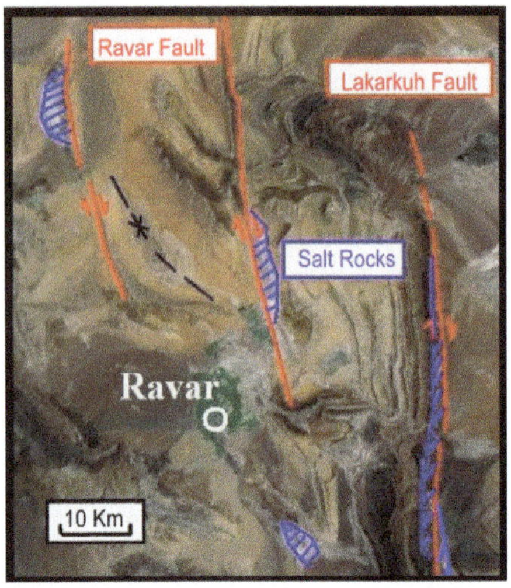

Figure 7. Interpreted ETM+ satellite image of Ravar region and its salt struc- tures (blue color). Map located in Figure 1.

Also, separated salt glacier (Figure 8) has been cropped out between Nayband and Lakarkuh faults, in exten- sional trend of resulting shear zone.

Kalot province

The salts of diapiric province of Kalot basin in northeast of Ardekan (Figure 3) is relate to the Qom basins. So, in this case, this province can be considered as the extent of southeastern part of the Qom diapiric province.

At Kalot region, the salt domes are mostly of oval shape; which at some regions, due to erosion function, are transformed to cabbage form flower. The salts of Kalot province are distinguished to be of old type and the external trigger has confirmed by shape of salt diapirs [22] .

Great Kavir Province

Great Kavir province (Figure 3), at south of Semnan can be considered as the most extensive salt basin of Central Iran. Because, based on the previous studies, such as [23] and [24] , there are old and new type salt diapirs. Now we encounter penetration of interrelated diapirs with old and new salts—side by side or inside each other. At this province, more than 50 salt domes are emergent. Erosion has resulted in the formation of 12 side-by- side diapirs (Figure 9), whose their appearance is similar to that of a canopy. The salts of this region are of old and new types. So the diapirism of this area can be assumed to have two separate origins. Also, a typical Crescentic folds (Figure 10) and a Vortex mushroom diaper (Figure 11) has been appeared in Great Kavir province.

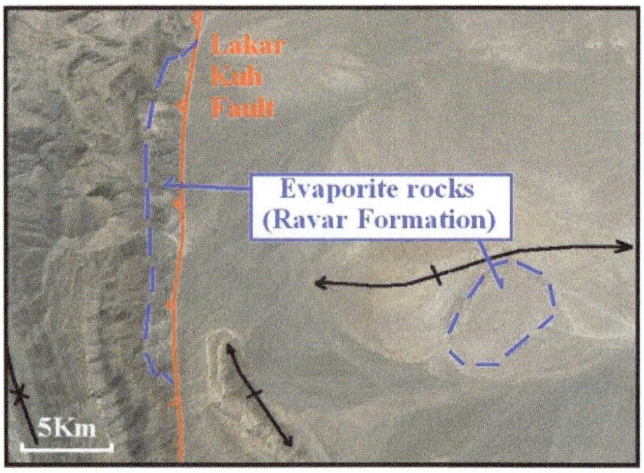

Figure 8. Interpreted ETM+ satellite image of a salt glacier (blue color) in the eastern part of Lakarkuh fault. Map located in Figure 1.

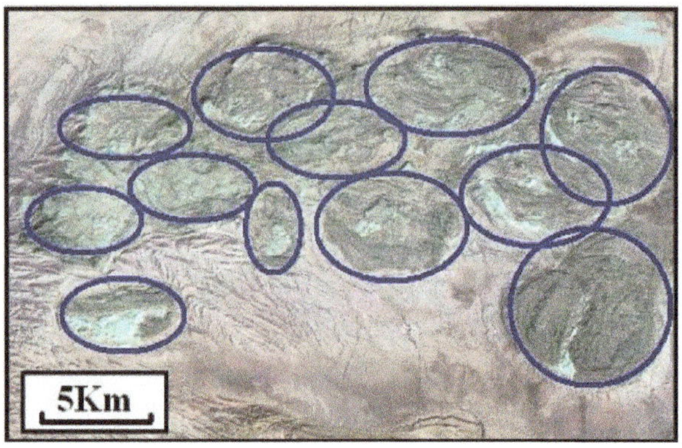

Figure 9. Interpreted ETM+ satellite image of 12 side-by-side salt diapirs (blue color) at south of Semnan. Map located in Figure 1.

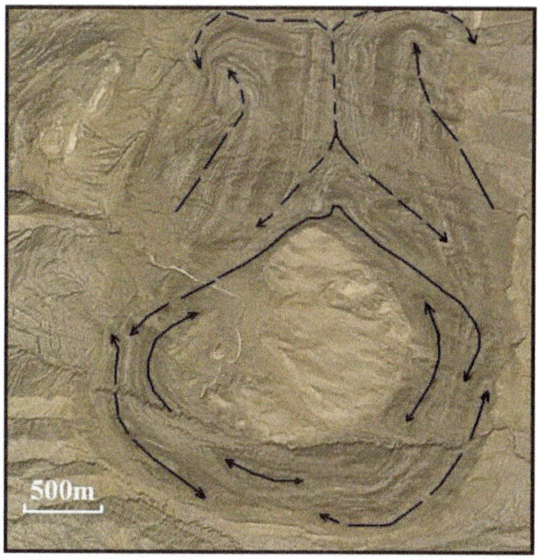

Figure 10. Interpreted ETM+ satellite image of a typical Cres- centic folds (black color) at southeast of Semnan. Map located in Figure 1.

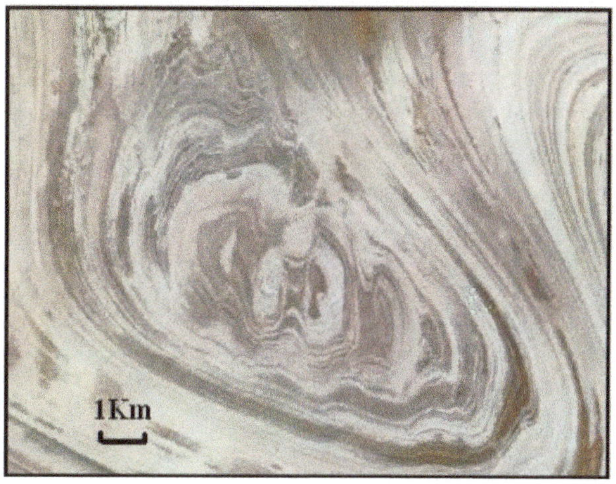

Figure 11. Interpreted ETM+ satellite image of a typical eroded Vor-tex mushroom diapir at Great Kavir province. Map located in Figure 1.

Garmsar Province

Garmsar province (Figure 3) belongs to depression of northwestern margin of Great Kavir. The salts of this province are of old type, in which presence of salt, gypsum and sulfur reserves has been resulted in increase of mining activities at this area. The origin of Garmsar region salt domes are distinguished as Lower Red formation and external trigger have confirmed by stretch shape of salt dome. There is a salt-tongue canopy with 130 km^2 areas that introduced as the biggest salt glacier in Iran (Figure 12) in front of an active region or Central Alborz [25] - [27] .

Qom Province

The diapiric province of Qom can be defined as a northwestern-southeastern depression. The maximum longitudinal extension of this province, is about 200 km, which extends from west of Saveh to south of Qom. The implemented studies indicate, that the salts of this province, are mostly of old type and their limited projections, relate to the presence of main structures at the region. There is a separate individual salt dome (Figure 13), at a relatively extensive plain. According to [28] , diapirism of this structure can be assumed to have two separate origins (a combination of old and new type).

Zanjan-Tasoj Province

This province is located from southwest of Zanjan to south of Tasoj (Figure 3). The implemented studies indicate that the salts of this province are of new type that intruded along main faults [29] .

DISCUSSION

Based on [30] , three diapiric provinces in the Southern Iran (Zagros and Persian Gulf) that they have concerned to the Hormuz, Nyriz-Jahrum and Shiraz-Kazerun salt basins, are related to the evaporate basins of thin-skinned folded and thrusted belt [31] and [32] . Two diapiric provinces in the Central Iran that they have concerned to the Great Kavir and Garmsar salt basins, are related to the evaporate basins of gravity gliding-related shortening at toe of a continental rift. Other diapiric provinces in the Central Iran that they have concerned to the Ravar, Kalot, Qom and Zanjan-Tasoj salt basins, are related to the evaporate basins of intracratonic inverted basin. Also, main characteristics of salt diapirism in the Southern Iran and Central Iran salt basins have summarized in table 1.

There are three mother layers in the Southern Iran salt basins. It seems that, salt combination of Hormuz and Dashtak mother layers (Late Triassic) caused to formation of Salt-Wing Intrusion [33] in Zagros and Persian Gulf. Also, thick evaporate Gachsaran formation (Late Miocene), formed a type of Raft Tectonics in Fars Group, especially in the Dezful Embayment. Diapirism process of Gachsaran mother layer is comparable to Upper Red mother layer in the Central Iran salt basins.

CONCLUSIONS

Based on the shape and orientation investigations of salt domes in Iran, nine diapiric provinces have introduced. Six diapiric provinces in Central Iran have concerned to the Ravar, Kalot, Great Kavir, Garmsar, Qom and Zanjan-Tasoj salt basins. Three diapiric provinces in Zagros and Persian Gulf have concerned to the Hormuz, Nyriz- Jahrum and Shiraz-Kazerun salt basins.

Salt diapirism of Central Iran provinces can be considered as originated from function of tectonic forces, effective in the foreland basins of Central Iran. It means that tectonic processes (external trigger) are predominant compared to Halokinesis (internal trigger) in Central Iran. But, Salt diapirism of Zagros and Persian Gulf has not triggered by tectonic forces, although has affected by it. It appears that Halokinesis are predominant compared to tectonic processes in Zagros that initiated by Neogene continental collision.

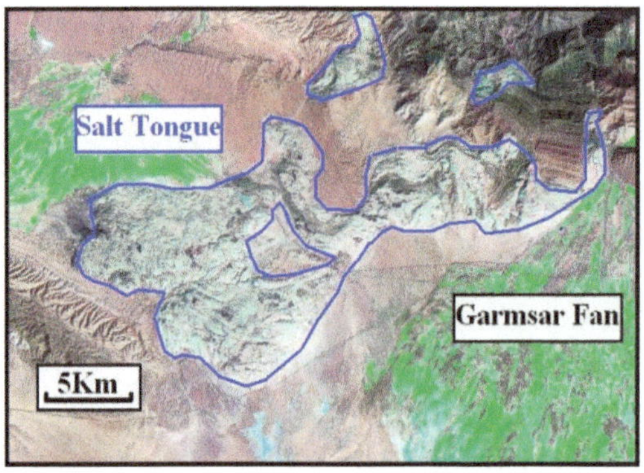

Figure 12. Interpreted ETM+ satellite image of a big salt-ton- gue (blue color) at Garmsar province. Map located in Figure 1.

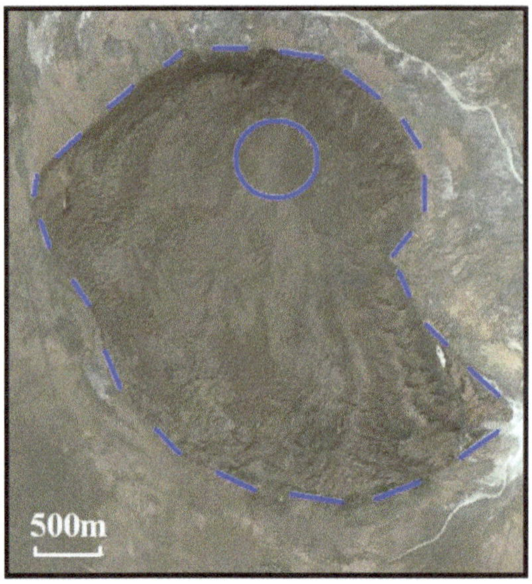

Figure 13. ETM+ Satellite image of a salt glacier (blue color) at northwest of Qom. Map located in Figure 1.

Table 1. Main characteristics of salt diapirism in the Southern Iran and Central Iran salt basins

Characteristics of salt diapirs	Southern Iran salt basins	Central Iran salt basins
Age of salt diapirs	Early Cambrian, Late Triassic	Early Cambrian, Late Jurassic, Eocene-Oligocene and Miocene
Name of diapiric formations or groups	Hormuz, Dashtak	Ravar, Mago, Lower and Upper Red
Thickness of salt rocks	Up to 1000 m	Down to 1000 m
Index structures	Salt Glacier and Diapiric Folds	Salt Glacier and Canopy
Salt deposition setting	Intercontinental Rift	Intracontinental Rift
Setting of salt diapirs evolution	Passive Margin and Collision Orogenic Belt	Retro arc Basin and Retro arc Foreland Basin
Shape of salt diapirs	Upward Narrowing Diapirs	Columnar and Upward Widening Diapirs
Regional shortening	Post-Diapirism	Pre & Syn-Diapirism
Diapirism as result of	Halokinesis	Tectonic Forces
Freeboard elevation	to 1000 m	to 500 m
Level of neutral buoyancy	Near to the Earth Surface	until 500 m Subsurface
Overburden behavior	More Brittle	More Ductile
Recent deformational rate	10 mm per year	2 mm per year
Seismicity frequency	High	Low
Folding type	Close to Open	Open to Gentle
Salt creep rate	More	Less
Humidity of the weather	High	Low
Diapirs position	Determined the shape and location of folds	Controlled by structures such as faults and folds
Number of diapiric provinces	3 big provinces	6 small and medium provinces

ACKNOWLEDGEMENTS

Mehran Arian, Hamideh Noroozpour This work has funded by the Department of Geology, Islamic Azad University, Science and Research Branch, Tehran, Iran. Also, special thanks to vice-president for research in Science and Research Branch, Tehran.

REFERENCES

1. Jackson, J.A. and Bates, R.L. (1997) Glossary of Geology. 4th Edition, American Geological Institute, Alexandria, 769 p.
2. Talbot, J.C. and Jackson, M.P.A. (1987) Salt Tectonics. Scientific American, Colorado, 70-79.
3. Qorashi, M. and Arian, M. (2011) Tectonics of Iran. Geologic Survey of Iran, Tehran, 336 p. (In Persian)
4. Arian, M. (2011) Basement Tectonics and Geology of Iran. Asar Nafis Press, Qum, 300 p. (In Persian)
5. Motiei, H. (2001) Simplified Table of Rock Unit in South West Iran. A Map Unpublished, KEPS Company, Tehran, 234 p.
6. Stocklin, J. (1968) Salt Deposits of the Middle East. Geological Society of America, Colorado, 181 p.
7. Arian, M. (2013) Physiographic-Tectonic Zoning of Iran's Sedimentary Basins. Open Journal of Geology, 3, 169-177.
8. Berberian, F., Muir, I.D., Pankhurst, R.J. and Berberian, M. (1982) Late Cretaceous and Early Miocene Andean-Type Plutonic Activity in Northern Makran and Central Iran. Journal of Geological Society of London, 139, 605-614.http://dx.doi.org/10.1144/gsjgs.139.5.0605
9. Stocklin, J. (1971) Stratigraphic Lexicon of Iran Part 1: Central, North and East Iran. Geologic Survey of Iran, Tehran, 338 p.
10. Jackson, M.P.A., Cornelius, R.R., Craig, C.H., Gansser, A., Stocklin, J. and Talbot, J.C. (1990) Salt Diapirs of the Great Kavir, Central Iran. Geological Society of America, Colorado, 139 p.
11. Talbot, C.J., Aftabi, P. and Chemia, Z. (2009) Potash in a Salt Mushroom at Hormoz Island, Hormoz Strait, Iran. Ore Geology Reviews, 35, 317-332.http://dx.doi.org/10.1016/j.oregeorev.2008.11.005
12. Arian, M. (2011) A Preface on Salt Diapirism of Iran. Asar Nafis Press, Qum, 309 p. (In Persian)
13. Asadian, F. and Arian, M. (2009) Identification of Diapiric Provinces of Central Iran through Geological and Geographical Analysis. International Journal of Agriculture Environment & Biotechnology, 2, 3443-3451.
14. Arian, M., Ahmadnia, A., Qorashi, M. and Pourkermani, M. (2002) Structural Analysis of Mengharak Transcurrent Fault System in Zagros, Iran. Special GEO 2002 Conference Issue Geoarabia, 7, 209-210.
15. Jahani, S., Callot, J.P., Frizon de Lamotte, D., Letouzey, J. and Leturmy, P. (2007) The Salt Diapirs of the Eastern Fars Province (Zagros, Iran): A Brief Outline of Their Past and Present. Frontiers in Earth Sciences, 2, 289-308. http://dx.doi.org/10.1007/978-3-540-69426-7_15
16. Arian, M. (2012) Clustering of Diapiric Provinces in the Central Iran Basin. Carbonates and Evaporites, 27, 9-18. http://dx.doi.org/10.1007/s13146-011-0079-9
17. Falcon, N. (1974) Southern Iran: Zagros Mountains. In: Spencer, A.M., Ed., Mesozoic-Cenozoic Orogenic Belts, Geological Society, London, Vol. 4, 199-211.
18. Arian, M. (2003) Seismotectonic and Earthquake-Fault Hazard Investigations in the Kerman Region. Research Institute of Applied Sciences, Tehran, 183 p. (In Persian)
19. Arian, M. (2010) Earthquake-Fault Hazard Investigations in the Kerman Quadrangle. Journal of Sciences (Islamic Azad University), 19, 176-182.
20. Arian, M., Pourkermani, M., Sistanipour, A. and Noroozpour, H. (2011) Kinematic Significance of Fold- and Fault- Related Fracture Systems in the Rafsanjan's Northeast Highlands (Central Iran). Journal of Basic and Applied Scientific Research, 1, 3398-3406.
21. Stocklin, J. (1961) Lagoonal Formations and Salt Domes in East Iran. Bulletin of Iranian Petroleum Institute, 3, 29-46.
22. Arfa-Nia, R. (1998) Diapirism in NE Ardekan, Kalot Region. M.Sc. Dissertation, Structural Geology, Islamic Azad University, North Tehran Branch, Tehran, 230 p. (In Persian)
23. Arian, M., Maleki, Z. and Noroozpour, H. (2011) Cenozoic Diastrophism and Deformational Events in the East-Central Alborz. Journal of Basic and Applied Scientific Research, 1, 2394-2400.
24. Feizi, F., Arian, A. and Rahmani, R. (2007) Seismotectonic Zoning in the Eastern Part of the Central Alborz. Journal of Sciences (Islamic Azad University), 17, 151-164. (In Persian)
25. Khavari, R., Arian, M. and Ghorashi, M. (2009) Neotectonics of the South Central Alborz Drainage Basin, in NW Tehran, N Iran. Journal of Applied Sciences, 9, 4115-4126.http://dx.doi.org/10.3923/jas.2009.4115.4126
26. Arian, M. and Bagha, N. (2012) Active Tectonics of Tehran Area, Iran. Journal of Basic and Applied Scientific Research, 2, 3805-3819.
27. Bagha, N., Arian, M., Ghorashi, M., Pourkermani, M., El Hamdouni, R. and Solgi, A. (2014) Evaluation of Relative Tectonic Activity in the Tehran Basin, Central Alborz, Northern Iran. Geomorphology, 213, 66-87. http://dx.doi.org/10.1016/j.geomorph.2013.12.041
28. Talbot, C.J. and Aftabi, P. (2004) Geology and Models of Salt Extrusion at Qum Kuh, Central Iran. Journal of Geological Society, 161, 321-334.http://dx.doi.org/10.1144/0016-764903-102
29. Pourkermani, M. and Arian, M. (2001) Structural Geomorphology of Northeastern Kurdistan, Sistan and Baluchestan University. Journal of Humanities, 7, 37-48.
30. Letouzey, J., Colleta, B., Vially, R. and Chermette, J.C. (1995) Evolution of Salt Related Structures in Compressional Settings. In: Jackson, M.P.A., Roberts, D.G. and Snelson, S., Eds., Salt Tectonics: A Global Perspective, American Association of Petroleum Geologists, Tulsa, Vol. 65, 41-60.

31. Arian, M. and Hashemi, A. (2008) Seismotectonic Zoning in the Zagros. Journal of Sciences, 18, 63-76.

32. Arian, M., Qorashi, M., Pourkermani, M. and Ahmadnia, A. (2006) The Structural Significance Kareh Bas Transcurrent Fault System in the Zagros Fold and Thrust Belt. Journal of Geosciences, 15, 126-133. Geological Survey of Iran, Tehran. (In Persian)

33. Hudec, M.R. and Jackson, M.P.A. (2006) Advance of Allochthonous Salt Sheets in Passive Margins and Orogens. American Association of Petroleum Geologist Bulletin, 90, 1535-1564.

CHAPTER 4

TECTONIC GEOMORPHOLOGY AND VOLCANO-TECTONIC INTERACTION

Engielle M. R. Paguican and Marcus I. Bursik

Keywords: volcano-tectonic, tectonic-geomorphology, Hat Creek Graben region, California, Southern Cascades, morphometry, spatial distribution, volcano alignments

The eastern boundary of the Southern Cascades (Hat Creek Graben region), California, USA, is an extensively faulted volcanic corridor between the Cascade Range and Modoc Plateau. The morphology of the region is a result of plate motions associated with different tectonic provinces, faulting, and recurring volcanic activity, making it an ideal place to study the interrelationship between tectonics, volcanoes, and geomorphology. We use the morphometry and spatial distribution of volcanoes and their interaction with regional structures to understand how long term regional deformation can affect volcano evolution. A database of volcanic centers and structures was created from interpretations of digital elevation models. Volcanic centers were classified by morphological type into cones, sub-cones, shields and massifs. A second classification by height separated the larger and smaller edifices, and revealed an evolutionary trend. Poisson Nearest Neighbor analysis showed that bigger volcanoes are spatially dispersed while smaller ones are clustered. Using volcano centroid locations, about 90 lineaments consisting of at least three centers within 6 km of one another were found, revealing that preferential north-northwest directed pathways control the transport of magma from the source to the surface, consistent with the strikes of the major fault systems. Most of the volcano crater and collapse scar openings are perpendicular to the north northwest-directed maximum horizontal stress, expected for extensional environments with dominant normal faulting. Early in the history of a volcano or volcano cluster, melt propagates to the surface using the easiest and most efficient pathway, mostly controlled by the pre-existing normal faults and near-surface stress fields, as indicated by the pervasive vent alignments. Volcano growth continues to be dependent on the regional structures as indicated by the opening directions, suggesting structural control on the growth of the volcanic edifices. The results present a particularly well-defined case in which extension of a volcanic region is accommodated mostly by faulting, and only partly by intrusion to form volcanoes. This is attributed to a low magma supply rate.

INTRODUCTION

Volcanoes occur in a wide variety of shapes and sizes as a result of the interaction of constructive and destructive geological and environmental processes. The quantitative description of volcano morphology, called volcano morphometry, is a systematic way to characterize the morphology of volcano edifices, allowing for the extraction of information on the processes that interact and factors that control growth history and evolution (Grosse et al., 2012, 2014). At the same time, describing a volcanic field in which volcanoes grow is essential for characterizing key controls on volcano growth in space, which reflects the pathway used by magma to reach the surface, and the existence of any dominantly controlling factor that may have created preferential pathways (Le Corvec et al., 2013). At the surface and through time, the growth or destruction of a volcano is influenced by the volcano-tectonic setting of the region. Studies looking at the influence of regional stress on the direction of sector-collapse and debris-avalanche amphitheater opening have been done for Southeast Asian volcanoes in tectonic settings dominated by strike-slip faults, particularly those in the Philippines, byLagmay and Valdivia (2006), Japan by Ui et al. (1986), and Indonesia byBahar and Girod (1983), which showed that the opening direction of such features is at an acute angle relative to the regional maximum horizontal stress. Their results are in disagreement with earlier models by Moriya (1980), Siebert (1984), and Nakamura (1977), which show volcano openings perpendicular to the maximum horizontal stress being more

Citation: Paguican EMR and Bursik MI (2016) Tectonic Geomorphology and Volcano-Tectonic Interaction. Front. Earth Sci. 4:76. doi: 10.3389/feart.2016.00076.

prevalent in extensional regions (Tibaldi, 1995), where tabular conduits and dikes that propagate along normal faults promote collapses perpendicular to the regional maximum horizontal stress. This difference in results may be due to differences in the underlying fault kinematics, the resulting geometry of large-scale sliding and debris avalanching, original crater opening direction due to temporal and topographic stress-field control within a volcanic body (Ui et al., 1986), a dipping substrate, or a spreading basement.

The eastern boundary of the Southern Cascades (herein referred to as the Hat Creek Graben region) in California is an extensively faulted volcanic corridor with spectacular, high, steep scarps in a bedrock of late Tertiary and Quaternary volcanic and sedimentary deposits (Wills, 1991; Muffler et al., 1994; Blakely et al., 1997). It is at the boundary between two distinct geologic and geomorphic provinces (Jenkins, 1938), the Cascade Range on the west and the Modoc Plateau on the east, between Mt. Shasta and Lassen Peak on the north and south, respectively. It is in the transition zone between several significant volcanic and tectonic provinces, including the east-west convergence and subduction of the Gorda Plate underneath the North American Plate, the current north-south shortening within the Klamath Mountain Region on the northwest, and the region of east-west extensional faulting that produced the dominant north- to northwest- striking normal faults, forming the horst and graben topography and voluminous basaltic volcanism (White and Crider, 2006). Volcano morphology is a result of recurring volcanic activity within this diverse tectonic setting, from more than 500 vents over the past 7 Ma (Muffler et al., 1994), making it an important place to study the long-term interrelationship between tectonics, geomorphology, and volcanic activity. Many studies have recently explored the importance of magma in shaping divergent plate boundaries (Wright et al., 2006; Ebinger et al., 2010; Sigmundsson et al., 2015; Acocella and Trippanera, 2016), and here we seek to discern whether some of these ideas also apply to intraplate magmatic provinces experiencing extension, such as the Hat Creek Graben region.

This study aims to provide a tectonic-geomorphic interpretation of the eastern boundary of the southern Cascades—the Hat Creek Graben region. It is hoped that the results will aid in producing a more refined understanding of regional volcano-tectonic interaction and history, which may help guide future hazard assessment. We explore the overall geomorphology and activity of individual volcanoes and the potential effects of regional stress regimes. We characterize local structures within each volcano, and compare the trend of volcano erosional and collapse scars with features in the basement faults, and regional maximum horizontal stress. We furthermore attempt to refine our understanding of the probable kinematics of underlying faults that may cause volcano instability.

REGIONAL TECTONIC SETTING AND FAULT SYSTEMS

The tectonics of the Hat Creek Graben region (Figure 1A) is dominated by the convergence of the Pacific, Juan de Fuca, Gorda and North American Plates; the rigid block of the Sierran Microplate; and the dominantly extensional western Basin and Range (Hammond and Thatcher, 2005). Subduction of the Juan de Fuca and Gorda Plates underneath the North American Plate forms the Cascadia Subduction Zone and the associated Cascade Arc volcanoes. To the south, the Sierran Microplate behaves as a rigid block that, due to the faster rate of right-lateral shear on its western side, accommodated by the right-lateral San Andreas Transform Fault and associated structures (Dixon et al., 2000; Faulds et al., 2005), is translated northward. This causes compression (north-south shortening) in the Klamath Mountains region on the northwest (McCaffrey, 2005). Furthermore, the motion of the Sierran Microplate contributes to the clockwise rotation of the Oregon forearc, where the Sierran Microplate and Cascadia forearc blocks meet (McCaffrey, 2005). On the eastern boundary of the Sierran Microplate, the Walker Lane accommodates some 15–25% of the total relative plate motion (Faulds et al., 2005), with GPS measured strain rates of about 11 mm yr^{-1} (Dokka and Travis, 1990; Argus and Gordon, 1991; Sauber et al., 1994; Bennett et al., 1999; Dixon et al., 2000; Hammond and Thatcher, 2007). Some of the right-lateral shear related to relative Pacific and North American plate motion is accommodated within the Hat Creek Graben region, and connects with poorly defined zones of deformation between the Gorda, Pacific, and North American plates near the Klamath Mountains (Miller et al., 2001). Crustal stresses related to the complex block motions and volcanic processes drive regional seismicity, and the observed patterns of faulting of the Hat Creek Graben region.

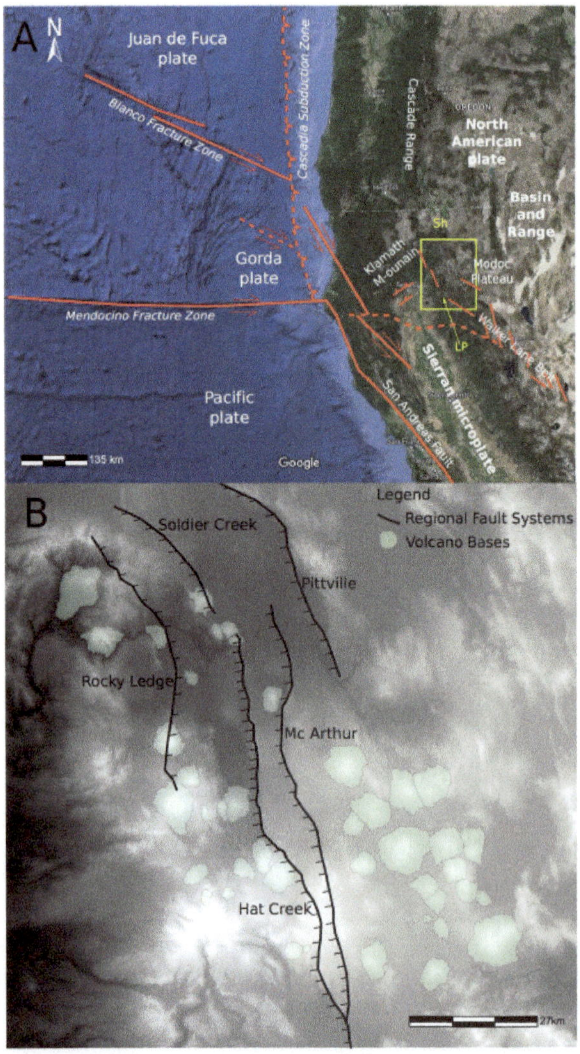

Figure 1. (A) Regional tectonic setting of the eastern boundary of the Southern Cascades (Hat Creek Graben region), California, is

dominated by the convergence of the Pacific, Juan de Fuca, Gorda, and North American Plates resulting in the formation of the Cascadia Subduction Zone and its associated Cascade Arc Volcano with Lassen Peak (LP) as its southernmost volcano; the rigid block of the Sierran Microplate that acts like a piston compressing the Klamath Mountain on the northern boundary and creating the Ink Creek Fold Belt southwest of Lassen Peak; the Oregon forearc where the Sierran Microplate and Cascadia forearc blocks meet; the dominantly extensional western Basin and Range Region and Walker Lane; and the Hat Creek Graben region (yellow box) that includes the area south of Mt. Shasta (Sh), Pit River Region, Mushroom Rock Anticline, Cinder Butte, Butte Creek, Bald Mountain, Big Lake, Brushy Butte, and Fall River Valley. Modified from Faulds et al. (2005) and Austin (2013). (B) General distribution of the regional fault systems and volcanoes.

Fault scarps are prominent in the Hat Creek Graben region, especially in places where they displace resistant Pliocene to early Pleistocene basalt, with up to millions of years of accumulated offset (Blakeslee and Kattenhorn, 2013). Major fault systems (Figure 1B) identified and described by previous workers include the Hat Creek, Rocky Ledge, McArthur, Soldier Creek, and Pittville Fault Systems.

The Hat Creek Fault System is a late Pleistocene to Holocene (Parrish, 2006;U.S. Geological Survey and California Geological Survey, 2006; Muffler et al., 2010), 47-km long, up to 490-m high, west-facing escarpment along the east side of Hat Creek Valley (Blakeslee and Kattenhorn, 2013). It has a slip rate of about 1-3.6 mm yr^{-1}, and is capable of producing an earthquake of moment magnitude about 6.7, with a recurrence interval of 667 ± 167 yr (Wills, 1991;Blakeslee and Kattenhorn, 2013). The Hat Creek Fault has a component of right-lateral offset along with normal offset (Wills, 1991), with segments connected by relay ramps and monoclines (Muffler et al., 1994). Due to sporadic volcanic activity, surface rupture along the Hat Creek fault has progressively migrated westward since the Late Pleistocene, with older scarps successively abandoned (Blakeslee and Kattenhorn, 2013). North of the Pit River, slip of the Hat Creek Fault probably transfers to the Soldier Creek Fault, which displaces a 1.2 Ma basalt by more than 260 m, giving a long-term vertical separation rate of 0.22 mm yr^{-1}. On the western edge of Hat Creek valley is the 9 km long, north-trending and east-facing Rocky Ledge Fault System. It has a component of right-lateral motion on a dominantly normal structure. The strike-slip component is consistent with the northwest propagation of the Walker Lane Belt, with suggested Holocene activity (Wills, 1991; Austin, 2013). The Rocky Ledge Fault System has an average slip rate of 0.4 ±0.2 mm yr^{-1}, an order magnitude less than the late Quaternary slip rate on the Hat Creek Fault (Sawyer and Ramelli, 2012). East of and parallel to the Hat Creek fault zone, there are normal faults with a right-lateral component, as well as the west-facing scarps of the McArthur Fault Zone. A steep scarp in alluvium with very weak soil development, a large closed depression that flanks a west-facing scarp in alluvium south of Big Lake, and tonal lineaments in plowed fields indicate Holocene offset along this fault (Wills, 1991). The more northwesterly Pittville Fault extends for about 48 km to the southeast, partly controlling the course of the Pit River and offsetting Brushy Butte, the volcanic center north of Big Lake (Peterson and Martin, 1980). Recent movements are indicated North of the Pit River Valley, where the fault forms a 1–3 m scarp in Late Pleistocene basalt, with open fissures up to 2 m deep on the upthrown side.

METHODOLOGY

Shuttle Radar Topography Mission (SRTM), higher-resolution, local LiDAR, and 10-m spacing National Elevation Data (NED) (Gesch et al., 2002) were used in the present study. NED data, covering the area of the Southern Cascades from 40.22 to 41.33 N and 122.08 to 120.22 W in decimal degrees, were acquired from the USGS National Map website (http://ned.usgs.gov), and were the primary dataset used for mapping and interpreting the overall tectonics, geomorphology and volcano distribution. In addition, Landsat satellite images, which have been used to study remote active volcanoes and to map basaltic volcanic fields (Francis and Wells, 1988), were downloaded. The elevation data were loaded in ArcGIS® software and saved as an ENVI compatible file for processing and topographic modeling.

Topographic Modeling, and Structural and Geomorphological Interpretation

Topographic modeling techniques were applied to the DEM using RSI ENVI®software generated surface parameters, including shaded relief, slope, and aspect maps. Landsat 7 Thematic Mapper Plus (ETM+) and panchromatic band 8 images were processed to combine bands and obtain maximum contrast to check surface features and deposits observed in the DEM. Generated images were compiled using ArcGIS® software.

Geomorphic features, including volcanic, tectonic, and volcano-tectonic features such as volcano edifices, cone and lava fields, and faults, as well as features resulting from gravitational instability and erosion were visually interpreted and described from the images. Interpretations were checked in the field. Structural analysis of the elevation dataset was based on the interpretation of rectilinear features or lineaments of regional scale. Lineaments consist of sharp tonal differences and alignments of geomorphological features like cones, streams and rivers, ridges and crests. Within tectonically active areas such as the Hat Creek Graben region, lineaments usually coincide with the fracture network and the major fracture patterns. Volcanoes are identified from descriptions in previous literature and visually interpreted by their positive relief with a circular or elongated shape, and may have a small crater near the summit. Volcano bases were manually delineated, by following the break in slope around the edifice base. Thus, only the objectively definable, visible edifice was traced and far-reaching fall and flow products were not considered in the delineation. Main lineament directions and volcano base elongations were then defined by the azimuth frequency distribution plotted on rose diagrams. Terrain modeling, geomorphological interpretation, and geomorphometry were then combined to interpret the relationships between the regional stress, basement faults, and volcanoes.

Volcano Morphometry

Individual volcano base and crater shapefiles were extracted as regions of interest from the DEM and processed using MORVOLC to generate a quantitative characterization of each individual edifice, and to develop a systematic way to compare the edifices. MORVOLC (Grosse et al., 2009) is an IDL-based code that generates a database of quantitative descriptions of the size (height, width, and volume), planview shape (ellipticity and irregularity), and profile shape (height-width ratio, summit width to basal width ratio, and slope) of stratovolcanoes and cones. Morphometric analysis can provide clues to the processes that interact during the growth history, development and evolution of each volcano, and can be used to draw an evolutionary trend for volcanoes in a region. MORVOLC calculates a 3D basal surface from the edifice outline using a least-squares criterion, which was used to estimate the volume and height parameters. Elevation contour lines were then generated from the DEM, and a summit region was defined at the elevation where the edifice starts flattening out.

Volcano Spatial Distribution and Alignments

MORVOLC generated the geographic coordinates of volcano summits that we used as the volcano centroids. These point data sets were used as input information for running the Geological Image Analysis Software (GIAS,www.geoanalysis.org) code in Matlab developed by Beggan and Hamilton (2010). GIAS evaluates the amount of randomness or any patterns of a spatial distribution of points by comparing the mean distance between all the nearest neighbors in the population with those defined by an idealized statistical model (Clark and Evans, 1954), Poisson Nearest Neighbor Analysis is preferred, in the present case. The nearest neighbor distances are determined both for the observed and expected distribution and then their mean distances are calculated to obtain the population-dependent R_0 and R_e(Baloga et al., 2007), respectively. Our observed population consists of the number of volcanoes per subgroup based on their morphometric class. The sample area is a convex hull generated by connecting the outermost points of the population (Hamilton et al., 2010; Le Corvec et al., 2013). To overcome the sample-dependent bias and properly assess the suitability of the PNN analysis, the R and c values were plotted with a confidence interval of 2σ. GIAS outputs a plot of the volcanic centers and the convex hull of the volcanic area, the R and c statistical values, and the skewness vs. kurtosis.

Alignment patterns found by PNN provided important constrains for understanding controls on regional volcanic evolution. Using the area of the convex hull from PNN analysis computed by GIAS, and the observed volcano population based on morphological characterization by MORVOLC, we calculated the normalized distance between the volcanoes used to create alignments, based on the density of the volcanoes. This takes into account the problem of volcanic fields covering a large area even though the mean distance between volcanic centers is generally small, due to magma propagating through new fractures or re-activated pre-existing structures (Valentine and Hirano, 2010). Due to differences in densities and mean distances of volcanic centers for each volcanic field within the Hat Creek region, we applied the relationship between the maximum length for the generation of alignments and the density of volcanic centers originally determined by Le Corvec et al. (2013) for monogenetic volcanic fields. This relationship was used to calculate the maximum length to generate lineaments in the Hat Creek volcanic region. A Three-point Alignments Matlab code used for volcanic alignment analysis (Le Corvec et al., 2013) was used to identify groups of points in a population of volcanic centers that form straight lines (Wadge and Cross, 1988; Connor et al., 1992; Baloga et al., 2007; Bleacher et al., 2009; Le Corvec et al., 2013), and to generate a list of alignment azimuths. Directions of alignments are presented in a rose diagram to find out if there is a a strong controlling factor that results in preferred alignment during volcano growth.

Volcano Morphology and Regional Stress Relationship

We compared crater morphology and regional stress relationships only for bigger volcanoes, to draw from a homogeneous population. Amphitheaters, breaches, scars or craters were our primary indicator for describing a part of the volcano's history that could be related to regional stress (Lagmay and Valdivia, 2006). We identified volcanoes that show amphitheaters, volcanic scars and breaches. A scar was used to describe craters deformed by major collapse events, including debris avalanches and large landslides. A crater breach is a scar-like deformation brought about by erosion such as by glaciers (cirques) or by meteoric and hydrological events forming gullies and gully heads.

To determine regional stress, we obtained the orientation of the nearest regional maximum horizontal stress node to each bigger volcano from the World Stress Map, a standard global compilation of contemporary tectonic stress (Heidbach et al., 2008). Stress orientation in the map is from various types of stress indicators, such as earthquake focal mechanisms; well bore breakout and drilling induced features; in-situ stress measurement such as overcoring, hydraulic fracturing, and borehole slotter; and young geologic data from fault slip analysis and volcanic vent alignments (Heidbach et al., 2008). We then measured the direction at which scars and breaches open using the angle measurement tool in GNU Image Manipulation Program®. The absolute difference between these two angles is the angle difference.

RESULTS

Topographic Modeling, and Structural and Geomorphological Interpretation

Elevation Modeling

The elevation range in the Hat Creek region is from 150 to 3200 m with a mean elevation of 1400 ± 400 m. To simplify initial exploration of terrain development, elevation density slicing was applied to the DEM, generalizing the topography. This process enhanced the regional detail of the DEM. Based on the original and density sliced DEM (Figures 2A,B), there are two major elevation profiles: the low elevation areas in cool colors and high areas in warm colors. The low elevation areas are subdivided into four areas based on the continuity of the low elevations compared to adjacent, higher areas: A, B1, B2, and C. The low areas decrease in mean elevation from east to west. The lowest mean elevation (A), darkest blue in Figure 2A and magenta in Figure2B, are the Sacramento Valley and Sacramento River and its tributaries from the Klamath Mountains in the north to the western side of Lassen Volcano National Park in the south. This area is separated from the lower area of Lake Britton (B1) on the east by Chalk and Hatchet Mountains, which are both part of the Klamath

Mountains. The Pit River, which is the northern main tributary of the Sacramento River, carves the divide between these two mountains. In lightest blue and highest in mean elevation among the three low areas is an area that includes lakes in the southern part of Brushy Butte lava field, and McArthur and Fall River Mills (B2). It is separated from the lower western area (B1), of Lake Britton and Hat Creek Valley, by the Hogback ridge and Saddle and Haney Mountains. The green, flat area consists of Lookout and Bieber areas and Ash Creek State Wildlife Area (C), possibly the easternmost tributary of the Sacramento River.

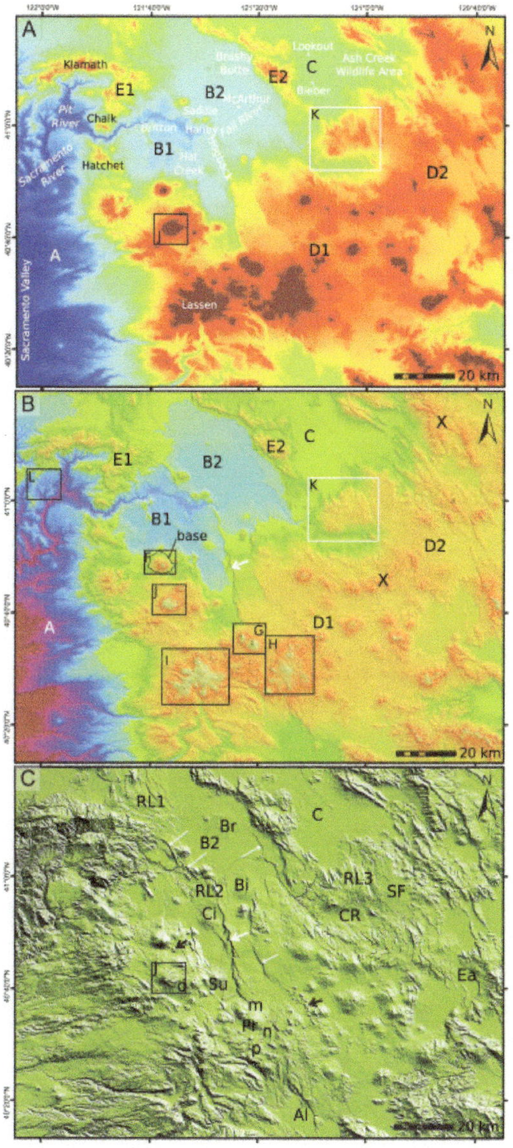

Figure 2. Density-sliced DEM and Shaded Relief Maps of the Hat Creek Graben and its vicinity (Gesch et al., 2002). (A) Highlights the different elevation regions high in warm colors and low in cold colors and (B) green as the boundary between the two main regions and red as the break in slope that marks the bases of volcanoes. Based on mean elevation and continuity of elevation vs. adjacent areas, the low elevation areas are subdivided into A: Sacramento Valley and Sacramento River including its main tributary on the north, the Pit River; B1: Hat Creek Valley and Lake Britton, B2: McArthur and Fall River, and C: Lookout, Bieber and Ash Creek Wildlife Area. High elevation areas are subdivided based on texture and intensity of warm colors and geographic location. D1: Cascade Range; D2: Modoc Plateau; E1: Klamath Mountains; and E2: Taylor, Jimmerson and Widow Mountains. Red linear areas marked with x and K (inset) are ridges, features in insets F, G, and J denote stratovolcanoes, and insets H and I denote complex volcano edifices. Inset L denotes areas of river erosion.(C) Shaded relief image with water bodies: lakes Almanor (Al) and Eagle (EA); reservoirs: Silva Flat (SF) and Catfish (CR); flood plains: B2 and C; volcanoes: Sugarloaf Mountain (Su), Crater Peak (inset J), Prospect Peak (Pr); lava fields: Cinder Butte (Ci), Big Cave (Bi), Brushy Butte (Br), m and n; lava flows (o and p); cones (black arrows); faults (white arrows); and possible lineaments as localized faults (RL1, RL2, RL3). Arrows are drawn perpendicular to the general strike of the lineaments.

The high elevation areas are subdivided based on the texture and intensity of warm colors in the DEM and the geographic location (D1, D2, E1, E2). The warm, geographically continuous colors in the southern part of the study area are divided into an area of smooth contours on the western side that coincides with the Cascade Range (D1) and the relatively irregular-edged Modoc Plateau (D2) on the east and northeast sides. Two other high elevation areas are geographically separated and thus are identified as separate profiles: the Klamath Mountains (E1) and Taylor, Jimmerson and Widow Mountains (E2). The elevation density map (Figure 2B) highlights the edges of high elevation areas (in green: D1, D2, E1, E2), and the break in slope that marks the base of volcano edifices (in red: insets F, G, H, I, J).

Features observed and mapped include the morphology of the edifices. Mountain ridges that form narrow and linear features (x) are contrasted against volcano edifices, observed as being more equant in planview (insets F, G, J). Stratovolcano bases are identified by the outermost, circular or equant, boundaries (insets F, G:1-2, J). The non-circular and very irregular edges of the base also identify the more complex volcanoes (such as those in insets H and I). "Piedmont" volcanoes (such as in inset G) are formed when the base of one volcano (1) is truncated by or buried under an adjacent edifice (2). Big volcanoes or broad areas of higher elevation relative to adjacent areas (insets H, I) are comprised of concentrations of (sometimes linear) cones (inset H), or are heavily eroded stratovolcanoes (inset I). Smaller stratovolcanoes (insets F, G, J) can also show irregularities of their bases, which can be evidence for collapse, sliding or unequal deposition and erosion. In inset J is an area that has experienced a deep-seated erosional or mass-wasting process.

Other high elevation areas (insets K, L) with differences in the regularity and smoothness of the edges (in green) can indicate areas that may be experiencing varying degrees of erosion. Inset K, showing a volcano base with a regular planview edge as indicated by smoother color contour, may indicate a relatively lesser amount and lower rate of erosion, while the volcano in inset L may be experiencing a higher and more stable rate of erosion by a river tributary.

Shaded Relief

The shaded relief map (Figure 2C) shows a 3D depiction of the earth's surface. The flat and very smooth surfaces represent: the water bodies such as lakes Almanor (Al) and Eagle (Ea); reservoirs such as Silva Flat (SF) and Catfish (CR); and flood plains such as B2 and C. Rough and sloping areas represent either mountains or mountain ridges if they are elongated, and volcanoes if they are more circular in planview. Volcanoes having rough surface textures but very few gullies or other linear erosional features, such as Sugarloaf Mountain (Su), have relatively fresh lava (lava is not easily erodible), which may translate into a young relative age of the volcano. Those with rough surface texture or lower slopes, such as Cinder Butte (Ci), Big Cave (Bi), and Brushy Butte (Br), or those being non-circular in planview, such as those in"m" and "n," are identified as lava fields, as long as there is a recognizable source. Lava flows ("o" and "p") are also recognizable as having levees on the narrower sides, and subsidiary flows on the edges of the main flow. Cones (black arrows) are easily recognizable as small mounds with relatively smooth and steep slopes. Relative ages of mountains are marked by the existence and density of gullies.

Other surface feature easily recognized on a shaded relief image are faults and linear features. Linear features that show continuity, and coincide with regional fault systems described in the literature, as well as those that show obvious vertical displacement of other features (white arrows), are identified as faults. Identification of linear arrays of cones (black arrows) and faults (white arrows) is a starting point in recognizing more subtle structures. Repeating lineaments (RL1, RL2, RL3) in the same area, and with the same strike, are also likely faults. In recognizing faults, the most obvious ones may not have the biggest offset, but may have the best preserved scarp, which is due to their age, offset, and erosion and deposition rates.

The hillshade image enhances some of what can be picked out in color mapping (Figure 2C). The addition of a 3D depiction behind the DEM enhances the surface features observed, while the simplified and generalized version of the shaded relief image confirms the existence of both the surface features and structures that may have been otherwise over-interpreted. Although (Figure 2C) is just an enhancement of Figure 2A, one of its advantages is that shadows generated by the particular sun angle considered in the generation of the shaded relief image are not emphasized. Shadow emphasis might be helpful for delineating faults (thick white arrow), but this is not always the case for delineating volcano bases like that of Crater Peak (inset J), as the surfaces are of much the same texture, generating the same types of shadows, in all directions.

Slope Modeling

Topographic modeling was also used to generate a slope map (Figures 3A,B). The dataset for the study area ranges from 0° to 81.5° with a mean slope of 11°. Steep slopes are emphasized in Figure 3A. This color mapping of the slope map is useful for confirming the delineations of faults.

Faults are identified as relatively straight lineaments, often members of a set with consistent strike. Faults in the study area have a considerable amount of vertical movement, as measured by the extent of the area colored yellow. The faults were classified according to the degree of curvature. Faults in RL1, RL2, RL3, for example, are grouped together. Faults that belong to the same set generally have the same orientation, maximum along-strike throw as indicated by maximum cross-strike width, planview curvature, and thus approximately the same age, as should be the case.

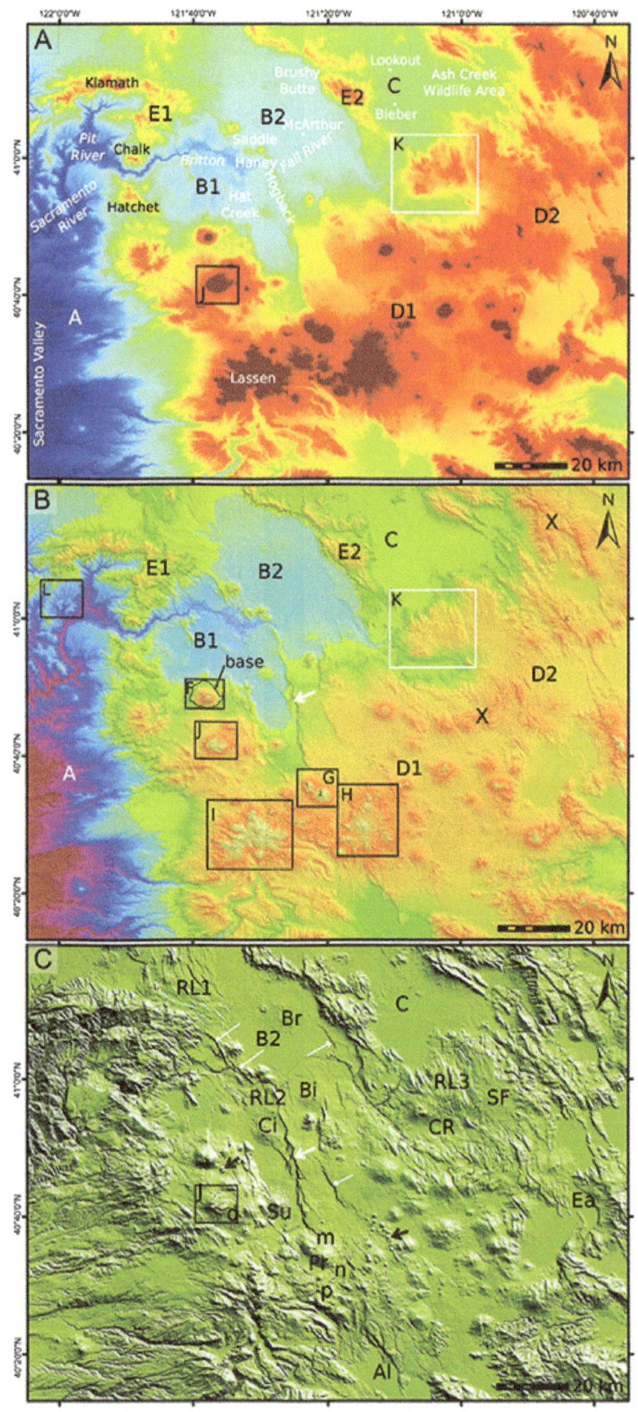

Figure 3. Slope and aspect maps (Gesch et al., 2002).(A) Density sliced slope map where steep areas including volcano and cone summits and fault scarps are yellow, edifices are pink, and relatively flat surfaces and water bodies are blue. The boundaries between the blue and pink areas are the break in slopes that mark a volcano base. Arrows are drawn perpendicular to the general strike of the lineaments. Fault groups based on orientation and strike, degree of curvature and sinuosity are RL1, RL2, RL3. Hat Creek Fault and its projected continuation is shown by thick white arrow. McArthur fault is indicated by a thin white arrow. Steep gorges are marked G1, G2, G3, and G4. Steeply sloping, wide areas are S1 and S2. Stratovolcanoes with known and observed deformed edifices are Crater Peak (CP), Chaos Crags (CC) and Lassen Peak (LP). Other edifices are Sugarloaf Peak (Su), Cinder Butte (Ci), Big Cave (Bi), Bald Mountain (Ba), and Burney Mountain (Bu) and lava fields (o) and Hermit Butte (He). (B) Emphasizes lava fields and their source areas labeled 1–11. (C) Aspect map of the Hat Creek Graben and adjacent areas. Polygon colors and azimuths: red (0–45°), green (45–90°), blue (90–135°), yellow (135–180°), cyan

(180–225°), magenta (225–270°), maroon (270–315°), and sea green (315–360°). Water bodies: lakes Almanor (Al) and Eagle (Ea) are in yellow. White and black arrows point at west- and south- facing fault controlled geomorphic features and linear discontinuities.

The main Hat Creek fault may well have a southern continuation (thick white arrow), which occurs along the northeastern bank of the northwestern lobe of Almanor Lake (Al). To the east of the Hat Creek Fault is the trace of the McArthur fault (thin white arrow), which may continue to the south or southeast as evidenced by the linear cones directly north of AL. Steep gorges, G1, G2, G3, and G4, are identifiable; they often indicate in this region that the stream segment has been uplifted. Steeply sloping, wide areas, such as S1 and S2 indicate landsliding or a high rate of erosion. A closer look at the steep summits of some volcanoes and domes, such as Chaos Crags (CC) and Crater Peak (CP), shows that these are relatively smooth in texture, compared to stratovolcanoes that have been deformed by violent activity, such as Lassen Peak.

Cones (black arrows) are defined by yellow circular regions that indicate steepness from the base to the summit, with a lower-slope angle, cool colored (blue) region in the middle. Because slope maps emphasize areas of steepness, they are good for delineating lava flow extent, since the edges of lava flows tend to be steep, irregular and narrow relative to adjacent areas. Because slope steepness is a factor in landslide susceptibility mapping, steeper regions on volcanic edifices are at increased risk for landslides and collapses.

Slope maps also indicate relative ages of volcanic edifices. If we compare Burney Mountain, Bald Mountain (Ba) and Hermit Butte (He), for example, Burney Mountain seems to have been covered by younger surface deposits and less affected by erosion than the other two volcanoes.

Another variation in slope map presentation (Figure 3B) emphasizes lava fields (labeled 1–11) and their source areas. Areas covered by lava flows, such as those near Sugarloaf Peak (Su), Cinder Butte (Ci), Big Cave (Bi), the area north of Burney Mountain (Bu) and the area southeast of Crater Peak (o), have rough texture in the slope map, but a mean slope near zero.

Aspect Modeling

Slope aspect mapping is classified as a continuous terrain segmentation scheme (Miliaresis and Kokkas, 2004). In the current approach, aspect was computed for every DEM grid point (Figure 3C). Then the aspect was standardized to eight major geographic directions: N, NE, E, SE, S, SW, W, and NW. Linearity in colored polygon in the figure is another type of evidence for faults (black arrows). Faults with dip-slip component of different orientation can be distinguished with this technique, e.g., those with the western block moving down relative to the eastern block (white arrows). Areas of deposition of rough materials, such as emplaced lava flows or avalanche and slide toes tend to have highly variable, specular color patterns, meaning that the area has slopes facing in many directions on a relatively fine spatial scale, such as at Cinder Butte (Ci), Brushy Butte (Br), and Crater Peak (CP).

Structural and Lineament Mapping

Regional fault systems (Figure 4A) were drawn and identified in consultation with previous work as: the Hat Creek, Rocky Ledge, McArthur, Soldier Creek, and Pittville Fault Systems. These fault systems are predominantly normal, with right-lateral components and evidence of activity up to Holocene time (Wills, 1991; Muffler et al., 1994, 2010; Parrish, 2006; U.S. Geological Survey and California Geological Survey, 2006; Sawyer and Ramelli, 2012; Blakeslee and Kattenhorn, 2013). We traced the possible extent of these structures, as well as other shorter, more curvilinear, and relatively minor, faults such as those that affect Burney Mountain (Bu), Crater Peak (CP), West Prospect Peak (WP), Prospect Peak (PP), Bogard Butte (Bo), Whaleback (Wh), and Fox Mountain (Fo) (Figures 4B–F). Some of the smaller faults, with strike directions that coincide with faults identified by U.S. Geological Survey and California Geological Survey (2006) to be <130 ka, <1600 ka and >1600 ka, are assumed to be of the same age as the larger faults, and are grouped with them. The smaller faults so mapped are possibly parts of the Walker Spring fault system, and the unnamed faults in the Susanville lake and Big Valley areas, while Other Primary Regional 1 faults are unnamed. Some of our other Primary Regional 2 traces also belong to the <1600 ka faults in the USGS Fault Map.

Most of the faults trend northwest (Figure 4A). However, in more detail, the fault strike evolves from north-northwest to more nearly west-northwest, from east to west near the northern edge, while it goes from west-northwest toward north in the central to southern part of the study area. The small faults are defined as short lineaments on volcanoes without clear continuity into nearby areas. Such lineaments include the trace just to the northeast of Crater Peak (CP) (Figure 4B), which continues northwest up to the northwest side of Burney Mountain (Bu). These traces can contain multiple, short lineaments, but are constrained areally. Another area of localized faulting is the corridor of northwest trending short faults that affect the northern sectors of Prospect (PP) and West Prospect Peaks (WP) (Figure 4E). On some volcanic edifices, such as Bogard Butte (Bo), Whaleback (Wh), and Fox Mountain (Fo) (Figure4F), small faults seem to split the edifices. In the case of Crater Peak, the small fault seems to have at least partly resulted from a major collapse. In these cases, the lineaments developed in response to a gravity-driven event that affected the volcano. However, in the case of Bogard Butte, Whaleback and Fox, it appears that the faults have influenced the shape and development of the volcano, suggesting that they are basement features that have propagated into the edifice.

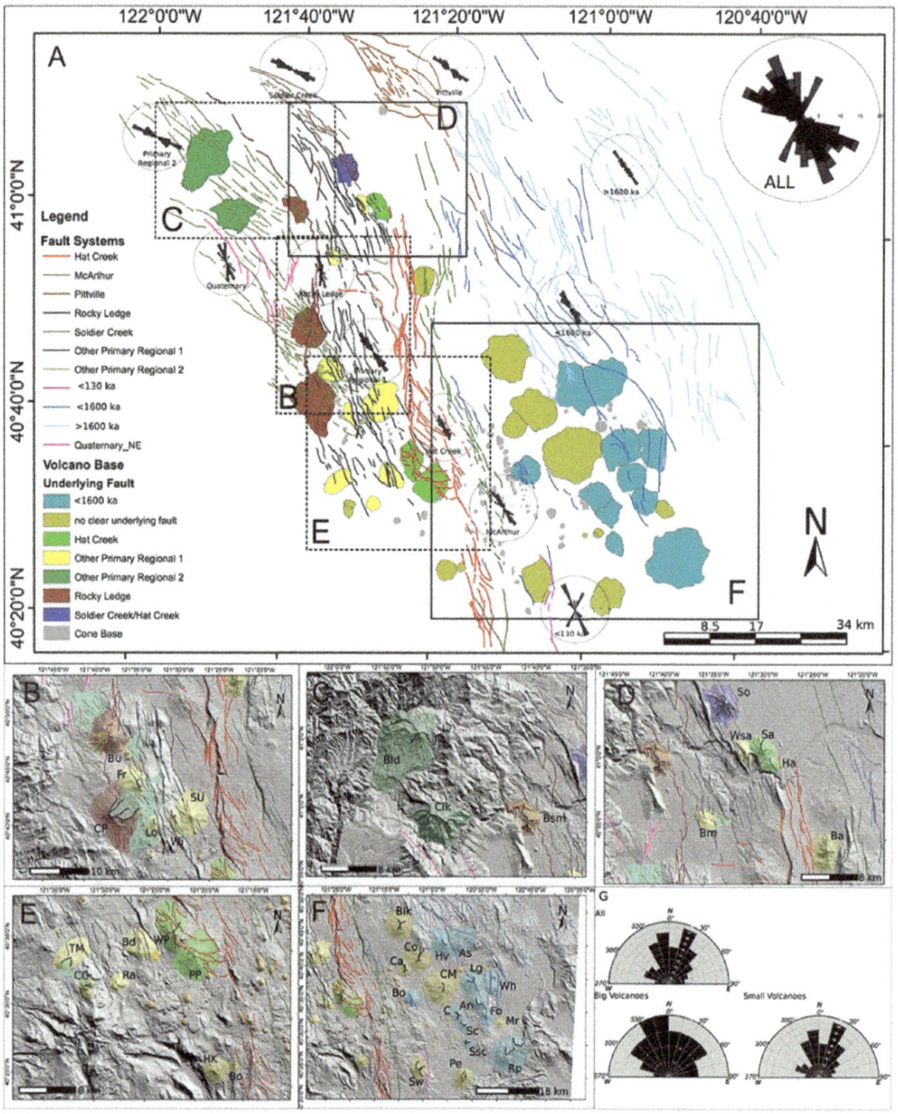

Figure 4. Volcano-tectonic relationships in Hat Creek Graben region. (A) Base map showing fault systems, volcano bases color coded on fault system that mostly affects edifice, and water bodies; rose diagrams of fault strikes by fault system; and locations (squares) of hill-shaded maps (B–F) showing fault traces and effects on edifice: edifice instability and collapse, edifice disintegration, and splitting. (G) Rose diagrams of volcano base elongations for both big (cones, sub-cones, shields, and massifs) and small (small cones, small sub-cones, and small shields) volcano sub-groups, and all volcanoes. Dataset available in Supplementary Materials.

Volcano Morphometry

Based on the MORVOLC output, we classified the volcanoes of the Hat Creek Graben region using primarily height to basal width ratio (Figure 5A). Those with a height to basal width ratio greater than 0.15 were classified as cones; those with intermediate values between 0.1 and 0.15 as sub-cones, and those with values less than 0.1 as shields or massifs. These last two were classified further based on height. A second classification based on height was done to differentiate small and large cones, sub-cones and shields. This classification was used to develop an evolutionary trend for volcanoes in the Hat Creek Graben region (Figure 5B).

The evolutionary trend begins with cones. Cones can either maintain their height to basal width ratio to increase in volume into larger cones, or enlarge in basal width to become sub-cones. Sub-cones can continuously increase in volume to become a bigger sub-cone, or evolve in basal width into a large but low volcano, a shield, and ultimately into a massif. Lastly, small volcanoes with height and basal width ratios of a shield can grow into massifs. Allmorphologically big volcanoes in the region are those with identified names, as discussed in the geomorphological classification of features in the region (Part 1 of this work). However, some of these previously named, big volcanoes needed to be reclassified as small cones, small sub-cones, or small shields, because their heights were less than 350 m (Figure 5B). Of the total of 185 drawn volcanic centers in the Hat Creek Graben region, 33 were classified as bigger stratovolcanoes (massifs), and 152 are morphologically smaller, mostly monogenetic cones (Tables 1, 2).

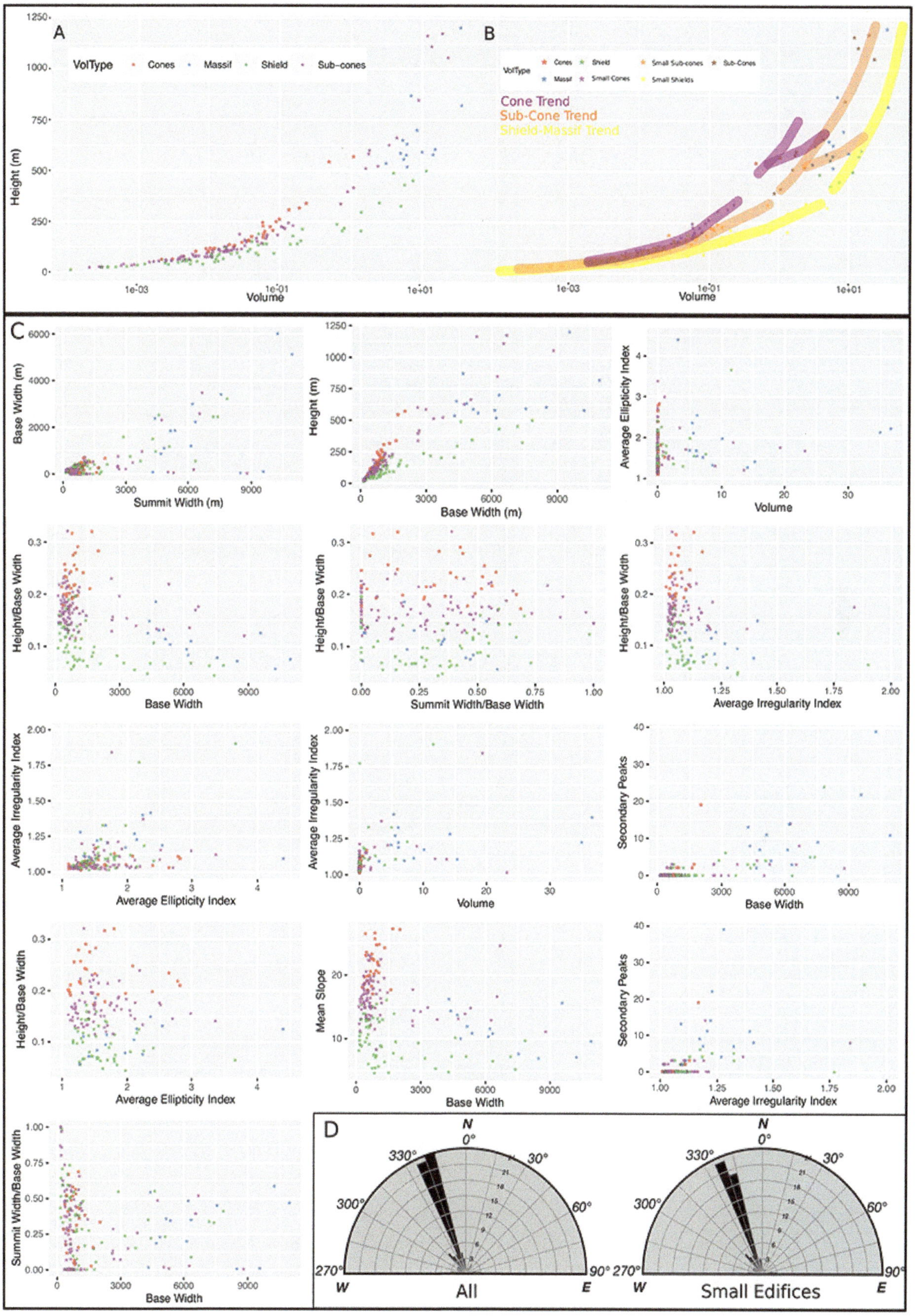

Figure 5. Volcano morphometry. **(A)** Volume vs. height plots for all volcanoes grouped by morphological classification: cones, sub-cones, shields, and massifs based on their height and basal width ratio; **(B)** A second morphological classification based on height separating bigger and smaller edifices on a volume vs. height plot to draw a volcano evolutionary trend; **(C)**X-Y plots of selected morphometric parameters grouped by morphological type (cones, sub-cones, shields and massifs); and**(D)** A rose diagram showing the preferred direction of alignments that can be formed by at least 3 volcano centroids within 6 km. Dataset available in Supplementary Materials.

Table 1. Morphometry and crater morphology of the Hat Creek Graben region stratovolcanoes.

Name	Type	HeightMax	VolMax	WidthBase	WidthSummit	σH_{max}	Crater direction
0-crater	C	84.61	0.05	1142.83	0.00		
1-crater	A	204.39	0.08	1041.97	374.30		
2-crater	B	183.54	0.09	1135.70	528.73		
3-crater	A	306.50	0.13	1055.52	449.16		
4-crater	C	139.37	0.06	1013.24	26.91		
5-crater	A	126.70	0.02	714.21	0.00		
6-crater	A	315.51	0.19	1308.55	500.73		
7-crater	B	242.07	0.13	1137.36	0.00		
8	A	117.98	0.01	507.83	206.89		
9-crater	A	224.08	0.07	877.83	464.97		
10-crater	B	229.48	0.15	1447.89	509.98		
11-crater	B	219.39	0.11	1060.86	692.81		
12-crater	B	214.63	0.12	1291.19	0.00		
13-crater	A	132.87	0.02	661.04	323.29		
14-crater	B	149.46	0.03	818.74	0.00		
15-crater	A	175.64	0.04	723.62	419.76		
16	A	114.79	0.01	447.11	51.27		
17-crater	B	101.96	0.01	457.38	256.21		
18-crater	A	261.68	0.10	939.22	410.03		
19-crater	B	102.62	0.02	623.90	267.48		
20-crater	A	171.83	0.06	994.90	682.40		
21-crater	B	177.72	0.07	1026.93	574.44		
22	B	52.45	0.00	290.10	83.82		
23-crater	A	154.56	0.04	733.61	378.59		
24-crater	B	76.64	0.01	576.39	79.77		
25-crater	C	47.43	0.00	487.56	0.00		
26-crater	C	55.34	0.00	474.03	227.70		
27-crater	A	135.25	0.02	625.51	0.00		
28	B	187.48	0.03	584.53	226.79		
29	B	108.22	0.02	643.34	47.04		
30-crater	C	139.33	0.07	1227.03	588.47		
31-crater	B	111.39	0.03	765.83	0.00		
32-crater	B	102.78	0.02	673.14	370.83		
33-crater	B	102.82	0.02	674.84	0.00		
34-crater	C	151.75	0.04	783.18	0.00		
35	C	88.47	0.01	629.07	111.51		
36	A	147.91	0.03	775.96	207.36		
37-crater	C	134.52	0.05	1069.24	484.42		
38-crater	B	112.97	0.03	863.77	383.09		
39-crater	B	165.27	0.06	1170.85	484.79		
40-crater	B	143.32	0.06	1125.81	487.76		
41-crater	A	105.69	0.01	528.68	352.49		
42	B	114.39	0.01	534.37	131.56		
43-crater	B	130.47	0.03	799.15	392.52		
44-crater	C	64.24	0.00	477.99	0.00		
45-crater	C	93.98	0.03	786.42	424.89		
46	B	126.67	0.02	591.44	175.94		
47-crater	B	93.58	0.01	508.92	0.00		
48	B	80.06	0.01	450.60	132.66		
49-crater	B	109.34	0.02	628.01	261.94		

Name	Type	HeightMax	VolMax	WidthBase	WidthSummit	σH_{max}	Crater direction
50-crater	C	120.30	0.02	636.28	0.00		
51-crater	B	182.88	0.07	936.56	360.90		
52-crater	B	167.79	0.05	822.82	0.00		
53-crater	B	108.30	0.02	675.58	0.00		
54-crater	B	224.42	0.11	1104.87	0.00		
55-crater	C	187.71	0.05	958.14	0.00		
56	B	158.98	0.03	664.97	97.53		
57	B	104.07	0.01	563.44	215.91		
58-crater	A	187.47	0.06	864.39	486.56		
59	C	51.58	0.01	852.92	322.74		
60	C	94.23	0.05	1310.23	215.51		
61	C	40.92	0.01	743.28	68.89		
62	C	37.97	0.00	430.59	221.30		
63	C	40.09	0.01	583.25	426.13		
64	B	77.89	0.01	682.35	86.88		
65	C	65.40	0.01	825.39	215.03		
66	C	56.35	0.01	885.72	343.24		
67	C	139.26	0.18	2107.28	401.35		
68	C	138.39	0.14	1721.25	251.66		
69	C	91.09	0.02	933.47	247.91		
70	C	113.38	0.04	1179.01	334.75		
71-crater	C	179.44	0.15	1823.35	96.44		
72	B	108.63	0.02	945.42	125.35		
73-crater	C	154.42	0.20	1766.18	0.00		
74	C	74.18	0.03	1040.01	395.14		
75	A	64.44	0.00	362.00	165.14		
76-crater	A	101.59	0.01	434.40	226.27		
77-crater	A	117.40	0.01	477.54	0.00		
78	B	102.70	0.02	605.24	356.17		
79-crater	B	127.52	0.02	564.57	0.00		
80	B	77.40	0.01	524.39	145.68		
81-crater	C	211.84	0.21	1425.68	0.00		
82-crater	A	157.53	0.04	818.09	52.04		
83-crater	B	114.51	0.02	650.72	357.50		
84	B	106.04	0.02	619.50	131.41		
85	C	57.50	0.00	565.37	239.89		
86	A	103.03	0.01	528.98	143.81		
87	B	49.49	0.00	263.73	164.77		
88	B	36.95	0.00	254.93	150.73		
89	A	77.47	0.00	390.64	68.46		
90	B	59.02	0.00	342.86	127.66		
91	B	65.75	0.01	455.40	169.45		
92	B	63.53	0.00	424.76	28.38		
93	B	54.93	0.00	340.69	160.95		
94	B	123.61	0.02	695.64	288.28		
95	B	89.32	0.01	387.70	170.94		
96	B	64.96	0.00	397.58	101.11		
97	B	60.23	0.00	455.70	126.14		
98-crater	B	79.23	0.01	475.31	0.00		
99	B	77.13	0.00	349.00	191.65		

Name	Type	HeightMax	VolMax	WidthBase	WidthSummit	σH_{max}	Crater direction
Burney Spring Mountain	Massif	536.63	3.10	4297.78	73.10	NF = 156°	293, 111
Cal Mountain	C	295.35	1.38	4423.38	963.10	NF = 156°	247, 66
Campbell Mountain	C	342.72	2.48	5110.78	1168.56	SS = 4°	10, 240
Chalk Mountain	Sub-cones	1174.27	19.33	6685.52	3465.11	NF = 153°	195, 193, 182, 230
Chaos Crags north	Cones	543.46	0.48	1698.41	262.63	NF = 156°	300
Chaos Crags south	Cones	569.94	1.07	2015.65	669.66		
Cone Mountain	C	324.32	4.99	7417.35	2446.71	SS = 4°	278
Crater Mountain-crater	Massif	578.82	14.22	10240.00	6012.92	SS = 4°	68,97
CraterPeak-crater	Sub-cones	1054.51	23.26	8809.16	21.25	NF = 156°	66
Fox Mountain	Massif	632.63	6.11	4850.22	1100.21		
FreanerPeak	Massif	869.30	6.99	4686.70	864.89	NF = 156°	76
Haney Mountain	B	337.23	0.71	2624.56	344.25	NF = 156°	44
Harkness Mountain	Sub-cones	521.40	2.39	3760.78	959.82		
Harvey Mountain	Massif	592.13	10.12	7563.22	3408.96	SS = 4°	25
Logan Mountain	Shield	450.59	7.50	7194.79	2435.96	SS = 4°	321
Logan Mountain	Sub-cones	600.12	1.81	2612.66	291.51	NF = 156°	65
Merrill Mountain	C	215.82	0.79	3194.44	751.84		
Mudleft	C	103.07	0.04	938.09	66.45		
Mudright	C	92.44	0.08	1501.66	410.10		
Pegleg Mountain	Massif	577.98	6.00	6148.08	1209.84	SS = 4°	89
ProspectPeak-crater	Massif	699.11	8.53	6352.69	2226.61		98
RakerPeak	Sub-cones	395.72	0.88	2637.22	746.77		
Roop Mountain	Massif	820.53	36.66	10924.98	5125.81		61
Saddle Mountain	Sub-cones	511.10	1.61	3332.40	492.58		14
SCampbell Mountain	C	203.86	1.36	4282.21	1491.79		345
SLongLake	C	67.64	0.02	912.56	255.52		221
SofSCampbell Mountain	C	202.66	0.75	3691.73	551.48		54
Soldier Mountain	Sub-cones	662.69	4.38	5045.36	84.61		40
SugarloafPeak	Sub-cones	847.47	8.96	6274.42	170.30		
Swain Mountain	Shield	429.74	5.62	6355.22	557.52	SS = 4°	1,290
Table Mountain	Shield	487.35	3.92	4590.87	2526.60	NF = 156°	7, 232, 195, 240, 306
WestProspectPeak	Sub-cones	1163.52	11.90	5306.43	46.43		
Whaleback Mountain	Massif	518.45	6.12	6219.10	2998.09	SS = 4°	62, 99
WilcoxPeak	Sub-cones	583.33	1.76	2985.07	205.76	NF = 156°	27, 108
WSaddle Mountain	Sub-cones	414.88	1.08	2674.36	199.76		

Height and width are in m; Volume is in m^3 ; σH_{max} is the maximum horizontal stress orientation (azimuth, in degrees) and tectonic regime (SS = strike-slip or NF = normal faulting) nearest to each volcano edifice with crater opening or collapse scar (stress data from Heidbach et al., 2008); crater directions are in degrees; volcano types A, B, and C refer to small cones, small sub-cones, and small shields that fall within the lower purple, orange, and yellow lines in Figure 5B. Dataset available in Supplementary Materials.

We present x-y scatter plots of volcanoes in the region based on morphological classification parameters (Figure 5C). Cones have narrow summit widths, while shields and massifs tend to have broader summits, wider bases and more peaks, and are sometimes more irregular in outline. Basically, shields and massifs are high volcanoes with wide bases. Sub-cones, shields and massifs are voluminous volcanoes compared to cones. Cones are mostly identifiable by their height and high mean slope angle, in which sub-cones follow them closely. Sub-cones, massifs and some shields show the most irregular or elliptical planview shapes (Figure 6A).

Volcano bases are elongated (Figure 4G), with two modes, north-northwest and northeast; most of this scatter is influenced by the trend of the large population of smaller volcanoes. Big volcanoes also generally trend north-northwest, but this trend is less clear. Volcano bases seem to be elongated consistent with the trend of the directly underlying local structures.

Table 2. Properties and results of the statistical analyses for each morphological subgroup (big and small), and for the all the volcanoes in the Hat Creek Graben.

		Smaller	Bigger	ALL
Measured nearest neighbor properties	Number of volcanic centers (N)	152	33	185
	Area Convex Hull (m^2)	3.26E+09	3.54E+09	4.28E+09
	Density (nb volcanic centers/m^2)	4.66E−10	9.33E−11	4.32E−10
	Min distance NN (m)	4.92E+02	4.29E+03	4.92E+02
	Max distance NN (m)	1.83E+05	1.65E+05	1.23E+05
	Mean distance NN (m)	1.60E+04	7.01E+04	2.01E+04
	Standard deviation (1σ)	2.17E+04	4.60E+04	2.36E+04
	Standard deviation (2σ)	4.33E+04	9.19E+04	4.72E+04
	Skewness	4.28	0.62	2.43
	Kurtosis	25.75	−0.80	6.40
Nearest neighbor results relative to the Poisson model	Poisson Re (m)	2.32E+04	5.18E+04	2.40E+04
	R	0.69	1.35	0.83
	Ideal R given N_i	1.04	1.08	1.03
	R negative thresholds at 1σ	1.08	1.19	1.07
	R positive thresholds at 1σ	1.13	1.29	1.12
	R negative thresholds at 2σ	0.99	0.98	0.99
	R positive thresholds at 2σ	0.94	0.87	0.95
	c	−7.25	3.89	−4.33
	Ideal c given N_i	0.84	0.90	0.84
	c negative thresholds at 1σ	1.93	2.07	1.92
	c positive thresholds at 1σ	3.02	3.24	3.00
	c negative thresholds at 2σ	−0.25	−0.27	−0.24
	c positive thresholds at 2σ	−1.34	−1.44	−1.33
Alignment analysis	Distance maximal for the generation of lineaments (m)	5800	9900	6000
	Number of lineaments (m)	85	0	93

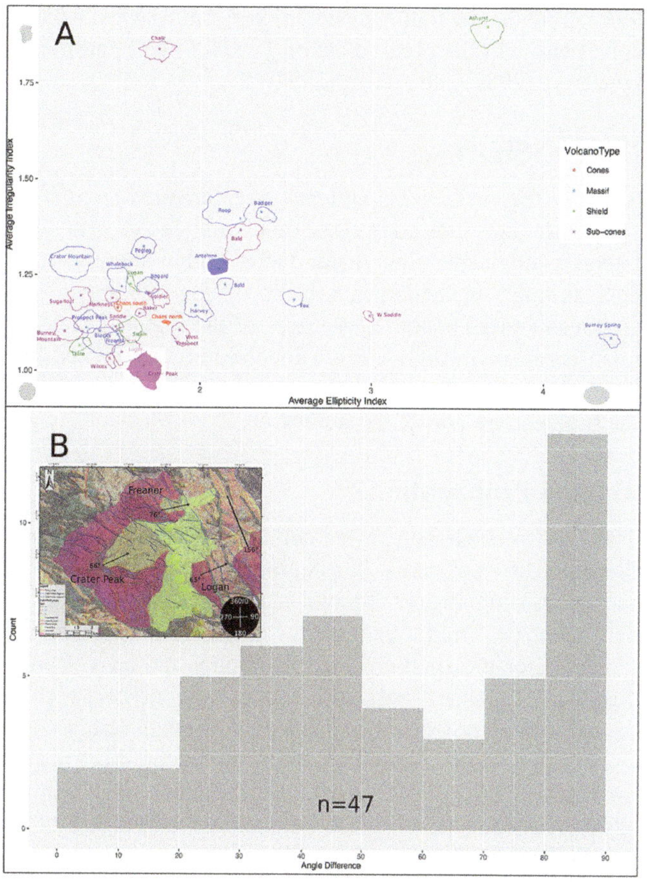

Figure 6. Volcano morphology and breaches. (A) Ellipticity and Irregularity Indices (not to scale) of bigger volcanoes in the Hat Creek Graben region grouped by morphological type as cones, sub-cones, shields and massifs. Filled polygons are those with evidence of collapse scars. **(B)**Frequency diagram showing the relationship between the opening direction of craters and the maximum horizontal stress. Inset is an example of how angle difference is calculated.

Volcano Spatial Distribution and Alignments

The Hat Creek Graben region appears to be a volcanic field with distinctive subgroups of volcanoes, having different spatial distributions. Poisson Nearest Neighbor Analysis was applied to the morphologically smaller and bigger volcanoes as subgroups, and separately, to all volcanoes (Table 2). The smaller volcano subgroup and the group of all volcanoes have c values outside the $\pm 2\sigma$ significance level but R less than -2σ, which means they are spatially clustered. The bigger volcano subgroup, however, has c values outside the $\pm 2\sigma$ significance level and R greater than $+2\sigma$, which means that volcanoes in this group are dispersed relative to the Poisson distribution. Based on the density of volcanoes per subgroup, and using the relationship established using the global database of monogenetic volcanic fields (Le Corvec et al., 2013), we used 5800, 9900, and 6000 m as the maximal distance needed for the recognition of lineaments for smaller and bigger subgroups, and the group of all volcanoes, respectively.

The three-point MATLAB script was used to automatically extract the different alignments of the volcanoes, and plot for each the direction of the alignment on a rose diagram (Figure 5D). Bigger volcanoes generated no preferred alignment, while the smaller volcano subgroup and the group of all volcanoes have a north-northwest preferred direction of alignments. Again, this quantitatively determined strike is consistent with the strike of local, tectonic faults.

Crater Morphology and Stress Field

Only the subgroup of bigger volcanoes was considered for investigating crater opening, either a volcano collapse scar or an erosional breach, because of the prevalence of these features on the bigger volcanoes. The regional maximum horizontal stress and crater opening directions, both in azimuth, were noted for each bigger volcano with crater openings in Table 1. For example, as in Figure 6B (inset), the maximum horizontal regional stress has an azimuth of 156° near Freaner, Logan and Crater Peak volcanoes, which have summit openings (erosional breaches for Freaner and Logan, and collapse scar for Crater Peak) of 76°, 65°, and 66°, respectively. The angle difference for each of these three volcanoes is the absolute value of the difference between 156°and the summit opening direction, which is 80°, 89°, and 90°, respectively (Table 1). A frequency diagram (Figure 6B) of the angular difference between the crater opening direction for the bigger volcanoes and the maximum horizontal stress has a mode at about 90°, although crater openings at an acute angle are also found, albeit at lower frequencies. Most of the crater openings are thus perpendicular to the maximum horizontal stress, which is consistent with the observation of Tibaldi (1995) and models byMoriya (1980) and Siebert (1984).

DISCUSSION

Several regional fault systems are oriented northwest-southeast; these are comprised of normal faults with right-lateral components. Individual faults within these systems pass through or underneath most of the volcanoes. The stratovolcanoes contain clear traces of the regional faults, or manifestations of collapse or erosional features on their edifices that can be related to the faults. Cones are more prevalent in the southeastern part of the study area.

Magma Propagation

The Hat Creek Graben region consists of volcano subgroups that have different spatial distributions; smaller volcanoes tend to cluster, while bigger volcanoes are dispersed. The density-dependent, preferred alignments reveal a single north-northwest direction for the small volcano subgroup, and all volcanoes considered together. These observations are consistent with the control of melt production at depth and magma propagation by pre-existing regional structures, and with the control of magma propagation in the near surface by the local stress regime. Melt production is related to the oblique subduction of the Gorda Plate underneath the Northern American Plate, and is manifested in the volcanism at the surface. Regional and near-surface stresses are influenced by Walker Lane and Cascades tectonism. Extension is accommodated by the pervasive faulting and magmatism. As the area extends, it produces weak regions that provide energy-efficient paths along which magma propagates to the surface.

Normal faults, dominant in the area, influence dike orientation and the generation of alignments at the surface, because dikes are more likely to intercept and follow dipping faults than vertical fractures (i.e., Le Corvec et al., 2013). Near-surface stress also affects magma propagation (i.e., Caputo, 1995) it is possible that high differential stress did not allow stress swap, therefore the intrusions followed the orientation of the earlier dikes. The generation of multiple orientations of volcanic alignments is perhaps attributable to the interaction of near-surface stress field influences and pre-existing fractures.

Magma pressure, which is also a factor in dike propagation (i.e., Jolly and Sanderson, 1997), may be low, so that magma only follows pre-existing fractures that are favorable for reactivation in the prevailing stress field. Although pathways may vary slightly according to stress field at time of propagation and eventual emplacement (i.e., Valentine and Krogh, 2006), in general, they have been consistent through the lifetime of Hat Creek region volcanism. The propagation of magma that formed the bigger volcanoes is also structurally controlled, because these volcanoes stand atop major regional structures, and the faults affect the stability of the volcanic edifices. The random, dispersed spatial distribution of the bigger volcanoes may be attributed to their long-lived plumbing systems, which are capable of capturing or re-focusing nearby dikes as they propagate to the surface, most of the time erupting from a single vent. Changes in the local stress field caused by the growing edifice may also have enhanced dike capture.

Edifice Evolution

Most volcanoes in the study area are small, with maximum heights about 250–300 m, and volumes generally less than 1 km^3. Based on the height and basal width ratio, some volcanoes can be classified as cones, which are relatively regular and more nearly circular in planview, have relatively smaller summit areas and narrower basal widths compared to sub-cones and to the bigger and more voluminous shields and massif volcanoes. Quantitative morphology furthermore suggests that volcanoes evolve along several paths: small cones continue to grow as cones, or they grow into sub-cones; small sub-cones continue to grow into bigger sub-cones, or they develop into shields and massifs; and shields can grow into massifs. Underlying major faults systems have greatly influenced the growth, partial destruction, and general evolution of volcano morphology.

Quantitative evaluation furthermore revealed volcanic basal elongations in two main directions, north-northwest and northeast. The existence and magnitude of the northeast mode in the elongations is consistent with spreading parallel to displacements on regional normal faults with right-lateral components, either by preferred feeder dike orientation or by deformation.

Deformed and Collapsing Volcanoes

Several types of volcanic edifice deformation are observed in the Hat Creek Graben region. Some volcanoes show collapse features, others erosional breaches, and some are being split by faulting.

Chaos Crags is a stratovolcano with a known collapse scar (Figure 4E). The collapse of the unstable, very steep, soft-cored, and unconsolidated plug dome of Chaos North resulted in a debris avalanche, called the Chaos Jumbles. The jumbles were emplaced as a high yield strength material capable of deforming and shearing in three separate rockfalls about 300 years ago Eppler et al. (1987), damming Manzanita Creek and forming Manzanita Lake. A projection of a northwest striking fault, here part of a Primary Regional 1 group of faults (Figures 4A,E), which splits Table Mountain to the northwest, may be masked underneath the Crags and Jumbles. Post-eruptive mass wasting and sedimentation on the slopes of Chaos Crags have masked any evidence for the mechanism of generation of the debris avalanche (Perez, 1998). Activity on the fault could have triggered instability and collapse.

Crater Peak is another stratovolcano that shows morphological evidence of collapse (Figure 4B). Crater Peak has a large and deep-seated scar, and a possible runout deposit elongated toward the northeast marked by a rough surface texture that would be expected in a debris avalanche deposit. The scar was glacially eroded after the collapse, as shown by the smoother, uppermost parts of the amphitheater, which thus resembles, or is in part, a cirque. Comparing the structures and scar elongation direction of Crater Peak to Figure 9E of Wooller et al. (2009), we speculate that the collapse could have been caused by a basement normal fault, with a right-lateral component, possibly the Rocky Ledge Fault (red line in Figures 4A,B, thicker black line Figure 7A), which passes through the center of the edifice. In this scenario, the Rocky Ledge Fault destabilized and eventually caused a collapse approximately normal to its strike direction.

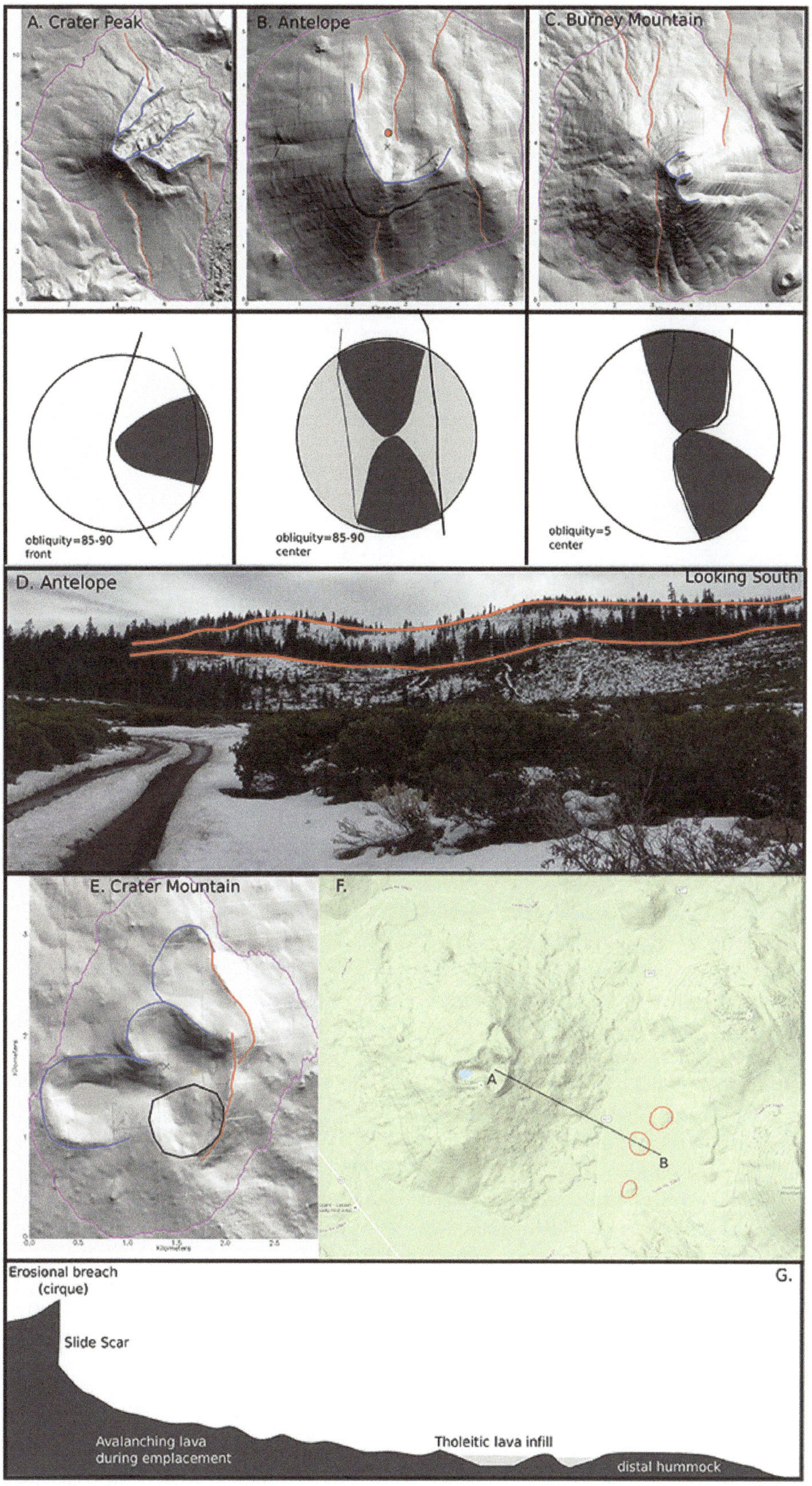

Figure 7. Deformed volcanoes in the Hat Creek Graben region. (A–C) Shaded relief delineated with volcano base (purple), collapse scars or erosional breaches (blue) and structural lineaments (red) for Crater Peak **(B)**, and Antelope**(C)**, and Burney **(D)** Mountains compared with their predictive model cones by Wooller et al. (2009) underneath, where the gray circles represent the edifice; solid lines are main normal faults; dotted polylines are antithetic and minor faults; and gray polygons are unstable sectors; **(D)** View looking at Antelope Mountain Slide scar from the northern sector of the volcano marked as red dot in **(B)**. The red lines are the slide scar (top), and terrace (bottom) that mark the top of the major listric faults that accommodate the sliding of the whole area. **(E)** Shaded relief and **(F)** google earth images, and **(G)** cross section of Crater Mountain Slide. Location of cross section A-B is drawn over google earth image **(F)**. Volcano base (purple); cirques (blue); scar and hummocks (red); possible crater trace (black) are marked.

Relations, confirmed in the field, showed that Antelope Mountain has also collapsed (Figures 4F, 7B,D), with deposits emplaced toward the northeast. From remote sensing, lineaments on the eastern base of the volcano can be traced (Figures 4A,F), of the same set that split Whaleback volcano to the northeast. Another set that perhaps influenced the stability of Antelope Mountain more strikes northwest, splits Logan Mountain to the north, and affects the western flank of Fox Mountain. The structures affecting Antelope Mountain are grouped with the <1600 ka faults, are possibly part of the Walker Spring fault system. These may be related to the faults in the Susanville and Big Lake Valley areas, which may have been active during the Holocene.

Based on remote sensing analysis, four more volcanoes notably show evidence of edifice collapse: Crater, Burney and Table Mountains; and Raker Peak. Burney and Table Mountains are deformed, and show morphological traces of edifice collapse structures (Figures 4B,E). The structures at Burney Mountain appear to be manifestations of ongoing deformation of the edifice by the Rocky Ledge Fault, the underlying basement fault (Figure 4B), which may ultimately result in sector collapse. In the field, Burney Mountain has two small craters that both open to the east. These are likely to have evolved into cirques by small, summit glaciers flowing to the east, leaving behind till, possibly in the form of a moraine, below the northern cirque (Figure 7C). The location of lineaments on the northwest and southeast sectors of Burney Mountain is consistent with underlying normal faults with little strike-slip component (Figure 7C). The lineaments (in yellow) perpendicular to the Rocky Ledge Fault trace may not be tectonic in origin, but rather lava flow ridges (Figure 4B).

Crater Mountain (Figures 4E–G) is a relatively "strong" and stable volcano, built mostly of lava flows, with no apparent hydrothermal alteration at the surface, nor any indication that it is built on deformable and weak substrata. At the summit, three craters have been traced, all elongated to the east, which may be modified by glaciers. There may be another, fourth, crater as well. The head-scarp morphology in the summit area, where the north-south scar is drawn (Figures 7E–G), suggests some type of gravity failure on the east flank. The scarp could have been made by multiple slides, affecting slightly different sectors. It is possible that a series of slides were emplaced without a thin, weak or wet sliding base, which resulted in slumping or rotation of relatively coherent blocks, forming a toe now exposed as three low mounds or distal hummocks (Paguican et al., 2014). Spaces between hummocks have been infilled by tholeiite lavas.

Raker Peak (Figure 4E) is another edifice with an apparent northeastward opening. This sector deformed relative to the others. Although there is no basement fault near the edifice, a river flows across the eastern piedmont, which could have cut the base of the edifice, causing its eventual collapse. It is also possible that a segment of a Primary Regional 2 fault is buried by the volcano.

Last, the southwestern sector of Table Mountain (Figure 4E) has a crater-like structure that opens toward the south. The difference in slope steepness between the upper and lower portions of the edifice within the scar is evident.

Volcanoes with Erosional Breaches and their Underlying Faults

Volcanoes with an erosional breach or particular sectors affected by relatively more gullies include Pegleg, Swain, Roop Mountain, Logan, Campbell, South of Campbell, South of South Campbell, and Harvey Mountain (Figure 4F); Chalk and Bald Mountains in the north (Figure 4C); Raker Peak (Figure 4E), Cone and Cal Mountains (Figure 4F), Soldier, Haney and, Saddle Mountains (Figure 4D), and Burney Spring Mountain (Figure 4C) in the east.

Harvey Mountain is bounded to the east by Ashurst volcano. These two stratovolcanoes are separated by a segment of a <1600 ka regional fault that also affects Logan, Antelope and Fox Mountains. On the northeast is a possible erosional feature most likely to develop under the influence of an active normal fault with a strike-slip component. Logan, Campbell and South of Campbell all have erosional features opening toward the northeast or northwest. Of these, only Logan Mountain is clearly traversed by a <1600 ka fault, the same fault that splits Fox Mountain and is related to the slide on Antelope Mountain. The volcano to the south of South Campbell has no apparent nearby fault to which we can relate the north and northwest scars. However, a river flows near the western base, which might have laterally undermined the volcano base, causing instability on the northwestern flank. The western flank of Roop Mountain has gullies. The northeastern sector is the most incised by deep gullies, with major gully heads. Based on its relatively bigger size and volume compared to other volcanoes, and the density and size of gullies, Roop Mountain is probably an old volcanic edifice.

Swain Mountain has traces of two fault on the southwestern and northeastern piedmonts. Two erosional scars are shaped like cirques and have gullies within their walls. Breaks in slope that create benches or terraces within the scar may indicate that this is the most deformed part of the edifice, causing it to be more easily affected by erosional processes. The locations and opening directions of the scars are consistent with deformed sectors of volcanoes affected by normal faults with a strike-slip component occurring at the base and facing toward the edifice (Wooller et al., 2009). Pegleg Mountain also has two scar traces but unlike Swain Mountain, there is no nearby fault that can be explicitly connected to it. Thus, all sectors of the volcano are affected by gullies.

In the northern portion of the Hat Creek Graben region, where both the Rocky Ledge and Hat Creek faults terminate, and more dense Primary Regional 1 and 2 faults, and other Quaternary faults, are delineated, Chalk, Bald, Soldier, Haney, Saddle and Burney Spring Mountains are observed to be heavily traversed by the regional structures. In addition, the volcanoes occur along a major river (the Pit River), so they have many more erosional features and gullies in almost all sectors. They are therefore more affected by both frequent sliding and gully formation than other volcanoes.

On the eastern side of the study area, two volcanic edifices, Cone and Cal Mountains, each, have two erosional scars opening in opposite directions. There is no clear trace of a regional fault or any manifestation of movement of possible faults except for these scars.

Splitting Volcanoes, and the Faults Splitting Them

A few volcanoes are either very young, composed of strong and consolidated materials, or constructed on relatively stable basements. Even though they stand atop regional faults, may remain intact, except for splitting, in which some parts of the edifice move vertically relative to adjacent areas. Splitting volcanoes are affected by through-going faults, deforming yet not causing major instability or collapse. Whaleback, Fox, and Bogard (Figure 4F) Mountains are examples of this. Freaner and Wilcox (Figure 4B) are splitting elliptical volcanoes, as shown by vertical differences between adjacent parts of the edifices. Some volcanoes, such as West Prospect Peak, Prospect Peak (Figure 4E), and Sugarloaf (Figure 4B), have retained their ellipticity and regularity, even though faults split them.

Fault vs. Dike Dominated Regional Extension

In magmatic slow and ultraslow spreading plate boundaries and rift zones, extension is often accommodated by magmatic activity, or diking with associated faulting (Acocella and Trippanera, 2016, and references therein). Normal faults that propagate from the surface downwards and create graben-like structures on the surface are recurrent structural features. Symmetric, minor grabens are observed in Lakagigar and Bardarbunga in Iceland, and Dallol in Afar, induced by single diking episodes. Asymmetric, larger grabens can be seen in Krafla, Iceland, and Fantale, Ethiopia, formed by repeated or distributed diking. At these spreading plate boundaries, extension at depth seems to be accommodated mostly by diking, with normal faulting playing a negligible role. In shallower layers, normal faults and extensional tectonic structures accommodate the deformation, alongside a few feeder dikes (Acocella and Trippanera, 2016, and references therein).

While we cannot rule out the idea that dikes play a major role in extension at depth in the Hat Creek Graben region, and that some of the shorter faults might have been generated by diking episodes, it seems that at shallower levels, regional normal faults accommodate most of the extension. The extent of the Hat Creek Graben region, its regional fault systems with total displacements as high as a few hundred meters, recurrence intervals of a few 100 years, and the influence of the regional fault systems on the development of the volcanoes furthermore suggest that it is unlikely that most accommodation is by diking episodes, even deeper in the crust. The Hat Creek Graben region and the East African Rift are spreading at similar rates of 2.9 ± 0.8 mm yr^{-1} (Zeng and Shen, 2014) and 3–4 mm yr^{-1} (Stamps et al., 2008), respectively. However, it appears that the Hat Creek Graben region has a lower magma supply rate because it has smaller volcanoes than does the East African Rift (Grosse et al., 2014). The Hat Creek Graben region presents as an example of a different mechanism of extension in a small rift system associated with a transform plate boundary.

CONCLUSIONS

Topographic mapping and modeling, based mostly on NED data, were used to describe the geomorphology and structures of the Hat Creek Graben region. In addition to the obvious, known major geomorphological features, such as Lassen Peak, Pit River and Hat Creek valleys, smaller-scale geomorphological features, including mountain ridges, volcanic edifices, and lava fields, were characterized in detail using different topographic modeling techniques. Initial assessment of the shape and regularity of the volcano bases gave an overview of the processes that have affected or are currently affecting the volcanoes. Erosional features and the density of structures delineated on volcanoes were also noted. The study can serve as a standard for other volcanic areas in the world with strong tectonic interaction.

Using the height and basal width ratios from the morphometric analysis allowed us to classify the individual volcanic edifices of the Hat Creek Graben region as cones, sub-cones, shields and massifs. Separating those less than 350 m high as smaller cones, sub-cones and shields revealed an evolutionary trend of volcanic forms in the region: smaller cones can grow into bigger cones if they increase their height in proportion to their basal width, or evolve into sub-cones if their basal width increases faster than their height. Smaller sub-cones can grow into bigger sub-cones, or grow into the wide variety of shields or massifs; and smaller shields can grow into larger shields or massifs. As volcanoes develop and grow on top of major regional faults, their activity and growth pattern are influenced by regional extension accommodated on north-northwest striking faults that are mostly normal with strike-slip component. The stress regime and fault orientation can cause edifice deformation by collapse, erosion, or splitting. The majority of the deformational features, often manifested as collapse scars or erosional breaches, have opening directions perpendicular to the maximum horizontal stress, similar to those in other extensional settings (Tibaldi, 1995) and may, therefore, be related to the normal component of movement on the underlying faults. A considerable number, however, open at an acute angle and therefore are affected by the right-lateral component of movement on the underlying faults.

Spatial distribution varies for the volcano subgroups in the Hat Creek Graben region. The smaller volcanoes subgroup, and the combined smaller and bigger volcanoes, are clustered with northwest-southeast alignments formed by at least three volcanoes within 6 km of one another, while the bigger volcanoes subgroup is random and dispersed. This implies that the initial and clustered growth of volcanoes in the Hat Creek Graben region is controlled by the pre-existing major structures. Thus, the regional structures and volcanoes within the Hat Creek Graben region are interacting. The spatial dispersion of the bigger volcanoes can be attributed to volcanogenic, near-surface pressure changes—as volcanoes grow and long-lived plumbing systems develop—capable of refocusing local dikes. Thus, in the case of the Hat Creek Graben region, the regional, pre-existing major faults within the crust greatly influenced the initial alignment and distribution of volcanoes, and their continued growth is closely related to local as well as regional strain.

Faults (by slip) and volcanoes (by increasing total crustal volume through dilation during intrusion and propagation of magma to the surface) both accommodate regional extension in the Hat Creek Graben region. However, extension is accommodated more by faulting than by dikes, which along with the small size of the stratovolcanoes, suggests that the magma supply rate is low, with low-pressure, as magma propagates to the surface using the most efficient, easiest, pre-existing fractures. Thus, the propagation of magma to the surface through a complex plumbing system has been insufficient to accommodate much of the extension, which is therefore dominated by normal and other extensional, purely tectonic structures. This is manifested by the dominantly fault-related topography of the Hat Creek Graben region, and the overwhelming control of the tectonic structures on the growth, development, and evolution of volcanoes. The tectonic structures should continue to have a great influence on the stability and deformation of the volcanoes into the future.

We recognize that a more comprehensive factoring in of temporal data, such as description of erosion rates, and dating of collapse events relative to the age of fault activity, will be important for a complete characterization of the morphological and morphometric evolution of the Hat Creek Graben region.

AUTHOR CONTRIBUTIONS

EP and MB both contributed to the conception and design of the work, acquired, analyzed, and interpreted the data for the work. EP executed the detailed analyses, and drafted the manuscript. Revisions were made by both EP and MB. We both approved the version of the manuscript we submitted and we will both be accountable for the accuracy and integrity of this work.

CONFLICT OF INTEREST STATEMENT

The authors declare that the research was conducted in the absence of any commercial or financial relationships that could be construed as a potential conflict of interest.

ACKNOWLEDGMENTS

We thank Laura Pioli and Gianluca Groppelli, and the Associate Editor, Dr. Roberto Sulpizio for their comprehensive and constructive reviews. We also thank the Editor in Chief, Dr. Valerio Acocella for his helpful comments. Dr. G. N. Fabbro provided essential help in formatting technicalities and improving the drafts. We are grateful for field assistance by Mr. G. Bursik and Ms. Leandra Marshall. This research was funded in part by NASA grant NNX12AQ10G, and by the State University of New York.

REFERENCES

1. Acocella, V., and Trippanera, D. (2016). How diking affects the tectonomagmatic evolution of slow spreading plate boundaries: overview and model. Geosphere 12:GES01271.1. doi: 10.1130/GES01271.1
2. Argus, D. F., and Gordon, R. G. (1991). Current Sierra Nevada-North America motion from very long baseline interferometry: implications for the kinematics of the western United States. Geology19, 1085–1088.
3. Austin, L. J. (2013). "Fault kinematics from LiDAR analysis of the Rocky Ledge Fault, Shasta County, California," in Geological Society of America Abstracts with Programs (Denver, CO: Geological Society of America), 546.
4. Bahar, I., and Girod, M. (1983). Controle structural du volcanisme indonesien (Sumatra, Java-Bali); application et critique de la methode de Nakamura. Bull. Soc. Geol. Fr. XXV, 609–614. doi: 10.2113/gssgfbull.S7-XXV.4.609
5. Baloga, S. M., Glaze, L. S., and Bruno, B. C. (2007). Nearest-neighbor analysis of small features on Mars: applications to tumuli and rootless cones. J. Geophys. Res. E Planets 112, 1–17. doi: 10.1029/2005JE002652
6. Beggan, C., and Hamilton, C. W. (2010). New image processing software for analyzing object size-frequency distributions, geometry, orientation, and spatial distribution. Comput. Geosci. 36, 539–549. doi: 10.1016/j.cageo.2009.09.003
7. Bennett, R. A., Davis, J. L., and Wernicke, B. P. (1999). Present-day pattern of Cordilleran deformation in the western United States. Geology 27, 371–374.
8. Blakely, R. J., Christiansen, R. L., Guffanti, M., Wells, R. E., Donnelly-Nolan, J. M., Muffler, L., et al. (1997). Gravity anomalies, Quaternary vents, and Quaternary faults in the southern Cascade Range, Oregon and California: implications for arc and backarc evolution. J. Geophys. Res. 102:22513. doi: 10.1029/97JB01516
9. Blakeslee, M. W., and Kattenhorn, S. A. (2013). Revised earthquake hazard of the Hat Creek fault, northern California: a case example of a normal fault dissecting variable-age basaltic lavas.Geosphere 9, 1397–1409. doi: 10.1130/GES00910.1
10. Bleacher, J. E., Glaze, L. S., Greeley, R., Hauber, E., Baloga, S. M., Sakimoto, S. E. H., et al. (2009). Spatial and alignment analyses for a field of small volcanic vents south of Pavonis Mons and implications for the Tharsis province, Mars. J. Volcanol. Geotherm. Res. 185, 96–102. doi: 10.1016/j.jvolgeores.2009.04.008
11. Caputo, R. (1995). Evolution of orthogonal sets of coeval extension joints. Terra Nov. 7, 479–490. doi: 10.1111/j.1365-3121.1995.tb00549.x
12. Clark, P. J., and Evans, F. C. (1954). Distance to nearest neighbor as a measure of spatial relationships in populations. Ecol. Soc. Am. 35, 445–453. doi: 10.2307/1931034
13. Connor, C. B., Condit, C. D., Crumpler, L. S., and Aubele, J. C. (1992). Evidence of regional structural controls on vent distribution: springerville volcanic field, Arizona. J. Geophys. Res. 97, 12349–12359. doi: 10.1029/92JB00929
14. Dixon, T. H., Miller, M., Farina, F., Wang, H., and Johnson, D. (2000). Present-day motion of the Sierra Nevada block and some tectonic implications for the Basin and Range province, North American Cordillera. Tectonics 19:1. doi: 10.1029/1998TC001088
15. Dokka, R., and Travis, J. (1990). Role of the eastern california shear zone in integrated shear along the to historic measurements mm yr present rate mm yr 4 determined. Geophys. Res. Lett. 17, 1323–1326. doi: 10.1029/GL017i009p01323
16. Ebinger, C., Ayele, A., Keir, D., Rowland, J., Yirgu, G., Wright, T., et al. (2010). Length and timescales of rift faulting and magma intrusion: the afar rifting cycle from 2005 to present. Annu. Rev. Earth Planet. Sci. 38, 439–

466. doi: 10.1146/annurev-earth-040809-152333

17. Eppler, D. B., Fink, J., and Fletcher, R. (1987). Rheologic properties and kinematics of emplacement of the chaos jumbles rockfall avalanche, Lassen Volcanic National Park, California. J. Geophys. Res.92:3623. doi: 10.1029/JB092iB05p03623

18. Faulds, J. E., Henry, C. D., and Hinz, N. H. (2005). Kinematics of the northern Walker Lane: An incipient transform fault along the Pacific-North American plate boundary. Geology 33, 505–508. doi: 10.1130/G21274.1

19. Francis, P., and Wells, A. (1988). LANDSAT Thematic Mapper observa- tions of debris avalanche deposits in the Central Andes. Bull. Volcanol. 50, 258–278. doi: 10.1007/BF01047488

20. Gesch, D., Oimoen, M., Greenlee, S., Nelson, C., Steuck, M., and Tyler, D. (2002). The national elevation dataset. Photogramm. Eng. Remote Sens. 68, 5–32.

21. Grosse, P., Euillades, P. A., Euillades, L. D., and van Wyk de Vries, B. (2014). A global database of composite volcano morphometry. Bull. Volcanol. 76:784. doi: 10.1007/s00445-013-0784-4

22. Grosse, P., van Wyk de Vries, B., Euillades, P. A., Kervyn, M., and Petrinovic, I. A. (2012). Systematic morphometric characterization of volcanic edifices using digital elevation models.Geomorphology 136, 114–131. doi: 10.1016/j.geomorph.2011.06.001

23. Grosse, P., van Wyk de Vries, B., Petrinovic, I. A., Euillades, P. A., and Alvarado, G. E. (2009). Morphometry and evolution of arc volcanoes. Geology 37, 651–654. doi: 10.1130/G25734A.1

24. Hamilton, C. W., Fagents, S. A., and Thordarson, T. (2010). Explosive lavawater interactions II: self-organization processes among volcanic rootless eruption sites in the 17831784 Laki lava flow, Iceland. Bull. Volcanol. 72, 469–485. doi: 10.1007/s00445-009-0331-5

25. Hammond, W. C., and Thatcher, W. (2005). Northwest Basin and Range tectonic deformation observed with the Global Positioning System, 1999-2003. J. Geophys. Res. Solid Earth 110, 1–12. doi: 10.1029/2005JB003678

26. Hammond, W. C., and Thatcher, W. (2007). Crustal deformation across the Sierra Nevada, northern Walker Lane, Basin and Range transition, western United States measured with GPS, 2000-2004. J. Geophys. Res. Solid Earth 112, 1–26. doi: 10.1029/2006JB004625

27. Heidbach, O., Tingay, M., Barth, A., Reinecker, J., Kurfeß, D., and Müller, B. (2008). The World Stress Map Database Release 2008. doi: 10.1594/GFZ.WSM.Rel2008

28. Jenkins, O. (1938). Geomorphic Map of California: California Division of Mines and Geology Map Sheet, Scale Approximately 1:2,000,000.

29. Jolly, R., and Sanderson, D. (1997). A Mohr circle construction for the opening of a pre-existing fracture. J. Struct. Geol. 19, 887–892. doi: 10.1016/S0191-8141(97)00014-X

30. Lagmay, A. M. F., and Valdivia, W. (2006). Regional stress influence on the opening direction of crater amphitheaters in Southeast Asian volcanoes. J. Volcanol. Geotherm. Res. 158, 139–150. doi: 10.1016/j.jvolgeores.2006.04.020

31. Le Corvec, N., Spörli, K. B., Rowland, J., and Lindsay, J. (2013). Spatial distribution and alignments of volcanic centers: clues to the formation of monogenetic volcanic fields. Earth-Sci. Rev. 124, 96–114. doi: 10.1016/j.earscirev.2013.05.005

32. McCaffrey, R. (2005). Block kinematics of the Pacific-North America plate boundary in the southwestern United States from inversion of GPS, seismological, and geologic data. J. Geophys. Res. Solid Earth 110, 1–27. doi: 10.1029/2004JB003307

33. Miliaresis, G., and Kokkas, N. (2004). Segmentation and terrain modeling of extraterrestrial chasmata. J. Spat. Sci. 49, 89–99. doi: 10.1080/14498596.2004.9635024

34. Miller, M., Johnson, D. J., Rubin, C. M., Dragert, H., Wang, K., Qamar, A., et al. (2001). GPS-determination of along-strike variation in Cascadia margin kinematics: implications for relative plate motion, subduction zone coupling, and permanent deformation. Tectonics 20, 161–176. doi: 10.1029/2000TC001224

35. Moriya, I. (1980). Bandaian Eruption and Landforms Associated with It: Collection of Articles in Memory of Retirement of Prof. K. Hishimura. Technical Report, Tohoku University.

36. Muffler, L., Blakely, R. J., and Clynne, M. (1994). Late Quaternary normal faulting of the Hat Creek Basalt, Northern California. Geol. Soc. Am. Bull. 106, 195–200.

37. Muffler, L., Robinson, J., Felgar, T., Dutton, D., and Clynne, M. (2010). Database for the Geologic Map of Lassen Volcanic National Park and Vicinity, California. Data to accompany Scientific Investigations Map 2899. Available online at: http://pubs.usgs.gov/sim/2899/database.html

38. Nakamura, K. (1977). Volcanoes as possible indicators of tectonic stress orientation-principle and proposal. J. Volcanol. Geotherm. Res. 2, 1–16. doi: 10.1016/0377-0273(77)90012-9

39. Paguican, E. M. R., van Wyk de Vries, B., and Lagmay, A. M. F. (2014). Hummocks: how they form and how they evolve in rockslide-debris avalanches. Landslides 11, 67–80. doi: 10.1007/s10346-012-0368-y

40. Parrish, J. G. (2006). Simplified Geologic Map of California. California Geological Survey, Map Sheet 57. Available online at:http://www.conservation.ca.gov/cgs/information/publications/ms/documents/ms057.pdf

41. Perez, F. L. (1998). Talus fabric, clast morphology, and botanical indicators of slope processes on the chaos crags (California cascades), U.S.A. Geogr. Phys. Quat. 52, 1–22. doi: 10.7202/004861ar

42. Peterson, J., and Martin, L. (1980). Geologic Map of the Baker-Cypress BLM Roadless Area and Timbered Crater Rare II Areas, Modoc, Shasta, and Siskiyou Counties, Northeastern California.U.S. Geological Survey, Miscellaneous Field Studies Map 1214-A. Available online at:https://pubs.er.usgs.gov/publication/mf1214A

43. Sauber, J., Thatcher, W., Solomon, S., and Lisowski,

M. (1994). Geodetic slip rate for the Eastern California shear zone and the recurrence time of Mojave Desert earthquakes. Nature 367, 264–266. doi: 10.1038/367264a0

44. Sawyer, T., and Ramelli, A. (2012). Structural Analysis, Evaluation, and Quaternary Activity of the Hat Creek and Rocky Ledge Faults, Shasta County, Northeastern California. Technical Report, Piedmont Geosciences Report, October 2012.

45. Siebert, L. (1984). Large volcanic debris avalanches: characteristics of source areas, deposits, and associated eruptions. J. Volcanol. Geotherm. Res. 22, 163–197. doi: 10.1016/0377-0273(84)90002-7

46. Sigmundsson, F., Hooper, A., Hreinsdóttir, S., Vogfjörd, K. S., Ófeigsson, B. G., Heimisson, E. R., et al. (2015). Segmented lateral dyke growth in a rifting event at BárÃřarbunga volcanic system, Iceland. Nature 517, 191–195. doi: 10.1038/nature14111

47. PubMed Abstract

48. Stamps, D. S., Calais, E., Saria, E., Hartnady, C., Nocquet, J. M., Ebinger, C. J., et al. (2008). A kinematic model for the East African Rift. Geophys. Res. Lett. 35, 1–6. doi: 10.1029/2007GL032781

49. Tibaldi, A. (1995). Morphology of pyroclastic cones and tectonics. J. Geophys. Res. 100, 24521–24535. doi: 10.1029/95JB02250

50. Ui, T., Yamamoto, H., and Suzuki-Kamata, K. (1986). Characterization of debris avalanche deposits in Japan. J. Volcanol. Geotherm. Res. 29, 231–243. doi: 10.1016/0377-0273(86)90046-6

51. U.S. Geological Survey California Geological Survey (2006). Quaternary Fault and Fold Database for the United States. Available online at: http://earthquake.usgs.gov/hazards/qfaults/

52. Valentine, G., and Krogh, K. (2006). Emplacement of shallow dikes and sills beneath a small basaltic volcanic center The role of pre-existing structure (Paiute Ridge, southern Nevada, USA).Earth Planet. Sci. Lett. 246, 217–230. doi: 10.1016/j.epsl.2006.04.031

53. Valentine, G. A., and Hirano, N. (2010). Mechanisms of low-flux intraplate volcanic fields - Basin and Range (North America) and northwest Pacific Ocean. Geology 38, 55–58. doi: 10.1130/G30427.1

54. Wadge, G., and Cross, A. (1988). Quantitative methods for detecting aligned points: an application to the volcanic vents of the Michoacan-Guanajuato volcanic field, Mexico. Geology 16, 815–818.

55. White, I., and Crider, J. (2006). Extensional fault-propagation folds: mechanical models and observations from the Modoc Plateau, Northeastern California. J. Struct. Geol. 28, 1352–1370. doi: 10.1016/j.jsg.2006.03.028

56. Wills, C. J. (1991). Active faults north of Lassen Volcanic National Park, Northern California. Calif. Geol. 44, 49–72.

57. Wooller, L., van Wyk de Vries, B., Cecchi, E., and Rymer, H. (2009). Analogue models of the effect of long-term basement fault movement on volcanic edifices. Bull. Volcanol. 71, 1111–1131. doi: 10.1007/s00445-009-0289-3

58. Wright, T. J., Ebinger, C., Biggs, J., Ayele, A., Yirgu, G., Keir, D., et al. (2006). Magma-maintained rift segmentation at continental rupture in the 2005 Afar dyking episode. Nature 442, 291–294. doi: 10.1038/nature04978

59. Zeng, Y., and Shen, Z. K. (2014). Fault network modeling of crustal deformation in California constrained using GPS and geologic observations. Tectonophysics 612–613, 1–17. doi: 10.1016/j.tecto.2013.11.030

CHAPTER 5 TECTONIC GEOMORPHOLOGY

Elahe Javadi Mosavi, Mehran Arian

Keywords: Atrak, Kope Dagh, Geomorphology, River, Basin, Iran

INTRODUCTION

Neotectonics is a major factor controlling landform development in quaternary tectonically active regions. The effect factors in morphology region are tectonics, lithology, climate and human. If tectonic has been most effective factor in region, it's related to tectonic geomorphology. The study area (Figure 1) of Atrak River basin is at north east of Iran [1] [2] . The river is located in the Kope dagh and Atrak basin is divided to 56 sub-basins (Figure 2).

This area belongs to Kope dagh province [3] . Dominant structural trend in Kope dagh province (Figure 3) is NW-SE. From tectonics view, it contains the Kope dagh hinterland or Kope dagh fold and thrust belt that formed in passive margin of Eurasian plate until late Triassic and then marine sedimentation on Kope Dagh proforeland basin has continued to Eocene. Kope dagh hinterland has uplifted related to Karakorum foreland basin in northeast along Eshghabad fault.

The western part of studied area has located in South Caspian foreland basin. Dominant structural trend in

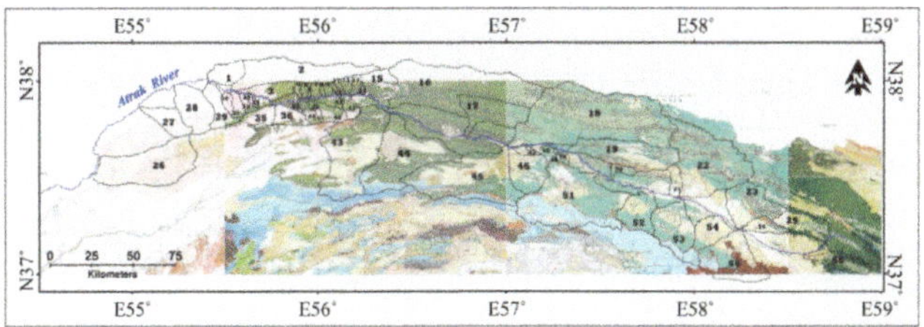

Figure 1. Location of the study area.

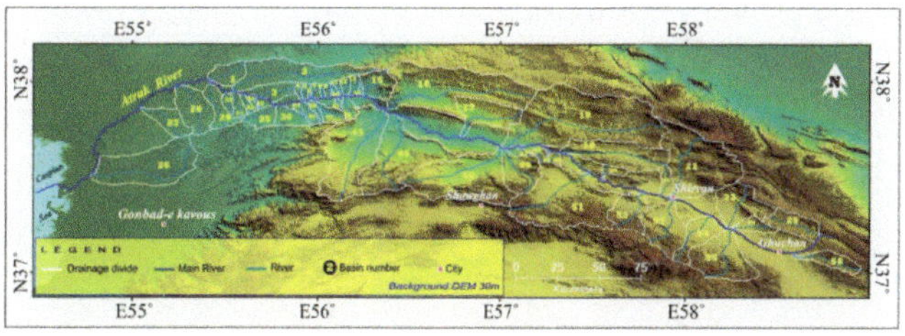

Figure 2. Fifty six sub-basins of the Atrak River basin.

South Caspian foreland basin province (Figure 3) is NW-SE. From tectonics view, it contains the northern fo- reland basin of West-Central Alborz and lesser Caucasus hinterland in the south margin of Eurasian plate since late Eocene. Although median part of South Caspian and Black sea basin has uplifted by collision between Eurasian and Cimmerian plates [3] .

This area has semi active tectonics regim in compared to Alborz [4] - [11] , Central Iran [12] - [18] and Zagros in the southern Iran [19] - [22] .

MATERIALS AND METHODS

This article applies a quantitative geomorphologic method to an area in the Zagros to evaluate rates of active tectonics. Considering the diversity of the morphotectonic features [23] [24] . Spatial tools including geographic information systems (GIS) and morphometric analyses may provide useful information on this subject, we have analyzed below geomorphic indices by [25] method (Table 1). Also, results were shown in Figure 4.

To study the indices, there is a formula which we turn to describe each one of indices; It is necessary to have some primary maps to calculate the indices, and the most important of which are: Digital Elevation Model (DEM), the drainage network and the sub-basins map of the Atrak basin that have been extracted from DEM. DEM (SRTM) extracted from a digitized topographic map (1:30,000).

1—the stream-gradient index (SL): $SL = (\Delta H/\Delta L) L$;

Citation: Elahe Javadi Mosavi, Mehran Arian, Tectonic Geomorphology, DOI: 10.4236/ojg.2015.53010.

2—drainage basin asymmetry (Af): Af = (Ar/At) 100;

3—hypsometric integral (Hi): $Hi = (\text{average elevation} - \text{min. elev.})/(\text{max. elev.} - \text{min. elev.})$;

4—valley floor width-valley height ratio (Vf): $Vf = 2Vfw/(Ald + Ard - 2Asc)$;

5—drainage basin shape (Bs): Bs = Bl/Bw;

6—mountain-front sinuosity index (Smf): J = Lj/Ls.

RESULTS AND DISCUSSION

We computed a single index (Iat) from the six indices to characterize relative active tectonics. Previous studies in Iran [26] - [42] tend to focus on a particular mountain front or area to discuss relative tectonic activity and geomorphic indices [42] - [47] . This study tried to evaluate tectonics in a wider area, using a number of omor- phologic parameters. The average of the six measured geomorphic indices (Iat) was used to evaluate the distribution of relative tectonic activity in the study area. The values of the index were divided into four classes to define the degree of active tectonics: 1—very high ($1.0 \le Iat < 1.5$); 2—high ($1.5 \le Iat < 2.0$); 3—moderate ($2.0 \le Iat < 2.5$); and 4—low($2.5 \le Iat$).

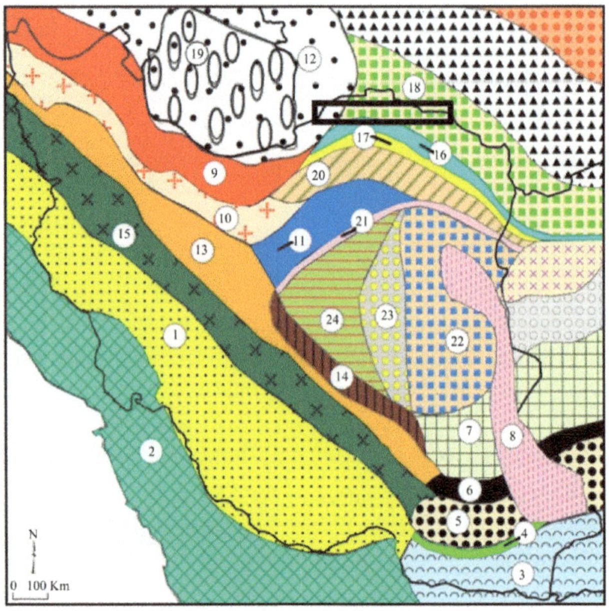

Figure 3. Physiographic-tectonic zoning map of Iran's sedimentary basins Iran, modified from [10] . Numbers in this figure are, 1: Zagros-East Taurus hinterland, 2: Persian Gulf-Mesopotamian foreland basin, 3: Makran accretionary prism, 4: Bashagard Mountains, 5: Jazmorian-Mashkel fore arc basin, 6: Shahsavaran-Soltan magmatic arc, 7: South Lut-South Helmand back arc basin, 8: East Iran Mountain belt, 9: West-Central Alborz and lesser Caucasus hinterland, 10: Great Kavir-Nor- then Urmieh lake foreland basin, 11: South Great Kavirfold and thrust belt, 12: South Caspian- Black sea foreland basin, 13: Urmieh-Dokhtar Magmatic Arc, 14: Naien-Kerman retro arc fore- land basin, 15: Sanandaj-Sirjanoverthrust belts, 16: East Alborz or Binalod hinterland, 17: Torbat-e am-Neyshabour retro arc foreland basin, 18: Kope Dagh hinterland, 19: South Caspian remnant basin, 20: Maiamay-Taibad Inverted back arc basin, 21: Khaf-Kavir Plain Magmatic Arc, 22: Lut Plain-Gonabad back arc basin, 23: Tabas hinterland, 24:Yazd-Khour Piggy back basin. The study area is shown in the black rectangle.

Table 1. Range of geomorphic indices, based on [25] .

Geomorphic index	Class 1	Class 2	Class 3
Vf	Vf < 0.3	0.3 < Vf < 1	Vf > 1
J	J < 1.1	1.1 < J < 1.5	J > 1.5
Bs	Bs > 4	3 < Bs < 4	Bs < 3
Af	Af < 35 or Af > 63	57 < Af < 65 or 35 < Af < 43	43 < Af < 57
Sl	Sl > 500	300 < Sl < 500	Sl < 300
Hi	Hi > 0.5	0.4 < Hi < 0.5	Hi < 0.4

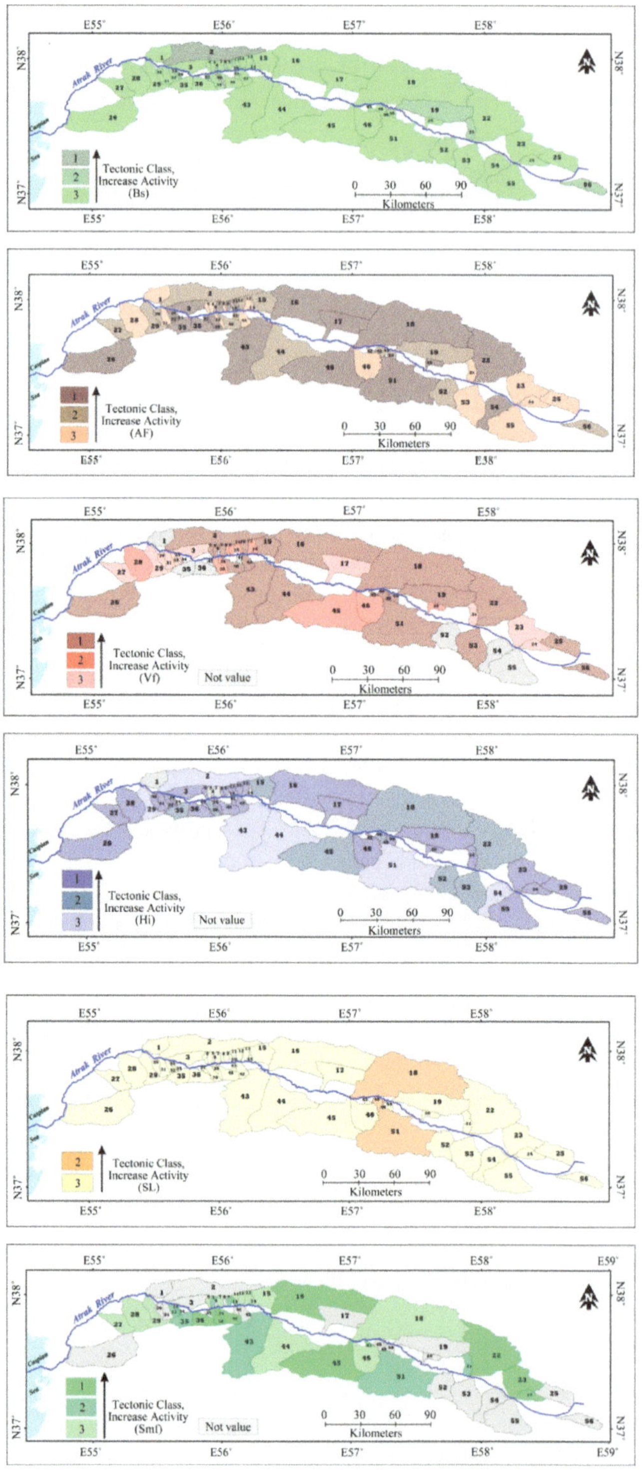

Figure 4. Distribution of six indices and classification of them into three classes.

The distribution of the four classes is shown in Figure 4. The values of the six geomorphic indices as well as Iat often change corresponding to the distribution of fault zones. Kope dagh, Ghochan, Kerivan, Esfraien and Chamanbid faults are the structural boundary that they have been changed in deformational history of study area. In study area, the low class of Iat is mainly in the sub-basins of 1, 14, 29 & 31 while the other parts of the study area has moderate active tectonics. Table 2 shows the result of the classification for each sub-basin.

CONCLUSIONS

Atrak River, north east Iran is a quaternary tectonically active region. The values of SL, Hi, and Bs were found to be high along major faults and folds. The values of Af show widespread drainage-basin asymmetry related to the tectonic tilting. The values of J suggest that mountain fronts are tectonically active, and the values of Vf show that many valleys are narrow and deep, suggesting a high rate of incision associated with tectonic uplift.

About the whole of the study area has Iat values of classes 2 and 3, indicating moderately to highly active tectonics. Class 2 of Iat, indicative of highly active tectonics, occurs mainly along the fault zone. These classes also correspond well to the areas with prominent scarps of tectonic origin, triangular facets, deformed alluvial fans, and deep narrow gorges.

ACKNOWLEDGEMENTS

This work has funded by the Department of geology, Islamic Azad University, Science and Research branch,

Table 2. Values of Lat.

Basin	Bs Tectonic Class	AF Tectonic Class	T Tectonic Class	Vf Tectonic Class	Hi Tectonic Class	SL Tectonic Class	Smf Tectonic Class	Iat Detail of Average	Iat Tectonic Class
1	3	3	3	_	_	3	_	3.00	3 (Low Activity)
2	1	2	3	1	3	3	_	2.17	2 (Moderate Activity)
3	3	1	3	3	1	3	_	2.33	2
4	3	3	3	1	1	3	3	2.43	2
5	3	3	2	1	_	3	2	2.33	2
6	2	2	2	_	_	3	_	2.25	2
7	3	1	2	1	_	3	3	2.17	2
8	2	3	2	2	1	3	_	2.17	2
9	3	1	1	2	1	3	2	1.86	2
10	2	2	3	3	1	3	1	2.14	2
11	3	1	1	2	1	3	3	2.00	2
12	3	2	1	1	1	3	3	2.00	2
13	3	3	2	2	1	3	3	2.43	2
14	3	3	2	2	_	3	3	2.67	3
15	3	2	1	1	2	3	3	2.14	2
16	3	1	1	1	1	3	1	1.57	2
17	3	1	2	3	1	3	_	2.17	2
18	3	1	2	1	2	2	3	2.00	2
19	2	2	2	1	1	3	_	1.83	2
20	3	1	2	2	1	3	_	2.00	2
21	2	3	2	3	1	3	2	2.29	2
22	3	1	1	1	2	3	1	1.71	2
23	3	3	3	3	1	3	1	2.43	2
24	3	3	2	3	1	3	1	2.29	2
25	3	3	3	1	1	3	_	2.33	2
26	3	1	1	1	1	3	_	1.67	2
27	3	2	1	3	1	3	3	2.29	2
28	3	3	2	2	1	3	3	2.43	2
29	3	2	2	3	3	3	3	2.71	3
30	3	2	2	3	1	3	_	2.33	2
31	3	3	2	3	1	3	_	2.50	3
32	3	1	2	3	1	3	2	2.14	2
33	2	1	1	3	1	3	2	1.86	2
34	1	2	2	3	1	3	2	2.00	2
35	3	1	2	_	2	3	2	2.17	2
36	3	1	1	_	1	3	1	1.67	2
37	2	1	1	_	1	3	_	1.60	2

38	3	1	1	2	2	3	1	1.86	2
39	1	3	2	2	1	3	1	1.86	2
40	3	2	2	1	1	3	1	1.86	2
41	3	2	3	_	1	3	_	2.40	2
42	3	3	2	1	1	3	_	2.17	2
43	3	1	2	1	3	3	2	2.14	2
44	3	2	1	1	3	3	3	2.29	2
45	3	1	1	2	2	3	1	1.86	2
46	3	3	2	2	1	3	3	2.43	2
47	3	2	2	1	1	3	1	1.86	2
48	3	2	2	1	2	2	_	2.00	2
49	2	3	2	1	1	2	_	1.83	2
50	3	2	2	1	1	3	_	2.00	2
51	3	1	3	1	3	2	2	2.14	2
52	3	2	2	_	2	3	_	2.40	2
53	3	3	2	1	2	3	_	2.33	2
54	3	1	1	_	3	3	_	2.20	2
55	3	3	2	_	1	3	_	2.40	2
56	2	2	3	1	1	3	_	2.00	2

Iat classification: 1 (High Activity), 2 (Moderate Activity), 3 (Low Activity).

Elahe Javadi Mosavi, Mehran Arian Tehran, Iran. Also, Special thanks to vice-president for research in Science and Research branch, Tehran.

REFERENCES

1. Qorashi, M. and Arian, M. (2011) Tectonics of Iran. Geologic Survey of Iran, 336 p.r reference:1]
2. Arian, M. (2011) Basement Tectonics and Geology of Iran. Asar Nafis Press, 300 per reference:1]
3. Arian, M. (2013) Physiographic-Tectonic Zoning of Iran's Sedimentary Basins. Open Journal of Geologper reference:2]
4. Arian, M., Maleki, Z. and Noroozpour, H. (2011) Cenozoic Diastrophism and Deformational Events in the East-Cen- tral Alborz. Journal of Basic and Applied Scientific Researcaper reference:1]
5. Feizi, F., Arian, A. and Rahmani, R. (2007) Seismotectonic Zoning in the Eastern Part of the Central Alborz. Journal of Sciences, Islamic Azad University, 17, 151-164. (In Persian)
6. Nouri, R., Jafari, M.R., Arian, M., Feizi, F. and Afzal, P. (2013) Correlation between Cu Mineralization and Major Faults Using Multifractal Modelling in the Tarom Area (NW Iran). Geologica Carpathica, 64, 409-416.
7. Nouri, R., Afzal, P., Arian, M., Jafari, M. and Feizi, F. (2013) Reconnaissance of Copper and Gold Mineralization Using Analytical Hierarchy Process (AHP) in the Rudbar 1:100,000 Map Sheet, Northwest Iran. Journal of Mining and Metallurgy A: Mining, 49, 9-19.
8. Poroohan, N., Pour Kermani, M. and Arian, M. (2013) An Assessment of Relationship in F-Parameter and Paleostress Fields in Heterogeneous Lithologies: Roudbar Area (Northwest of Iran). Australian Journal of Basic & Applied Sciences, 7, 933-942.
9. Poroohan, N., Kermani, M.P. and Aryan, M. (2009) An Assessment on Correlations of Seismotectonic Parameters Preceding and Following Roudbar-Manjil Earthquake (Gilan, North of Iran). Australian Journal of Basic & Applied Sciences, 3, 2643-2652.
10. Arian, M. and Pourkermani, M. (2004) Tectonic Elements of South Flank in the East-Central Alborz Mountain. Journal of Sciences, Teacher Training 8. [Paper reference:1]
11. Arian, M. and Qorashi, M. (2006) The Movement Potential Evaluation of the Major Quaternary Faults in Alborz-Central Iran Border Zone, from the East of Tehran to the East of Semnan. Journal of Geosciences, Geological Surve8. [Paper reference:1]
12. Pourkermani, M. and Arian, M. (2001) Structural Geomorphology of Northeastern Kurdistan, Sistan and Baluchestan University. Journal 48. [Paper reference:1]
13. Arian, M. (2012) Clustering of Diapiric Provinces in the Central Iran Basin. Carbonates and Evaporites, 27, 9-18. http://dx.doi.org/10.1007/s13146-011-0079-9
14. Pourkermani, M. and Arian, M. (1997) Salt Domes of Central Iran. Journal of Humanities University of Sistan and Balouchestan, 3, 29-41. (In Persian)
15. Arian, M., Pourkermani, M., Sistanipour, A. and Noroozpour, H. (2011) Kinematic Significance of Fold- and Fault- Related Fracture Systems in the Rafsanjan's Northeast Highlands (Central Iran). Journal of Basic and Applied Scientific Research, 1, 3398-3406.
16. Arian, M., Pourkermani, M., Sistanipour, A. and Noroozpour, H. (2011) Seismicity and Fault Segmentation of Bafq- Baghin Fault System (Central Iran). Journal of Applied Environmental and Biological Sciences, 1, 382-396.
17. Asadian, F. and Arian, M. (2009) Identification of Diapiric Provinces of Central Iran through Geological and Geographical Analysis. International Journal of Agriculture Environment & Biotechnology, 2, 3443-3451.
18. Arian, M. (2010) Earthquake—Fault Hazard Investigations in the Kerman Quadrangle. Journal

of Sciences, Islamic Azad Univers Persian) [Paper reference:1]

19. Rahimi, N. and Arian, M. (2014) Tectonic Geomorphplogy of Hamedan-Sosangerd Region, West Iran. Advances in Env, 119-124. [Paper reference:1]
20. Arian, M. and Hashemi, A. (2008) Seismotectonic Zoning in the Zagros. Journal of Sciences, 18, 63-76. (In Persian)
21. Arian, M., Ahmadnia, A., Qorashi, M. and Pourkermani, M. (2002) Structural Analysis of Mengharak Transcurrent Fault System in Zagros, Iran. GeoArabia, 7, 209-210.
22. Arian, M., Qorashi, M., Pourkermani, M. and Ahmadnia, A. (2006) The Structural Significance Kareh Bas Transcurrent Fault System in the Zagros Fold and Thrust Belt. Journal of Geosciences, Geological Survey (In Persian) [Paper reference:1]
23. Burbank, D.W. and Anderson, R.S. (2001) Tectonic GeomorphologyOxford, 274 p. [Paper reference:1]
24. Keller, E.A. and Pinter, N. (1996) Active Tectonics: Earthquakes, Uplift, and Landscnglewood Cliff. [Paper reference:1]
25. El Hamdouni, R., Irigaray, C., Fernandez, T., Chacón, J. and Keller, E.A. (2007) Assessment of Relative Active Tectonics, Southwest Border of Sierra Nevada (Southerngy, 96, 150-173. [Paper reference:2]
26. Arian, M. and Aram, Z. (2014) Relative Tectonic Activity Classification in the Kermanshah Area, Western Iran. Solid Earth, 5, 1277-1291. ht94/se-5-1277-2014 [Paper reference:1]
27. Mashal, M., Kermani, M.P., Charchi, A., Almasian, M. and Arian, M. (2013) Pattern of Structural Geology Underground in Eastern of North Dezfol Embayment. Advances in Environmental Biology, 7, 260-268.
28. Housini Toudeshki, V. and Arian, M. (2011) Morphotectonic Analysis in the Ghezel Ozan River Basin, NW Iran. Journal of Geography and Geology, 3, 258-260.
29. Pazhoohan, M., Arian, M., Ghorashi, M. and Khosrotehrani, K. (2014) A Study of Drainage Pattern Responses to Active Tectonics in Tadvan Region, SW Iran. Geodynamics, 1, 36-41.
30. Arian, M., Bagha, N., Khavari, R. and Noroozpour, H. (2012) Seismic Sources and Neo-Tectonics of Tehran Area (North Iran). Indian Journal of Science and Technology, 5, 2379-2383.
31. Arian, M., Pourkermani, M., Sistanipour, A. and Noroozpour, H. (2011) Seismicity and Fault Segmentation of Bafq-Baghin Fault System (Central Iran). Journal of Applied Environmental and Biological Sciences, 1, 382-396.
32. Bagha, N., Arian, M., Ghorashi, M., Pourkermani, M., El Hamdouni, R. and Solgi, A. (2014) Evaluation of Relative Tectonic Activity in the Tehran Basin, Central Alborz, Northern Iran. Geomorphology, 213, 66-87. http://dx.doi.org/10.1016/j.geomorph.2013.12.041
33. Arian, M. and Feizi, F. (2005) Application of Geomorphic Indices to the Assessment of Relative Tectonic Activity Levels in the Alborz—Central Iran Border Zone (From the East of Varamin to the East of Semnan). Journal of Sciences (Islamic Azad University), 15, 378-403.
34. Khavari, R., Arian, M. and Ghorashi, M. (2009) Neotectonics of the South Central Alborz Drainage Basin, in NW Tehran, N Iran. Journal of Applied Sciences, 9, 4115-4126.http://dx.doi.org/10.3923/jas.2009.4115.4126
35. Arian, M. and Bagha, N. (2012) Active Tectonics of Tehran Area, Iran. Journal of Basic and Applied Scientific Research, 2, 3805-3819.
36. Arian, M., Toudeshki, V.H. and Noroozpour, H. (2011) Active Tectonics of Qezel Ozan River Basin, NW Iran. Journal of Applied Environmental and Biological Sciences, 1, 291-295.
37. Housini Toudeshki, V., Pourkermani, M., Arian, M. and Khosrotehrani, K.H. (2011) Influence of Structures on the Ghezel Ozan River. Geosciences, 21, 55-60.
38. Mardani, Z., Ghorashi, M. and Arian, M. (2011) Geomorphic Signatures of Active Tectonics in the Talaghan Rud, Shah Rud and Sefidrud Drainage Basins in Central Alborz, N Iran. Geosciences, 20, 159-166.
39. Sorbi, A., Arian, M. and Pourkermani, M. (2011) The Application of Geomorphic Indices to the Assessment of Relative Tectonic Activity Levels in Tehran Quadrangle. Journal of the Earth, 6, 1-9.
40. Khavari, R., Ghorashi, M., Arian, M. and Khosrotehrani, K. (2010) Geomorphic Signatures of Active Tectonics in the Karaj Drainage Basin in South Central Alborz, N Iran. Geosciences Scientific Quarterly Journal, 19, 67-74.
41. Arian, M. and Noroozpour, H. (2015) The Biggest Salt-Tongue Canopy of Central Iran. Open Journal of Geology, 5, 55-60. http://dx.doi.org/10.4236/ojg.2015.52005
42. Arian, M. and Noroozpour, H. (2015) Tectonic Geomorphology of Iran's Salt Structures. Open Journal of Geolo/dx.doi.org/10.4236/ojg.2015.52006 [Paper reference:2]
43. Azor, A., Keller, E.A. and Yeats, R.S. (2002) Geomorphic Indicators of Active Fold Growth: South Mountain-Oak Ridge Ventura Basin, Southern California. Geological Society of America Bulletin, 114, 745-753. http://dx.doi.org/10.1130/0016-7606(2002)114<0745:GIOAFG>2.0.CO;2
44. Bull, W.B. and McFadden, L.D. (1977) Tectonic Geomorphology North and South of the Garlock Fault, California. Proceedings of the Eighth Annual Geomorphology Symposium, Binghamton, 23-24 September 1977, 115-138.
45. Molin, P., Pazzaglia, F.J. and Dramis, F. (2004) Geomorphic Expression of Active Tectonics in a Rapidly-Deforming Forearc, Sila Massif, Calabria, Southern Italy. American Journal of Science, 304, 559-589. http://dx.doi.org/10.2475/ajs.304.7.559
46. Farrokhnia, A., Pirasteh, S., Pradhan, B., Pourkermani, M. and Arian, M. (2011) A Recent Scenario of Mass Wasting and Its Impact on the Transportation in Alborz Mountains, Iran Using Geo-Information Technology. Arabian Journal of Geosciences, 4, 1337-1349.

47. Rockwell, T.K., Keller, E.A. and Johnson, D.L. (1985) Tectonic Geomorphology of Alluvial Fans and Mountain Fronts near Ventura, California. In: Morisawa, M., Ed., Tectonand Unwin Publishers, Boston, 183-207. [Paper reference:1]

CHAPTER 6 EXTENSIONAL SEISMOGENIC STRESS AND TECTONIC MOVEMENT

Jiren Xu and Zhixin Zhao

INTRODUCTION

The continental crust of the Indian plate collides with and subducts underneath Eurasia along great thrusts at the southern edge of the Himalayas. The compressive tectonic force due to the collision along the Himalayas between the Indian and Eurasian plates results in orogenic motions in the Himalayas, Tibetan plateau, and Altyn and Qilian mountains [1, 2]. Such violent crustal movements and frequent and strong earthquakes extend to the Tianshan Mountains [3, 4]. The compressive tectonic force affects crustal motions in the west of China and as far as Mongolia [5, 6]. Such motion and deformation make the Tibetan plateau and its surrounding areas one of the most active regions in terms of tectonics and seismicity in the world [7, 8]. The synchronous temporal variations of the seismic activities on the Tibetan plateau region and its vicinity are related to the compressional collision between the Indian and Eurasian plates [9, 10]. Many papers have reported significant findings in the fields of tectonics and seismicity in this region [11, 12]. Besides the compressive tectonic motions, some extensional tectonic motions have been reported on the Tibetan plateau. The normal faulting type motions on the Tibetan plateau is attributable to the gravitational collapse of the high plateau [13]. Extensional crevices and grabens have been detected on the Tibetan plateau [14]. LANDSAT image data shows that extensional activity appears in the Tibetan region [15]. A Late Cenozoic E-W extension occurred on the Tibetan p1ateau [16]. The southern part of the plateau is cut by N-S trending normal faults that are the surface expression of regional stretching in an ESE direction, approximately parallel to the Himalayas [17].

Citation: Jiren Xu and Zhixin Zhao, "Extensional Seismogenic Stress and Tectonic Movement" International Journal of Geophysics, vol. 2009, Article ID 897424, 8 pages, 2009. doi:10.1155/2009/897424.

In this analysis, the focal mechanisms for studying the tectonic characteristics in and around the plateau are analyzed. The extensional lithosphere motion on the Tibetan plateau and tectonic implications are investigated in detail, based on the analyses of types of earthquake occurrences and the stress field, geothermal as well as GPS data.

Figure 1 shows the tectonic scheme of the Tibetan plateau and surrounding regions as a background map, including topographic information elevation and some faults. The elevation chart images the outline of the plateau.

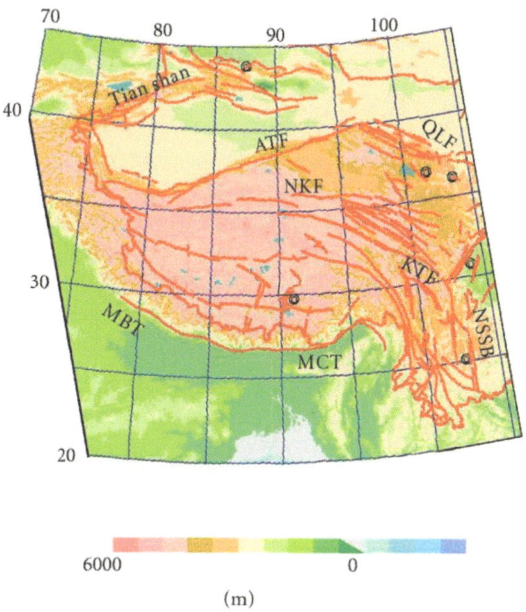

Figure 1: Topographic and tectonic map of the Tibetan Plateau and surrounding regions. Solid lines show major faults (ATF: Altyn Tagh Fault; NKF: North Kunlun Fault; KTF: Kang Ding Fault; MCT: Main Central Thrust; MBT: Main Boundary Thrust; QLF: Qilian fault). NSSB: southern segment of the North-South Seismic belt.

DATA OF FAULT PLANE SOLUTIONS

Focal mechanism solutions of 924 earthquakes with magnitude equal to or greater than M4.5 on the Tibetan plateau and its vicinities during the period between 1933 and 2003 were analyzed and used in this study. In which, the selected events greater than or equal to 5.0 (Mb) are more than 900. The depths of the events are less than 87 km in the lithosphere. The majority of events used in this analysis occurred mainly in the crust.

The 217 fault plane solutions of events were determined form the distribution of the polarities of initial P-wave motion on the focal hemisphere by the authors referring to data from the Bulletin of the International Seismological Center (ISC). Another 253 solutions determined by P-wave first motions are selected from the other authors referring to data from the Chinese Seismic Net and the Bulletin of the International Seismological Center (ISC) using P-wave first motions. (e.g., [4]). The above 470 solutions were determined using the first polarities of P-waves of long- and short- period waveforms by the analog and digital records. The earthquake data from the Chinese Seismic Net have been collected from digital records since 2000.

All the solutions used in the analysis are well selected based on the contradictory rate of the plus to minus for the first polarity of P-wave motion. The focal mechanism solutions with symbolic contradictory ratio of P-wave onsets less than 10 percent are of about 89 percent of the total solutions. Solutions with symbolic contradictory ratios between 11 and 15 percent are of 11 percent of the total solutions.

The others 454 CMT solutions (determined by the Centroid Moment tensor solutions) were selected from the Harvard University and USGS. CMT solutions for events with magnitude greater than 5.0 from 1977 to 2003 were also used in this study (e.g., [18, 19]).

The characteristics of crust and lithosphere motions were analyzed based on the investigations of fault motions of events in and around the plateau. Earthquakes are divided into three types—normal, thrust, and strike-slip faulting events—depending on the plunge angles of the compressive stress P- and extensive stress T-axes. The statistical effect of characteristics of fault plane solutions were employed for analyses of the tectonics in and around plateau.

ACTIVITY OF NORMAL FAULTING EARTHQUAKES

Earthquakes can generally be understood as brittle rock ruptures in the crust and lithosphere. So the study of faulting types of earthquakes may help us to understand the characteristics of lithosphere motion. The form of occurrence of earthquakes reflects the stress regime in a region. It is necessary to analyze fault types of earthquakes for investigation of tectonic motion characteristics on the Tibetan plateau and its surroundings. Figure 2(a) shows the fault plane solutions of events. Figure 2(b) visualizes the distributions of normal, thrust, and strike-slip faulting events that are divided according to the plunge angles of their compressive P- and extensional T-axes. It can be seen from Figure 2 that many events with normal faulting are clearly concentrated in the region with altitudes greater than about 4000 m on the Tibetan plateau. Normal faulting type regions AB and CD encircled by dashed line in Figure 2 are zoned based on the azimuths of T-axes.

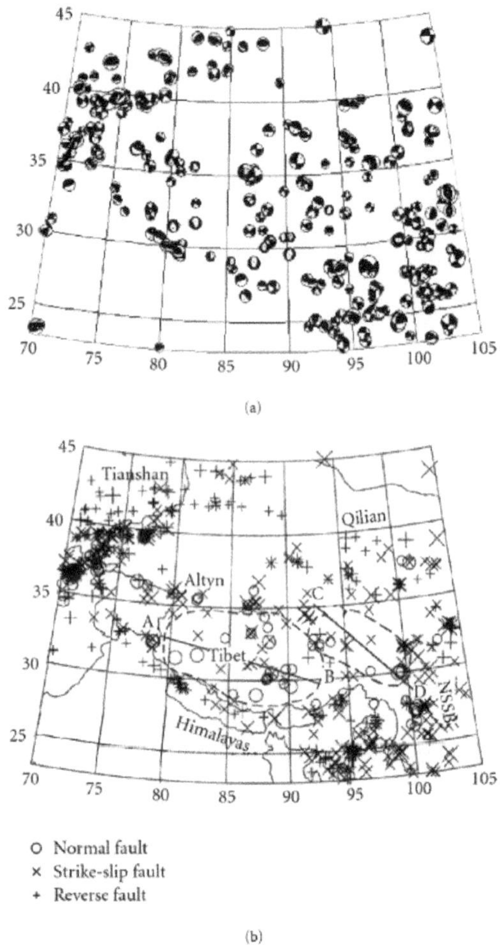

Figure 2: (a) Fault plane solutions of events. (b) Distribution of normal, thrust, and strike-slip faulting events. The Tibetan plateau is surrounded by the Himalayas, Altyn, Qilian, and NSSB. Open circle: normal faulting event; Cross: strike-slip faulting event; Plus: reverse faulting event. Regions AB (the western part of the normal faulting event region, Lhasa block) and CD (the eastern part of the normal faulting event region, Qiangtang-Changdu region) enclosed with dashed lines.

In the surroundings of the normal faulting event region, the region in Tibet characterized by abundant normal faults is bounded on the north by the discontinuous zone of strike-slip and thrust faults in the Altyn and Qilian mountains. A number of large strike-slip faulting events occurred along the lines of the Karakorum and Altyn mountains. In the Qilian Mountains many thrust faulting events occurred besides the strike-slip faulting events. Seismic activity with normal faulting events disappears on the northern boundary of the plateau, where the style of seismic activity differs from that in the high-altitude region of the plateau. The thrust faulting events are clearly predominant in the Tianshan Mountains. The thrust faulting events occurred mainly along the western part of the Himalayan front to the Kashmir region. A number of strike-slip faulting events occurred along the eastern Himalayan front, except for some thrust faulting type events. The focal mechanism solutions show mainly strike-slip faulting events in the southern segment of the NSSB (North-South Seismic Belt), the eastern boundary of the plateau. Many reverse and strike-slip faulting events occurred in the low surrounding regions of the Tibetan plateau. It may related with the strong compressive stress from the relative movement between the Indo-Australian and Eurasian plates acting on the surroundings of the plateau [20].

DISTRIBUTIONS OF THE HORIZONTAL COMPONENTS OF THE P- AND T-AXES

The stress field in and around the Tibetan plateau was investigated by analyzing the distributions of the P- and T-axes of focal mechanism solutions for analyses of style of earthquake occurrence. Figures 3 and 4 show the horizontal projections of the P- and T-axes, respectively, in the last 70 years or so. It can be seen from the figures that the P-axes of events lie mainly in N-S to NE-SW directions on the plateau and its surrounding regions, with a few exceptional events. On the Himalayan fronts, the P-axes are almost in an NNE-SSW direction and are perpendicular to the Himalayan arc. The horizontal projections of some T-axes are very small there. The horizontal components of the P-axes are on average greater than those of the T-axes. This shows that the tectonic movements in the Himalayan front region are characterized by compressive stress in an NNE-SSW direction.

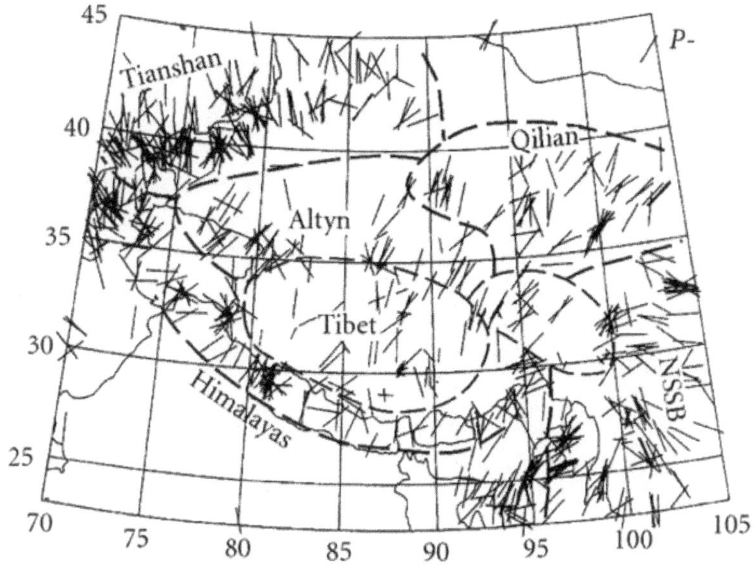

Figure 3: Distribution of compressive axes (P-) in and around the Tibetan plateau.

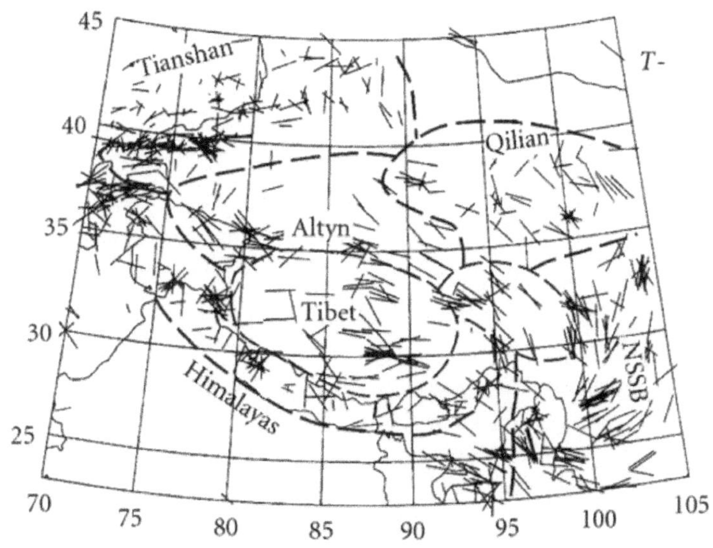

Figure 4: Distribution of extensional axes (T-) in and around the Tibetan plateau.

In the region north of the Himalayas and on the central Tibetan plateau (regions AB and CD enclosed with dashed lines in Figure 2) the horizontal components of the P-axes as a whole become less than those of the T-axes, although the P-axes basically do not change their orientations compared with those in the southern Himalayas. Most of T-axes lie nearly in an E-W direction in the AB part. T-axes are nearly in an NW-SE or NNW-SSE direction in the CD part. The horizontal projections of the T-axes of some events are markedly large, especially in the regions of high elevation on the Tibetan plateau.

Figure 5 shows the statistical charts of azimuths and dips of the compressive (P) and extensional (T) axes in the Lhasa block (AB part in Figure 2). The extensive stress in the nearly E-W direction dominates the seismic fault activities. The characteristics of the tectonic stress field in the regions of high elevation on the Tibetan plateau are clearly different from those in the southern margin of the Himalayan front.

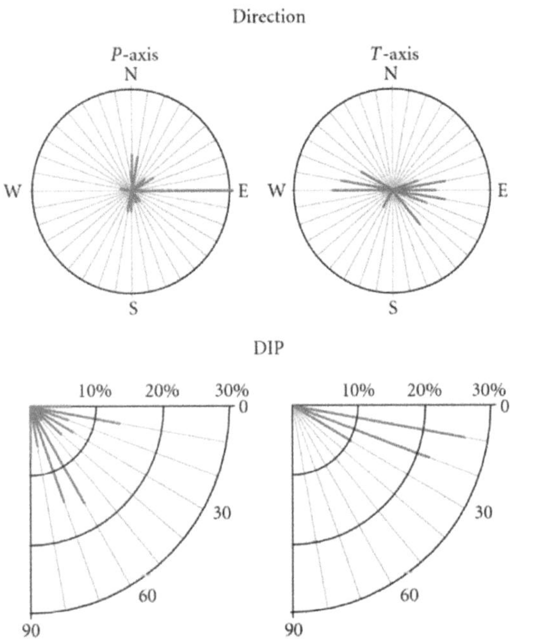

Figure 5: Statistical charts of azimuths and dips of compressive (P) and extensional (T) axes in the southern and central part of the Tibetan plateau (AB region in Figure 2).

The P-axes are mainly in an N-S to NE-SW direction on the northern borders of the Tibet plateau, Karakorum, Altyn, and Qilian Mountains. The orientations of the T-axes are in the general direction E-W to NNW-SSE. The horizontal projections of the P- and T-axes are great and have almost same size as the whole of the Altyn Mountains. The horizontal components of some T-axes are somewhat small in the Qilian Mountains region, while those of the P-axes are great. Strong compressive stress has been identified in the region north of the Qilian Mountains. The horizontal projections of the P-axes are large and lie in a general N-S direction in the Tianshan Mountains region. The horizontal components of the T-axes are quite small there.

In the western part of the southern segment (south of 33°N) of the NSSB (along about 96–104°E) east of the Tibetan plateau, the horizontal components of the P-axes lie almost in an NNE-SSW direction in the southern segment, changing to nearly NE-SW in the northern region of about latitude 33°N. The P-axes lie in a general E-W direction in the regions around 33°N in the NSSB. The horizontal components of the T-axes lie in and around the NW-SE direction in the southern segment of the NSSB, which forms the eastern boundary of the Tibetan plateau. The horizontal components of the P- and T-axes are almost the same in the eastern boundary.

PROJECTIONS FOR P- AND T-AXES ALONG PROFILES

Tectonic stress characteristics in the AB region of the central Tibetan plateau and its eastern vicinity, CD region, encircled with dashed lines as shown in Figure 2 are analyzed in detail for a further understanding of the tectonic implications of normal fault earthquakes on the plateau altitudes. Figures 6 and 7 show the projections for T-axes and stereonet diagrams of focal mechanism solutions on the vertical section beneath the lines AB and CD in the region in Figure 2. The maximum hypocentral depths of events are 78 km and 50 km in Figures 6 and 7, respectively. In the top profile of Figure 6, most of the T-axes lie in a nearly horizontal direction, with large components along the line AB, in a general E-W direction. These characteristics suggest that a horizontal extensional stress in a general E-W direction probably dominates the seismic fault motions in this region. Most of events occurring in this region are normal-faulting-type. The characteristics of normal-faulting events can also be identified in the bottom profile. This suggests that the horizontal motions of earthquake faults extend mainly in a general E-W direction.

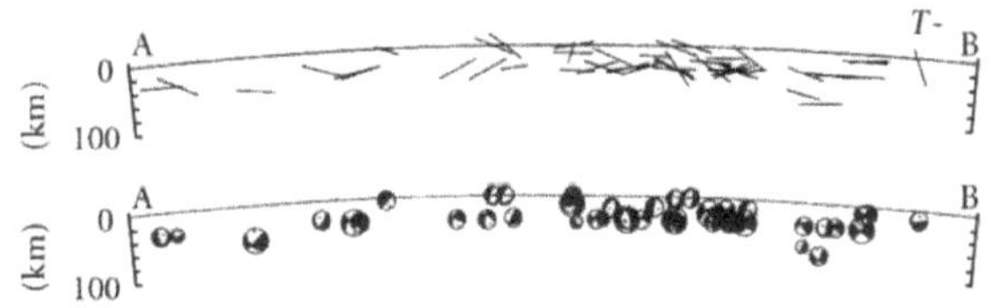

Figure 6: Projections of the vertical section along the line AB in a general E-W direction for the T-axes and equal area projection diagrams of focal mechanism solutions of events (1933–2003,) in the region encircled by a dashed line in Figure 2.

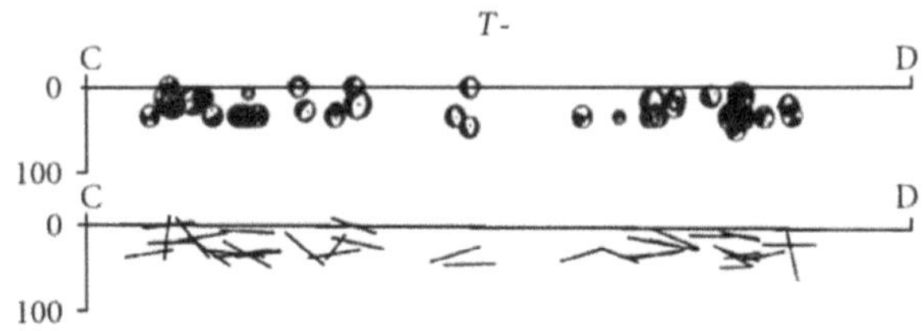

Figure 7: Projections of the vertical section along the line CD near the NW-SE direction for equal area projection diagrams and T-axes of focal mechanism solutions of events (1933–2003,) in the region encircled by the dashed line in Figure 2.

With regard to the characteristics of the T-axes of focal mechanism solutions of events in profile CD in Figure7, most of the T-axes lie in a generally horizontal direction, with large components, along the line CD in a nearly NW-SE direction in Figure 7. These characteristics suggest that a horizontal extensional stress in the NW-NE direction probably dominates the seismic fault motions in the region CD in Figure 2. The characteristics of the stereonet diagrams of events can also be identified in the top profile.

NORMAL-FAULTING EARTHQUAKE REGION

The motional characteristics of normal-faulting-type events in Figure 2 are further analyzed by investigating the focal mechanism solutions. Figure 8 shows the stereonet diagrams of focal mechanism solutions only for normal faulting events with magnitudes greater than M5 in and around the Tibetan plateau. It can be seen that many normal-faulting-type events are concentrated on the Tibetan plateau. The strikes of many fault planes seem to be in a general N-S direction on high altitudes of the Tibetan plateau, corresponding to the region encircled by dashed line AB in Figure 2. While the strike directions of the fault planes of normal faulting events occurring in the Tianshan Mountains and southern segment of the BSSN are scattered in different directions.

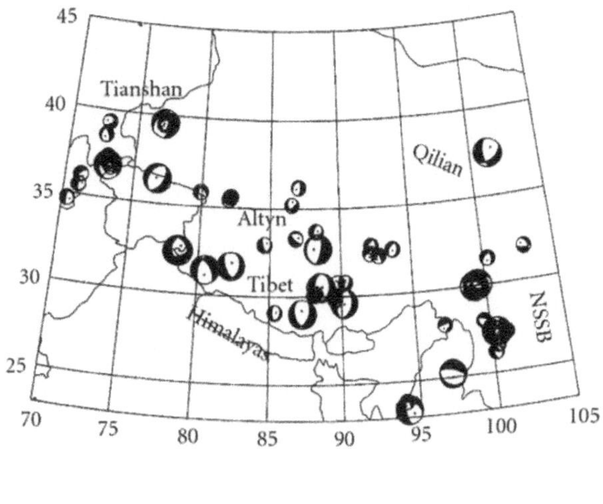

Figure 8: Distribution of the fault plane solutions for the normal-fault-type events (1933-2003,) in and around the Tibetan plateau.

The characteristics of tectonic motions due to earthquakes can be further determined by analyzing in detail the strike directions of the fault planes of the events. Figure 9 shows the equal area projection diagrams of focal mechanism solutions of events in the AB region in Figure 2. The total number of events is 54 in the region enclosed by a dashed line in Figure 9. Few of them are thrust-fault-type events; some belong to the category of strike-slip-faulting events.

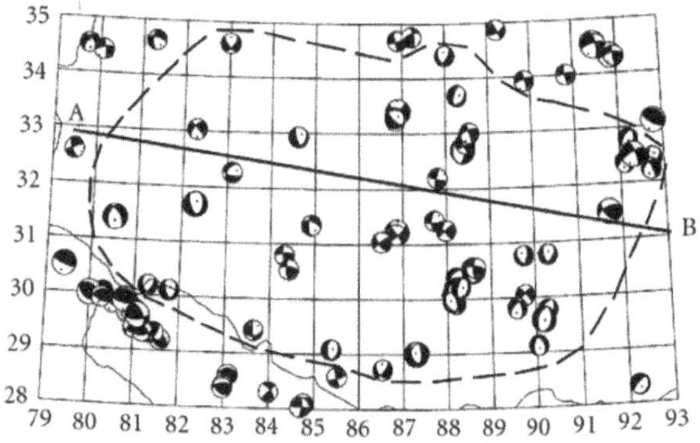

Figure 9: Distribution of the fault plane solutions of earthquakes in and around the AB region of the Tibetan plateau in Figure 2

The majority of them are normal-fault-type earthquakes, and 25 of them are almost perfect normal faulting events in which the strike directions of the two node planes lie in a general N-S direction. Their P-axes are almost located at centers of the equal area projection diagrams. This implies that the P-axes are almost perpendicular to the surface, and the slip vectors of earthquake faults have major components in the E-W direction. Such events perhaps result in major horizontal extensions in the E-W direction whichever of the two node planes is taken as the source fault. Similarly, major horizontal extensions appeared in the NW-SE or NNW-SSE direction in the CD region in Figure 2.

Based on the above analyses, we can conclusively state that the crust extends in a general E-W direction in the AB region and NW-SE or NNW-SSE in the CD region at the altitudes of the Tibetan plateau due to the normal-faulting-type earthquakes.

CRUSTAL MOVEMENTS FROM GPS AND GEOTHERMAL DATA

The crustal east-west extensional motions are also confirmable by analyzing GPS and geothermal data. The rates of horizontal displacement lie in a general N-S direction on the western plateau and the western edge of the Kunlun Mountains. The eastward annually rates are between 15 and 20 mm/a in the central plateau [8]. The annually rates of displacements lie identically in a generally easterly direction on the eastern boundary of the Tibetan plateau, in the southern segment of the NSSB, almost same as those in the central plateau. The eastward horizontal displacement components increase on average from the western plateau to the eastern margin of the plateau, in the northern segment of the NSSB, and through the central plateau. The rapid clockwise flow occurs around the syntaxis of the Himalayas on the eastern margin of the plateau [22]. The above analyses based on GPS findings at the surface imply that there exists an eastward crustal movement on the central and east plateau. This coincides with the extensional motion in the E-W and NW-SE directions due to the normal faulting events in the present analysis. Deformation of a continuous medium at depth best describes the present-day tectonics of the Tibetan plateau.

Geological tectonic and geothermal activities in the Lhasa block of the plateau are shown in Figure 10. Large mountains and early major compressive or compressive-shear faults mainly lie in an E-W direction in the plateau. The north boundary of the Lhasa block is the Bangong-Nujiang suture zone, and the southern boundary is the Yarlung Zangbo Jiang suture zone. A number of faulted-depression zones with nearly N-S or NNE strike direction, however, occur in the plateau. The faulted-depression zones with middle or small scales were probably formed by extensions in a generally east-west direction. Many extensional faulting activities along a general NEE direction exist on the northern part of the plateau also [23]. The Tibetan plateau is also a geothermally active zone. Many hot springs and geysers are located on the plateau. It is interested that many hydrothermal activities are gathered along the above extensional faulted depression zones. According to a discussion of the lithospheric thickness and hypocentral depth the faulted-depression zones with hydrothermal activities probably indicate that cracks occur in the depth due to extensional motion in the plateau also. The hypocenters of some normal faulting events mentioned in this paper extend also as deep as about 87 km. The eastward extensional motions due to the normal faulting events probably extend to near the mantle.

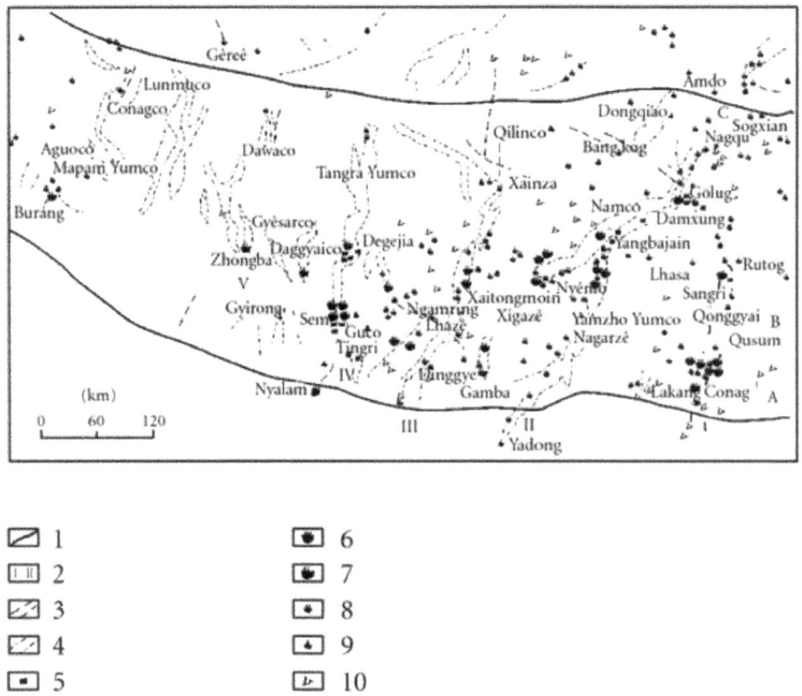

Figure 10: Distribution of hydrothermal manifestations in the Lhasa block of the Tibetan plateau (about 80–94°E, 27–33°N). 1: Early major compressive or compressive-shear fault: the upper one is the Bangong-Nujiang suture zone, and the lower one is the Yarlung Zangbo Jiang suture zone; 2: the number of active tectonic zone; 3: boundary fault in active tectonic zone; 4: faulted-depression zone in active tectonic zone; 5: site of sampling; 6: boiling spring; 7: hydrothermal explosion, geyser and boiling fountain; 8: thermal spring; 9: hot spring; 10: untested hydrothermal area (according to [21]).

Figure 11 shows the eastward extension regions on the central Tibetan plateau as the result. The extensional-motion regions marked by open rhombs and the direction of the eastward extension motions marked by open arrow on the plateau altitudes were substantiated based on the present analyses. The other tectonic properties are referenced the study of Avouac and Tapponnier [24].

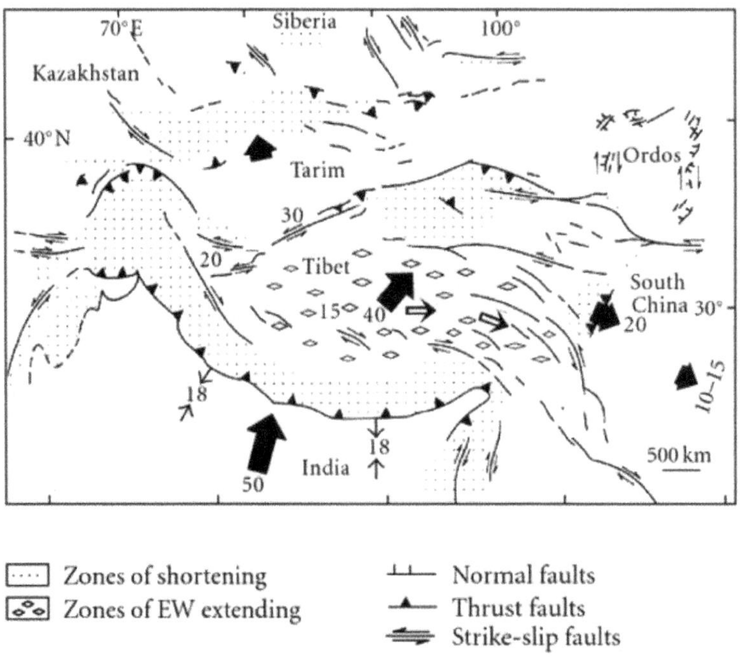

Figure 11: Eeastward extension regions on the central Tibetan plateau. Open rhomb area: extensional motion region obtained from the present analysis; open arrows: extensional direction of the plateau altitudes obtained in the present analysis (other tectonic properties are referenced the study of Avouac and Tapponnier [24]).

DISCUSSION

The characteristics and mechanism of extensional motions at the high altitudes of the Tibetan plateau have been analyzed [25–29]. The activities of normal faulting earthquakes concentrated at high altitudes were investigated based on distribution characteristics of stress field, GPS data, as well as geothermal data in this paper. The extensional-stress-predominating earthquake activity at these altitudes superposes on the compressive stress predominating seismic activities on the edges of the plateau. The eastward extensional motion at high altitudes becomes an obviously important tectonic activity regime, although shortening deformation and movement in a nearly N-S direction are revealed there too.

E-W extension shown by the normal faulting events behind the Himalayas may be related to tectonic stresses due to collision causing a wide distribution of P-axes in the general NNE-SSW or N-S directions in western China [13, 30]. The strong compressive stress due to the northward movement of the Indian plate also causes the Tibetan plateau rise gradually [31]. The weight of a high plateau and the buoyancy acting on the bottom of the plateau crust from the mantle may create an additional vertical compressive stress which can result in the relaxation of the topography and corresponding thinning of the crust together with the normal faulting activity. The altitude of the Tibetan plateau is so high with respect to its surroundings that active normal faulting may also be caused by gravitational collapse toward the plateau margins [25, 32, 33]

A large number of thrust and strike-slip fault events occur in the front of the Himalayan Mountains. Many strike-slip fault events occur in the Karakorum and Altyn mountains. Thrust-faulting events occur frequently in regions of the Qilian and Tianshan mountains [34]. These findings show that strong compressive stresses exist on the southern, western, and northern boundaries of the plateau. Moreover, the GPS data show that the northward annual slip rates on the northern margin of the plateau, for example, in the Altyn and Qilian mountains, are one tenth of those on the Himalayan front [22, 35–37]. These findings imply that the northward motion of the Tibetan plateau shortened in the N-S direction probably encounters strong obstructions at the western and northern margins [38].

A stress field in an NWW-SEE direction appears on the CD region in Figures 2 and 6. On the eastern margin of the Tibetan plateau, however, in the southern end of the NSSB, events are mainly of the strike-slip faulting type. The annually rates of crustal horizontal displacements are eastward and southward there. The extensional motion due to the relaxation of topography and/or gravitational collapse on the high plateau may easily become an eastward motion. The E-W stretching due to the normal fault events on the plateau is also consistent with eastward "escape-block" tectonics [24, 39].

Seismic velocity is rapid beneath the western Himalayas and western Tibet and is less rapid beneath north-central Tibet [40, 41]. The Tarim, Qaidam, Turfan, and Junggar basins on the western and northern sides of the plateau are underlain by large, rapid-velocity anomalies at depths of 100 km and 300 km [42]. The rapid-velocity anomalies are probably related to the ancient cold subduction materials beneath the Eurasian lithosphere of the Tethyan oceanic lithosphere [43]. It is a younger subduction history that dip angle of subduction is about ° [44]. The younger subduction motions encounter strong resistance from the front plate [45, 46]. The seismic velocity beneath the NSSB is less than that beneath the central plateau [47]. The image implies that the mantle materials under the plateau thrust beneath the Yangtze block. This may imply that hotter materials lie beneath the eastern part of the plateau. So the hotter materials possibly modulate continuously northward convergence movements from the Indian plate in the Tibetan plateau by the eastward extrusion of mantle material. The latter may drag the lithosphere eastwards.

ACKNOWLEDGMENTS

The authors are most grateful to the two reviewers for their helpful comments. This study was supported partly by the Nature Science Foundation of China (Grant no. 40674026) and Commonweal Special Science Foundation of China (Grant no. 200811037).

REFERENCES

1. P. Molnar and P. Tapponnier, "Cenozoic tectonics of Asia: effects of a continental collision," Science, vol. 189, no. 4201, pp. 419–426, 1975. View at Google Scholar
2. P. Molnar, P. England, and J. Martinod, "Mantle dynamics, uplift of the Tibetan Plateau, and the Indian monsoon," Reviews of Geophysics, vol. 31, no. 4, pp. 357–396, 1993. View at Google Scholar · View at Scopus
3. C. Qin, C. Papazachos, and E. Papadimitriou, "Velocity field for crustal deformation in China derived from seismic moment tensor summation of earthquakes," Tectonophysics, vol. 359, no. 1-2, pp. 29–46, 2002. View at Publisher · View at Google Scholar · View at Scopus
4. Z. Xu, "Modern tectonic stress in East Asia," Acta Seismologica Sinica, vol. 23, pp. 492–501, 2001. View at Google Scholar
5. R. Van der Voo, W. Spakman, and H. Bijwaard, "Tethyan subducted slabs under India," Earth and Planetary Science Letters, vol. 171, no. 1, pp. 7–20, 1999. View at Publisher · View at Google Scholar ·View at Scopus
6. E. T. Brown, R. Bendick, D. L. Bourlès, et al., "Slip rates of the Karakorum fault, Ladakh, India, determined using cosmic ray exposure dating of debris flows and moraines," Journal of Geophysical Research B, vol. 107, no. 9, p. 2192, 2002. View at Publisher · View at Google Scholar
7. Y. Wang and H. Xu, "Convergence rate of the India plate to Eurasia subduction beneath Qing-Zang plateau—inversion result from the GPS observed data," Chinese Journal of Geophysics, vol. 46, pp. 185–190, 2003. View at Google Scholar
8. Q. Wang, P.-Z. Zhang, J. T. Freymueller, et al., "Present-day crustal deformation in China constrained by global positioning system measurements," Science, vol. 294, no. 5542, pp. 574–577, 2001. View at Publisher · View at Google Scholar · View at Scopus
9. Z. Zhao, K. Oike, K. Matsumura, and Y. Ishikawa, "Stress field in the continental part of China derived from temporal variations of seismic activity," Tectonophysics,

vol. 178, no. 2–4, pp. 357–372, 1990. View at Google Scholar · View at Scopus

10. Z. Zhao, J. Xu, and S. Horiuchi, "Differentiation operation in the wave equation for the pseudospectral method with a staggered mesh," Earth, Planets and Space, vol. 53, no. 5, pp. 327–332, 2001. View at Google Scholar · View at Scopus
11. Z. Zeng, Z. Ding, and Q. Wu, "Seismological evidence for the mulitiple incomplete crustal subductions in Himalaya and south Tibet," Chinese Journal of Geophysics, vol. 43, pp. 780–797, 2000. View at Google Scholar
12. Z. Fu, J. Xu, and J. Huang, "Numerical simulation of the compression uplift of the Qinghai-Xizang plateau," Chinese Journal of Geophysics, vol. 43, no. 3, pp. 346–355, 2000. View at Google Scholar
13. J. Xu, Z. Zhao, Y. Ishikawa, and K. Oike, "Properties of the stress field in and around west China derived from earthquake mechanism solutions," in Bulletin of the Disaster Prevention Research Institute, vol. 38, pp. 49–78, Kyoto University, Kyoto, Japan, 1988. View at Google Scholar
14. J. F. Deway and J. M. Bird, "Mountain belts and new global tectonics," Journal of Geophysical Research, vol. 75, no. 14, pp. 2625–2647, 1970. View at Google Scholar · View at Scopus
15. P. Molnar, "A review of the seismicity and the rates of active underthrusting and deformation at the Himalaya," Journal of Himalayan Geology, vol. 1, pp. 131–154, 1990. View at Google Scholar
16. J. Ni and J. E. York, "Late Cenozoic tectonics of the Tibetan plateau," Journal of Geophysical Research, vol. 83, no. B11, pp. 5377–5384, 1978. View at Google Scholar
17. R. Armijao, P. Tapponnier, J. L. Merccier, and T.-L. Han, "Quaternary extension in southern Tibet: field observations and tectonic implications," Journal of Geophysical Research, vol. 91, pp. 13083–13872, 1986. View at Google Scholar
18. A. M. Dziewonski and J. H. Woodhouse, "An experiment in systematic study of global seismicity: centroid-moment tensor solutions for 201 moderate and large earthquakes in 1981," Journal of Geophysical Research, vol. 88, no. 4, pp. 3247–3271, 1983. View at Google Scholar · View at Scopus
19. A. M. Dziewonski, A. Friedman, D. Giardini, and J. H. woodhouse, "A experiment in systematic study of Geobal seismicity of 1982: centroid-moment tensor solution fot 308 earthquake," Physics of the Earth and Planetary Interiors, vol. 33, pp. 76–90, 1983. View at Google Scholar
20. C. Froidevaux and Y. Ricard, "Tectonic evolution of high plateaus," Tectonophysics, vol. 134, no. 1–3, pp. 227–238, 1987. View at Google Scholar · View at Scopus
21. H. Tonglin, "Relationship of the active structural system to geothermal activity in southern xizang (Tibet)," in Himalayan Geology, pp. 45–58, Geological Press, Beijing, China, 1981. View at Google Scholar
22. P.-Z. Zhang, Z. Shen, M. Wang, et al., "Continuous deformation of the Tibetan Plateau from global positioning system data," Geology, vol. 32, no. 9, pp. 809–812, 2004. View at Publisher · View at Google Scholar
23. L. Yalin, W. Chengshan, Y. Haisheng, et al., "Characteristics of the shuanghu graben and cenozoic extension in northern Tibet," Science in China, Series D, vol. 44, supplement 1, pp. 285–291, 2001. View at Google Scholar · View at Scopus
24. J.-P. Avouac and P. Tapponnier, "Kinematic model of active deformation in central Asia," Geophysical Research Letters, vol. 20, no. 10, pp. 895–898, 1993. View at Google Scholar
25. P. M. Blisniuk, B. R. Hacker, J. Glodny, et al., "Normal faulting in central Tibet since at least 13.5 Myr ago," Nature, vol. 412, no. 6847, pp. 628–632, 2001. View at Publisher · View at Google Scholar
26. H. Z. Wu, P. Ye, and D. Hu, Evolvement for Crustal Deformation and Tectonic Physiognomy, Geological Press, Beijing, China, 2003.
27. J. F. Dewey, "Extensional collapse of orogens," Tectonics, vol. 7, no. 6, pp. 1123–1139, 1988. View at Google Scholar · View at Scopus
28. P. Molnar and P. Tapponier, "Active tectonics of Tibet," Journal of Geophysical Research, vol. 83, no. B11, pp. 5361–5375, 1978. View at Google Scholar
29. S. L. Klemperer, "Crustal flow in Tibet: geophysical evidence for the physical state of Tibetan lithosphere, and inferred patterns of active flow," in Channel Flow, Ductile Extrusion and Exhumation in Continental Collision Zones, R. D. Law, M. P. Searle, and L. Godin, Eds., vol. 268, pp. 39–70, Geological Society, London, UK, 2006. View at Google Scholar
30. J. Xu and K. Oike, "Stress characteristics in the southern segment of the North-South Seismic Belt," Acta Seismologica Sinica, vol. 17, pp. 31–40, 1995 (Chinese). View at Google Scholar
31. R. S. Yeats and R. J. Lillie, "Contemporary tectonics of the Himalayan frontal fault system: folds, blind thrusts and the 1905 Kangra earthquake," Journal of Structural Geology, vol. 13, no. 2, pp. 215–225, 1991.View at Google Scholar · View at Scopus
32. S. Robert, K. S. Yeats, and R. A. Clarence, The Geology of Earthquakes, Oxford University Press, Oxford, UK, 1997.
33. J. Lavé, J. P. Avouac, R. Lacassin, P. Tapponnier, and J. P. Montagner, "Seismic anisotropy beneath Tibet: evidence for eastward extrusion of the Tibetan lithosphere?" Earth and Planetary Science Letters, vol. 140, no. 1–4, pp. 83–96, 1996. View at Publisher · View at Google Scholar
34. T. M. Harrison, P. Copeland, W. S. F. Kidd, and A. Yin, "Raising tibet," Science, vol. 255, no. 5052, pp. 1663–1670, 1992. View at Google Scholar · View at Scopus
35. R. Bendick, R. Bilham, J. Freymueller, K. Larson, and G. Yin, "Geodetic evidence for a low slip rate in the Altyn Tagh fault system," Nature, vol. 404, no. 6773, pp. 69–72, 2000. View at Publisher · View at Google Scholar
36. R. Hetzel, S. Niedermann, M. Tao, et al., "Low slip rates

and long-term preservation of geomorphic features in Central Asia," Nature, vol. 417, no. 6887, pp. 428–432, 2002. View at Publisher · View at Google Scholar · View at Scopus

37. Z.-K. Shen, M. Wang, Y. Li, et al., "Crustal deformation along the Altyn Tagh fault system, western China, from GPS," Journal of Geophysical Research B, vol. 106, no. 12, pp. 30607–30621, 2001. View at Google Scholar · View at Scopus

38. Z. Chen, B. C. Burchfiel, Y. Liu, et al., "Global positioning system measurements from eastern Tibet and their implications for India/Eurasia intercontinental deformation," Journal of Geophysical Research B, vol. 105, no. 7, pp. 16215–16227, 2000. View at Google Scholar · View at Scopus

39. P. Tapponnier, "The Ailao Shan-Red River metamorphic belt: tertiary left-lateral shear between Indochina and South China," Nature, vol. 294, pp. 405–410, 1990. View at Google Scholar

40. L. Bourjot and B. Romanowicz, "Crust and upper mantle tomography in Tibet using surface waves,"Geophysical Research Letters, vol. 19, no. 9, pp. 881–884, 1992. View at Google Scholar · View at Scopus

41. R. Zhou, S. P. Grand, F. Tajima, and X.-Y. Ding, "High velocity zone beneath the southern Tibetan plateau from P-wave differential travel-time data," Geophysical Research Letters, vol. 23, no. 1, pp. 25–28, 1996. View at Google Scholar · View at Scopus

42. Y. Zhang and T. Tanimoto, "High-resolution global upper mantle structure and plate tectonics," Journal of Geophysical Research, vol. 98, pp. 9793–9823, 1993. View at Google Scholar

43. P. Molnar, "The geologic evolution of the Tibetan plateau," American Scientist, vol. 77, no. 4, pp. 350–360, 1989. View at Google Scholar · View at Scopus

44. W. Zhao, K. D. Nelson, and Project INDEPTH Team, "Deep seismic reflection evidence for continental underthrusting beneath southern Tibet," Nature, vol. 366, no. 6455, pp. 557–559, 1993. View at Publisher· View at Google Scholar · View at Scopus

45. J. Xu and K. Yoshiteru, "Geometry of slab, intraslab stress field and its tectonic implication in the Nankai trough, Japan," Earth, Planets and Space, vol. 54, no. 7, pp. 733–742, 2002. View at Google Scholar · View at Scopus

46. J. Xu, Z. Zhao, Y. Kono, and H. Kinoshita, "Regional characteristics of stress field and its dynamics in and around the Nankai Trough, Japan," Acta Geophysica Sinica, vol. 46, no. 4, pp. 488–494, 2003. View at Google Scholar · View at Scopus

47. Y. Zhang, "Three dimension upper mantle structure beneath East Asia and its tectonic implication," inMantle Dynamics and Plate Interaction in East Asia, M. E. J. Flower, S.-L. Chung, C.-H. Lo, and T.-Y. Lee, Eds., pp. 11–23, American Geophysical Union, Washington, DC, USA, 2000. View at Google Scholar

CHAPTER 7

TECTONIC-GEOPHYSICAL MAPPING: IMPLICATIONS FOR HYDROCARBON PROSPECTING

Lev Eppelbaum, Youri Katz

Keywords: Eastern Mediterranean, Isopach Maps, Hydrocarbon Potential, Satellite Gravity, Airborne Magnetics

INTRODUCTION

The eastern Mediterranean is a tectonically complex region evolving in the long term located in the midst of the progressive Afro-Eurasian collision. Its geological-geophysical structure has been studied for years, but is still not completely known. The eastern Mediterranean represents a classic area for the emergence of plate tectonics [2-7]. Recent discoveries of significant hydrocarbon deposits in this region (e.g., [8-10]) have heightened the need for a robust tectonic-structural description to assist in prospecting for hydrocarbon structures. This article reports on a combined geophysical (potential geophysical fields and seismic) and geological (formational, facial, tectonic-structural and paleogeographical) investigation conducted for this purpose.

In the eastern Mediterranean there are three main types of crust: (a) platforms with a Precambrian crust, (b) regions with an oceanic crust, and (c) a folded belt of the transition type with a thinned continental crust [4]. However, the exact boundaries of these zones are not reliably known.

Integrated geophysical data on crust thicknesses [2,4, 11,12] indicate that the folded belt is composed of several terranes (Cyprus, Eratosthenes, the southern part of the Levant Basin (Pleshet), Galilee-Lebanon, JudeaSamaria and Negev) (Figure 1A). Figure 1B shows a velocity-depth distribution of compressional seismic velocities for these terranes (after [4] and [13]). It is obvious that these graphs exhibit different behaviors.

Alongside geophysical investigations, the Mesozoic formations were also mapped. The presence of a large number of deep wells in this region makes it possible to carry out geological mapping of the terranes and their boundaries. For this purpose a series of maps of the Early Mesozoic sedimentary rocks was constructed (here, the two most significant paleogeological maps are presented). These maps confirm an earlier model of continental accretion [2]. They show in particular that abrupt changes in the trend and thickness of the Early Mesozoic formations coincide with the terrane boundaries [14].

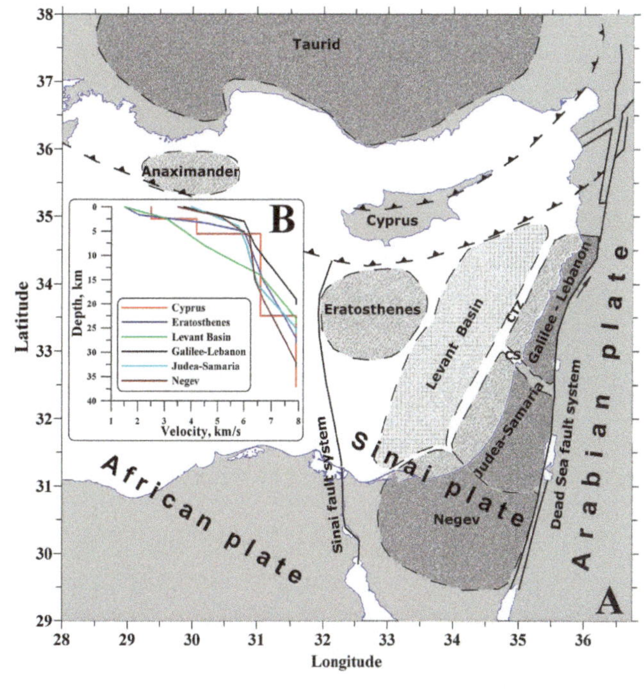

Figure 1. Tectonic scheme of the eastern Mediterranean (A) with generalized seismic sections (B). (A) after [11] and (B) after [4] and [12]. CS, Carmel structure; CTZ, crustal transition zone.

In this region numerous seismostratigraphic investigations have been carried out (e.g., [9,10,15-21]). These works have mainly been oriented toward the study of prospective hydrocarbons structures.

Citation: L. Eppelbaum and Y. Katz, "Tectonic-Geophysical Mapping: Implications for Hydrocarbon Prospecting," Positioning, Vol. 2 No. 1, 2011, pp. 36-54. doi: 10.4236/pos.2011.21004.

A BRIEF GEOLOGICAL OVERVIEW

According to modern geological-geophysical constructions, the eastern Mediterranean region covers four geotectonic plates: the African, Sinai, Arabian and AegeanAnatolian (Figure 1). The first two plates consist of heterogeneous blocks of continental and oceanic crust and each of these plates is circumscribed by deep faults. According to Robertson et al. [22], the oceanic crust of the eastern Mediterranean is a remnant of the Neotethys. The Sinai plate [22] is bounded from the west and the east by a system of Late Cenozoic deep faults [3,24]. The eastern part of the eastern Mediterranean is formed by the Dead Sea Fault System (Figure 1) along which about 100 km of left-lateral strike-slip motion has developed [25]. Within this fault system numerous horsts and grabens have emerged. These grabens usually represent pull-apart basins [26] filled by terrigenous evaporite molasses and basic rocks.

Between the oceanic and continental crust is the Syrian fold belt (Syrian arc) (e.g., [5]). This belt is the eastern part of the larger Alpine-Mediterranean fold belt and basin system [6,27]. The structural inhomogeneity of the Syrian arc was revealed by analysis of subsurface Paleogene-Upper Cretaceous sediments [28,29]. However, the tectonic characteristics of the deeper structural stages have been studied to a lesser extent. Geophysically it was shown [2] that a thinned continental crust composes the Syrian arc structure. Tectonically, this belt is made up of a sequence of extensional independent crystal units [11].

Precambrian, Paleozoic, Mesozoic and Cenozoic associations form the geological section of the African and Arabian platforms within the eastern Mediterranean. The Precambrian is known to contain metamorphic and igneous rocks, and in the upper part of Precambrian there is molassa which passes to the Cambrian base. Cambrian rocks are extensive in the Sinai, but are not found in the borehole sections of the Negev desert. In the Negev the Permian deposits overlie the Precambrian arkosic sandstone. In the Heletz block there is no molasses and the Permian deposits are in direct contact with the crystalline schist.

More completely than in the Sinai, Lower Paleozoic sequences are present to the east of the eastern Mediterranean coast in Syria, Jordan and Iran. Analysis of geophysical data testifies to the absence of Lower Paleozoic strata in the Levant Basin and in the Pleshet terrane [12,17,18]. Devonian and Carbon deposits have not been detected in the borehole sections of Israel. The Carbon and Permian continental associations were studied in Sinai [30]. Deep boreholes drilled in the Negev, the coastal plain of Israel, the Judean mountains and western Jordan have revealed marine Permian deposits [31,32]. Thus, the Precambrian and Paleozoic associations of the Arabian plate and southern Sinai differ significantly from the folded zone of the Syrian arc.

The Mesozoic and Paleogene associations make up the carbonate platform of the eastern Mediterranean (including reefs and other organogenic constructions). These carbonate associations also include small terrigeneous complexes and traps. The Miocene and Quaternary associations are represented by terregeneous molassa with Early Messinian evaporates, sometimes generating salt domes and diapirs. These associations are developed mainly in the pull-apart basins within the Dead Sea transform zone and Levant Basin [33]. Besides sedimentary deposits, in the eastern Mediterranean there are widely developed Mesozoic and Late Cenozoic magmatic rocks [29,34,35].

MATERIALS AND METHODS

The tectonic-geophysical investigation of such a complex region requires careful analysis of numerous materials and data sources such as deep boreholes sections and major outcroppings, geological and geophysical maps, seismic regional constructions and seismostratigraphic examinations, as well as the results from potential geophysical fields.

Here, we utilized and examined more than 200 deep boreholes (isopach constructions were made from more than 120 boreholes) and about 10 of the most significant outcroppings (for instance, such as Jebel Maghara, Makhtesh Ramon and Mt. Hermon) (Figure 2). Data on borehole sections and outcroppings were taken from numerous publications and reports [8,17,19,21,30-32,36-49].

The geological materials included the following data: geological maps of Israel [29], Jordan [50], Egypt [46] and photomaps of Israel and adjacent areas [51]. The following materials on the dating of Mesozoic and Cenozoic magmatic associations were employed: [34,35,47,52,53].

The geophysical data used here encompassed a wide spectrum of geophysical fields. Of primary importance are the geophysical maps that included gravity and magnetic maps of Israel and the Levant [54-56], satellite gravity data [57], a map of the Moho discontinuity for the eastern Mediterranean [58], a map of the Curie discontinuity for Israel [58], the geothermal gradient map of Israel [59] and maps of earthquakes epicenters [60,61]. In addition we drew on the results of regional analyses of seismic, gravity, magnetic and thermal data in this region [2-4,11,13,20,62,63], seismostratigraphic examinations [10, 11,15-21], and detailed geophysical-geological studies in separate areas [14,64-74].

For the first time in this region we applied a regional integrated geological-geophysical zonation of deep-seated complexes. A facial-paleogeographical analysis and thickness analysis were conducted mainly for the localization of geological structures of the Early Mesozoic. This analysis was also employed to study both sedimentary and magmatic complexes. Paleobiogeographical examinations were utilized to assess initial positions of the terranes [75]. This integrated methodology was accompanied by a broad examination of previous geological-geophysical research in the region.

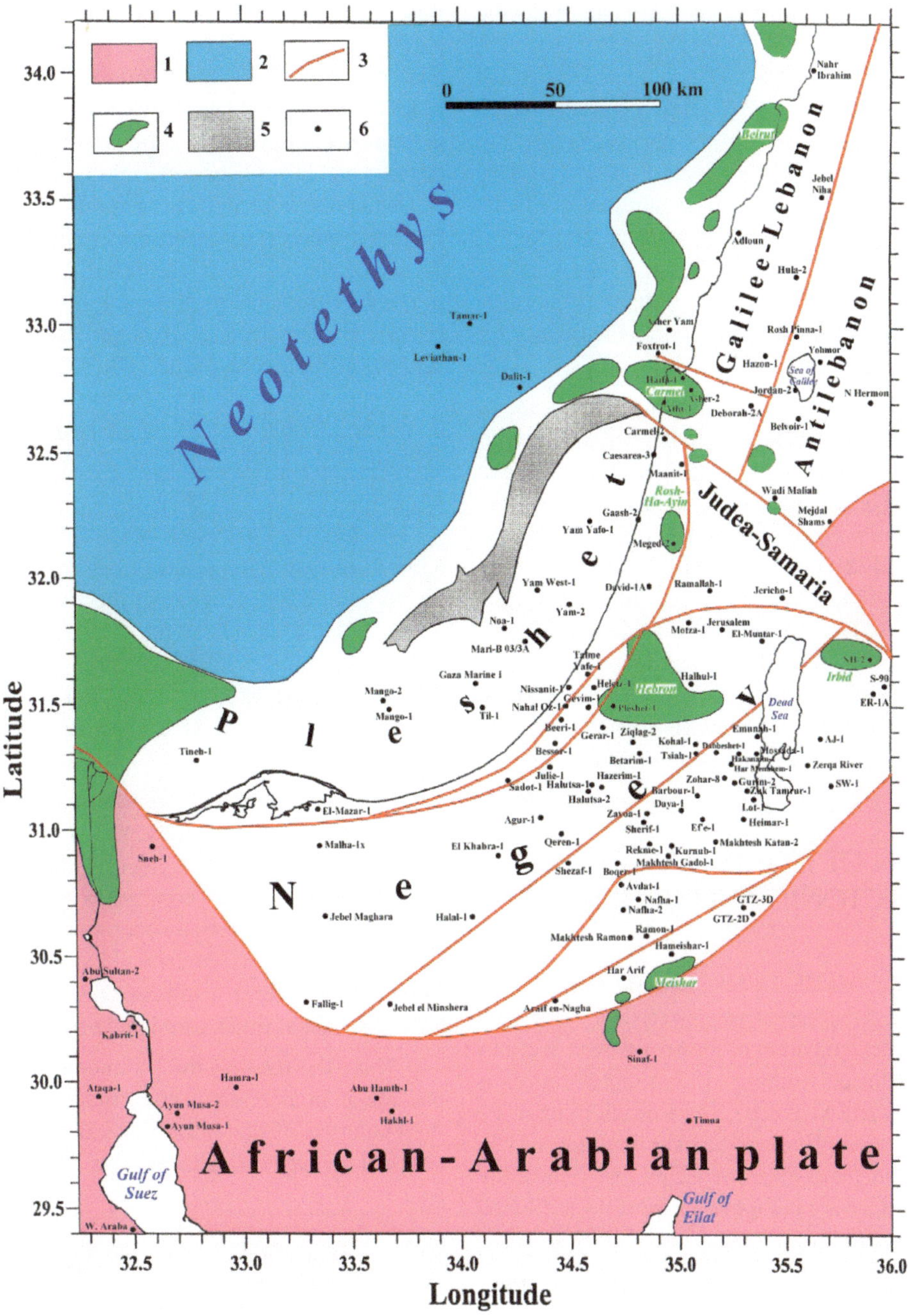

Figure 2. Map of Lower Mesozoic wells and outcrop locations in the eastern Mediterranean with some geological-geophysical zonation. (1) African-Arabian Precambrian plate, (2) Neotethys, (3) main faults, (4) contours of significant magnetic anomalies, (5) marginal erosional zone (Jonah Ridge), (6) deep boreholes and distinctive outcroppings.

ANALYSIS OF GEOPHYSICAL FIELDS

Previous divisions of the area under study for the abovementioned terranes have been made on the basis of comprehensive analyses of seismic, gravity and magnetic data (e.g., [2,4,11,62]). The first Curie map of Israel [58] also correlates with the position of the terranes (e.g., [11]). The sources of the two most significant gravity-magnetic anomalies in Israel – Hebron and Carmel – were classified as phenomena associated with tectonically weakened zones between the terranes [11,68,69].

The satellite gravity data were obtained from the World Gravity DB as retracked from Geosat and ERS-1 altimetry [57]. A highly positive factor is that these observations were made with regular global 1-minute grids [56] and the error of gravity data computation was estimated at 2-3 mGals. The compiled gravity map (**Figure 3**) shows the intricate gravity pattern of the investigated area (isoline interval is 5 mGal; "zero" isoline is dashed and bolded). This figure indisputably proves that in specific cases it is useful to display the gravity field map without any reduction. The selected positive and negative gravity anomalies (**Figure 3**) clearly reflect the main structural-geotectonical units of the region. The geomorphologic names of the corresponding gravity anomalies were taken from [11,18,32,76,77].

A map of the magnetic field horizontal gradients (Figure 4) was computed (a sliding interval of 2 × 2 km was applied) on the basis of an airborne total magnetic field analysis [54]. Some peculiarities of the horizontal magnetic gradient behavior indicate the presence of regional tectonic faults which mark crystal units (terranes and Precambrian platform) in northern

Israel. The Cretaceous and Neogene-Quaternary local volcanic eruptions also follow these faults (**Figure 4**). Magnetoactive rocks producing high gradients generally correspond to surface and subsurface trap complexes, the thicknesses of which depend on the disjunctive tectonics in the region [53,71, 74]. This map indicates that the sharpest gradients (125- 750 nT/km) correspond to the Precambrian plate. The minimal gradients (0-12.5 nT/km) correspond to northern part of the Negev, Pleshet, Judea-Samaria and the northern part of the Galilee-Lebanon terranes. The location of the Neogene-Quaternary local volcanic eruptions was used as an additional indicator for the detection of significant faults.

In order to define the position of tectonic boundaries of the main structures in the eastern Mediterranean, an analysis of the regional magnetic anomalies (on the basis of [54] and [56]) was carried out. The main magnetic features it reveals are the high-amplitude anomalies of Hebron [54,68,78], Carmel and Beirut [62], Irbid [78] and Meishar [74] (see **Figure 2**). Furthermore a lowamplitude anomaly was contoured on this map in RoshHa-Ayin [64] and several unnamed low-amplitude magnetic anomalies associated with the boundaries of terranes. This map shows the positions of the deep boreholes and the most distinctive outcroppings of the Mesozoic sediments utilized for the construction of the series of isopach maps (see below).

CONSTRUCTION OF ISOPACH MAPS: THEIR RELATION TO TECTONIC AND GEOPHYSICAL PATTERNS

Isopach maps of the Middle-Upper Jurassic and Lower Cretaceous (Figures 5 & 6, respectively) were constructed on the basis of a detailed examination of numerous well sections and the most significant outcroppings (a list of references is presented in Materials and Methods). The isopach maps were compiled on classical principles described, for instance, in [79]. A thorough analysis of these maps indicates that practically all the aforementioned terranes (**Figure 1**) are reflected in the MiddleUpper Jurassic and Lower Cretaceous isopach maps.

The compiled maps also capture significant distinctions between the Arabian and Sinai plates on the one hand and the Syrian arc on the other. Mesozoic associations in the ancient African-Arabian plate are developed to a lesser extent than in the Syrian arc and sometimes are lacking. From a facial point of view [80,81] thin terrigenous marine and continental associations are presented here mainly comparatively. The isopach map of the MiddleUpper Jurassic (**Figure 5**) indicates that the Cenozoic folded belt of the Syrian arc was formed within the complex system of paleotectonic blocks. These blocks correspond completely to the terranes identified earlier. The isopach map permits, for instance, to define terrane boundaries and to single out some block structures within these terranes. This map indicates that the axes of the terrane and platform paleostructures are located discordantly to each other. The axes of the platform Early Mesozoic eastern Desert [46] and Jebel Druze depression [31] are located at an angle of 90 to the axes of Syrian Arc paleostructures (**Figure 5**). Analyses of facies and thicknesses of Jurassic deposits show that the Late Cenozoic Dead Sea Transform in the southern part of the region dissects the Anti-Lebanon, Judea-Samaria and Negev terranes and displays their western and eastern blocks for 100 km. The distribution of the Lower Cretaceous thicknesses (**Figure 6**) testifies to the lack of influence of the terranes' geotectonic boundaries on sedimentation. Rather, it is caused by the influence of erosion relief forms and denudation processes on the post-accretional sedimentation. As can clearly be seen from the map, the relief of the denudation surface is inclined from SE to NW at 1600 m to the studied region (**Figure 6**). The depth of erosional cutting reaches 500 m, and in the transition zone of the Tethys Ocean erosional processes have shaped the canyons to a depth exceeding 1000 m. This is an important criterion for regional oil-and-gas prospecting (an example is the Heletz oil deposit – the first oil field discovered in the eastern Mediterranean [82]).

Besides thickness analyses, a facial examination was also carried out. Given the series of bauxite iron-pisolith redbeds of the Jurassic weathering crust, the Negev and Pleshet terranes as well as the northern and southern Palmyride and Heletz blocks are joined to a common tectonic-paleogeographic structure. At the same time the Galilee-Lebanon terranes and Anti-Lebanon as well as Aleppo belong to another tectonic-paleogeographical zone.

The facial analysis highlights that the terranes were isolated structures until the end of Jurassic and the beginning of the Early Cretaceous. Evidently, their movements triggered the formation of dyke complexes with an estimated age of 140 m.y. [47]. The terrane accretion was estimated to be the Lower Cretaceous (Hauterive) [75] after which post-collision traps began to form in this region [74]. Besides these traps, terrigenic and at times carbonate associations were generated.

Thus, the data on the sedimentary thickness distribution provide an unambiguous confirmation of the terrain model of the evolution of the eastern Mediterranean derived from the regional geophysical data analysis [2].

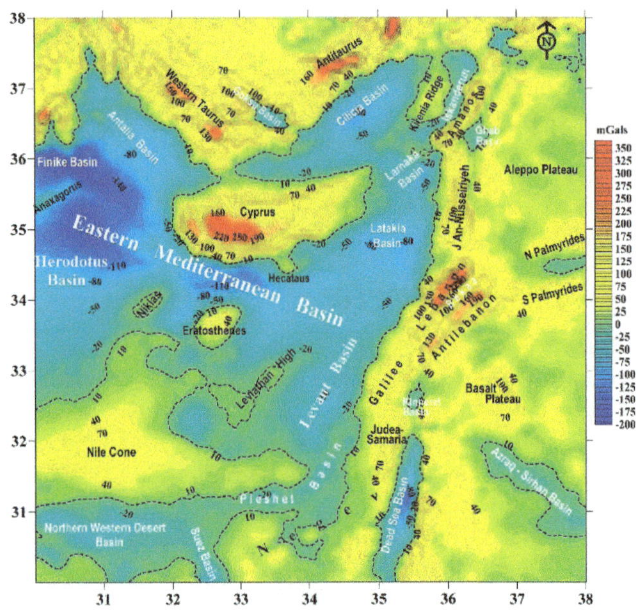

Figure 3. Map of significant gravity anomalies of the eastern Mediterranean and adjacent areas (isolines are given in mGals).

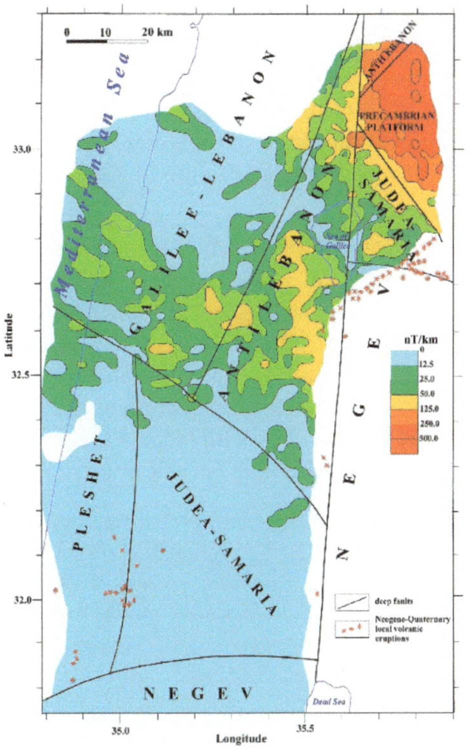

Figure 4. Magnetic gradient map of Israel and some adjacent areas with elements of tectonic analysis.

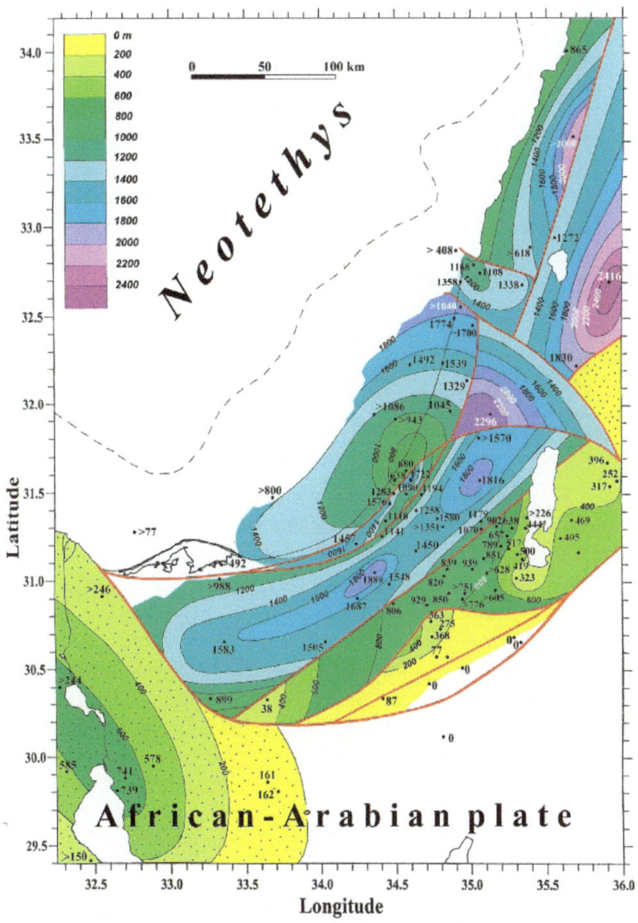

Figure 5. Isopach map of the Middle-Upper Jurassic for the Levant color isopach column is presented in the upper left corner. Black lines with numbers in italics represent isopachs. Numbers next to solid black dots designate determined Bajocian – Upper Jurassic thicknesses. Dotted areas show terrigeneous (predominantly continental and shallow marine) deposits (other captions are the same as in **Figure 3**).

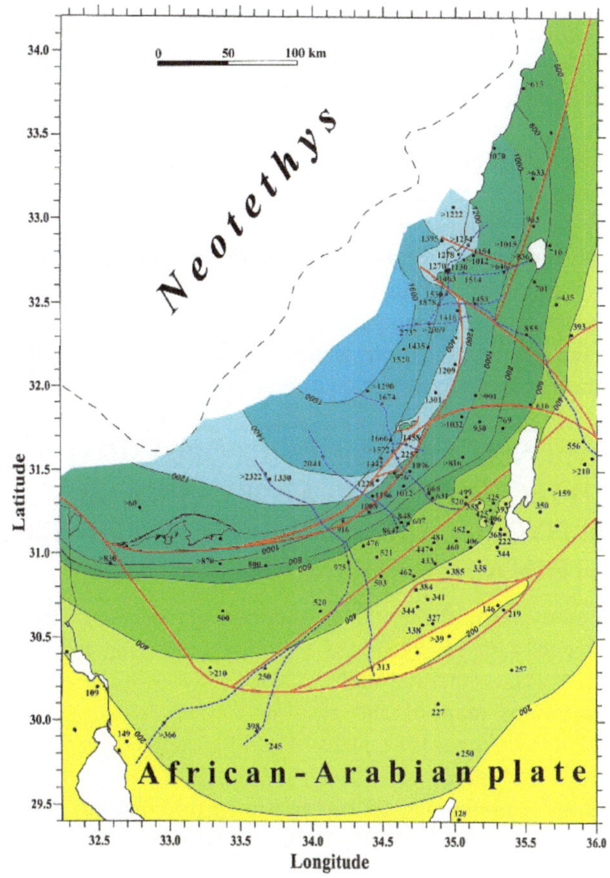

Figure 6. Isopach map of the Lower Cretaceous for the Levant Dashed blue lines designate the position of the ancient erosional system (other captions are the same as in Figures 3 and 4).

APPLICATION OF PALEOTECTONIC ZONATION IN THE HELETZ-ASHDOD OIL FIELD (CENTRAL ISRAEL)

Deep paleotectonic mapping (zonation) supported by gravity-magnetic examination was used to investigate the hydrocarbon fields of the central plain of Israel (Figures 7-9). In the detail drilled (about 100 boreholes) of the Heletz-Ashdod oil and gas area, paleogeographic and paleotectonic mapping based on the facies analysis was applied and the Jurassic and Cretaceous horizons were studied. **Figure 7** shows the Early Cretaceous erosion section at a depth of up to 1000–1300 m, splitting the Jurassic thicknesses down to the Sherif formation of the Late Bajocian. The post-accretional Gevar formation of the Late Hauterivian – Barremian is well known as regards alluvial, deltaic and deep marine channel facies [83].

The structural differentiation of this erosional zone appears both in differences in the integrated thicknesses of Jurassic rocks (**Figure 7**) and in thicknesses of their separate horizons (**Figure 8**). The examined geological structures are mainly inherited ones. The Heletz oil field is a triangular depression within which thicknesses of the Upper Jurassic and Lower Cretaceous are sharply increased. The Lower Cretaceous thicknesses in this area offshore Israel reach maximal eastern Mediterranean values in the Heletz-33 (2359 m), Heletz-31 (2257 m) and Heletz-32 (2290 m) boreholes [42]. The Ashdod gas field may be described as an anomalous uplifted zone where thicknesses of the Jurassic and Lower Cretaceous horizons abruptly decrease.

Gravity and magnetic anomalies in the area under study are significantly influenced by some regional anomalies. Therefore, for better visualization, the gravity and magnetic maps are overlaid and also show the geological structural elements (**Figure 9**). In this integrated map the local structural elements of the oil and gas HeletzAshdod area stand out well. The clearest delineations are the Heletz triangle, Yinnon High, the Beeri depression and Gevim High (**Figure 9**). Note that the Talme Yafe arch, within which the Ashdod High is located, is characterized by a heterogeneous structure that calls for further detailed geological-geophysical investigation for prospective oil and gas structure detection in this area.

COMBINED ANALYSIS OF GEOPHYSICAL AND TECTONIC INDICATORS AS A BASIS FOR HYDROCARBON PROSPECTING IN THE EASTERN MEDITERRANEAN

The analysis of the regional gravity field (**Figure 2**) was combined with an integrated examination of the magnetic field [4,84], a thermal field analysis [4,63,66], the Moho discontinuity map of the eastern Mediterranean and the Curie map of Israel [58], maps of earthquake epicenters (e.g., [50,61]), and seismic data analyses (e.g., [4,11]) to construct a modified tectonic map of the Eastern Mediterranean (**Figure 10**).

The satellite gravity map displays the terranes, the main and intraplate tectonic faults, the African-Arabian Precambrian basement with the continental crust, areas occupied by oceanic crust, Mesozoic terranes and the Alpine tectonic belt (**Figure 10**). To improve readability, the gravity map was smoothed and the isoline interval was increased to 20 mGals.

This map (**Figure 10**) was developed on the basis of regional tectonic reconstructions of the eastern Mediterranean [2-4,11-14,22,75,85] and the present investigation. This map depicts the main plate tectonic structures of this region: the African, Arabian, Sinai and Aegean-Anatolian plates. The latter plate is part of the Alpine mobile belt. Associations of the Precambrian continental crust, oceanic crust and Mesozoic accretional complexes compose the African, Arabian and Sinai plates. It should be noted that the earlier (e.g., [86]) Pelusium Line (some 60 km off the coast of Israel) in the northern part mainly coincides with the boundary between the oceanic crust and the Galilee-Lebanon terrane. Thus, the Pelusium Line cannot be a marker of the northern part of the Sinai plate (**Figure 10**). At the same time it agrees with the conceptual model proposed by Ben-Avraham [3] where the northern boundary of the Sinai plate collides with the Aegean-Anatolian (Turkish) plate.

In the region under study, several hundred deep boreholes were drilled. Among these roughly 40 known borehole hydrocarbon reserves have been discovered [8, 19,36,40,42,45]. Interestingly, borehole Tamar-1 (where a deposit was discovered with total reserves of more than 200 billion m^{3}of gas condensate) and the promising prospective borehole Leviathan [9] (the most recent estimates, in January of 2011 report about 450 billion m^3 of gas condensate), are located in the vicinity of a positive gravity anomaly (**Figure 10**). The last U. S. Geological Survey estimate indicates the presence of 1.7 billion barrels of recoverable oil and more than 4 trillion m^3 of recoverable gas in the Levant Basin Province [1].

Integrated methodology of the regional-structural, facial and paleo-geodynamical analysis can serve to estimate the overall hydrocarbon potential in the eastern Mediterranean. For the first time three structural-prognosis geotectonic zones for hydrocarbon prospection were revealed: 1) a province located on the ancient Nubian-Arabian Precambrian platform, 2) a province of a terrane belt with thinned Precambrian crust, and 3) a province of basins with oceanic crust.

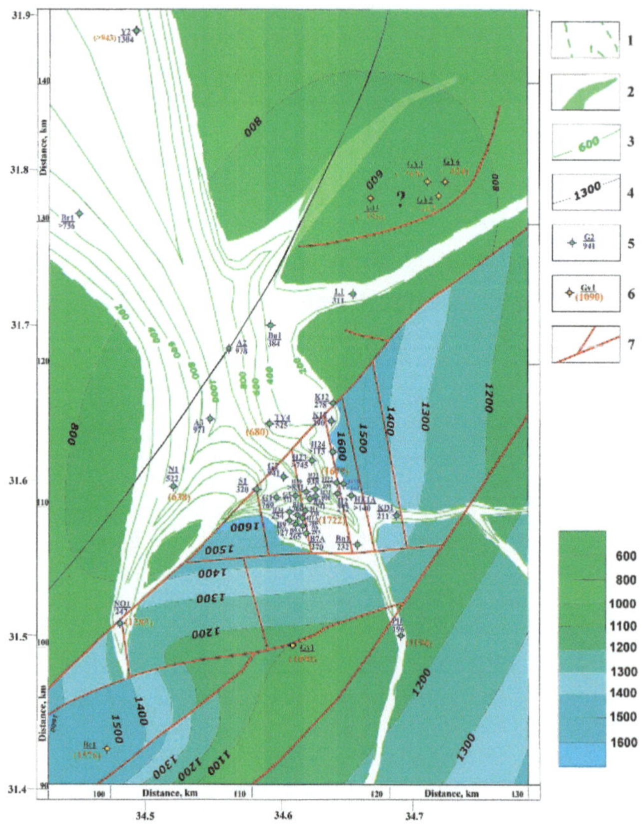

Figure 7. Paleogeomorphological map of GevarAm erosional stage (Hauterivian – Barremian) of the Heletz-Ashdod oil and gas field and adjacent areas. (1) Gevar Am deltaic, alluvial and marine canyon zone, (2) zone of posterior Gevar Am denudation, (3) isopachs of the Gevar Am formation (in m), (4) isopachs of the Upper – Middle Jurassic (Halutza – Upper Inmar formations) (in m), (5) boreholes penetrating Gevar Am formation (numerator – borehole name and number, denominator – thickness in m, (6) boreholes penetrating Jurassic rocks (numerator – borehole name and number, denominator – thickness in m), (7) faults. Borehole abbreviations: A – Ashkelon, Ad – Ashdod, B – Brur, Ba – Barnea, Be – Beeri, Bn – Binyamin, Br – Bravo, BW – Beeri West, G – Gevar Am, Gi – Givati, Gv – Gevim, GY – Gan Yavne, H – Heletz, Hd – Hodiya, Ho – Hoga, HE – Heletz East, K – Kokhav, Kr – Karmon, KD – King David, L – lior, MY – Massuot Yizhak, N – Nissanit, Ng – Negba, NA – Nir Am, NO – Nahal Oz, Pl – Pleshet, Rv – Revaha, S – Shimon, Sa – Saad, Sh – Shuva, SA – Sedot Akiva, TY – Talme Yafo, Y – Yam, Ya – Yakhini, Yi – Yinnon Color columns show the Middle – Upper Jurassic isopach intervals (in m).

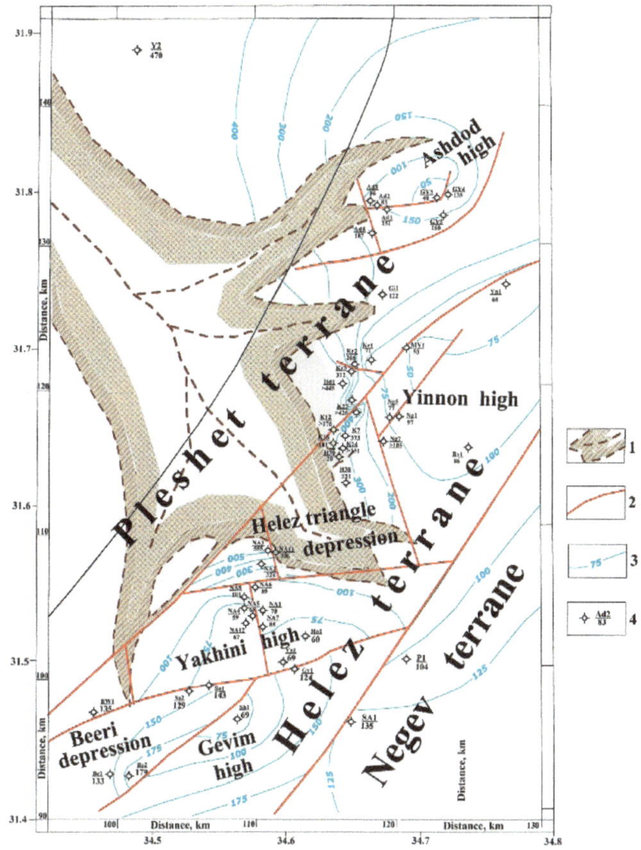

Figure 8. Late Hauterivian – Barremian structural-paleogeographic map of the Heletz-Ashdod oil-gas field and adjacent areas. (1) The Early Cretaceous Gevar Am) erosoinal zone, (2) isopachs of the Middle Oxfordian (Beer Sheva formation) (in m), (3) boreholes penetrating the Beer Sheva formation (numerator – borehole name and number (see caption to**Figure 7**), denominator – thickness in m), (4) faults.

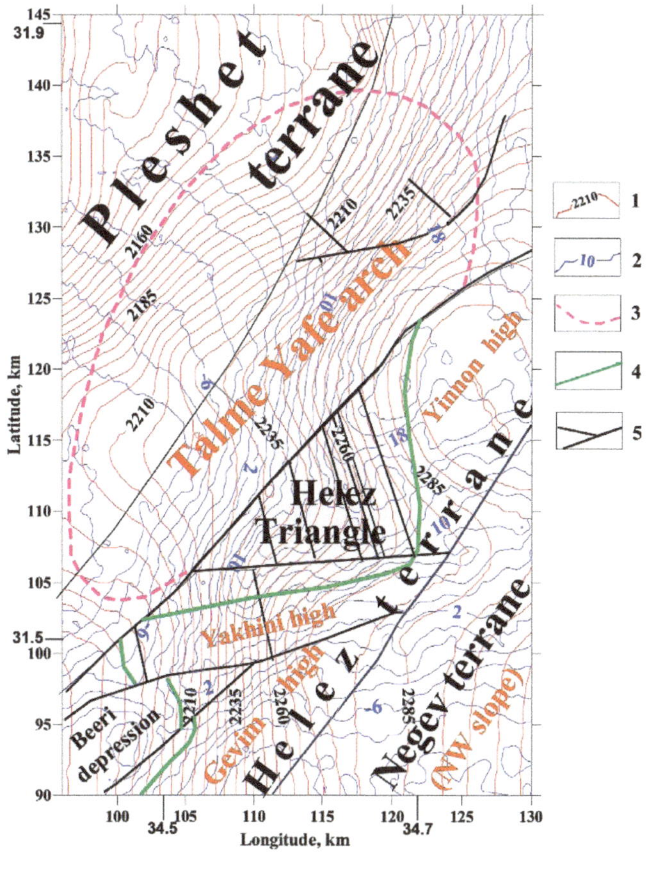

Figure 9. Combined gravity-magnetic map overlaid on the Late Jurassic-Early Cretaceous paleotectonic structure of the Heletz-Ashdod oil and gas field and adjacent areas. (1) isolines of Bouguer anomalies (in mGal), (2) isolines of magnetic anomalies (in nT), (3) isopach 800 m of the Middle - Upper Jurassic rocks in the Pleshet terrane, (4) isopach 1300 m of the Middle - Upper Jurassic rocks in the Heletz terrane, (5) faults. The gravity and magnetic maps were constructed using the Gravity-Magnetic Database compiled by the Geophysical Institute of Israel.

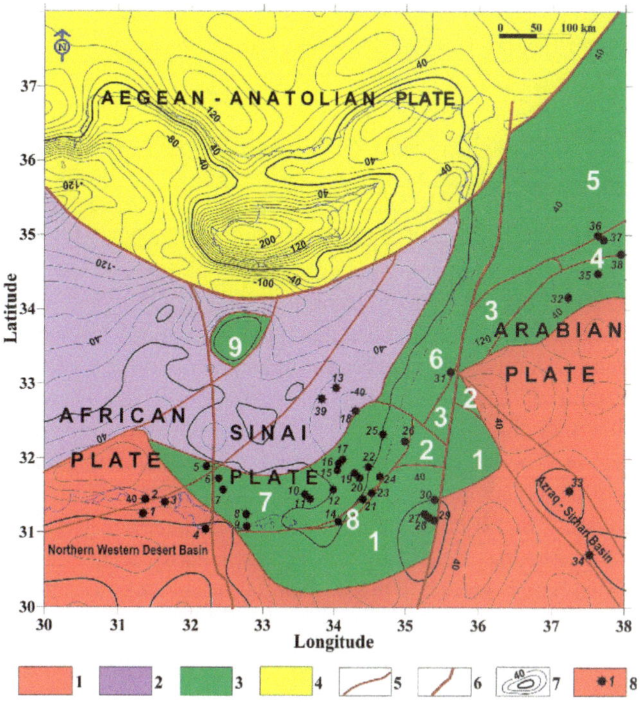

Figure 10. Tectonic map of the eastern Mediterranean overlaid on the smoothed gravity map. Terranes position (white numbers): 1 – Negev, 2 – Judea-Samaria, 3 – Antilebanon, 4 – Palmyride, 5 – Aleppo, 6 – Galilee-Lebanon, 7 – Pleshet, 8 – Heletz, and 9 – Eratosthenes. (1) Precambrian plates with continental crust, (2) oceanic crust, (3) Mesozoic terranes, (4) Alpine tectonic belt, (5) main tectonic faults, (6) intraplate tectonic faults, (7) gravity isolines, mGal, (8) boreholes with discovered hydrocarbon reserves (black numbers): 1 – Khelala-1, 2 – El-Qara-1, 3 – Wastani-1, 4 – Qnantara-1, 5 – Temsah-1, 6 – Kersh-1, 7 – Port Fouad-1, 8 – Tineh-1, 9 – Bougaz-1, 10 – Mango-1, 11 – Mango-2, 12 – Gaza Marine, 13 – Tamar-1, 14 – Sadot-1, 15 – Noa South-1, 16 – Noa-1, 17 – Or-1, 18 – Dalit-1, 19 – Mari-1,2, 20 – Nir-1,2, 21 – Shiqma-1, 22 – Yam-2, 23 – Heletz, 24 – Ashdod, 25 – Yam-Yafo-1, 26 – Meged-2, 27 – Zahar-8, 28 – Gurim-1, 29 – Zuk-Tamrur-4, 30 – Emunah-1, 31 – Hula, 32 - Qaruateine-1, 33 – Hamza, 34 – Wadi Sirhan-4, 35 – Cheriffe-1, 36 & 37 – boreholes with unknown names, 38 – Al Shaier-1, 39 – Leviathan-1.

The hydrocarbon basins of the first province are narrow, intraplate marginal graben-like troughs. Their formation was initiated in the Late Cretaceous era. However, the main stage of their development and hydrocarbon migration are associated with new Neotectonics epochs (30 m.y.a. – to present) and are connected to the Red Sea system riftogenesis. There are known oil deposits (October and other oil fields in the Suez Basin) [87], ozocerites (Dead Sea Basin) [88] and small gas deposits (Hula Basin) [89]. The prospects for oil and gas discovery are associated with the presence of a thick series of the Early Messinian salt.

In the second province, three structural floors have developed in a sedimentary cover: (a) Late Precambrian molasses, (b) the Pre-accretional carbonate platform of the Late Permian – Early Mesozoic, and (c) the Postaccretional Late Mesozoic-Cenozoic predominantly terrigenic complex. Oil and gas deposits were found in floors (b) and (c). The carbonate platform (b) has thickness up to 5-6 km and hydrocarbon searching is oriented here to discovering reef traps. The present investigation helps define a search strategy for these objects based on a combined historical-geodynamical and paleogeographical examination. Taking into account that the thick Mesozoic carbonate series were formed in terranes of the peripheral zones of the Tethys Ocean, the presence of reef complexes and the backreef sediments in this region is obvious. A predominantly terrigenic complex (c) mostly developed in the coastal plain of the eastern Mediterranean and directly in the shelf zone (its thickness reaches 5-6 km). Hydrocarbon collectors in this area are the Late Cenozoic terrigeneous rocks associated with Late Messinian salt. The Early Cretaceous erosional channel complexes are of special interest. The long-term Heletz oil deposit (central Israel) (see Section 6 and Figures 7-9) belongs to just this type. In the Early Cretaceous paleogeographical map (**Figure 6**) for the first time a complex system of ancient erosional channels is shown. Therefore, this map could be used for hydrocarbon investigation in this region. The presence of a majority of hydrocarbon deposits and a geotectonic basis in the Pleshet terrane (**Figure 10**) is confirmed by the present work. In this terrane terrigene-carbonaceous rocks and salt predominantly occur. Such a facial type is known in many oil and gas provinces worldwide.

The third province consists of many basins, but our analysis focuses solely on the largest Levant basin (Figures 1,2,10). The tectonic-geophysical analysis was used to designate these basin boundaries with respect to the zone of terrane belt and other eastern Mediterranean basins of this province. The western part of this basin is bounded by the Leviathan high and eastern part – by the continental-ocean transition zone (**Figure 10**). Note that in the central part of the Levant basin a Moho discontinuity lowering-in was detected previously (see **Figure 3** in [58]). The Early Mesozoic-Cenozoic facial characteristics of the sedimentary cover are similar to the corresponding cover of the Pleshet terrane. The available data indicate that structural zones have developed here of the oceanic type as found in the Gulf of Mexico, the Pricaspian basin and the southern Caspian basin. Ben-Avraham [3] claims that there are intensive geodynamic movements at the boundary of the Sinai and Aegean-Anatolian plates allow us to suggest that in the Levant basin numerous prospective local structures and swells may occur. Hence, it is likely that the high-potential hydrocarbon deposits in Tamar-1 and Leviathan (**Figure 10**) discovered recently in this region are not isolated features.

CONCLUSIONS

This article presented the results of combined structural isopach map compilations and geophysical field analyses of the deep floors of the eastern Mediterranean region, thus constituting the first deep geological-geophysical mapping ever produced. For the construction of the isopach maps of the Middle-Upper Jurassic and Lower Cretaceous a regional structural analysis method was applied (previously employed mainly for mapping subsurface structures in this region). The same analysis for zoning the Heletz-Ashdod oil field was successfully applied. It revealed that the Syrian arc is an accretional belt of the Mesozoic terranes, which was formed in the Early Cretaceous. It was shown that in the Early Cretaceous post-accretional stage numerous erosional channels, which may be used for searching for hydrocarbon collectors, were formed. An integrated methodology of regional-structural, paleogeographical and paleogeodynamical analysis was used to estimate the overall hydrocarbon potential of the eastern Mediterranean. Three structural-prognosis geotectonic zones for hydrocarbon prospecting were delineated: 1) a province located on the ancient Nubian-Arabian Precambrian platform, 2) a province of a terrane belt with a thinned Precambrian crust, and 3) a province of basins with oceanic crust.

ACKNOWLEDGEMENTS

The authors would like to thank Prof. Zvi Ben-Avraham (Tel Aviv University) for useful discussions and comments that were very helpful in preparing the manuscript. Comments of Dr. A. Pilchin (Canada) and two anonymous reviewers were essentially practical ones. The authors also thank the Geophysical Institute of Israel for the permission to use the Gravity-Magnetic Database of Israel and the eastern Mediterranean.

REFERENCES

1. C. J. Schenk, M. A. Kirschbaum, R. R. Charpentier, T. R. Klett, M. E. Brownfield, J. K. Pitman, T. A. Cook and M. E. Tennyson, "Assessment of Undiscovered Oil and Gas Resources of the Levant Basin Province, Eastern Mediterranean," U. S. Geological Survey Fact Sheet, 2010- 3014, 2010.
2. Z. Ben-Avraham and A. Ginzburg, "Displaced Terranes and Crustal Evolution of the Levant and the Eastern Mediterranean," Tectonics, Vol. 9, No. 4, 1990, pp. 613-622.doi:10.1029/TC009i004p00613
3. Z. Ben-Avraham, "The Structure and Tectonic Setting of the Levant Continental Margin, Eastern Mediterranean," Tectonophysics, Vol. 46, No. 3-4, 1978, pp. 313-331. doi:10.1016/0040-1951(78)90210-X
4. Z. Ben-Avraham, A. Ginzburg, J. Makris and L. Eppelbaum, "Crustal Structure of the Levant Basin, Eastern Mediterranean," Tectonophysics, Vol. 346, No. 1-2, 2002, pp. 23-43. doi:10.1016/S0040-1951(01)00226-8
5. Z. Garfunkel, "Constrains on the Origin and History of the Eastern Mediterranean Basin," Tectonophysics, Vol. 298, No. 1-3, 1998, pp. 5-35. doi:10.1016/S0040-1951(98)00176-0
6. V. E. Khain, "Regional Geodynamics. Alpine Mediterranean Belt," Nedra Publishing, Moscow, 1984.
7. A. H. F. Robertson and J. E. Dixon, "Introduction: Aspects of the Geological Evolution of the Eastern Mediterranean," In: J. E. Dixon and A. H. F. Robertson, Eds., The Geological Evolution of the Eastern Mediterranean, Geological Society Special Publishing Vol. 17, 1984, pp. 1-74.
8. "Tamar Oil Find Fires Interest in East Mediterranean Data," First Break, Vol. 27, No. 6, 2009, p. 47.
9. "Noble Identifies Giant Prospect Offshore Israel,"

Offshore, Vol. 70, No. 5, June 2010. http://www.offshoremag.com/index/article-display/6855731723/articles/off-shore/field-development/middleeast/2010/06/noble-identifies_giant.html

10. "Eastern Mediterranean Prospects Buoyed by USGS Assessment," First Break, Vol. 28, No. 6, 2010, p. 23.

11. Z. Ben-Avraham, U. Schattner, M. Lazar, J. K. Hall, Y. Ben-Gai, D. Neev and M. Reshef, "Segmentation of the Levant Continental Margin, Eastern Mediterranean," Tectonics, Vol. 25, No. TC5002, 2006, p. 17.

12. Z. Cohen, V. Kaptsan and A. Flexer, "The Tectonic Mosaic of the Southern Levant: Implications for Hydrocarbon Prospects," Journal of Petroleum Geology, Vol. 13, No. 4, 1990, pp. 437-462. doi:10.1111/j.1747-5457.1990.tb00858.x

13. J. Mechie, K. Abu-Ayyash, Z. Ben-Avraham, R. El-Kelani, I. Qabbani and M. Weber, "Crustal Structure of the Southern Dead Sea Basin Derived from Project DESIRE Wide-Angle Seismic Data," Geophysical Journal International, Vol. 178, No. 1, 2009, pp. 457-478. doi:10.1111/j.1365-246X.2009.04161.x

14. Y. Katz, Z. Ben-Avraham and L. Eppelbaum, "Early Mesozoic Facial Distribution and Tectonic-Geophysical Setting in Israel," Transactions of the Conference of the Israel Geology Social Annual Meeting, Neve Zohar, Dead Sea, 2007, p. 59.

15. A. A. Aal, A. El Barkooky, M. Gerrits, H. Meyer, M. Schwander and H. Zaki, "Tectonic Evolution of the Eastern Mediterranean Basin and Its Significance for Hydrocarbon Prospectivity in the Ultra-Deep Water of the Nile Delta," The Leading Edge, Vol. 19, 2000, pp. 1086-1102. doi:10.1190/1.1438485

16. Y. Druckman, D. Gill, L. Fleischer, E. Gelbermann and O. Wolff, "Subsurface Geology and Structural Evolution of the Northwestern Negev, Southern Israel," Israel Journal of Earth Science, Vol. 44, 1995, pp. 115-135.

17. M. A. Gardosh, Z. Garfunkel, Y. Druckman and B. Buchbinder, "Tethyan Rifting in the Levant Region and Its Role in Early Mesozoic Crustal Evolution," Geological Society of London, Vol. 341, 2010, pp. 9-36. doi:10.1144/SP341.2

18. M. A. Gardosh, Y. Druckman, B. Buchbinder and M. Rybakov, "The Levant Basin Offshore Israel: Stratigraphy, Structure, Tectonic Evolution and Implications for Hydrocarbon Exploration," GII 429/328/08, Geological Institute of Israel, 2008, pp. 1-118.

19. G. Roberts and D. Peace, "Hydrocarbon Plays and Prospectivity of the Levantine Basin, offshore Lebanon and Syria from Modern Seismic Data," GeoArabia, Vol. 12, No. 3, 2007, pp. 99-124.

20. H. Shulman, M. Reshef and Z. Ben-Avraham, "The Structure of the Golan Heights and Its Tectonic Linkage to the Dead Sea Transform and the Palmyrides Folding," Israel Jour. of Earth Science, Vol. 53, No. 3, 2004, pp. 225-237. doi:10.1560/MWVC-CGPU-65KU-FFPY

21. J. Steinberg, Z. Gvirtzman, H. Gvirtzman and Y. Ben-Gai, "Late Tertiary Faulting along the Coastal Plain of Israel," Tectonics, Vol. 27, No. TC4014, 2008.

22. A. H. F. Robertson, P. D. Clift, P. J. Degnan and G. Jones, "Palaeogeographic and Palaeotectonic Evolution of the Eastern Mediterranean Neotethys," Palaeogeography, Palaeoclimatology, Palaeoecology, Vol. 87, 1991, pp. 289- 343. doi:10.1016/0031-0182(91)90140-M

23. D. P. Mckenzie, D. Davies and P. Molnar, "Plate Tectonics of the Red Sea and East Africa," Nature, Vol. 226, No. 5242, 1970, pp. 243-248. doi:10.1038/226243a0

24. Y. P. Malovitskiy, E. M. Emelyanov, O. V. Kazakov, V. N. Moskalenko, G. V. Osipov, K. M. Shimkus and I. S. Chumakov, "Geological Structure of the Mediterranean Sea Floor (Based on Geological-Geophysical Data)," Marine Geology, Vol. 18, 1975, pp. 231-261. doi:10.1016/0025-3227(75)90015-8

25. L. Dubertret, "Les Formes Structurales de la Syrie et de la Palestine," Comptes rendus de l›Académie des Sciences, Paris, Vol. 195, 1932, pp. 66-68.

26. S. Cloetingh, Z. Ben-Avraham, W. Sassi and F. Horvath, "Dynamics of Basin Formation and Strike-Slip Tectonics," Tectonophysics, Vol. 449, No. 1-4, 1996, pp. 1-10.doi:10.1016/S0040-1951(96)00179-5

27. W. Bosworth, A. S. El-Hawat, D. E. Helgelson and K. Burke, "Cyrenacian 'Shock Absorber' and Associated Inversion Strain Shadow in the Collision Zone of Northeast Africa," Geology, Vol. 36, No. 9, 2008, pp. 695-698. doi:10.1130/G24909A.1

28. A. Flexer, "Late Cretaceous Palaeogeography of Northern Israel and Its Significance for the Levant Geology," Palaeogeography, Palaeoclimatology, Palaeoecology, Vol. 10, 1971, pp. 293-316. doi:10.1016/0031-0182(71)90053-8

29. A. Sneh, Y. Bartov and M. Rozensaft, "Geological Map of Israel, Scale 1:200,000," Geological Survey of Israel, Ministry of National Infrastructure, Jerusalem.

30. Y. Druckman, T. Weissbrod, M. Goldberg and S. Boehm, "Lithological and Stratigraphic Data on the Precambrian, Paleozoic, Triassic and Jurassic Subsurface Sequences Penetrated in the Negev and Northern Sinai," Geological Survey of Israel, Mapping, Stratigraphy and Oil Division, 1983.

31. I. J. Andrews, "Permian, Triassic and Jurassic Lithostratigraphy of the Subsurface of Jordan," Subsurface Geology Bulletin, Vol. 4, 1992, pp. 1-60.

32. J. K. Hall, V. A. Krasheninnikov, F. Hirsch, C. Benjamini and A. Flexer, Eds., "Geological Framework of the Levant," Volume II: The Levantine Basin and Israel, Jerusalem, 2005.

33. Z. Ben-Avraham, "Development of Asymmetric Basins along Continental Transform Faults," Tectonophysics, Vol. 215, 1992, pp. 209-220. doi:10.1016/0040-1951(92)90082-H

34. Z. Garfunkel, "Tectonic Setting of Phanerozoic Magmatism in Israel," Israel Journal of Earth Science, Vol. 38, No. 2-4, 1989, pp. 51-74.

35. A. Segev, "Synchronous Magmatic Cycles during the Fragmentation of Gondwana: Radiometric Ages from the Levant and Other Provinces," Tectonophysics, Vol. 325, No. 3-4, 2000, pp. 257-277. doi:10.1016/S0040-

1951(00)00122-0

36. A. S. Alsharhan and M. G. Salah, "Geologic Setting and Hydrocarbon Potential of North Sinai, Egypt," Bulletin of Canadian Petroleum and Geology, Vol. 44, No. 4, 1996, pp. 615-631.

37. I. J. Andrews, "Cretaceous and Paleogene Lithostratigraphy of the Subsurface of Jordan," Subsurface Geology Bulletin, No. 5, 1992, pp. 1-60.

38. Y. Bartov, Z. Lewy, G. Steinitz and I. Zak, "Mesozoic and Tertiary Stratigraphy, Paleogeography and Structural History of the Gebel Areif en Naqa Area, Eastern Sinai," Israel Journal of Earth Science, Vol. 29, 1980, pp. 114-139.

39. Z. R. Beydoun, "Some Open Questions Relating to the Petroleum Prospects of Lebanon," Journal of Petroleum Geology, Vol. 3, No. 3, 1981, pp. 303-314. doi:10.1111/j.1747-5457.1981.tb00932.x

40. Z. R. Beydoun, A. R. Futyan and A. H. Jawzi, "Jordan Revisited: Hydrocarbons Habitat and Potential," Journal of Petroleum Geology, Vol. 17, No. 2, 1994, pp. 177-194.doi:10.1111/j.1747-5457.1994.tb00125.x

41. Y. Druckman, "The Triassic in Southern Israel and Sinai. A Sedimentological Model of Marginal Epicontinental Marine Environments," Ph.D. Thesis, Hebrew University, Jerusalem, 1976, p. 188.

42. L. Fleischer and A. Varshavsky, "A Lithostratigraphic Data Base of Oil and Gas Wells Drilled in Israel," Ministry of National Infrastructures of Israel, 2002-2010.

43. V. A. Krasheninnikov, J. K. Hall, F. Hirsch, H. Benjamini and A. Flexer, "Geological Framework of the Levant," Volume 1: Cyprus and Syria, Jerusalem, 2005.

44. Z. Lewy and M. Raab, "Mid-Cretaceous Stratigraphy of the Middle East," Annales de Museum d'Histoire National de Nice, Vol. 4, No. 32, 1976, pp. 1-19.

45. F. H. Nader and R. Swennen, "Petroleum Prospects of Lebanon: Some Remarks from Sedimentological and Diagenetic Studies of Jurassic Carbonates," Marine and Petroleum and Geology, Vol. 21, No. 4, 2004, pp. 427-441. doi:10.1016/S0264-8172(03)00095-3

46. R. Said, Ed., "The Geology of Egypt," A. A. Balkema/ Rotterdam/Brookfield, 1990.

47. A. Segev, T. Weissbrod and B. Lang, "$^{40}Ar/^{39}Ar$ Dating of the Aptian-Albian Igneous Rocks in Makhtesh Ramon (Negev, Israel) and Its Stratigraphic Implications," Cretaceous Research, Vol. 26, 2005, pp. 633-656. doi:10.1016/j.cretres.2005.03.003

48. F. Shaaban, R. Lutz, R. Littke, C. Bueker and K. Odisho, "Source-Rock Evaluation and Basin Modelling in NE Egypt (NE Nile Delta and Northern Sinai)," Journal of Petroleum Geology, Vol. 29, No. 2, 2006, pp. 103-124. doi:10.1111/j.1747-5457.2006.00103.x

49. C. Walley, "Some Outstanding Issues in the Geology of Lebanon and Their Importance in the Tectonic Evolution of the Levantine Region," Tectonophysics, Vol. 298, 1998, pp. 37-62. doi:10.1016/S0040-1951(98)00177-2

50. F. Bender, "Geology of Jordan," Berlin, Gebruder Borntraeger, 1974.

51. Y. Bartov, "Geological Photomap of Israel and Adjacent Areas, 1:750,000," The Geological Survey of Israel, Jerusalem, 1974.

52. B. Lang and G. Steinitz, "K–Ar Dating of Mesozoic Magmatic Rocks in Israel: A Review," Israel Journal of Earth Science, Vol. 38, 1989, pp. 89-103.

53. G. Shaliv, "Stages in the Tectonics and Volcanic History of the Neogene Basin in the Lower Galilee and the Valleys," Ph.D. Thesis, Hebrew University, Jerusalem, 1991, p. 94. (in Hebrew, English summary)

54. Y. Folkman and Z. Yuval, "Aeromagnetic Map of Israel, 1:250,000," Israel Institute for Petroleum Research and Geophysics, Holon, Israel, 1976.

55. A. Ginzburg, Y. Folkman, M. Rybakov, Y. Rotstein, Y. Assael and Z. Yuval, "Israel: Bouguer Gravity Map, Scale 1:250,000," Survey of Israel, Jerusalem, 1993.

56. M. Rybakov, V. Goldshmidt and Y. Rotstein, "New Compilation of the Gravity and Magnetic Maps of the Levant," Geophysical Research Letters, Vol. 24, 1997, pp. 33-36.

57. D. T. Sandwell and W. H. F. Smith, "Global Marine Gravity from Retracked Geosat and ERS-1 Altimetry: Ridge Segmentation versus Spreading Rate," Journal of Geophysical Research, Vol. 114, No. B01411, 2009. doi:10.1029/96GL03617

58. L. V. Eppelbaum and A. N. Pilchin, "Methodology of Curie Discontinuity Map Development for Regions with Low Thermal Characteristics: An Example from Israel," Earth and Planetary Sciences Letters, Vol. 243, No. 3-4, 2006, pp. 536-551.doi:10.1016/j.epsl.2006.01.003 [Paper reference:5]

59. D. Levitte and A. Olshina, "Isotherm and Geothermal Gradient Maps of Israel," Geological Survey of Israel, 1985.

60. A. Shapira and G. Shamir, "Seismicity Parameters of Seismogenic Zones in and around Israel," 1994.

61. G. Vannucci, S. Pondrelli, A. Argnani, A. Morelli, P. Gasperini and E. Boschi, "An Atlas of Mediterranean Seismicity," Annals of Geophysics, Vol. 47, No. 1, 2004, pp. 247-306.

62. Z. Ben-Avraham and A. Ginzburg, "Magnetic Anomalies across the Levant Continental Margin," Marine and Petroleum Geology, Vol. 3, 1986, pp. 220-233. doi:10.1016/0264-8172(86)90046-2

63. A. N. Pilchin and L. V. Eppelbaum, "Determination of Magnetized Bodies Lower Edges by Using Geothermal Data," Geophysical Journal International, Vol. 128, No. 1, 1997, pp. 167-174. doi:10.1111/j.1365-246X.1997.tb04077.x

64. A. Berkovitch, I. Binkin, L. Eppelbaum, N. Scharff and E. Guberman, "Integration of Advanced Multifocusing Seismics with Potential Field Analysis: Heletz Oil Field (Central Israel) Example," Journal of the Balkan Geophysical Society, Vol. 8, No. (Suppl. 1), 2005, pp. 593- 596.

65. L. V. Eppelbaum, "Combined Interpretation of Gravity and Magnetic Fields over the Rosh-Ha'Ayin Structure, Israel," Transactions of the Conference of the Geological

Society of America, North-Central Section, Iowa, Vol. 28, No. 6, 1996, p. 38.

66. L. V. Eppelbaum, M. M. Modelevsky and Pilchin A. N., “Thermal Investigation in Petroleum Geology: The Experience of Implication in the Dead Sea Rift Zone, Israel,” Journal of Petroleum Geology, Vol. 19, No. 4, 1996, pp. 425-444. doi:10.1111/j.1747-5457.1996.tb00448.x

67. L. V. Eppelbaum, Y. A. Livshits, A. Flexer and Z. BenAvraham, “Integrated Geological-Geophysical Analysis of Ring Structures Phenomenon in the Eastern Mediterranean,” Transactions of the Conference of Israel Geological Society, Annual Meeting, Mizpe-Ramon, 1998, p. 25.

68. L. Eppelbaum, Z. Ben-Avraham and Y. Katz, “Does Mantle Diapir Produce Hebron Magnetic Anomaly?” Transactions of the Conference of Israel Geological Society, Annual Meeting, Moshabim Country Logging (Negev), 2005, p. 26.

69. L. Eppelbaum, Y. Katz and Z. Ben-Avraham, “Mt. Carmel Structure as a Plate Tectonics Phenomenon,” Transactions of the Conference of the Israel Geological Society, Annual meeting, Beit-Shean, 2006, p. 31.

70. L. V. Eppelbaum, V. L. Vaksman, S. V. Kouznetsov, L. M. Sazonova, S. A. Smirnov, A. V. Surkov, B. Bezlepkin, Y. Katz, N. N. Korotaeva and G. Belovitskaya, “Discovering of Microdiamonds and Minerals-Satellites in Canyon Makhtesh Ramon (Negev Desert, Israel),” Doklady Earth Sciences (Springer), Vol. 407, No. 2, 2006, pp. 202- 204.doi:10.1134/S1028334X06020097

71. L. V. Eppelbaum, Z. Ben-Avraham and Y. I. Katz, “Structure of the Sea of Galilee and Kinarot Valley Derived from Combined Geological-Geophysical Analysis,” First Break, Vol. 25, No. 1, 2007, pp. 21-28.

72. A. Ginzburg and L. Eppelbaum, “A Combined 3-D Interpretation of the Carmel Gravity and Magnetic Anomalies,” Report Prepared on Behalf of Oilfields Ltd., Israel, 1993.

73. A. Ginzburg and L. Eppelbaum, “A 3-D Reinterpretation of the Ma’Anit Gravity Anomaly,” Report Prepared on Behalf of Oilfields Ltd., Israel, 1994.

74. Y. I. Katz and L. V. Eppelbaum, “Preliminary Results of Basin Mapping of the Lower Cretaceous Traps in Northern Israel,” Transactions of the Conference of Israel Geological Society, Annual Meeting, Dead Sea, 1999, p. 40.

75. Y. I. Katz and L. V. Eppelbaum, “Levantine Phase of Tectonic-Thermal Activity in the Eastern Mediterranean,” Transactions of the Annual Meet of Geological Society of American, Section: Planetary Geology, Vol. 31, No. 7, 1999, p. A119.

76. Z. Ben-Avraham, J. Woodside, E. Lodolo, M. Gardosh, M. Grasso, A. Camerlenghi and G.-B. Vai, “Eastern Mediterranean Basin Systems,” In: D. G. Gee and R. A. Stephenson, Eds., European Lithosphere Dynamics, Geological Society, London, Vol. 32, 2006, pp. 263-276.

77. A. Segev and M. Rybakov, “Effects of Cretaceous Plume and Convergence, and Early Tertiary Tectonomagmatic quiescence on the Central and Southern Levant Continental Margin,” Journal of the Geological Society, London, Vol. 167, 2010, pp. 731-749.doi:10.1144/0016-76492009-118

78. M. Rybakov, “What is the Reason for the Dead Sea Transform Fault System Segmentation?” Israel Journal of Earth Science, Vol. 52, 2003, pp. 65-75.doi:10.1560/JNQ0-UWCX-PAVN-XJT7

79. A. I. Levorsen, “Geology of Petroleum,” Freeman & Co., San-Francisco/London, 1967.

80. F. Hirsch and L. Picard, ”The Jurassic Facies in the Levant,” Journal of Petroleum Geology, Vol. 11, No. 3, 1988, pp. 277-308. doi:10.1111/j.1747-5457.1988.tb00819.x

81. F. Hirsch, ”Jurassic of Israel,” In: J. K. Hall, V. A. Krasheninnikov, F. Hirsch, C. Benjamini and A. Flexer, Eds., Volume II: The Levantine Basin and Israel, Jerusalem, 2005, pp. 361-391.

82. H. J. Tschopp, “The Oilfield of Heletz, Israel,” Bulletin of Swiss Association on Petroleum, Geology & Engineering, Vol. 22, No. 63, 1956, pp. 41-54.

83. Z. Cohen, “Early Cretaceous Buried Canyon: Influence on Accumulation of Hydrocarbons in Heletz Oil Field, Israel,” AAPG Bulletin, Vol. 60, No. 1, 1976, pp. 108-114.

84. L. V. Eppelbaum, “Methodology of 3-D Combined Modeling of Magnetic and Gravity Fields in the Eastern Mediterranean,” Proceedings of the 3rd EUG Meeting, Vol. 8, 2006, p. 4.

85. J. Mascle, J. Benkhelil, G. Bellaiche, T. Zitter and L. Loncke, “Marine Geologic Evidence for a Levantine-Sinai Plate, New Piece of the Mediterranean Puzzle,” Geology, Vol. 28, No. 9, 2000, pp. 779-282. doi:10.1130/0091-7613(2000)028<0779:MGEFAL>2.3.CO;2

86. D. Neev, “The Pelusium Line – A Major Transcontinental Shear,” Tectonophysics, Vol. 38, No. 3-4, 1977, pp. 114-123. doi:10.1016/0040-1951(77)90207-4

87. M. A. El-Ghamri, I. C. Warburton and S. D. Burley, “Hydrocarbon Generation and Charging in the October Field, Gulf of Suez, Egypt,” Journal of Petroleum Geology, Vol. 25, No. 4, 2002, pp. 433-464. doi:10.1111/j.1747-5457.2002.tb00094.x

88. A. Nissenbaum and M. Goldberg, “Asphalts, Heavy Oils, Ozocerite and Gases in the Dead Sea Basin,” Organic Geochemistry, Vol. 2, No. 3-4, 1980, pp. 167-180.doi:10.1016/0146-6380(80)90007-8

89. P. R. May, “Oil and Gas Potential of the Galilee, Israel,” Rep. 87/12, Oil Exploration of Israel Ltd, Tel Aviv, 1987, p. 79.

CHAPTER 8

DYNAMICS OF AN EARTHQUAKE UNDER MAGMA THRUST STRENGTH

L. Y. Kagho, M. W. Dongmo, and F. B. Pelap

INTRODUCTION

Many research works related to earthquake dynamics were focused on the study of a single spring-block model [1–3]. This single block model is deduced from the simplification of models of several blocks [4–8]. In these models, the motion of the fault is only stimulated by the strength due to the motion of the tectonic plates. As far as the one spring-block model is concerned, few studies have shown the appearance of chaotic behavior within the system when its motion is unidirectional [3] or bidirectional [2]. Moreover, Vasconcelos has investigated the phase transition in this single-block model and demonstrated that when the friction characteristic velocity is greater than 0.5, the block displacement vanishes after a slip event and the transition occurs [1]. In this study, the only force that provokes the rupture of the fault is induced by the movement of tectonic plates. One feature of our work is to examine this transition in the presence of a supplementary action referred to as the magma thrust strength.

In this paper, we consider a modified 1D spring-block model subjected to the strength due to the movement of the tectonic plates during subduction movement and to the magma thrust strength that appears during a volcanic activity. Therefore, we plan to present the origin and investigate the effects of this new force on the system and thereby contribute to the better understanding of earthquake dynamics.

This paper is outlined as follows. In Section 2, we present the earthquake model under consideration and determine the equation of motion. Section 3 is devoted to the analytical analysis of the motion equation exploiting the direct integration method. These investigations are done for two cases which deal with the closeness (or not) between the frequency of the magma and the fundamental frequency of the spring block. In Section 4, the magnitude and the event probability of an earthquake are examined. The last section is devoted to discussions and concluding remarks.

Citation: L. Y. Kagho, M. W. Dongmo, and F. B. Pelap, "Dynamics of an Earthquake under Magma Thrust Strength," Journal of Earthquakes, vol. 2015, Article ID 434156, 9 pages, 2015. doi:10.1155/2015/434156.

EARTHQUAKE MODEL AND EQUATION OF MOTION

The mechanical model under consideration (Figure 1) consists of a block of mass m attached to a spring of stiffness k_p that moves the upper line with a constant velocity V which represents the speed of the tectonic plate. This block rests on a rough surface and is connected by a harmonic spring of stiffness k_c which stands for the elasticity of rocks separating the block of mass m and the loading point of the strength F(t) that represents the magma thrust.

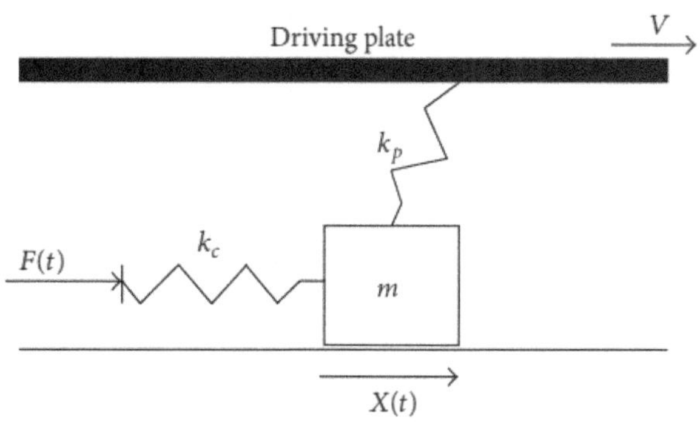

Figure 1: Spring-block model of earthquakes under consideration. Forces that intervene in this system are, respectively, the strength $F(t)$ due to the magma upflow, the elastic force $f_1(X) = -k_cX$ due to the contraction or dilatation of the spring of stiffness k_c, the external force $f_2(X) = -k_p(X-Vt)$ due to the motion of the driving plate, and the frictional force $f(dX/dt)$ that deals with the asperities between the lips

It is well known that the subduction movement is responsible of many earthquakes (tectonic event). However, many events (volcano-tremor, volcano-tectonic) are often recorded around active volcanoes. It is also known that volcano-tectonic events are due to fault's sliding under the volcano activity. The interior of a volcano is modeled by many authors [9–13] who showed that active volcano is one of the main sources of earthquakes particularly when the plates are initially loaded (Figure 2). They presented the interior of a volcano consisting of a shallow magma reservoir fed by a deeper source and a cylindrical-shaped conduit that transports the fluid upwards and may be connected to a network of other conduits [9]. Generally, the reservoir is considered to behave elastically under variable stress conditions imposed by fluid accumulation or outflow. On the other hand, the cylindrical conduit is believed to exhibit more complicated viscous behavior so that a high driving fluid pressure from below results in deformations transmitted periodically as expansions or contractions of its walls [11, 13]. After a relative long time, these periodic deformations induced a periodic strength that expanded to the rocks nearby during the motion of the magma. Hereafter, we refer to this force as the magma thrust strength F(t) defined by

$$F(t) = F_M \cos(\omega_e t) \tag{1}$$

where F_M represents the amplitude of the force and depends on the magma characteristics: magma viscosity (η_1), rock viscosity (η_2), the radius of the magmatic room (R), gravity (g), that is, $F_M = F(\eta_1, \eta_2, R, g)$, and ω_e which is the external frequency. Hereafter, (t) could be referred to as the external force acting on the block of mass m. During the investigation of the dynamics of our system, we assume that the fracturing zone where the magma outbreaks is orthogonal to the fault [14]. Then, the forces created by the thrust of the magma on the earth crust are parallel to the fault during the volcanic emission. Initially (t=0), we consider that the system is at rest, and the elastic energy accumulated in the horizontal spring kc is due to the force $f_1(X) = -{}_{kcX}$ that yields by the contraction or dilation of this spring. The motion of the driving plate also participates in the movement of the block by exerting an external force of the form $f_2(X) = -kp(X - Vt)$ [1]. The asperities which exist between the lips generate the frictional force (dX/dt) which is opposed to the motion of the block.

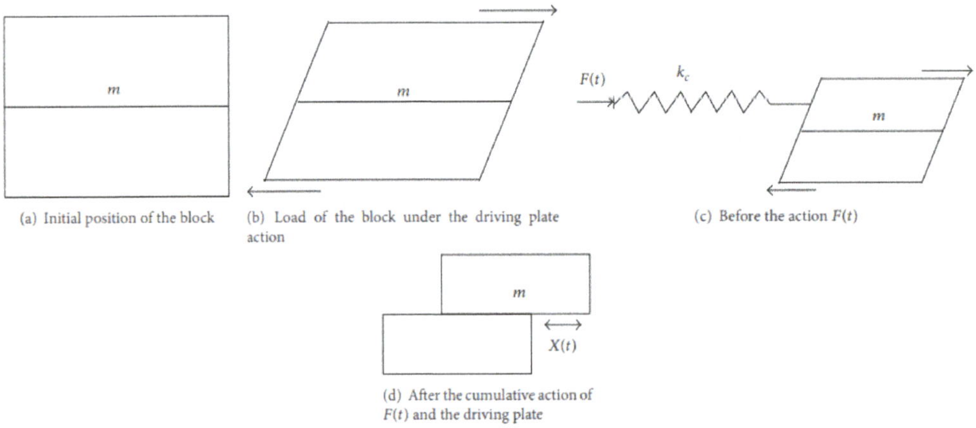

Figure 2: Action on the magma thrust strength on the block of mass.

During volcano emission, the system is also subjected to a force F(t) due to the upflow of the magma. Therefore, the system of Figure 1 could be described by the upcoming equation in terms of real variables:

$$m\frac{d^2X}{dt^2} = -k_p(X - Vt) - k_cX - f\left(\frac{dX}{dt}\right) + F(t) \tag{2}$$

It should be stressed that there exist several models of frictional force [1, 8, 15, 16]. If we suppose that the speeds are low during the sliding, then we can consider the Vasconcelos frictional model [1] defined by

$$f\left(\frac{dX}{dt}\right) = F_0\Phi\left(\frac{1}{v_f}\frac{dX}{dt}\right) \tag{3}$$

wherein F_0 is the amplitude of frictional force, v_f stands for the characteristic velocity of friction, and $\Phi(x)$ represents a continuous function for $x \geq 0$ that vanishes for large values of x and satisfies the following conditions:

$$\Phi(0) = 1, \qquad \Phi'(0) = -1 \tag{4}$$

in which the prime denotes differentiation with respect to the argument. The second condition in (4) expresses the velocity weakening effects of the friction and implies that (dX/dt) will be a decreasing function of dX/dt. Substitution of the mechanical actions (1) and (3) into (2) leads to

$$m\frac{d^2X}{dt^2} = -k_p(X - Vt) - k_cX - F_0\Phi\left(\frac{1}{v_f}\frac{dX}{dt}\right) + F_M\cos(\omega_e t). \tag{5}$$

We transform this equation into dimensionless form by defining new variables $U = k_pX/F_0$ and $\tau = t\sqrt{k_p/m} = \omega_p t$ so that (5) becomes as follows:

$$\frac{d^2U}{d\tau^2} = -\left(1+l^2\right)U - \Phi\left(\frac{1}{v_f}\frac{dU}{d\tau}\right) + v\tau + F\cos(\omega\tau) \tag{6}$$

where $l = \sqrt{kc/kp}$ defines the sound velocity, $v_f = V_f/V_0$ is the dimensionless characteristic velocity of friction, $v = V/V_0$ designates the dimensionless driven velocity, $V_0 = F_0/\sqrt{k_p m}$ is the maximum velocity attained by the block that experiences friction as it moves, $\omega = \omega_e/\omega_p$ defines the dimensionless external frequency, and $F = F_M/F_0$ stands for the dimensionless amplitude of the external force. At this level, if we assume that $(dU(\tau)/d\tau) \ll v_f$, then the function $\Phi((1/v_f)(dU/d\tau))$ can be developed in the linear approximation in terms of Taylor series according to Langer and Tang [17]. Therefore, (6) takes the following form:

$$\frac{d^2U}{d\tau^2} = -\left(1+l^2\right)U - \left(1 - \frac{1}{v_f}\frac{dU}{d\tau}\right) + v\tau + F\cos(\omega\tau) \tag{7}$$

Now, let us redefine the origin of displacement so as to eliminate the unit constant that would otherwise appear on the right-hand side of (7). Hence, the motion equation is put in the simplest form:

$$\frac{d^2U}{d\tau^2} - 2\alpha\frac{dU}{d\tau} + \omega_0^2 U = v\tau + F\cos(\omega\tau) \tag{8}$$

where $\alpha = 1/2v_f$ is the friction coefficient and $\omega_0 = \sqrt{1+l^2}$ defines the natural frequency. Equation (8) governs the slip of the block subjected to both the movement of the driving plate and the action of the magma thrust strength. It satisfies the initial conditions as below:

$$U(0) = \frac{dU(0)}{d\tau} = 0 \tag{9}$$

Expression (8) is the equation of motion that characterizes the movement of the block. If $\omega_0 = 1$ (i.e., when the sound velocity is zero) and $F=0$, we recover the Vasconcelos results [1]. In the upcoming sections, we carry out the analytical analysis of (8).

ANALYTICAL ANALYSIS

In this section, we intend to solve (8) and examine the properties of its solutions. Observation of this equation shows that one can distinguish two subcases that deal with the sign of the discriminant $\alpha^2 - \omega_0^2$ of the characteristic equation associated with (8). Hereafter, we will investigate separately the behavior of the system in each case.

Case Where $\alpha \geq \omega_0$. This case corresponds to the physical situation in which the frictional coefficient is greater than the natural frequency of the block. Mathematically, the general solution of (8) is sought under the form:

$$U(\tau) = U^*(\tau) + \overline{U}(\tau) \tag{10}$$

in which $U^*(\tau)$ is the solution of the homogeneous equation and $\overline{U}(\tau)$ is a particular solution of the entire equation (8). Computations show that the solution $U^*(\tau)$ has the explicit form:

$$U^*(t) = \exp(\alpha\tau)\left[A_1\cosh(\tilde{\omega}\tau) + B_1\sinh(\tilde{\omega}\tau)\right] \tag{11}$$

wherein A_1 and B_1 are integration constants to be determined by using the initial conditions. The last part of solution (10) can be put under the form:

$$\overline{U}(t) = at + d + b\cos(\omega\tau) + c\sin(\omega\tau) \tag{12}$$

where a, d, b, and c are constants to be determined. While exploiting (12) and (8), we find that

$$a = \frac{v}{\omega_0^2}, \qquad d = \frac{2\alpha v}{\omega_0^4},$$
$$b = \frac{F\left(\omega_0^2 - \omega^2\right)}{\left(\omega_0^2 - \omega^2\right)^2 + 4\alpha^2\omega^2},$$
$$c = -\frac{2\alpha\omega F}{\left(\omega_0^2 - \omega^2\right)^2 + 4\alpha^2\omega^2}. \tag{13}$$

Hence, we substitute (11) and (12) into (10) and use the initial conditions (9). Therefore, the displacement (τ), solution of (8), is written as

$$U(\tau) = \exp(\alpha\tau)\left[A_1\cosh(\tilde{\omega}\tau) + B_1\sinh(\tilde{\omega}\tau)\right] + at + d + b\cos(\omega\tau) + c\sin(\omega\tau) \tag{14}$$

With

$$A_1 = -\left[\left(\frac{2\alpha v}{\omega_0^4}\right) + \frac{F\left(\omega_0^2 - \omega^2\right)}{\left(\omega_0^2 - \omega^2\right) + 4\alpha^2\omega^2}\right],$$
$$B_1 = \frac{1}{\tilde{\omega}}\left[v\left(\frac{2\alpha^2}{\omega_0^4} - \frac{1}{\omega_0^2}\right) + \frac{F\alpha\left(\omega_0^2 + \omega^2\right)}{\left(\omega_0^2 - \omega^2\right) + 4\alpha^2\omega^2}\right],$$
$$\tilde{\omega} = \sqrt{\alpha^2 - \omega_0^2}. \tag{15}$$

For this case, where $\alpha \geq \omega_0$, we use solution (14) and examine the impact of the amplitude of the magma thrust strength F and that of the sound velocity l on the behavior the block. Numerical results are plotted in Figure 3.

Figure 3 shows that the slip increases with the time and the amplitude of the external force but decreases with the growth of the sound velocity. Curves of Figure 3 which deal with the parameters $l=0$ and $F=0$ correspond to the results established by Vasconcelos [1]. On these curves, the slip is very low (of 10^{-7} order) compared to the case where the magma strength is not zero. The decay of the slip with the growth of the sound velocity explains the role played by the horizontal spring of our model. Moreover, when the horizontal spring is removed, the external force acts directly on the block. Then, the slip which occurs is greater than the results obtained in the presence of the horizontal spring.

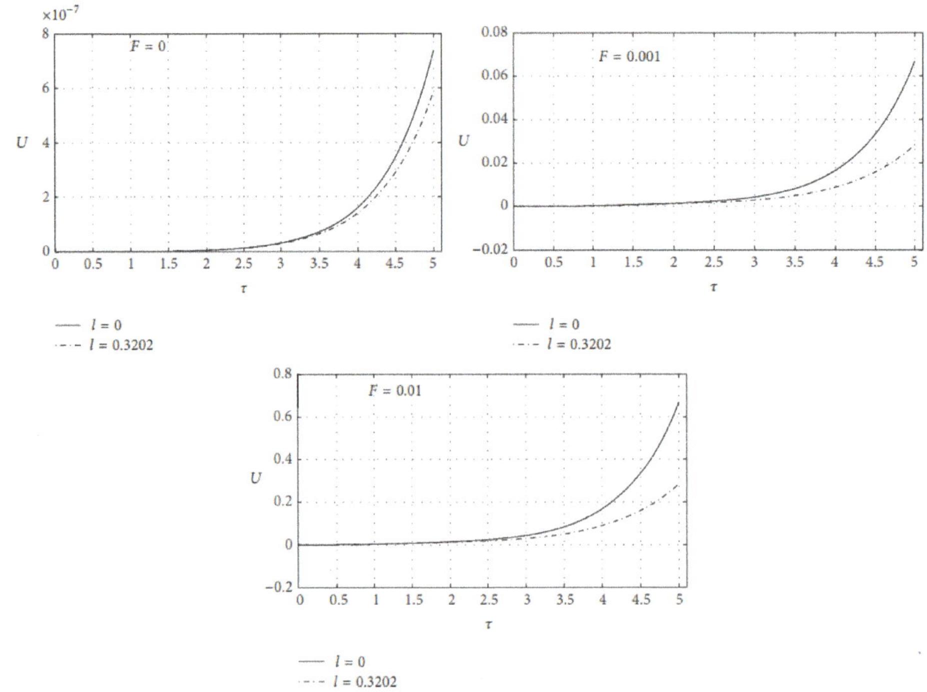

Figure 3: Influence of the magma thrust strength and the sound velocity on the slip for the parameters $\alpha = 1.06$ and $\omega = 1.5$.

Case Where $\alpha<\omega_0$: In this subsection, we focus our attention in the physical situation where the frictional coefficient is less than the natural frequency of the block. Here, the displacement of the block (τ), solution of the motion equation (8), is given by

$$U(\tau) = \exp(\alpha\tau)\left[A_1\cos(\tilde{\omega}\tau) + B_1\sin(\tilde{\omega}\tau)\right] + at + d + b\cos(\omega\tau) + c\sin(\omega\tau) \quad (16)$$

with $\tilde{\omega} = \sqrt{\omega_0^2 - \alpha^2}$

In order to measure the effects of the magma thrust strength on the dynamics of the system, it is necessary to investigate the temporal evolution of the displacement (τ) when those parameters vary.

Temporal Evolution of the Displacement. Here, we examine the impact of the amplitude and the frequency of the external force as well as the effects of the sound velocity on the dynamics of the system. We believe necessary to examine the two different cases dealing with ω=0 and $\omega = \not{0}$, respectively.

When ω=0, the external force exerted by the magma on the block becomes constant. Figure 4 shows that the displacement of the block is oscillatory. It also appears from these graphs that the amplitude of this displacement increases with the growth of the amplitude of the external force but decays while the sound velocity increases.

When $\omega = \not{0}$, one notes from Figure 4 that the displacement of the block is still oscillatory and its amplitude increases with the growth of the magma thrust strength amplitude. These curves also exhibit the increment of the displacement with the sound velocity for nonzero values of F. Figures 4 and 5 enable us to appreciate the effects of the external frequency on the dynamics of the system. One notices that, for ω=0, the displacement of the block presents a simple oscillatory behavior while its profile becomes complex for nonnull values of the external frequency

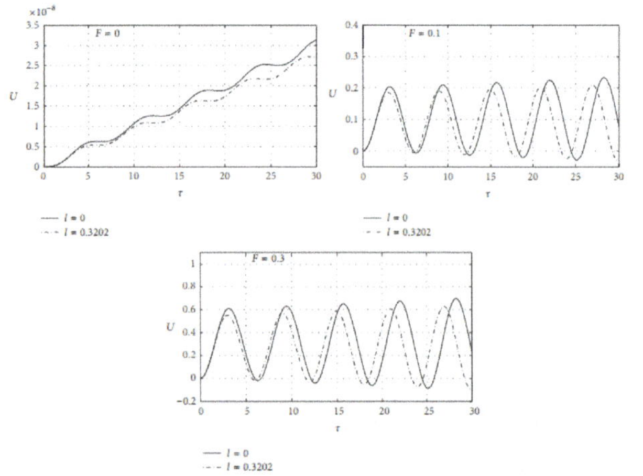

Figure 4: Temporal evolution of the block displacement when the amplitude of the magma thrust strength and the sound velocity vary for $\alpha = 0.01$ and ω=0.

Displacement of the Block at the End of the Earthquake. At the end of the slip, the block occupies an undefined position that can be identical or not to his initial position. The length between the final and initial positions of the block depends on the forces which have generated the slip. Vasconcelos [1] showed that when the displacement of the block is provoked only by the movement of the tectonic plates, the block exhibits a creep motion since its initial and final positions after the slip are identical. Therefore, its displacement is null when] → 0. In this subsection, we wish to evaluate this position when the block is subjected to cumulative actions created by the motion of the tectonic plates and the magma upflow. Indeed, when the block ceases to move, it occupies a position whose determination requires the knowledge of the time $\tau 0$ at which the block stopped to slip. Concretely, when the block stops, its speed vanishes. Then, the corresponding time $\tau 0$ is the solution of the following equation:

$$[(A_1\alpha + \tilde{\omega}B_1)\cos(\tilde{\omega}\tau_0) + (B_1\alpha - \tilde{\omega}A_1)\sin\tilde{\omega}\tau_0]\exp(\alpha\tau_0) + c\omega\cos(\omega\tau_0) - b\omega\sin(\omega\tau_0) + \frac{\nu}{\omega_0^2} = 0. \quad (17)$$

Now, we plan to solve analytically this equation. For simplification matters, we consider the particular case where the pseudo frequency of the block equals the frequency of the magma thrust strength (i.e., $\omega = \tilde{\omega}$). Therefore, expression (17) is reduced to

$$[(A_1\alpha + \tilde{\omega}B_1)\exp(\alpha\tau_0) + \tilde{\omega}c]\cos(\tilde{\omega}\tau_0) + [(B_1\alpha - \tilde{\omega}A_1)\exp(\alpha\tau_0) - \tilde{\omega}b]\sin(\tilde{\omega}\tau_0) + \frac{\nu}{\omega_0^2} = 0 \quad (18)$$

Making use of relations (18) and (16) to find the final position of the block, we obtain that it depends on the duration of the slip event τ_0:

$$\Delta = (B_1\exp(\alpha\tau_0) + c)\,\xi(\tau_0) + (A_1\exp(\alpha\tau_0) + b)\sqrt{1 - \xi^2(\tau_0)} + \nu\left(\frac{\tau_0}{\omega_0^2} + \frac{2\alpha}{\omega_0^4}\right), \quad (19)$$

wherein $\xi(\tau_0) = (\nu\zeta_2(\tau_0) - \zeta_1(\tau_0)\sqrt{\omega_0^2\zeta_3(\tau_0) - \nu^2})/\zeta_3(\tau_0)$, $\zeta_1(\tau_0) = (A_2\exp(\alpha\tau_0) + \tilde{\omega}c)$, $\zeta_2(\tau_0) = B_2\exp(\alpha\tau_0) - \tilde{\omega}b$, $\zeta_3(\tau_0) = \omega_0^2(\zeta_1^2(\tau_0) + \zeta_2^2(\tau_0))$, $B_2 = \alpha A_1 + \tilde{\omega}B_1$, $A_2 = BA_1 - \tilde{\omega}A_1$.

While exploiting expression (19) in the limit case, where $v \to 0$, we note that the position of the block at the end of the earthquake is not null ($\Delta = 0\!\!/$). Physically, this result means that the slip which occurs is a stick-slip motion. Consequently, the fault remains seismologically active. It is not useless to remember that when the motion of the block is only provoked by the subduction movement, its displacement at the end of the slip is zero (Δ=0) for $v \to 0$ and the block experiences a creep motion. In this case, the fault is said to be seismologically inactive [1].

These results present the great role played by the magma thrust strength F in the dynamical behavior of the earthquake. First of all, Vasconcelos [1] demonstrated that the aperiodic movement generates a stick-slip motion while the periodic oscillation induces a creep motion, and the transition occurs. In the present study, Figure 6 shows that when F=0, the position occupied by the block at the end of earthquake is zero (Δ=0). Therefore, the movement created by the oscillatory behavior of the block is a stick-slip. On the other hand, it appears from this figure that when $F = 0\!\!/$, the position occupied by the block at the end of the earthquake is nonnull ($\Delta = 0\!\!/$). Then, the block experiences a stickslip motion. These results show that when the rupture of a fault occurs, the motion generated is a stick-slip irrespective of the nature of the movement (periodic or aperiodic); consequently, there is no transition. In comparison to the results of [1], the transformation of the creep to stick-slip occurs because of the magma thrust strength action. Equally, we note from Figure 6 that Δ increases lineally with F. It arises from this result that the more the action of this external load is considerable, the more the probability of the transition occurrence is weak. Thus, if an external force amplifies the movement of the tectonic plate to destabilise a fault, then phase transition will not be observed. In what follows the effects of F on the power and the probability of an event are investigated.

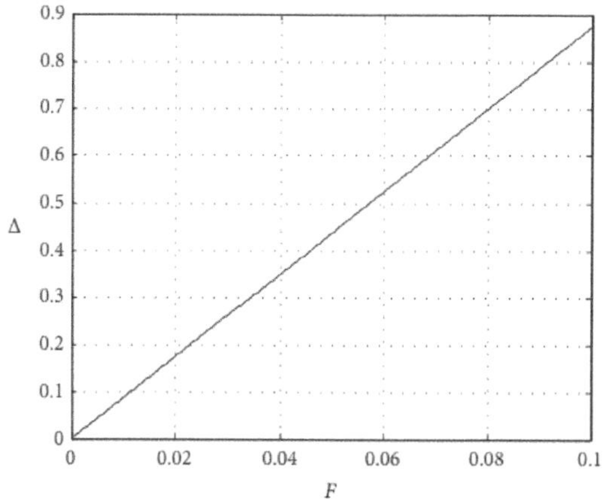

Figure 6: Position of the block at the end of the slip versus the amplitude of the external force for the parameters $\alpha = 0.01$, $\omega_0 = 2$, and $\tau_0 = 30$.

MAGNITUDE OF THE EARTHQUAKE AND PROBABILITY OF AN EVENT

In this section, we evaluate the power of the earthquake (i.e., the magnitude) and the probability of an event when the system under consideration is defined by Figure 1. As the magnitude is concerned, various expressions have been proposed within which some consider the physical properties of the rocks [18–21] and others take into account the parameters characterizing the seismic waves [22, 23]. Hereafter, we work in the former case with the magnitude expression given by

$$M = \frac{2}{3}\left[\log_{10}\left(\frac{M_0}{1N\cdot m}\right) - 9.1\right] \quad (20)$$

and the probability of an event defined in Gutenberg and Richter [24] by

$$W = 10^{(e - h\times\log M_0)} \quad (21)$$

In these quantities, M_0 represents the seismic moment introduced by [25, 26]: $M_0 = \mu\cdot S\cdot\Delta$ in which μ is the rocks rigidity, S characterizes the surface broken during the earthquake, and Δ designates the displacement of the block during the motion or its position after the seism. Since the broken surface and the rock rigidity are unknown, we consider the explicit forms of expressions (20) and (21) defined in [27]:

$$M = \frac{2}{3}\left[\log_{10}\left(K\cdot\Delta^3\right) - 9.1\right] \quad (22a)$$

$$W = 10^{e - h\times\log(K\cdot\Delta^3)} \quad (22b)$$

in which e and h are arbitrary constants and K is a constant which can be determined by exploiting the experimental curve giving the magnitude as a function of displacement [28].

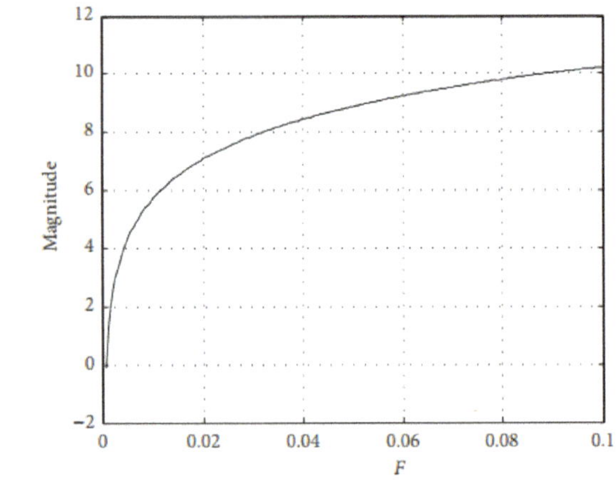

Figure 7: Magnitude of the seism as a function of the amplitude of the external force for the parameters $\alpha = 0.01$, $\omega_0 = 2$, $\tau_0 = 30$, and $k = 1.2589254 \times 10^{16}$.

To appreciate the influence of the external force on these two quantities, we draw each of them as a function of F. Figure 7 shows that the magnitude of the earthquake increases with the amplitude of the external force while the probability of an event decreases with the growth of F (Figure 8). We could also note from these plots that, for small values of F, the appearing probability of the event is significant, but earthquakes which occur are not powerful since their magnitudes are very small.

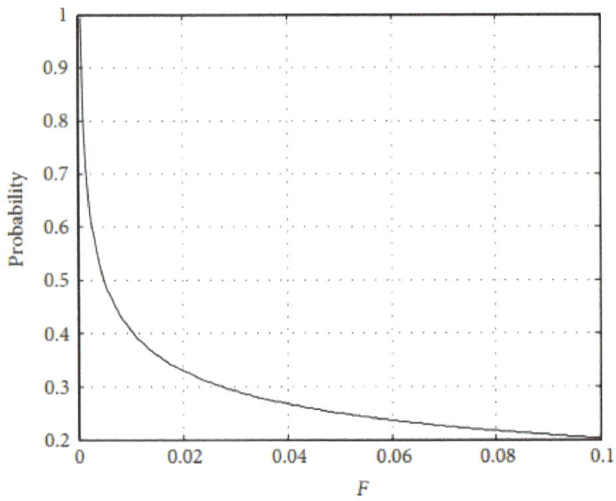

Figure 8: Probability of the slip versus the amplitude of the magma thrust strength for the parameters of Figure 7 with $e = 0.9$ and $h = 0.1$.

DISCUSSIONS AND CONCLUSION

In this paper, we have examined the dynamics of a 1D spring-block model for earthquakes subjected to cumulative strengths due to the motion of the tectonic plates and the upflow of magma during volcanism. The presence of the horizontal spring during the modeling has allowed for the introduction of a new parameter called the sound velocity. Through the direct integration method, we have explained the impacts of the magma thrust strength (amplitude F, frequency ω) on the dynamics of the earthquakes. Indeed, we have shown that the displacement of the block is oscillatory and its amplitude increases with the increment of F and the sound velocity when $\omega = \emptyset$. We have also demonstrated that when the external frequency ω equals the pseudo frequency of the block's oscillation, the slip which occurs during this periodic movement is a stick-slip motion instead of a creep motion like that previously established [1]. It has appeared from our analytical investigations that the magma thrust strength deletes the first order phase transition observed in [1]. We have also noted that the power of the earthquake increases with the amplitude of the magma load while the corresponding event probability decreases with it. The model studied in this paper represents a generalization of previous models about the behavior of a springblock model subjected to the strengths due to the relative motion between two plates since it takes into account the flow of magma during volcano emission. Based on the physical interpretations arising from this study on the mechanisms of the earthquakes' dynamics under magma thrust strengths, we hope the results established within this paper could be also valuable as an interesting application that can be used to illustrate this sort of concepts to graduate students. In solving (17), we have limited investigations to the resonance case in which the frequency of the magma upflow equals the fundamental frequency of the spring block. In the corresponding nonresonance case, this equation cannot be solved analytically. Numerical integration of this relation is under consideration.

CONFLICT OF INTERESTS

There is no conflict of interests concerning the work done in this paper.

REFERENCES

1. G. L. Vasconcelos, "First-order phase transition in a model for earthquakes," Physical Review Letters, vol. 76, no. 25, pp. 4865–4868, 1996.
2. R. Montagne and G. L. Vasconcelos, "Complex dynamics in a one-block model for earthquakes," Physica A: Statistical Mechanics and its Applications, vol. 342, no. 1-2, pp. 178–185, 2004.
3. B. Erickson, B. Birnir, and D. Lavallée, "A model for aperiodicity in earthquakes," Nonlinear Processes in Geophysics, vol. 15, no. 1, pp. 1–12, 2008.
4. R. Burridge and L. Knopoff, "Model and theoretical seismity," Bulletin of the Seismological Society of America, vol. 57, pp. 341–371, 1967.
5. G. L. Vasconcelos, M. de Sousa Vieira, and S. R. Nagel, "Phase transitions in a spring-block model of earthquakes," Physica A: Statistical Mechanics and Its Applications, vol. 191, no. 1–4, pp. 69–74, 1992. · ·
6. M. De Sousa Vieira, G. L. Vasconcelos, and S. R. Nagel, "Dynamics of spring-block models: tuning to criticality," Physical Review E, vol. 47, no. 4, pp. R2221–R2224, 1993.
7. P. G. Akishin, M. V. Altaisky, I. Antoniou, A. D. Budnik, and V. V. Ivanov, "Burridge-Knopoff model and self-similarity," Chaos, Solitons & Fractals, vol. 11, no. 1–3, pp. 207–222, 2000. · · View at MathSciNet ·
8. E. Brittany, Bjorn, and D. Lavallée, Periodicity, Chaos

and Localisation in Burridge-Knopoff Model of an Earthquake with Dieterich-Ruina Friction, Center for Complex and Nonlinear Science, Santa Barbara, Calif, USA, 2010.

9. S. A. Fedotov, "Magma rates in feeding conduits of different volcanic centres," Journal of Volcanology and Geothermal Research, vol. 9, no. 4, pp. 379–394, 1981.
10. B. A. Chouet, "Excitation of a buried magmatic pipe: a seismic source model for volcanic tremor," Journal of Geophysical Research, vol. 90, no. 2, pp. 1881–1893, 1985.
11. Y. Ida and M. Kumazawa, "Ascent of magma in a deformable vent," Journal of Geophysical Research, vol. 91, no. B9, pp. 9297–9301, 1986. ·
12. B. A. Chouet, "A seismic model for the source of long period events and harmonic tremor," in Volcanic Seismology, P. Gasparini, R. Scarpa, and K. Aki, Eds., vol. 3 of IAVCEI Proceedings in Volcanology, pp. 133–156, Springer, 1992.
13. Y. Ida, "Cyclic fluid effusion accompanied by pressure change: implication for volcanic eruptions and tremor," Geophysical Research Letters, vol. 23, no. 12, pp. 1457–1460, 1996.
14. M. Kwékam, J.-P. Liégeois, E. Njonfang, P. Affaton, G. Hartmann, and F. Tchoua, "Nature, origin and significance of the Fomopéa Pan-African high-K calc-alkaline plutonic complex in the Central African fold belt (Cameroon)," Journal of African Earth Sciences, vol. 57, no. 1-2, pp. 79–95, 2010.
15. C. H. Scholz, The Mechanics of Earthquakes and Faulting, Cambridge University Press, New York, NY, USA, 1990.
16. P. Hähner and Y. Drossinos, "Nonlinear dynamics of a continuous spring-block model of earthquake faults," Journal of Physics A: Mathematical and General, vol. 31, no. 10, pp. L185–L191, 1998.
17. J. S. Langer and C. Tang, "Rupture propagation in a model of an earthquake fault," Physical Review Letters, vol. 67, no. 8, pp. 1043–1046, 1991.
18. H. Kanamori and D. L. Anderson, "Theoretical basis of some empirical relations in seismology," Bulletin of the Seismological Society of America, vol. 65, pp. 1073–1095, 1975.
19. T. C. Hanks and H. Kanamori, "A moment magnitude scale," Journal of Geophysical Research B: Solid Earth, vol. 84, no. 5, pp. 2348–2350, 1979.
20. H. Kanamori, H.-K. Thio, D. Dreger, E. Hauksson, and T. Heaton, "Initial investigation of the Landers, California, earthquake of 28 June 1992 using TERRAscope," Geophysical Research Letters, vol. 19, no. 22, pp. 2267–2270, 1992.
21. T. Utsu, "Relationships between magnitude scales," in International Handbook of Earthquake, Engineering Seismology, W. H. K. Lee, H. Kanamori, P. C. Jennings, and C. Kisslinger, Eds., vol. 81-A of International Geophysics, pp. 733–746, Academic Press, Division of Elsevier, 2002.
22. C. F. Richter, "An instrumental earthquake magnitude scale," Bulletin of the Seismological Society of America, vol. 25, pp. 1–32, 1935.
23. B. Gutenberg, "Amplitudes of surface waves and magnitude of shallow earthquakes," Bulletin of the Seismological Society of America, vol. 35, pp. 3–12, 1945.
24. B. Gutenberg and C. F. Richter, "Earthquake magnitude, intensity, energy, and acceleration," Bulletin of the Seismological Society of America, vol. 46, pp. 105–146, 1956.
25. A. V. Vvendenskaya, "The determination of displacement fields by means of dislocation theory," Izvestiya Akademii Nauk SSSR, Seriya Geograficheskaya, pp. 227–284, 1956.
26. A. Keiti, "Estimation of earthquakes moment, released energy and stress-strain drop from wave spectrum," Bulletin of the Earthquake Research Institute, vol. 44, pp. 23–88, 1966.
27. M. W. Dongmo, L. Y. Kagho, F. B. Pelap, G. B. Tanekou, Y. L. Makenne, and A. Fomethe, "Water effects on the first-order transition in a model of earthquakes," ISRN Geophysics, vol. 2014, Article ID 160378, 7 pages, 2014. ·
28. P. Pascal, Qu›est-ce qui fait trembler la terre ?EDP Science, 2003.

CHAPTER 9

VOLCANO-TECTONIC EARTHQUAKE FORECASTS BASED ON A MOVING MEAN RECURRENCE TIME ALGORITHM

Alicia García, Servando De la Cruz-Reyna, José M. Marrero, and Ramón Ortiz

INTRODUCTION

Volcano-tectonic (VT) earthquakes are considered to be lowmagnitude compared to tectonic earthquakes, even though in some cases (three in the twentieth century) they have reached M $\geq$ 7 (Zobin, 2001). Volcanic earthquakes occurring near active subduction zones, which may often reach magnitudes in the range 4–6, rarely cause the same degree of damage or concern among the population that stronger tectonic earthquakes cause. However, in volcanic regions with low tectonic activity, and consequently with a higher vulnerability to strong ground motions, events exceeding M4 may produce significant damage. Such a situation may become quite serious when that type of VT activity occurs in volcanic islands with unstable topographies, prone to landslides, as is the case of El Hierro, the westernmost, youngest (< 2 Ma, Ancochea et al., 1994), and smallest (269 km2) of the Canary Islands. In general, earthquake forecasting requires answers to four questions: where, when, how likely, and how large? In VT earthquake forecasting, the answer to where the earthquake is is constrained by the extent of the volcanic system. The answer to the second question on timing is related to the onset and evolution of a volcanic crisis. Such crises are generally characterized by VT seismic swarms, lasting an average of 5.5 days in the case of El Hierro, concentrated in regions within or around the volcanic system, and generally defined as "a sequence of events closely clustered in time and space" (Benoit and McNutt, 1996). Since such swarms may include

Citation: García, A., De la Cruz-Reyna, S., Marrero, J. M., and Ortiz, R.: volcano-tectonic earthquake forecasts based on a moving mean recurrence time algorithm, Nat. Hazards Earth Syst. Sci., 16, 1135-1144, doi:10.5194/nhess-16-1135-2016, 2016.

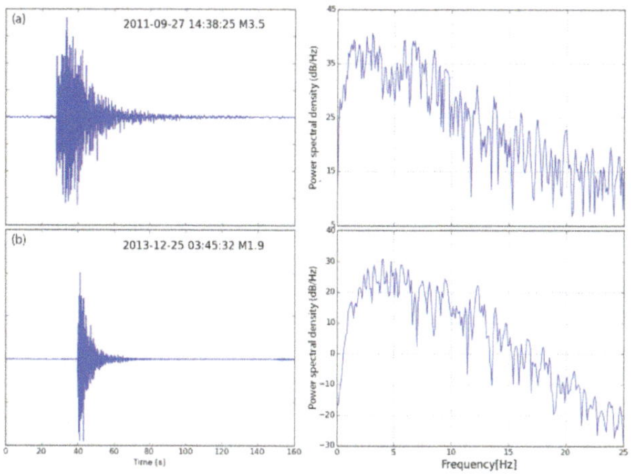

Figure 1. Characteristic VT earthquakes of the El Hierro seismo-volcanic process and their spectra. (a) Event recorded on 27 September 2011, before the eruption, at ~ 6.3 km to the SE of the island centre. (b) Event recorded on 24 December 2013, after the eruption at ~ 10.4 km SW of the island centre.

events of damaging magnitudes, it is important to detect the non-random components of the swarm evolution, as it contains factors linking different stages of the process that may allow major VT events to be forecast. We assume that such "causal factors" are related to the nature of the magmatic intrusion process at depth that involves an increasing accumulation of localized stress, which produces the VT earthquake when released. In particular, the magmatic intrusions may produce clustered sources of stress in which pressure increases over relatively short timescales.

We picture the situation at El Hierro as voluminous bodies of dense magma intruding into the mantle under the crust from a deeper magma generation region (Martínez-Arevalo et al., 2013). That magma is dense enough to remain in almost isostatic equilibrium at the base of the crust in a process of magmatic underplating (Leahy et al., 2010). As it evolves under the local conditions at mantle-crust boundary depths, the diffusive assimilation of crustal minerals causes an increase in the magma buoyancy over the timescale of centuries as the assimilation diffusivity is at least 1 order of magnitude higher than the thermal diffusivity (McLeod and Sparks, 1998; Portnyagin et al., 2008). The increasing buoyancy causes additional stresses and deformations in the overlying crust. Such processes may induce a causative evolution of the seismicity, reflected as an increased rate of seismic energy release, and a changing distribution in the magnitudes of the VT events, as discussed in the following sections. However, at this point, it is important to note that the analysis of such evolution may enable scientists to forecast the occurrence of major VT events on the short-term timescale; nevertheless, in some cases this analysis may not foretell the occurrence of an eruption, as the magma injection and migration process that causes the stress concentration may never reach the surface (Bell and Kilburn, 2012), although an increase in the probability of such an event must be considered. Next we present this methodology, developed and used during the El Hierro volcanic process (2011–2015), which is aimed at detecting the precursory behaviour of the seismicity before the occurrence of major, potentially damaging VT earthquakes.

EL HIERRO VOLCANIC PROCESS

The morphology of El Hierro island (27.7◦ N; 18.0◦ W) has been attributed to a triple volcanic rift with the NE, NW, and S axes of these rifts diverging by about 120◦ from each other (Carracedo et al., 1999). The island landscape displays abundant dikes and fissures (Gee et al., 2001), as well as major landslides (Masson et al., 2002). The subaerial island displays a high concentration of volcanoes with a spatial distribution that may be explained by the model proposed by Stroncik et al. (2009), in which the volcanic activity at El Hierro is controlled by a complex array of magma pockets. Magma injections from the pockets through dikes or sills may generate VT seismicity. In July 2011, a sudden increase of seismicity was detected at El Hierro island, prompting several Spanish institutions to deploy recording instrumentation. In addition to the official seismic network operated by the Instituto Geográfico Nacional (IGN), in September 2011, the IGEO-CSIC Institute (see authors' affiliations) installed, for research purposes, a seismic network consisting of three permanent, real-time, vertical, short-period stations, and five similar, temporary, manual-access stations.

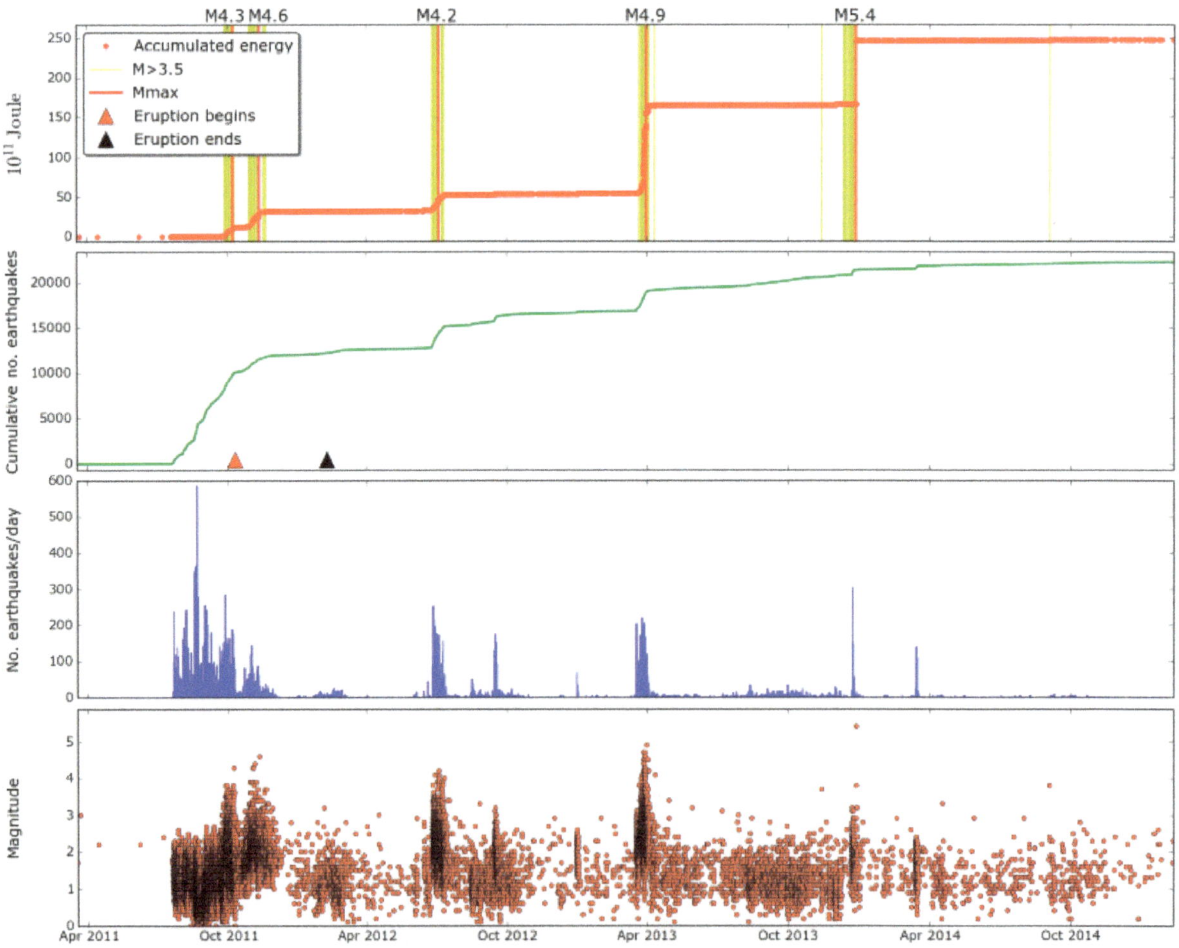

Figure 2. Evolution of the seismic activity at El Hierro island between 18 March 2011 and 12 February 2015. Data are from the IGN public earthquake catalogue and the IGEO-CSIC seismic research network.

Seismicity, deformations and diffuse gas emissions increased persistently until October 2011, when a submarine eruption at 27◦37.180 N; 17◦59.580 W was first detected (López et al., 2012; Ibáñez et al., 2012; Prates et al., 2013). The eruptive activity persisted for months, until March 2012. However, neither the seismic activity nor the deformations stopped, and have in fact continued to the time of this submission (García et al., 2014b, a). During the whole episode (2011–2015), only two types of seismicity have been observed: short-period volcano-tectonic (VT) earthquakes, with well-defined P and S phases, and a syn-eruptive long-period tremor, patently related to the submarine magma effusion. No long-period (LP) earthquakes have been observed at any time. Here, we analyse the time evolution of the VT earthquakes occurrence rates, as they seem to have similar origins, related to rapid stress concentrations caused by magma intrusions.

The VT events do not show significant changes in their waveforms or spectral contents along the duration of the whole episode, as illustrated in Fig. 1. The evolution of the seismic activity at El Hierro island for the whole period of unrest (2011–2015) is summarized in Fig. 2, in terms of the time distribution of earthquake magnitudes, the number of earthquakes exceeding M1.5 per day, the cumulative number of events, and the cumulative seismic energy released, estimated from the IGN public earthquake catalogue (http://www.ign.es/ign/layoutIn/ sismoFormularioCatalogo.do as downloaded on 29 August 2015), and from records of the above-mentioned IGEOCSIC seismic research network. The uncertainties in calculating the duration and magnitude of very low-amplitude signals recorded by the IGEO-CSIC have been treated according to Del Pezzo et al. (2003). Inspection of Fig. 2 reveals that the seismicity, deformations, and eruption evolution observed during the 2011 unrest episode not only are consistent with the model of Stroncik et al. (2009) (García et al., 2014b), but also reveal some other interesting features of the volcanic unrest process

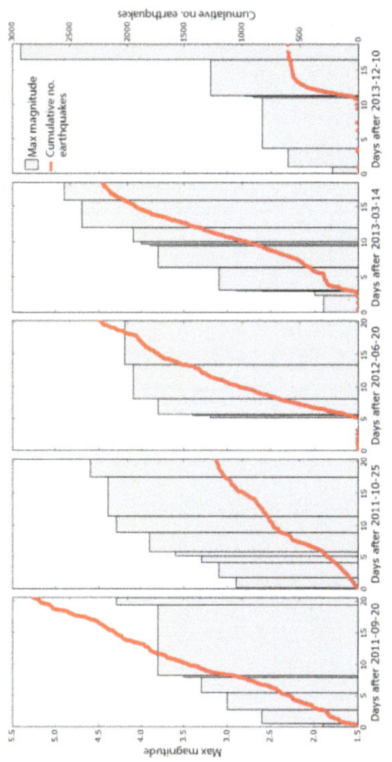

Figure 3. Time evolution of the daily count of VT events (red dots) and maximum magnitudes (gray bars) during the earthquake swarms of the El Hierro seismo-volcanic process.

dynamics. First, as expected, a marked rise of the VT event occurrence rate preceded the onset of the 10 October 2011 eruption. However, during the continuing seismic unrest, the event rate and the cumulative number of events have also markedly increased before the occurrence of major VT earthquakes; but it is the cumulative release of seismic energy, shown in the uppermost plot of Fig. 2, that is the parameter that best illustrates the precursory character before major VT events. Additionally, the M–T diagram in the lower plot of Fig. 2 shows a peculiar behaviour. The minimum magnitude rises as the number of higher-magnitude events increases. This may be caused by completeness fluctuations of the catalogue, derived from the difficulties in counting very small events when larger earthquakes are concurrent, or it may be an actual condition related to the physical process of seismic energy release.

Several other cases in which the completeness is maintained at the lowest measurable value during the duration of the VT swarm suggest that the latter condition may be real. For example, Benoit and McNutt (1996) report an increase of the minimum magnitude when the VT activity approaches a mean magnitude of 3. Likewise, De la CruzReyna et al. (2008) and Hurst et al. (2014) report a similar behaviour for the VT seismicity of Popocatépetl and Tongariro volcanoes respectively. To date, five seismo-volcanic crises or VT earthquake swarms have been identified at El Hierro (García et al., 2014b, a), showing an increasing trend of the maximum detected magnitudes in all cases, as shown in Fig. 3. It is significant that the acceleration of the seismic energy release rate begins in all cases when the cumulative seismic energy exceeds the 10^{11}J energy threshold criterion of Yokoyama (1988).

Hours to days after the onset of this acceleration stage, the swarm sequences culminated with earthquakes exceeding a magnitude of 4 (Fig. 4). Currently, VT earthquake swarms are classified according to three family types proposed by Benoit and McNutt (1996), and later updated by Farrell et al. (2009) and Brumbaugh et al. (2014): type A – mainshock and aftershocks; type B – foreshocks, mainshock, and aftershocks; and type C – swarms without a mainshock. Further examination of the lower part of Fig. 3 reveals that the evolution of the daily rate of seismicity does not correspond to any of these established earthquake families. We thus introduce a fourth family, type D, characterized by an increasing number of growing magnitude foreshock events culminating with a mainshock and followed by a rapid decay of seismicity lasting just a few hours. The differences between these models are illustrated in Fig. 5. Therefore, a type D sequence does not follow an ETAS model (Ogata, 1992), nor does it follow the Omori law (Utsu and Ogata, 1995), but it reveals a causal process that allows major VT events to be forecast.

FORECASTING MAJOR VT EARTHQUAKES AT EL HIERRO

The Gutenberg-Richter law (GRL, Gutenberg and Richter, 1944) scales the seismic activity, with respect to its magnitude, as

$$\log_{10}(N(M)) = a - bM \tag{1}$$

where N (M) is the number of earthquakes with magnitude M or larger detected within a given region, in a certain time interval (Bender, 1983). In a process that remains stationary over a time interval ΔT, the Gutenberg-Richter parameters (GRP) a and b remain constant. Here, we use a timedependent GRP to estimate the mean recurrence time (MRT) of volcanic earthquakes having or exceeding given magnitudes, and thus the likelihood of major events occurring on the shorter timescale (hours to days). The MRT (τ_T) between events with magnitude equal to or greater than M may thus be estimated from Eq. (1) as

$$N(M) = \frac{\Delta T}{\tau_T} = 10^{(a-bM)} \tag{2}$$

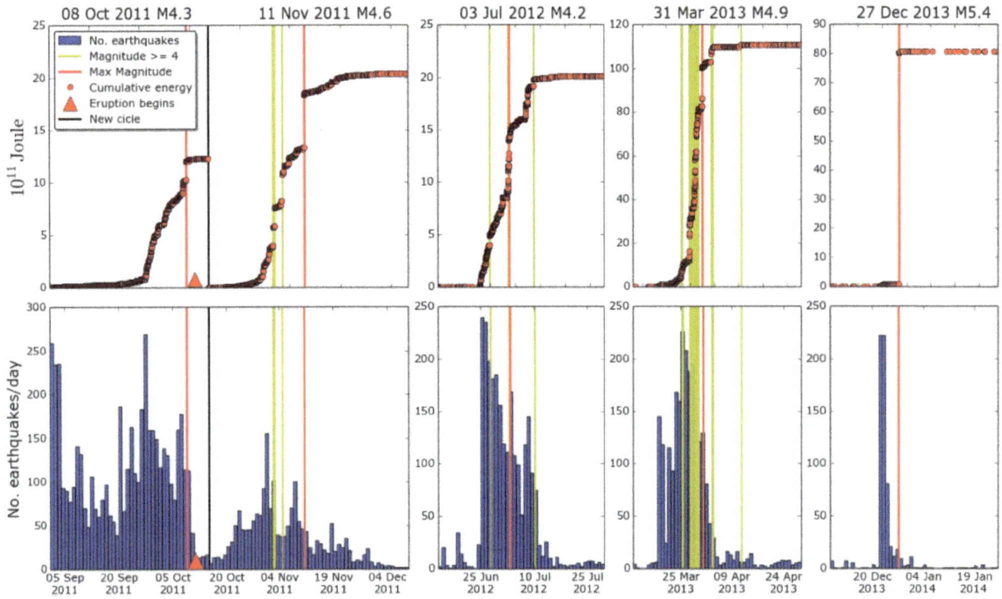

Figure 4. Seismo-volcanic crises identified at El Hierro island (2011–2014). The upper plots show the evolution of seismic energy for each of the cycles of unrest. The lower plots show the daily event count. The plots illustrate how the seismic activity drops rapidly after the largest magnitude earthquakes. Particularly significant is the absence of aftershocks after the largest (M5.4) earthquake of 27 December 2013. Data are from IGN public earthquake catalogue and the IGEO-CSIC seismic research network.

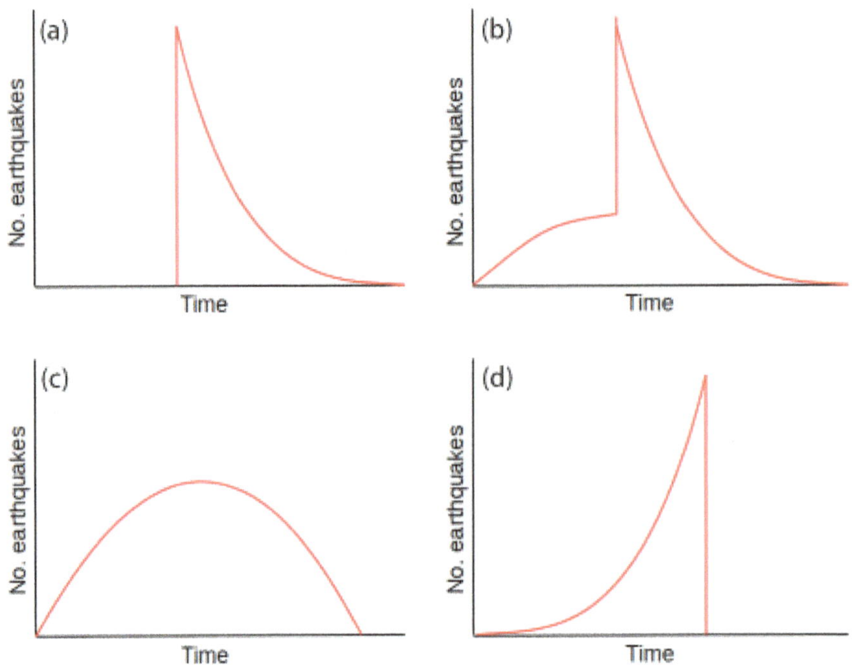

Figure 5. Seismic swarms are usually classified in three family types (A – panel a, B – panel b, C – panel c, Farrell et al., 2009; Brumbaugh et al., 2014). From our results at El Hierro in 2011– 2015, we propose a fourth family type D (panel d), characterized by an increasing number of VT earthquakes and their magnitudes with time, and by a quick decay of the seismicity, in a matter of hours after a mainshock usually exceeding M4.

The MRT may thus be calculated as

$$\tau_T = \Delta T \times 10^{(bM-a)} \tag{3}$$

Increasing likelihood of major events may thus be estimated from the increasing rates of occurrence of smaller earthquakes. In practice this may prove to be a difficult task since the available seismic catalogues frequently present completeness uncertainties derived from technical factors such as the stability of the seismic networks and links, weather conditions, background noise, event counting, and magnitude assignment routines, among others. Problems of catalogue completeness in general mean an undercounting of the lowest magnitudes (Bell and Kilburn, 2012) that define a lowend magnitude cutoff (MC) above which the GRL holds. A second major problem is the bias introduced by the use of binned magnitudes and errors of the magnitudes (Bender, 1983; Tinti and Mulargia, 1985, 1987; Marzocchi and Sandri, 2003). These problems have been amply discussed in the literature about seismicity from different origins (e.g. Wiemer et al., 1998; Lombardi, 2003; Marzocchi and Sandri, 2003; Helmstetter et al., 2007; Amorese et al., 2010; BengoubouValerius and Gibert, 2013; Alamilla et al., 2015; MárquezRamírez et al., 2015). In our case, an additional problem is that the development of the VT seismicity in the El Hierro crisis shows strong and rapid variations, and the estimates for the MC and the GRP may vary in time and space, causing temporary distributions of shocks quite different from those of the full catalogue. We have addressed these problems considering a moving and variable time window 1T in Eqs. (2) and (3). In most cases, an acceleration of the seismicity occurs when the cumulative energy released by a VT swarm reaches a value near the 1011 J energy threshold of Yokoyama (1988). Experience has shown that at El Hierro such swarms precede a stage of accelerated seismicity with increasing magnitudes that culminate in major (M > 4) VT events

Table 1. Characteristics of El Hierro swarms

Mainshock date	Magnitude		Swarm duration (days)	Number of events
	Mean	Max		
8 Oct 2011	2.33	4.3	10	934
11 Nov 2011	2.27	4.6	8	354
3 Jul 2012	2.35	4.2	5	635
31 Mar 2013	2.75	4.9	4	651
27 Dec 2013	2.09	5.4	4	466

The MRT algorithm starts a forecast when a swarm of at least 200 earthquakes with $M \geq 1.5$ are detected in a time span of 5 days. Then, a completeness magnitude MC is calculated using the maximum curvature method (MCM, Wiemer and Wyss, 2000). The GRP are estimated for each time window using a maximum likelihood estimator (MLE) (Peishanet al., 2003; Bengoubou-Valerius and Gibert, 2013). In the final standard-C version of the software, we implemented afast subroutine based on the equations of Marzocchi and Sandri (2003) to calculate the b value)

$$b = \frac{1}{\ln(10)\Delta M}\ln(p) \tag{4}$$

where

$$p = 1 + \frac{\Delta M}{\mu - M_c} \tag{5}$$

μ is the sampling average of the magnitudes and Mc is the cut-off (threshold) magnitude. We used a magnitude binning interval 1M = 0.1. If the seismicity rate drops below 200 earthquakes in 5 days, the MRT algorithm would stop running. In successive swarms, some patterns appear as the time to a mainshock approaches: M_c may increase from the initial 1.5 to as much as 3 or more, and the time window shortens. Algorithm 1 shows a schematic description (pseudocode) of the MRT calculation. The algorithm has been implemented in a batch processing mode that is fast and relatively simple to program. Table 1 shows the main features of the swarms detected at El Hierro, and Fig. 6 shows the evolution of the MRT for the initial four cycles of El Hierro activity.

Algorithm 1 Pseudocode of the MRT algorithm developed to forecast mainshocks applied during the El Hierro seismo-volcanic process (2011–2015).

```
Get seismic catalog ← download from IGN website
Count number of events ← In the last 5 days
if earthquakes M1.5 > 200 events then
  Compute current MC ← Magnitude Cutoff, using
  Maximum Curvature Method (MCM)
  Compute Time window ← Use constant MC
  Compute number of events where M>MC
  if Number of events (M>MC) > 200 events then
    Select all events where M>MC
    Compute Gutenberg-Richter Parameters ← Using
    Maximum Likelihood Estimator (MLE)
    Compute Mean-Recurrence-Time (MRT) for M4 events
    if MRT for M4 events < 10 days then
      Activate Warning Window
    end if
  end if
end if
```

A warning window for the likely occurrence of an earthquake exceeding M4 is issued when the calculated MRT for that magnitude drops below 10 days, and a statement saying that there is an increased likelihood of an M4 (or higher) earthquake in the following 10 days is delivered to the decision makers. The warning remains active until the MRT returns to a value exceeding 10 days. This method has been incorporated into a volcanic alert system, (described in Garcia et al., 2014a), where the continuous evolution of the MRT over the whole unrest process is described. An 8-month lull followed those four cycles, and as frequently happens, a belief that the seismovolcanic crisis had ended spread throughout the island's population. Additional factors, such as a particularly strong winter storm and the Christmas holidays, caused failures of the IGN official seismic network that translated into a rather incomplete seismic catalogue. When a new seismic swarm began on 20 December 2013, the MRT could only be calculated predominantly using data from the IGEOCSIC seismic research network. Figure 7 shows the MRT evolution using magnitudes from both catalogues. On 24 December 2013 a new warning was issued communicating to decision makers the increased likelihood of an earthquake M5 or larger occurring in the following days. On 27 December 2013, at 17:46:02 UTC, the largest earthquake of the entire episode, with magnitude 5.4, caused numerous landslides across the island, fortunately without casualties. It is interesting to note that no M4 earth-quakes occurred in this stage.

DISCUSSION AND CONCLUSIONS

Earthquake forecasting is a problem that has been addressed in the past from various different perspectives. Significant advances were obtained after the introduction of complexity theory (Lorenz, 1963; Mandelbrot, 1977) and the development of predictive algorithms (Gabrielov et al., 1986; Molchan, 1990; Keilis-Borok, 1996; Turcotte, 1997). Currently, the need to implement operational short-term earthquake forecasting based on predictive and forecasting algorithms has been stated by Jordan et al. (2011). The MRT is one of such algorithms, based on a property that is typical of complex systems: seismic events cluster according to a simple scaling law, the Gutenberg-Richter relationship (Eq. 1). This empirical relationship contains information about the occurrence of large magnitudes embedded in the number of small-magnitude events (Helmstetter et al., 2007), and the time variations of the distribution of the small-magnitude events provides information about the timescale of largemagnitude occurrences. In the case of VT earthquakes, two additional factors help to simplify the forecasting process: one is the limited geographic extent of the volcanic source of seismicity, confining the spatial extension of the forecast. The other factor is the short timescale, measured in days, at which orderly fluctuations of the GRPs evolve reflecting stress accumulations induced by the magma injections, and allowing rapid recognition of significant changes of the MRT as defined in Eq. (3), and therefore viable short-term forecasts (García et al., 2014a). In contrast, a similar method for tectonic earthquakes would require recognizing orderly evolution patterns of the GRPs over timescales of years to decades, making short-term forecasting non-viable (e.g. Jordan et al., 2011).

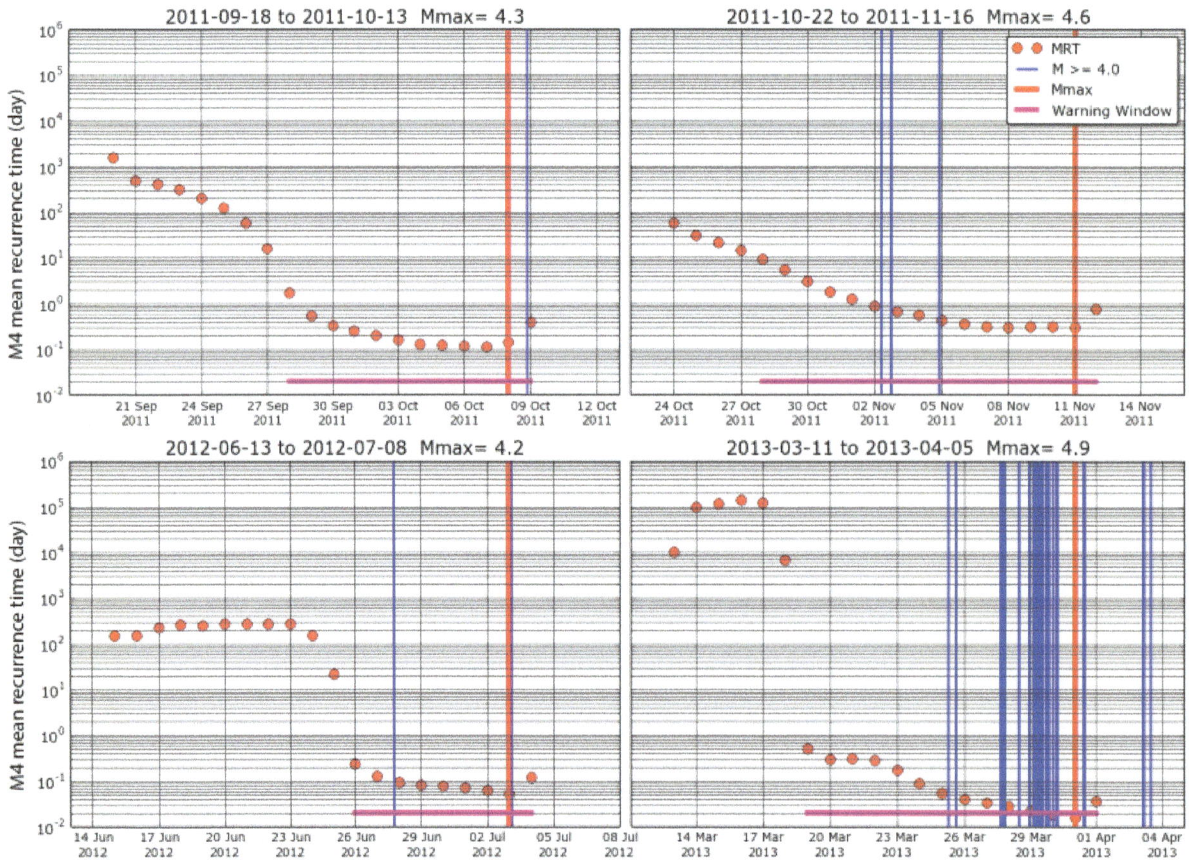

Figure 6. MRT evolution (computed using MLE) for four periods: 18 August 2011 to 10 October 2011; 22 October 2011 to 13 November 2011; 13 June 2012 to 5 July 2012; and 11 March 2013 to 2 April 2013. The warning windows were triggered when MRT dropped below 10 days.

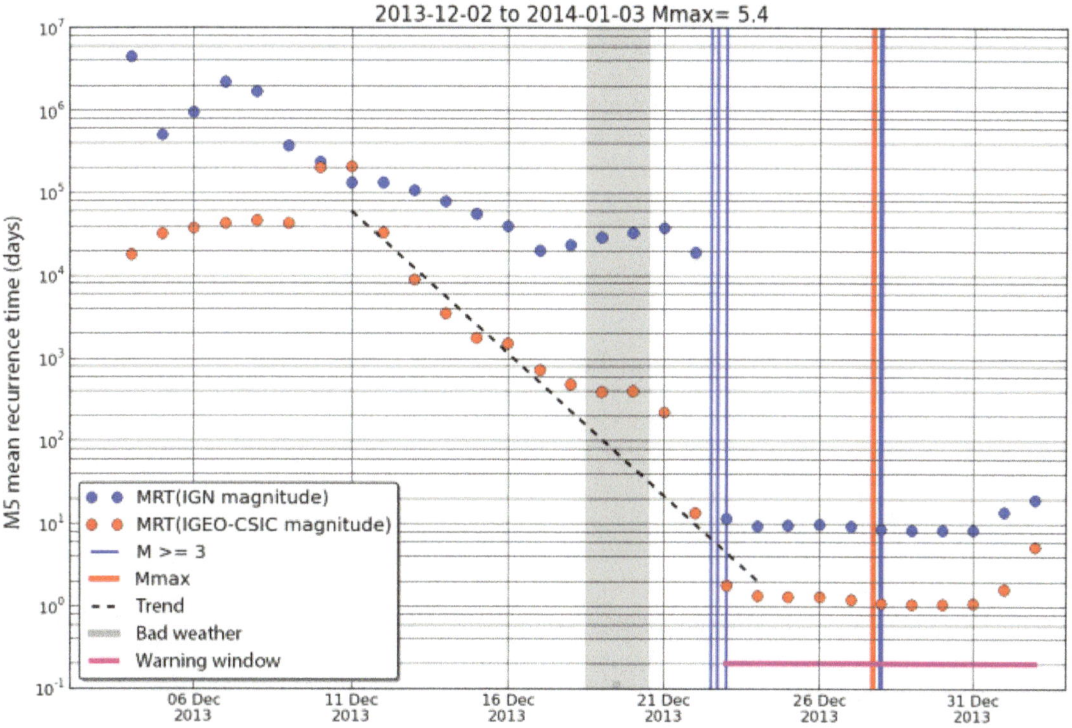

Figure 7. MRT evolution (computed using MLE) in December 2013 calculated from the IGN official catalogue (blue) and from the IGEOCSIC catalogue (red). The magnitude of the earthquake on 27 December 2013 17:46:02 was estimated by IGN as 5.1, and as 5.4 by the IGEO-CSIC network and other international catalogues.

We envision the El Hierro process as the result of magma injections causing rapid stress concentrations capable of triggering seismic swarms. Such swarms culminate with mainshocks that release stress concentrations, and are thus not followed by aftershock sequences. Further magma migrations cause new swarms and mainshocks, in repeating, but nonperiodic cycles of activity. For example, the largest earthquakes of the March and December 2013 swarms (see Table 1) were located less than 20 km west of the westernmost shore of the island, suggesting a nearby localized stress source possibly resulting from subsurface magma injections under that region (Fig. 8). From the initial stage of unrest in July 2011, five earthquake swarm cycles have been identified, but only the first of these culminated in an eruption. The subsequent cycles show two main features: first, the swarms are separated by increasingly longer lull intervals, and second, they tend to culminate in mainshocks of increasing magnitudes (see Fig. 3).

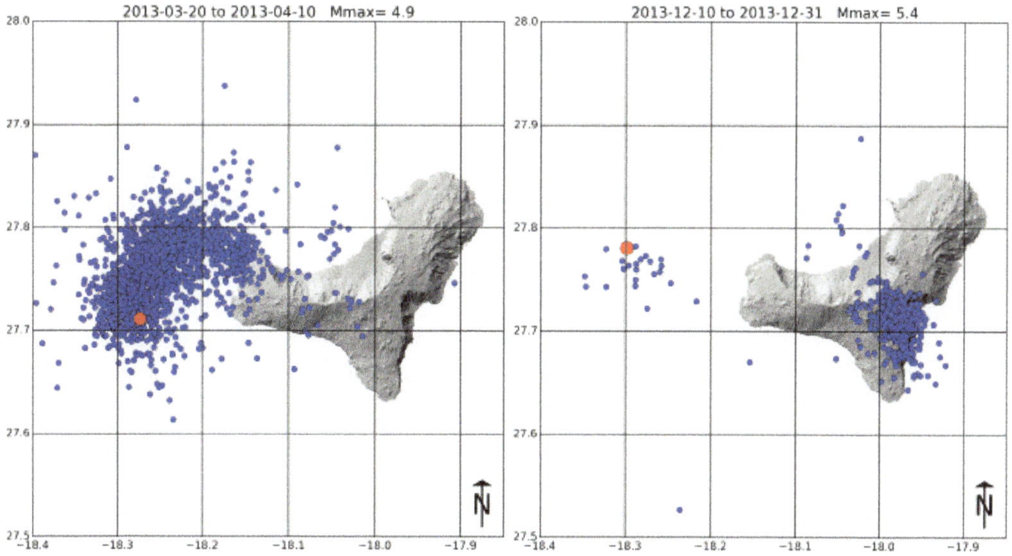

Figure 8. Swarms and mainshock epicentres of the March and December 2013 seismicity cycles. The M5.4 mainshock of 27 December 2013, at 17:46:02 UTC, is located in the same region of the March 2013 cycle.

The forecasting algorithm we are proposing here is based on the above-mentioned envisioning of the process that is causing the seismic swarms, and begins when the background seismic energy exceeds the 1011 J energy threshold criterion of Yokoyama (1988). The acceleration of the earthquake rate (Fig. 5d) is interpreted in terms of piecewise estimations of the Gutenberg-Richter parameters. The variations of those parameters are, in turn, translated as time windows of increased likelihood of a large-magnitude earthquake. The MRT algorithm may be, and has been, used in real time. It took the initial VT swarm crises to interpret the process and to recognize and develop the procedures and set the parameters of the MRT algorithm, which was first tested during the third crisis, when a significant reduction of the MRT was detected on 5 May 2012. A M4.2 VT earthquake occurred on 3 July 2012. From then on, the MRT algorithm ran continuously in the background. It was not until the fourth crisis that a formal warning was issued on 26 March 2013 communicating to the island authorities and to the Civil Protection officials the increased probability of a M5 earthquake for the following 8 days, as described in García et al. (2014a).

On 31 March 2013 a M4.9 earthquake occurred. In the next volcano-tectonic seismic crisis another formal warning was issued on 22 December 2013, communicating the increased probability of an earthquake exceeding M5 in the next 15 days. Further formal warnings were issued daily afterwards also indicating the increased probability of landslides and other eruption. On 27 December 2013, an M5.4 caused numerous landslides (García et al., 2014a). In the absence of swarms, the GRPs are calculated with a long database of the seismic catalogue, and updated on a daily basis (Algorithm 1), usually yielding fluctuating, but always very long values of the MRT (that may exceed 106 days). At the onset of a seismic swarm, the MRT drops rapidly to values of the order of 103 days. When this happens the GRPs are calculated from shorter time windows of variable duration, provided that they contain a large enough number of events (at least 200) to yield stable values of the GRPs. When the MRT for events of magnitude 4 or higher is less than 10 days, a warning is issued to the decision makers, expressed as a 10-day window of increased likelihood for larger, potentially damaging earthquakes. Up to the time of this submission, all the recent earthquakes of $M \geq 4$ at El Hierro have been forecasted several days in advance. Application of the MRT to other volcanoes may be relatively simple to implement, provided that the observed seismicity corresponds to swarms of type D (Fig. 5): increasing number of VT events with increasing magnitudes followed by a mainshock, and by a rapid decay of the seismicity afterwards. The algorithm may be readily adapted to any real-time monitoring system. In addition, the concept of time windows for the increased likelihood of a large-magnitude earthquake should be easily understood and accepted by governmental decision makers, making the management of risk more effective.

ACKNOWLEDGEMENTS

This research has been funded by projects from the CSIC (2011-30E070) and MINECO (CGL2011-28682- C02-01). The authors are grateful to the DGAPA-PAPIIT-UNAM program for their support. We used seismic data from the IGN public website (http://www.02.ign.es/ign/layout/sismo.do, public website, ©Instituto Geográfico Nacional). S. Malone, an anonymous reviewer and the editor, G. Macedonio, provided valuable comments that helped to clarify and strengthen various parts of the manuscript. We are also indebted to the Cabildo Insular de El Hierro, and its three municipalities (Valverde, El Pinar de El Hierro, and Frontera) for their support. We wish to thank all of the people living on El Hierro for their encouragement and understanding of our scientific work.

REFERENCES

1. Alamilla, J. L., Vai, R., and Esteva, L.: Completeness assessment of earthquake catalogues under uncertain knowledge, J. Seismol., 19, 27–40, doi:10.1007/s10950-014-9448-x, 2015.
2. Amorese, D., Grasso, J.-R., and Rydelek, P. A.: On varying b-values with depth: results from computer-intensive tests for Southern California, Geophys. J. Int., 180, 347–360, doi:10.1111/j.1365- 246X.2009.04414.x, 2010.
3. Ancochea, E., Hernán, F., Cendrero, A., Cantagrel, J. M., Fúster, J. M., Ibarrola, E., and Coello, J.: Constructive and destructive episodes in the building of a young Oceanic Island, La Palma, Canary Islands, and genesis of the Caldera de Taburiente, J. Volcanol. Geoth. Res., 60, 243–262, doi:10.1016/0377- 0273(94)90054-X, 1994.
4. Bell, A. F. and Kilburn, C. R.: Precursors to dyke-fed eruptions at basaltic volcanoes: insights from patterns of volcano-tectonic seismicity at Kilauea volcano, Hawaii, B. Volcanol., 74, 325– 339, doi:10.1007/s00445-011-0519-3, 2012.
5. Bender, B.: Maximum likelihood estimation of b values for magnitude grouped data, B. Seismol. Soc. Am., 73, 831–851, 1983.
6. Bengoubou-Valerius, M. and Gibert, D.: Bootstrap determination of the reliability of b-values: an assessment of statistical estimators with synthetic magnitude series, Nat. Hazards, 65, 443–459, doi:10.1007/s11069-012-0376-1, 2013.
7. Benoit, J. P. and McNutt, S. R.: Global volcanic earthquake swarm database 1979–1989. Open-file report 96-69, US Department of the Interior, US Gological Survey, 1996.
8. Brumbaugh, D., Hodge, B., Linville, L., and Cohen, A.: Analysis of the 2009 earthquake swarm near Sunset Crater volcano, Arizona, J. Volcanol. Geoth. Res., 285, 18–28, doi:10.1016/j.jvolgeores.2014.07.016, 2014.
9. Carracedo, J. C., Day, S. J., Guillou, H., and Pérez-Torrado, F. J.: Giant Quaternary landslides in the evolution of La Palma and El Hierro, Canary Islands, J. Volcanol. Geoth. Res., 94, 169–190, doi:10.1016/S0377-0273(99)00102-X, 1999.
10. De la Cruz-Reyna, S., Yokoyama, I., Martinez-Bringas, A., and Ramos, E.: Precursory seismicity of the 1994 eruption of Popocatepetl Volcano, Central Mexico, B. Volcanol., 70, 753– 767, doi:10.1007/s00445-008-0195-0, 2008.
11. Del Pezzo, E., Bianco, F., and Saccorotti, G.: Duration magnitude uncertainty due to seismic noise: inferences on the temporal pattern of GR b-value at Mt.

Vesuvius, Italy, B. Seismol. Soc. Am., 93, 1847–1853, doi:10.1785/0120020222, 2003.

12. Farrell, J., Husen, S., and Smith, R. B.: Earthquake swarm and b-value characterization of the Yellowstone volcanotectonic system, J. Volcanol. Geoth. Res., 188, 260–276, doi:10.1016/j.jvolgeores.2009.08.008, 2009.
13. Gabrielov, A., Dmitrieva, O., Keilis-Borok, V., Kossobokov, V., Kutznetsov, I., Levshina, T., Mirzoev, K., Molchan, G., Negmatullaev, S., Pisarenko, V., Prozorov, A., Rinheart, W., Rotwain, I., Shelbalin, P., Shnirman, M., and Schreider, S.: Algorithms of long-term earthquakes' prediction, international school on research oriented to earthquake prediction–algorithms, software and data handling, Centro Regional de Sismología para América del Sur, Lima, Peru, 1986.
14. García, A., Berrocoso, M., Marrero, J. M., Fernández-Ros, A., Prates, G., De la Cruz-Reyna, S., and Ortiz, R.: Volcanic alert system (VAS) developed during the 2011–2014 El Hierro (Canary Islands) volcanic process, B. Volcanol., 76, 825, doi:10.1007/s00445-014-0825-7, 2014a.
15. García, A., Fernández-Ros, A., Berrocoso, M., Marrero, J., Prates, G., De la Cruz-Reyna, S., and Ortiz, R.: Magma displacements under insular volcanic fields, applications to eruption forecasting: El Hierro, Canary Islands, 2011–2013, Geophys. J. Int., 196, 1–13, doi:10.1093/gji/ggt505, 2014b.
16. Gee, M. J., Watts, A. B., Masson, D. G., and Mitchell, N. C.: Landslides and the evolution of El Hierro in the Canary Islands, Mar. Geol., 177, 271–293, 2001.
17. Gutenberg, B. and Richter, C. F.: Frequency of earthquakes in California, B. Seismol. Soc. Am., 34, 185–188, 1944.
18. Helmstetter, A., Kagan, Y. Y., and Jackson, D. D.: Highresolution time-independent grid-based forecast for M 5 earthquakes in California, Seismol. Res. Lett., 78, 78–86, doi:10.1785/gssrl.78.1.78, 2007.
19. Hurst, T., Jolly, A. D., and Sherburn, S.: Precursory characteristics of the seismicity before the 6 August 2012 eruption of Tongariro volcano, North Island, New Zealand, J. Volcanol. Geoth. Res., 286, 294–302, doi:10.1016/j.jvolgeores.2014.03.004, 2014.
20. Ibáñez, J. M., De Angelis, S., Díaz-Moreno, A., Hernández, P., Alguacil, G., Posadas, A., and Pérez, N.: Insights into the 2011-2012 submarine eruption off the coast of El Hierro (Canary Islands, Spain) from statistical analyses of earthquake activity, Geophys. J. Int., 191, 659–670, doi:10.1111/j.1365- 246X.2012.05629.x, 2012.
21. Jordan, T. H., Chen, Y. T., Gasparini, P., Madariaga, R., Main, I., Marzocchi, W., Papadopoulos, G., Sobolev, G., Yamaoka, K., and Zschau, J.: Operational Earthquake Forecasting. State of Knowledge and Guidelines for Utilization, Ann. Geophys.-Italy 54, 315–391, doi:10.4401/ag-5350, 2011.
22. Keilis-Borok, V.: Intermediate-term earthquake prediction, P. Natl. Acad. Sci. USA, 93, 3748–3755, 1996.
23. Leahy, G. M., Collins, J. A., Wolfe, C. J., Laske, G., and Solomon, S. C.: Underplating of the Hawaiian Swell: evidence from teleseismic receiver functions, Geophys. J. Int., 183, 313–329, doi:10.1111/j.1365-246X.2010.04720.x, 2010.
24. Lombardi, A. M.: The maximum likelihood estimator of bvalue for mainshocks, B. Seismol. Soc. Am., 93, 2082–2088, doi:10.1785/0120020163, 2003.
25. López, C., Blanco, M. J., Abella, R., Brenes, B., RodríguezCabrera, V. M., Casas, B, Cerdeña, I. D., Felpeto, A., de Villalta, F., del Fresno, C., García, O., García-Arias, M., GarcíaCañada, L., Gomis-Moreno, A., González-Alonso, E., GuzmánPérez, J., Iribarren, I., López-Díaz, R., Luengo-Oroz, N., Meletlidis, S., Moreno, M., Moure, D., Pereda de Pablo, J., Rodero, C., Romero, E., Sainz-Maza, S., Sentre-Domingo, M., Torres, P., Trigo, P., and Villasante-Marcos, V.: Monitoring the volcanic unrest of El Hierro (Canary Islands) before the onset of the 2011-2012 submarine eruption, Geophys. Res. Lett., 39, L13303, doi:10.1029/2012GL051846, 2012.
26. Lorenz, E. N.: Deterministic nonperiodic flow, J. Atmos. Sci., 20, 130–141, 1963. Mandelbrot, B. B.: Fractals: Form, Chance and Dimension, W. H. Freeman and Co., San Francisco, 1977.
27. Márquez-Ramírez, V., Nava, F., and Zúñiga, F.: Correcting the Gutenberg–Richter b-value for effects of rounding and noise, Earthquake Science, 28, 129–134, doi:10.1007/s11589-015- 0116-1, 2015.
28. Martínez-Arevalo, C., Mancilla, F. d. L., Helffrich, G., and García, A.: Seismic Evidence of a Regional Sublithospheric Low Velocity Layer beneath the Canary Islands, Tectonophysics, 608, 586– 599, doi:10.1016/j.tecto.2013.08.021, 2013.
29. Marzocchi, W. and Sandri, L.: A review and new insights on the estimation of the b-value and its uncertainty, Ann. Geophys.-Italy, 46, 1271–1282, 2003.
30. Masson, D., Watts, A., Gee, M., Urgeles, R., Mitchell, N., Bas, T. L., and Canals, M.: Slope failures on the flanks of the western Canary Islands, Earth-Sci. Rev., 57, 1–35, 2002.
31. McLeod, P. and Sparks, R. S. J.: The dynamics of xenolith assimilation, Contrib. Mineral. Petr., 132, 21–33, doi:10.1007/s004100050402, 1998.
32. Molchan, G.: Strategies in strong earthquake prediction, Phys. Earth Planet. In., 61, 84–98, doi:10.1016/0031-9201(90)90097- H, 1990.
33. Ogata, Y.: Detection of precursory relative quiescence before great earthquakes through a statistical model, J. Geophys. Res.-Sol. Ea., 97, 19845–19871, doi:10.1029/92JB00708, 1992.
34. Peishan, C., Tongxia, B., and Baokun, L.: The b-Value and Earthquake Occurrence Period, Chinese J. Geophys., 46, 736–749, doi:10.1002/cjg2.3393, 2003.
35. Portnyagin, M., Almeev, R., Matveev, S., and Holtz, F.: Experimental evidence for rapid water exchange between melt inclusions in olivine and host magma, Earth Planet. Sc. Lett., 272, 541–552, doi:10.1016/j.epsl.2008.05.020, 2008.

36. Prates, G., García, A., Fernández-Ros, A., Marrero, J. M., Ortiz, R., and Berrocoso, M.: Enhancement of sub-daily positioning solutions for surface deformation surveillance at El Hierro volcano (Canary Islands – Spain), B. Volcanol., 75, 724, doi:10.1007/s00445-013-0724-3, 2013.
37. Stroncik, N., Klügel, A., and Hansteen, T.: The magmatic plumbing system beneath El Hierro (Canary Islands): constraints from phenocrysts and naturally quenched basaltic glasses in submarine rocks, Contrib. Mineral. Petr., 157, 593–607, doi:10.1007/s00410-008-0354-5, 2009.
38. Tinti, S. and Mulargia, F.: Effects of magnitude uncertainties on estimating the parameters in the Gutenberg-Richter frequencymagnitude law, B. Seismol. Soc. Am., 75, 1681–1697, 1985.
39. Tinti, S. and Mulargia, F.: Confidence intervals of b-values for grouped magnitudes, B. Seismol. Soc. Am., 77, 2125–2134, 1987.
40. Turcotte, D. L.: Fractals and chaos in geology and geophysics, Cambridge University Press, 1997.
41. Utsu, T. and Ogata, Y.: The centenary of the Omori formula for a decay law of aftershock activity, J. Phys. Earth, 43, 1–33, 1995.
42. Wiemer, S. and Wyss, M.: Minimum magnitude of completeness in earthquake catalogs: examples from Alaska, the western United States, and Japan, B. Seismol. Soc. Am., 90, 859–869, doi:10.1785/0119990114, 2000.
43. Wiemer, S., McNutt, S. R., and Wyss, M.: Temporal and threedimensional spatial analyses of the frequency–magnitude distribution near Long Valley Caldera, California, Geophys. J. Int., 134, 409–421, doi:10.1046/j.1365-246x.1998.00561.x, 1998.
44. Yokoyama, I.: Seismic energy releases from volcanoes, B. Volcanol., 50, 1–13, doi:10.1007/BF01047504, 1988. Zobin, V. M.: Seismic hazard of volcanic activity, J Volcanol. Geoth. Res., 112, 1–14, doi:10.1016/S0377-0273(01)00230-X, 2001.

CHAPTER 10

CONNECTION OF LARGE EARTHQUAKES OCCURRING MOMENT WITH THE MOVEMENT OF THE SUN AND THE MOON AND WITH THE EARTH CRUST TECTONIC STRESS CHARACTER

M. K. Kachakhidze, R. Kiladze, N. Kachakhidze, V. Kukhianidze, and G. Ramishvili

INTRODUCTION

It is necessary to assume that if it is not the variable stress in the earth (inter earth, meteorological or space origin) the earthquake will occur at the moment, when tectonic stress in the zone of its preparation reaches the certain critical meaning – the limit of the geological medium strength.

The presence of the variable stress in the Earth considerably changes this situation. These stresses being added on the smooth increasing tectonic stress can prematurely bring a total stress up to critical meaning and thus provoke earthquake at a rather early stage of its preparation. From this point of view we may think that earthquakes are not "ripen" up to the end and do not occur themselves. The moment of the earthquakes occurring depends on the joint action of the factors causing the variable stress in the Earth.

By their nature these initiative factors can be divided into two kinds – internal (endogenous) and external (exogenous). It is acceptable that internal forces oriented along tectonic stresses may have the most important influence on the seismicity of region. In order to check this suggestion we studied this work. Many scientific papers are devoted to the problem of the role of the cosmic factors in the earthquakes occurring process (Harris, 2000; Lopes et al, 1990; Nikolaev, 1995; Nikolaev et al, 1996; Rydelek et al, 1992; Tanaka et al, 2006; Tanaka, 2007; Tsuruoka et al, 1995; Van Ruymbeke et al, 2007).

In some cases there are shown the Sun perturbations and in other cases – the Moon perturbation components in- fluence on the earthquakes occurring moments. There are papers where the Sun and the Moon perturbations influence on the earthquakes are not fixed at all (Davidson, 2002; Davidson, 2005). The task of determination of the exogenous action on the Earth from the main celestial bodies must be solved for any point and any moment of the time. The fact that tides and deformations are depended on the geographic coordinates must be taken into account (Kiladze et al, 1997, 2005).

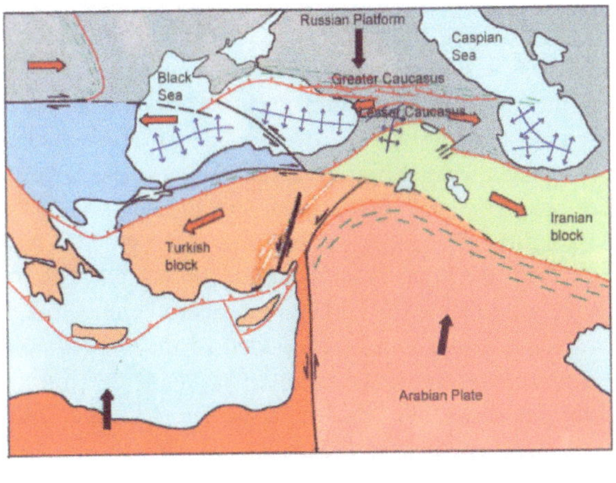

Figure 1. Tectonic stress map of Caucasus and neighbouring regions. Large red arrows are the approximate directions of motion of different parts of Caucasus and neighbouring regions relative to Eurasia the black arrow correspond to the compression field.

Citation: Kachakhidze, M., Kiladze, R., Kachakhidze, N., Kukhianidze, V., & Ramishvili, G. (2010). Connection of large earthquakes occurring moment with the movement of the Sun and the Moon and with the Earth crust tectonic stress character. Nat. Hazards Earth Syst. Sci., 10(7), 1629-1633. http://dx.doi.org/10.5194/nhess-10-1629-2010.

We must note that the task of determination of direction of tectonic stress action is solved with less accuracy. The field of the tectonic stress has the hierarchical structure. As it is known the different total tectonic stress acts in the different regions because of plates movements particularities. It is established that in the young folded areas like the Caucasus the field of tectonic stress is characterized by the sharp anisotropy with the predominance of the compression perpendicular to the trend of folding. Spatial location of the main positive and negative geotectonic morphostructures of the Caucasus shows the existence of the wavy tectonic movements in the region. They are caused by the horizontal compression, provoked evidently by advancement of the Arabian lithosphere plate to the North and its reaproachment with the Euroasian plate (Fig. 1). All these cause considerable deformation of the lithosphere of the Caucasian region and it is breaking up in separate blocks. This, in its turn, causes the concentration of stress along the boundaries of the blocks and rising of earthquakes focuses there (Jackson, 1992; Kachakhidze et al., 2001)

Basing on the geological and geophysical data we can pick out in Caucasus:

1. Interzonal deep faults of the Caucasian orientation, which divide the main tectonic units of the region.
2. Interzonal longitudinal and transversal faults.
3. Transzonal faults of the Anticaucasian and submeridional trend.

At present these faults determine heterogeneity of the Caucasian stress field, which is mainly characterized by the shear stress, directed diagonally towards the main structures and finally-by the tensile stress, directed subaltitudinally, parallel with the main structures.

As we mentioned above, in every region, including the Caucasus, besides of geological factors it may be the exogenous (cosmic) force of certain direction, which takes role of triggering factor in the occurring process of large earthquakes, if tectonic stress is near to its critical value. It is known that any planet is not only under the Sun attraction but under other bodies of the Solar system also. As the planets of the Solar system have the mass many times less than the Sun, the Sun and the Moon are considered as the "main bodies" of perturbation. Thus, the stress acted on the Earth from the solar system and, in our case, on the Caucasian region, can play a role of the triggering factor only in case when its direction helps to increase tectonic stress. At the same time this additional stress acts on the neighboring region of course (for instance on Turkey and Arabian Peninsula) but tectonic stress has different direction there. Because of it the force acted on the Caucasus region, from the Solar system bodies, can not play the role of the triggering factor for Turkey or Arabian Peninsula. On the contrary, the exogenous stress added to the Turkey tectonic stress appears as earthquakes initiative factor for Turkey and can not play the role of the triggering effect for the Caucasus and Arabian Peninsula earthquakes.

THE METHOD

Unfortunately, it is not possible to determine value of the force caused by plate's movement and the task of N – body is not solved analytically yet. Consequently, we are made to investigate earthquakes exogenous triggering factors by numerical methods. We consider earthquakes occurring moments connection with the Sun, the Moon attractive and gravity perturbation components of tide which will be called in future as "initiative agents" of the earthquakes. Vertical (Z) and horizontal (D) components of an attraction are following:

$$F_Z = G\frac{mM}{R^2}\cos Z \qquad (1)$$

$$F_D = G\frac{mM}{R^2}\sin Z\cos(A-D) \qquad (2)$$

where G – is gravitational constant, R – distance up to the celestial body, Z and A – its zenith distance and azimuth and D – azimuth of a horizontal projection of the force. Zenith distance and azimuth of the celestial body are calculated by the formulas:

$$\cos Z = \sin\phi\sin\delta + \cos\phi\cos\delta\cos t \qquad (4)$$

$$\tan A = \frac{\cos\delta\sin t}{\sin\delta\cos\phi - \cos\delta\sin\phi\cos t} \qquad (5)$$

where φ – is latitude of the terrain, δ – declination of celestial body and t – its hour angle.

Table 1. Azimuths of the perturbation forces

Triggering factor	Azimuth A of the force direction	$\frac{n-\bar{n}}{\sigma}$	Validity of the phenomena
Perturbation from the Sun	193°–206°	2.35	98%
Perturbation from the Moon	295°–310°	3.28	99.8%
Total perturbation	290°–307°	1.72	91.5%

Tides caused by the celestial body, are calculated by the formulas:

$$\Delta F_z = \frac{G\,r\,m\,M}{R^3}\left(1 - 3\cos^2 Z\right) \quad (6)$$

$$\Delta F_D = -3\frac{G\,r\,m\,M}{R^3}\sin Z\cos Z\cos(A - D) \quad (7)$$

where r – is the radius of the Earth.

DISCUSSION

We considered all 393 Caucasus earthquakes with $M \geq 4.5$. Histogram of earthquakes recurrence frequencies were constructed based on these observed frequencies at different intervals of azimuths (we considered frequencies of recurrence of azimuths in intervals of tides possible directions). The interval of possible azimuths was divided into 18 parts. In addition to it three histograms are created for the Sun, the Moon tides and for their total perturbations (Fig. 2a, b, and c). In Table 1 there are shown azimuths of the force directions (for the Sun, the Moon and total) deviation from the average meaning of which is more then 2σ. There are given: triggering factor, the meanings of the azimuths of the celestial bodies at the moment of the occurring of the significant quantity of the earthquakes, the relative detour from the average meaning and validity of the cases. Such four directions are revealed in all. In these cases the quantity of the earthquakes are more than their average significances. Sectorial effect is obviously shown in the Moon gravity perturbations and in the total gravity perturbations of the Sun and the Moon (Fig. 2b and c). The Moon gravity perturbation effect is given in the Fig. 2. It is shown the heterogeneity in the azimuth sector 295°–310° , but the total perturbation effect – in the sector 290°–307° (Fig. 2c). It is obvious the great share of the Moon in the total effect. It is clear that the each possible "agent" has only one azimuth, to which the increased quantity of earthquakes corresponds (Table 1). In order to more detail analysis of task the Moon, the Sun and their total perturbation directions were calculated and in- flicted on the Caucasus main faults map for each earthquake separately. Only three of them are shown in the Fig. 3 like examples (Fig. 3a, b, and c).

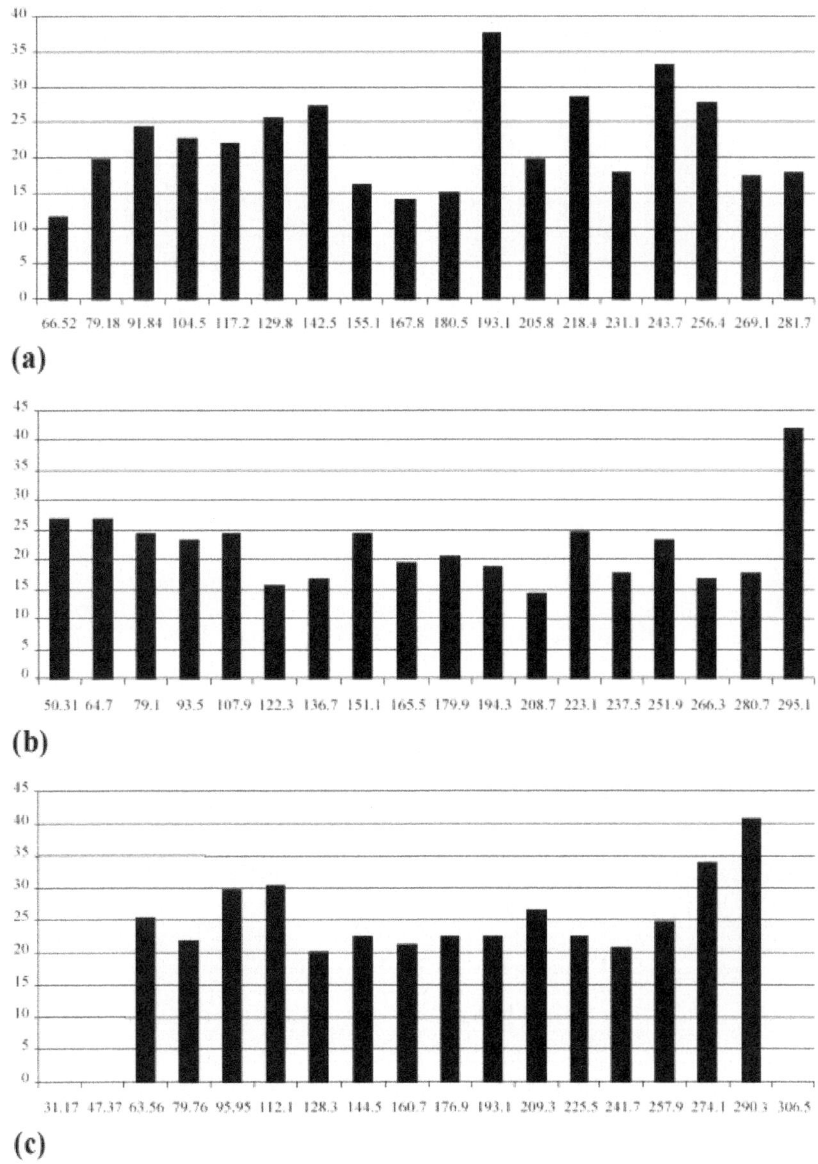

Figure 2. Histograms: (a) Sun gravity perturbation; (b) Moon gravityg.2. Histograms: a) Sun gravity perturbation; b) Moon gravity perturbation; c) the total (the sun and the moon) gravity perturbation.

CONCLUSIONS

By comparing of getting results with Caucasus tectonic map we may do conclusion that the majority quantity of Caucasus M≥4.5 earthquakes occur when:

1. Gravitational field and the Sun perturbation are perpendicular to the plane of Russian platform (i.e., they are perpendicular to fault along Caucasian mountain ridge) and they are directed from the North to the South, in particular, create 20◦–30◦ angles with 40◦ meridian
2. Lunar and total perturbations coincide with the direction of the tectonic stress, approximately, from SouthEast to North-West (along the above mentioned fault in the Caucasian seismoactive region) and is known as the stretching stress. They create 60◦ angle with 40◦ meridian.

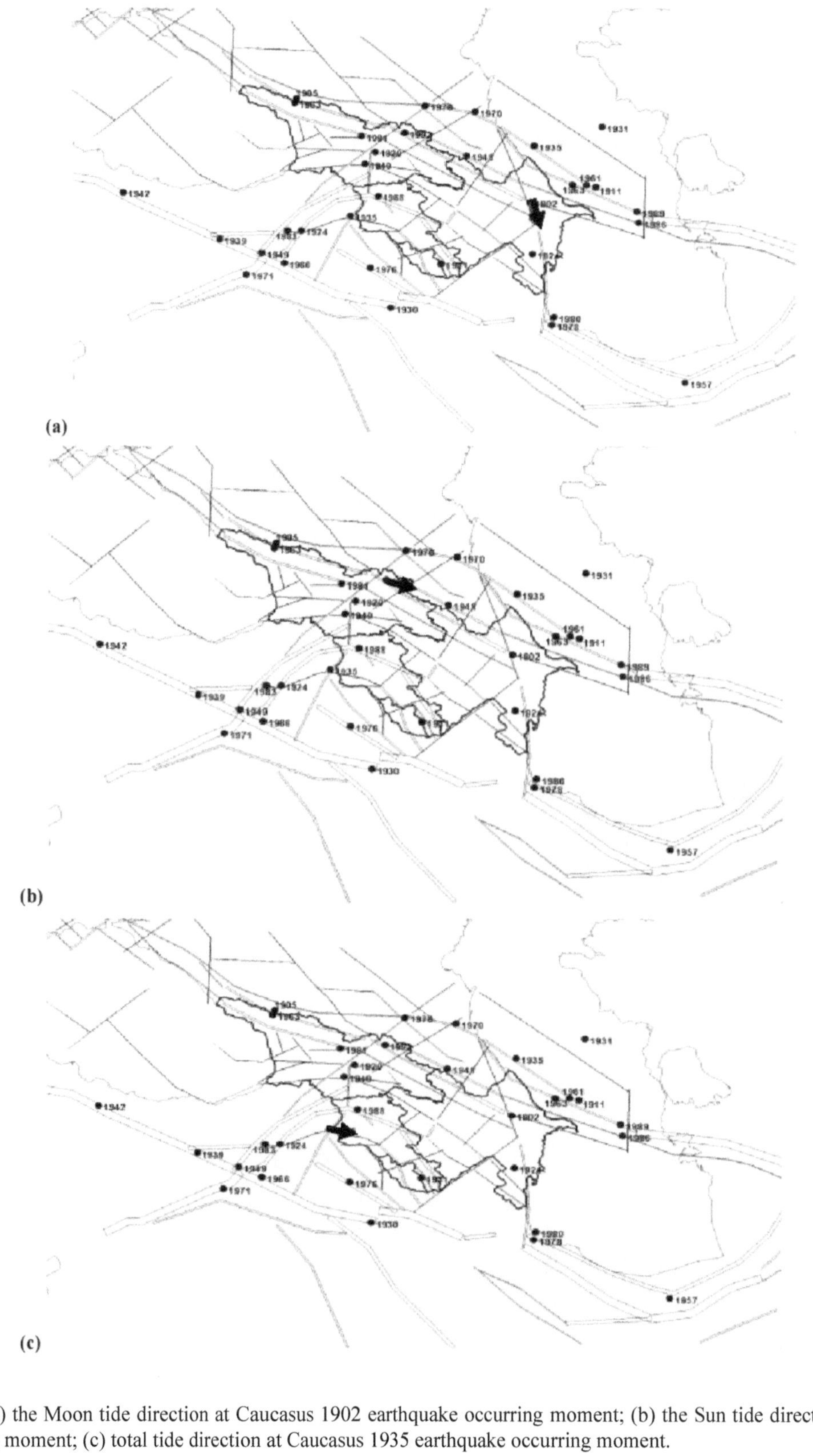

Figure 3. Arrows: (a) the Moon tide direction at Caucasus 1902 earthquake occurring moment; (b) the Sun tide direction at Caucasus 1992 earthquake occurring moment; (c) total tide direction at Caucasus 1935 earthquake occurring moment.

Thus, based on Caucasus region earthquakes, we conclude that, in case of consideration of the regional tectonic stress picture, we must take into account above described distributions of the celestial bodies on the heaven and relevant time intervals as the periods of increasing danger of large earthquakes occurring in the region.

REFERENCES

1. Davidson, S.: Solar and Lunar Triggers on Earthquakes and Volcanic Eruptions, Gravity Concepts, Sec. 70, 175–181, 2 March 2002.
2. Davidson, S.: High Tides Trigger Earthquakes, Gravity Concepts, Sec. 30, Rev. 1, 2005. Harris, R. A.: Earthquake stress triggers, stress shadows, and seismic hazard, in: Special Section: Seismology, Current Science, 79(9), 1215–1225, 2000.
3. Jackson, J.: Partitioning of Strike – Slip and Convergent Motion Between Eurasia and Arabia in Eastern Turkey and the Caucasus, J. Geophys. Res., 97(89), 12.471–12.473, 10 August 1992.
4. Kachakhidze, M., Kachakhidze, N., Kiladze, R., Kukhianidze, V., and Ramishvili, G.: Tectonic sources of caucasus strong earthquakes, Nat. Hazards Earth Syst. Sci., 4, 53–58, doi:10.5194/nhess-4-53-2004, 2004.
5. Kiladze, R., Kachakhidze, M., Kukhianidze, V., Kachakhidze, N., and Ramishvili, G.: Study of possible relationships between strong earthquakes and astronomical phenomena for seismoactive regions of the Caucasus, Turkey and Greece, Bulletin of Georgian Academy of Sciences, 155(N3), 367–369, 1997.
6. Kiladze, R., Kachakhidze, M., Kukhianidze, V., Kachakhidze, N., and Ramishvili, G.: A Search of Possible Relations Between Large Earthquakes and Astronomic Phenomena: the Caucasian Seismic region, J. Vulkanol. Seismol., (N3), 75–80, 2005.
7. Lopes, R. M. C., Malin, S. R., Mazzarella, A., and Palumbo, A.: Lunar and solar triggering of earthquakes, Phys. Earth. Planet. In., 59, 127–129, 1990.
8. Nikolaev, A. V.: On the earthquake triggering by the Earth's tide and underground nuclear explosions, in: Earthquakes induced by underground nuclear explosions, NATO ASI Series, 2-Environment, 255–266, 1995.
9. Nikolaev, A. V. and Nikolaev, V. A.: Lithosperic stress state in South Americans inferred from tidal triggering of erthquakes, Geophys. J. Int., 35(N3), 329–339, 1996.
10. Rydelek, P. A., Sacks, I. S., and Scarpa, R.: On tidal triggering of earthquakes at Campi Flegrei, Italy, Geophys. J. Int., 109, 125– 137, 1992.
11. Tanaka, S., Sato, H., Matsumura, S., and Ohtake, M.: Tidal triggering of earthquakes in the subducting Philippine Sea plate beneath the locked zone of the plate interface in the Tokai region, Japan, Tectonophysics, 417, 69–80, 2006.
12. Tanaka, S.: Tidal triggering of earthquakes precursory to the recent Sumatra megathrust earthquakes of 26 December 2004 (Mw9.0), 28 March 2005 (Mw8.6), and 12 September 2007(Mw8.5), Geophys. Res. Lett., 37(2), L02301, doi:10.1029/2009GL041581, 2010.
13. Tsuruoka, H., Ohtake, M., and Sato, H.: Statistical test of the tidal triggering of earthquakes: contribution of the ocean tide loading effect, Geophys. J. Int., 122, 183–194, 1995.
14. van Ruymbeke, M., Zhu, P., Cadicheanu, N., and Naslin, S.: Very Weak Signals (VWS) detected by stacking method according to different astronomical periodicities (HiCum), Nat. Hazards Earth Syst. Sci., 7, 651–656, doi:10.5194/nhess-7-651-2007, 2007.

CHAPTER 11

NEOTECTONICS, FLOODING PATTERNS AND LANDSCAPE EVOLUTION

U. Lombardo

INTRODUCTION

Palaeo-ecological reconstructions are a fundamental step in order to understand the origins of modern landscapes and the potential changes they could undergo due to climate change. Understanding how a particular region or ecosystem responded to past human disturbances and changes in past climate can tell us much about its resilience (Tinner and Ammann, 2005; Bush and Silman, 2007; Dearing, 2008). At the same time, understanding the processes behind the formation of modern landscapes and ecosystems can help assess to what extent these are "natural" or anthropogenic, eventually contributing to better informed conservation policies (Angermeier, 2000; Willis and Birks, 2006; Vegas-Vilarrúbia et al., 2011). These considerations are particularly true for the Llanos de Moxos (LM) (Fig. 1), which cover most of the savannah of southern Amazonia. Since the beginning of the Holocene this region has experienced both climate and human induced environmental changes (Mayle et al., 2000; Whitney et al., 2011; Urrego et al., 2013; Lombardo et al., 2013a, b). The LM, located between the central Andes and the Brazilian Shield, is one of the largest seasonally flooded savannahs of South America. This region covers most of the southern part of the Amazon Basin and represents about 10 % of the wetlands of South America (Junk, 2013). The forest–savannah ecotone is largely determined by seasonal floods (Mayle et al., 2007). The patchwork of savannahs and forests that covers the LM is a key element for the survival of its rich biodiversity, including rare and threatened species (Herzog et al., 2012; Wallace et al., 2013). The recent designation of three new protected areas has made the LM the world's largest Ramsar site (http://www.worldwildlife.org/press-releases/ bolivia-designates-world-s-largest-protected-wetland). In addition to its biodiversity, the LM is also of great palaeo

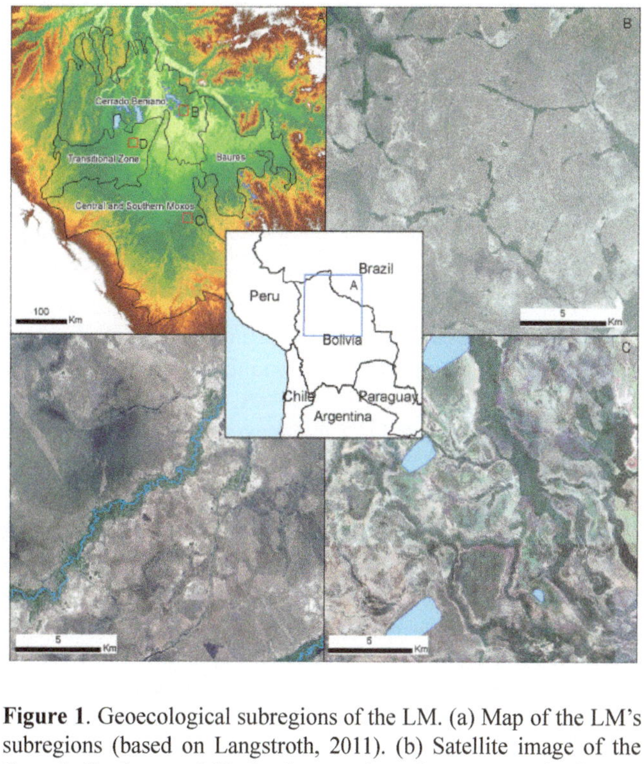

Figure 1. Geoecological subregions of the LM. (a) Map of the LM's subregions (based on Langstroth, 2011). (b) Satellite image of the Cerrado Beniano, a hilly landscape where forests grow in depressions. (c) Satellite image of the central and southern Moxos, a flat landscape where forests grow on elevated fluvial levees. (d) Satellite image of the transitional zone, where forests are relegated to the floodplains of underfit rivers.

ecological importance, as this region constitutes the southern border of the Amazonian rainforest, hence a preferential area to study past forest–savannah dynamics (Mayle et al., 2000). Inundation patterns, or flood pulses, have important effects on the biogeochemistry and the ecology of floodplain ecosystems,

Citation: Lombardo, U. (2014). Neotectonics, flooding patterns and landscape evolution. Earth Surface Dynamics Discussions, 2(2), 635-679. http://dx.doi.org/10.5194/esurfd-2-635-2014

as they determine the occurrence and distribution of plants and animals, they affect primary and secondary production and influence decomposition and nutrient cycles in water and soils (Junk et al., 1989; Junk, 2013).

Nowadays, the LM is constituted by different ecoregions which are characterized by specific vegetation assemblages which depend on the interplay between precipitation, drainage and soils (Navarro, 2011). Recent research shows that the LM was first inhabited by hunter-gatherers at the beginning of the Holocene (Lombardo et al., 2013b). Its landscape has since then been significantly changed during the late Holocene by the so-called "Earthmover" societies, who built an extensive array of earthworks (Erickson, 2008; Mann, 2008; Lombardo and Prümers, 2010; Lombardo et al., 2011b). We know very little about the way early hunter-gatherers dealt with the seasonal floods during the early to mid-Holocene, but evidence suggests that they had to abandon the LM around 4000 yr BP (before present) because of a large-scale environmental change, which also caused the burial of the early archaeological sites with fluvial sediments (Lombardo et al., 2013b).

Inundation patterns also played an important role in shaping the way in which the late Holocene "Earthmovers" adapted to their local environment and reshaped the landscape, determining the spatial distribution of different earthworks (Lombardo et al., 2011b); especially in order to improve the drainage of agricultural land (Lombardo et al., 2011a, 2012). Therefore, the reconstruction of past environments, landscapes and human–environment interactions in southern Amazonia throughout the Holocene greatly depends on our ability to reconstruct past inundation patterns. Up to now, with the exception of a recent study which reconstructs 6 kyr of palaeovegetation in the Baures subregion (Fig. 1) (Carson et al., 2014), palaeo-ecological reconstructions in the Bolivian Amazon have been mostly based on pollen and charcoal archives from lakes located outside the margins of the LM (Mayle et al., 2000; Burbridge et al., 2004; Whitney et al., 2011; Urrego et al., 2013).

These archives have been interpreted as the result of changes in climate and pre-Columbian human impacts, implicitly assuming that other factors, such as neotectonics, did not play an important role in the evolution of the landscape during the Holocene. However, several authors have stressed that the LM shows manifold evidence of neotectonic activity during this period (Hanagarth, 1993). In particular, the existence of hundreds of rectangular and oriented lakes have been attributed to the propagation to the surface of bedrock faults (Plafker, 1964, 1974; Allenby, 1988); the formation of ria lakes in the northern part of the LM and the changes in the meandering behaviour of the Mamoré River at Puerto Siles have been associated with uplift and tilting movements in the northern LM (Hanagarth, 1993; Dumont and Fournier, 1994) and the counter clockwise movement of the Beni and Grande rivers has been associated with the tilting of the basement (Dumont and Fournier, 1994; Dumont, 1996).

A ria lake forms when the lower part of a tributary is ponded at the margin of the trunk river because of a quick depositional event of the main river, a raise of the base level due to faulting or tilting, or because of regional subsidence (Schumm et al., 2002). These studies suggest that neotectonics had a significant impact on the formation of the modern landscape of the LM and, hence, on past inundation patters. However, because of the lack of chronological data, digital elevation models (DEMs), remote sensing imagery, and the difficulty of performing fieldwork in the region at the time of these early studies, published work to date does not provide quantitative measures of the neotectonic movements and do not de- fine a chronological framework for the tectonic activity. Furthermore, previous observations based on remote sensing imagery have not been corroborated by field evidence.

The extent and the chronology of neotectonic activity are extremely important in order to assess the potential impact that neotectonics could have had on the evolution of southern Amazonia. Neotectonics could have brought about changes in the drainage of the LM during the Holocene which, in turn, could have altered the flooding regime, vegetation assemblages, landscape evolution, and human–environment interactions in the region. In the present paper, previous studies are revisited in light of new remote sensing imagery, digital elevation models and extensive field work. Based on this evidence, I propose a general framework to explain the contribution of neotectonics to the evolution of the landscape and inundations patterns in southern Amazonia.

The paper describes the physical setting of the LM and the geomorphological evidence that supports the existence of neotectonic activity. The paper is organized in three main sections. The first part, Sect. 2, describes the study area, focusing on three aspects: the general landscape, the modern flooding patterns and the geology. The second part, Sect. 4, assesses the evidence of neotectonic activity. This is subdivided into two parts: rivers/palaeorivers and lakes. Here, the first paragraphs review the existing literature; the following paragraphs present new data and the results of the study. The third part, Sect. 5, addresses the following questions: 1. What could be the general mechanism behind neotectonics in southern Amazonia? 2. What was the extent and timing of neotectonic activity? 3. How did neotectonics affect the LM hydrology during the Holocene?

THE STUDY AREA

The LM and its subregions

The LM, in southern Amazonia, is a large seasonally flooded savannah criss-crossed by strips of forest. However, important differences exist between its subregions (Fig. 1). The northern LM, also called Cerrado Beniano, is covered by reddish, relatively well drained lateritic crusts, which host Cerrado-like savannahs (Langstroth, 2011; Navarro, 2011). Forests grow in the lower part of the landscape, mostly in the small valleys between lateritic hills where sediment accumulation provides enough humidity and room for tree roots to grow (Fig. 1b). Compared to the rest of the LM, which are almost totally flat, this region is hillier and shows noticeable topographic changes.

As it can be seen in the DEM in Fig. 1a, most of the Cerrado Beniano is well above the level of the seasonal floods. Hence, here, the forest–savannah ecotone is determined primarily by soil properties and local precipitation, and is relatively independent from the LM's flood regime. Some of the largest lakes in the LM are located here. In the central and southern LM (Fig. 1c) the topographic control on the savannah–forest assemblage is inverted with respect to the northern LM, in that forested areas tend to be concentrated on high ground rather than valleys.

The central and southern parts of the LM are an extremely flat floodplain, mostly covered by seasonally flooded savannahs, which, in general, are fluvial backswamps. Savannahs are interspersed by strips of forests, which mainly grow on the fluvial deposits that remain above the level of the seasonal floods (Mayle et al., 2007). Most of the soils are relatively fertile because they generally form on recently deposited, clay-rich, Andean sediments. The most common soil types are Gleysols in the savannahs, Fluvisols in the recent fluvial levees and Luvisols and Cambisols in the older palaeolevees (Langstroth, 1996; Boixadera et al., 2003; Lombardo et al., 2014). Here soil properties and the forest–savannah ecotone along fluvial splays and levees are primarily determined by the inundation patterns. This is the part of the LM where hundreds of rectangular and oriented lakes formed.

In the western LM, the Cerrado Beniano is separated from the central LM by a transitional zone (Fig. 1d). This encompasses the savannahs north and west of Santa Ana de Yacuma. This area is very flat and crossed by underfit rivers. These rivers flow on ancient courses of the Beni River (Dumont and Fournier, 1994) and do not overflow during the rainy season (Walker, 2004). The savannah is seasonally flooded, but floods are caused by local precipitation alone, with no contribution from river overflow. The lack of fluvial sediment deposition, combined with strong redoximorphic conditions, results in highly hydromorphic soils, very poor in nutrients, often with very low pH values and toxic levels of exchangeable aluminium (Navarro, 2011; Lombardo et al., 2013a).

The northeastern part of the LM, the Baures subregion, is characterized by a very poorly drained savannah dotted with several forested areas. These forests grow on 2–3 m high elevated platforms that constitute outcrops of tertiary rocks (Pitfield and Power, 1987) (Fig. 2b). This is the less accessible and therefore less studied area of the LM. With an average slope of 0.014 %, the LM is extremely flat and subject to seasonal floods. Seasonal floods and backswamp sedimentation along the basins of the Mamoré and Beni rivers are primarily controlled by El Niño/La Niña cycles (Aalto et al., 2003) and the South Atlantic sea surface temperature anomalies (Ronchail et al., 2005). Precipitation in the LM range from 1350 to 2450 mm yr−1 (yearly average precipitation between 1971 and 2000 at Trinidad; Ronchail and Gallaire, 2006), while the flooded area is in the range of 30 000–92 000 km2 (Hamilton et al., 2004; Melack and Hess, 2011).

This shows the high sensitivity of the LM and its flooding regime to relatively small changes in precipitation. Flood dynamics are also influenced by the Mamoré River, which is the base level for most of the rivers in the region (Bourrel and Pouilly, 2004). The duration and extent of the inundations depend on the combination of precipitation in the upper catchment of the Mamoré River, the impeded drainage of local precipitation (Bourrel et al., 2009) and the elevation of the water table (Ronchail et al., 2005; Fan et al., 2013). A lag of 1–3 months is observed between the time of maximum precipitation in the upper catchment of the Mamoré and the peak in floodplain inundation (Hamilton et al., 2004). This results from the movement of river runoff from upland catchments to the floodplains, as well as from the delay in the drainage of inundated areas to the rivers (Hamilton et al., 2004). On an interannual basis, consecutive years with large inundations are caused by groundwater storage (Ronchail et al., 2005).

The geology of the Llanos de Moxos

The Andes–Amazon foreland basin system is not a typical foreland basin system (DeCelles and Giles, 1996); it is divided into two different basins by the uplifting Fitzcarrald Arch, a 600 m high bulge (Espurt et al., 2007, 2010). It has been hypothesized that the Fitzcarrald Arch is the projection to the surface of a buoyant flat slab resulting from the subduction of the Nazca Ridge (Espurt et al., 2008). The formation of the flat slab about 4 Myr ago would have brought about the back-arc uplift of the Fitzcarrald Arch (Espurt et al., 2007, 2008; Regard et al., 2009). The LM constitutes the foredeep of the southern Amazonian foreland basin. It is located between the central Andes, to the W–SW, and the Brazilian Shield, to the E. The Brazilian Shield dips gently towards the Andes underlying unconsolidated sediments (Plafker, 1964; Hanagarth, 1993). Very little is known about the stratigraphy and thickness of these foreland deposits. Geophysical surveys in the southern part of the LM indicate the presence of thick layers of fluvial sediments covering the Brazilian Shield (Plafker, 1964). The sediments overlying the crystalline bedrock thicken gradually from the margin of the Brazilian Shield towards the west. The thickness of the sediments is more than 5500 m near the Andean foothills and approximately 800 m at a distance of 150 km from the Andes, in the village of Perú (Plafker, 1964) (Fig. 2b). In the northern part of the LM, recent surveys have shown the presence of metamorphic rock at 165 m below Santa Ana de Yacuma and at 65 m below the town of Magdalena (UNASBVI-JICA, 2009). A 200 m long and 50 m high rocky exposure, belonging to the Brazilian Shield, outcrops in the location of El Cerro, 60 km north of Santa Ana de Yacuma. Several authors have identified an uplifting region in northern Beni; the so-called Linea Bala–Rogaguado in Fig. 2b (Allenby, 1988; Dumont, 1996; Dumont and Fournier, 1994; Plafker, 1964; Hanagarth, 1993). The Linea Bala–Rogaguado, also called Reyes-Puerto Siles axis (Dumont, 1996), has been interpreted as a SW–NE faulting system which represent the south-western extension of the Tapajó fault in central Brazil (Dumont and Fournier, 1994). More recently, it has been proposed that the Puerto Siles uplift is part of the SE–NW oriented Andean forebulge axis (Aalto et al., 2003; Roddaz et al., 2006) (Fig. 2b).

METHODS

The study examines the evidence of neotectonic activity in southern Amazonia and its impact on flooding patterns and landscape evolution during the Holocene. It focuses on the analysis of modern and palaeorivers, ria lakes, palaeosols and landscape topography. For this it combines the analysis of remote sensing images, extensive fieldwork in the LM and a review of existing studies. ArcGis 10.1 has been used to produce the topographic maps shown in Figs. 1a, 2, 3a, 6a and 9 and the topographic profiles shown in Figs. 5, 6a and 10; these are based on the hole-filled Shuttle Radar Topography

Mission (SRTM) imagery, where elevation values for water bodies are interpolated (Jarvis et al., 2008). Images of landscape features used in Figs. 1b–d, 3b–d, 7a and 10b are retrieved from Google Earth. Images used in Figs. 4 and 10a

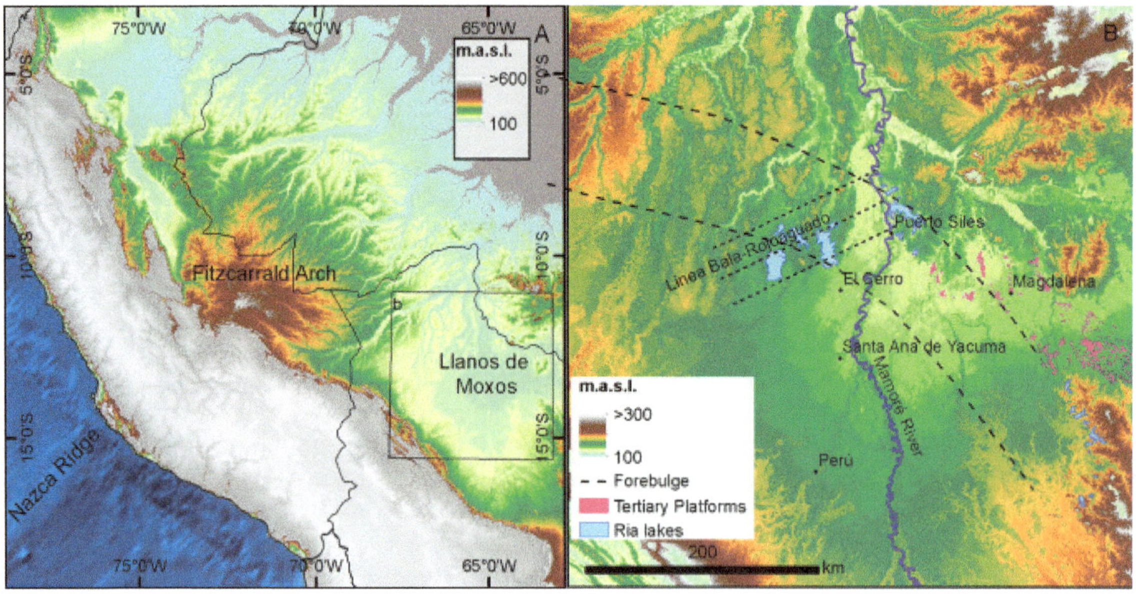

Figure 2. Geological setting of the LM. (a) Topographic map of central western South America showing how the south-eastern part of the Fitzcarrald Arch constitutes the LM's north-western border. This, together with the Brazilian Shield in the north-east, forms a continuous barrier which impedes the drainage of the LM. (b) Topographic map of the LM showing the location of the Linea Bala–Rogaguado (as in Dumont and Fournier, 1994) and the Andean forebulge (as proposed by Aalto et al., 2003).

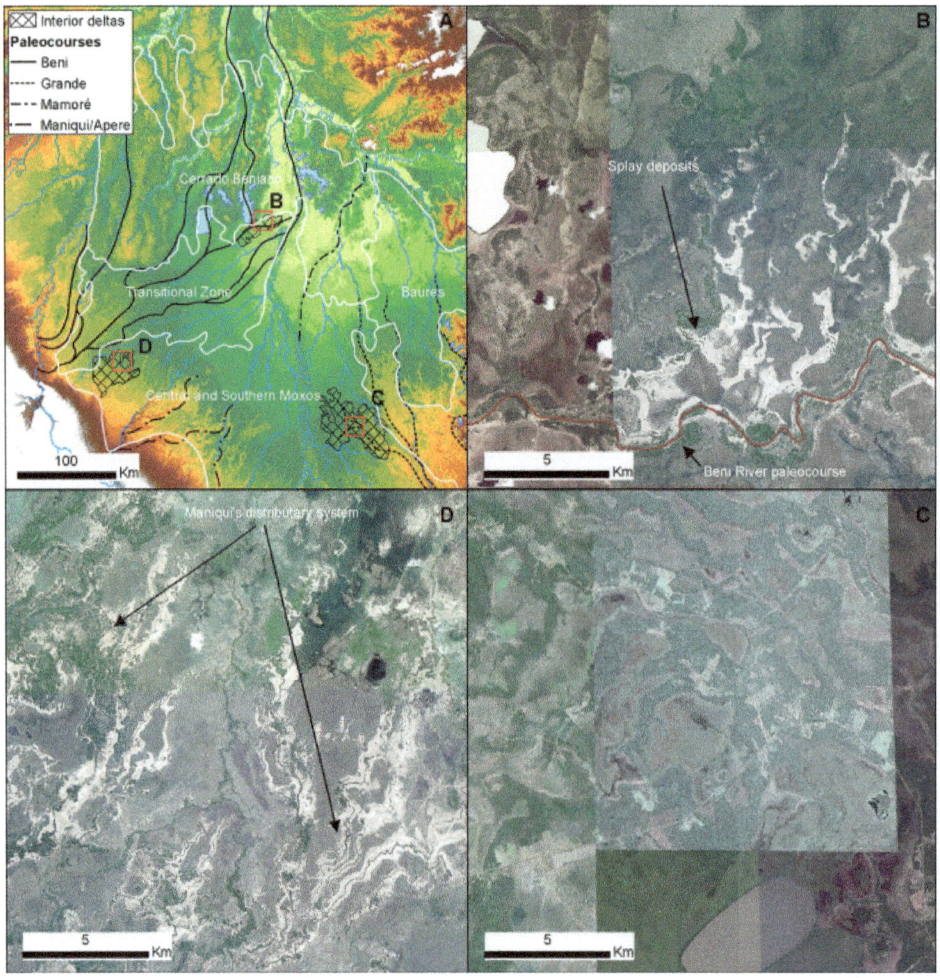

Figure 3. Map of rivers and palaeorivers in the LM. (b) An example of crevasse splays at the distal part of a palaeocourse of the Beni River. (c, d) Interior deltas formed by repeated avulsions of the Grande and Maniqui rivers, respectively.

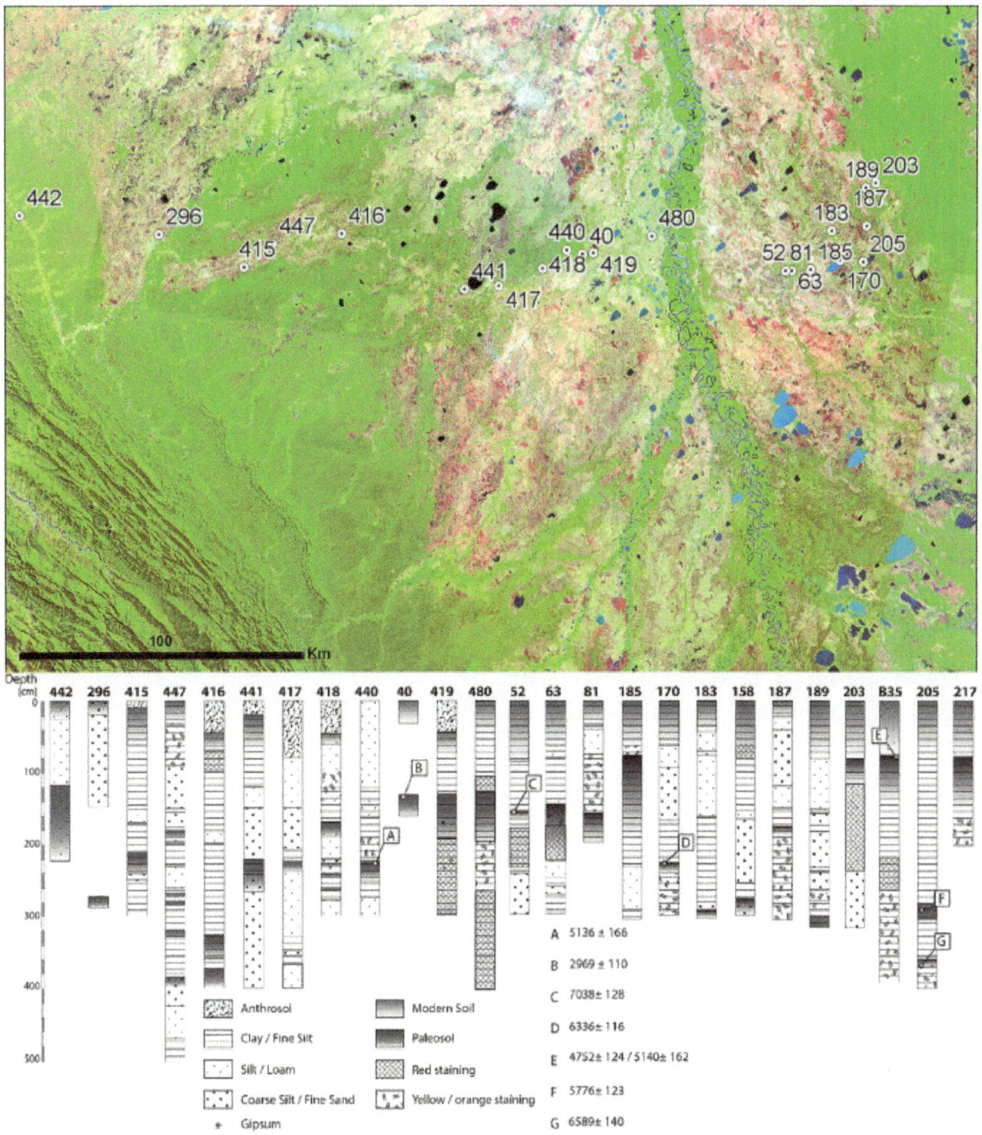

Figure 4. Cored profiles along a 300 km long transect crossing the central and southern LM. Palaeosols are always found below alluvia. Sequences of several organic horizons separated by alluvial deposits (for example core 447) indicate frequent avulsions. Radiocarbon ages indicate that the central and southern LM were covered with fluvial sediments during the mid- to late Holocene. AMS (applied mass spectrometry) radiocarbon ages of palaeosols in profile 40 (uncalibrated age 2900 ± 35 yr BP) and 440 (uncalibrated age 4520 ± 40 yr BP) were measured at the Poznan Radiocarbon Laboratory. The rest of the radiocarbon ages have been taken from Lombardo et al. (2012).

are Landsat ETM (Enhanced Thematic Mapper) and have been downloaded from GLOVIS (http://glovis.usgs.gov/). The Mamoré River has been manually digitalized over the hole-filled SRTM. Elevation points within the Mamoré River channel have been extracted from the hole-filled SRTM at each vertex of the digitalized line and plotted against distance downstream (Fig. 5b) and against their latitude (Fig. 5a, c). The latter facilitates the direct comparison between the slope and the morphology of the river. River reaches have been identified on the basis of the remote sensing imagery and after a visual analysis of the river geometry and slope.

The average slope of each reach is represented as a line (Fig. 5a, c). Sinuosity was measured for each river reach (1–9 in Fig. 5) as the river length divided by the valley length. In order to investigate the origin and time of formation of Lake Rogaguado, the lake has been cored in four places; however, only the data from two of the cores is presented in this paper, cores 398 and 397, as the other two cores did not reach the bottom of the lacustrine sediments. Lake cores have been taken with a gravimetric corer. In order to reach the sediments below the lacustrine phase, location 398 has been cored deeper using a Livingstone corer. The radiocarbon age from the base of the lacustrine phase of core 378 comes from a fragment of wood; it has been calibrated with Calib 7.0 (http://calib.qub.ac.uk/calib/calib.html) using the SHcal13 calibration curve (Hogg et al., 2013). In order to assess the sedimentary regime across the central and southern LM, a stratigraphic transect has been made from the Andes to the eastern LM (Fig. 4). Stratigraphic cores along this transect have been taken with a Waker motor vibracorer and described in the field. The description and radiocarbon ages of the cores taken east of the Mamoré River are based on Lombardo et al. (2012).

THE EVIDENCE OF NEOTECTONIC ACTIVITY IN SOUTHERN AMAZONIA

Rivers and palaeorivers

Tropical alluvial rivers can change in response to different factors, notably changes in climate, neotectonics or their intrinsic evolution over time (Schumm et al., 2002; Bridge, 2009; Burbank and Anderson, 2011; Gupta, 2011). In particular, river avulsions take place when the main channel becomes infilled with sediments, triggering the diversion of the river into a new course (Slingerland and Smith, 2004; Buehler et al., 2011). River avulsions are the main mechanism behind the formation of distributary fluvial systems (DFS), which fill alluvial basins (Hajek and Edmonds, 2014; Weissmann et al., 2010; Hartley et al., 2010). High aggradation rates in the main river channel can result from a variety of causes such as (i) an increase in the sedimentary load of the river; (ii) an increase in the peak water discharge of the river or (iii) a decrease in the channel gradient, which can be caused by a rise of the base level or downstream tectonic uplift (Slingerland and Smith, 2004). Therefore, the interpretation of environmental and palaeo-environmental fluvial forms and palaeoforms is not always straightforward. Avulsive processes and the formation of DFS within the LM have been described in the case of the Beni River (Plafker, 1964, 1974; Hanagarth, 1993; Dumont, 1996; Hartley et al., 2010) the Grande River (Hanagarth, 1993; Dumont, 1996; Lombardo et al., 2012; Plotzki, 2013) and the Mamoré River (Plotzki et al., 2011, 2013) (Fig. 3a) but the mechanisms behind these avulsions are controversial. The northward migration of the Beni River deflection point has been associated with a fault connected to the Andean foothill margin (Dumont, 1996; Plafker, 1964; Hanagarth, 1993; Allenby, 1988). Based on the assumption that the Grande River changed its course following a counter clockwise movement, the tilting of the basement, and the consequent uplift of its southeastern margin, has been considered responsible for its avulsions (Hanagarth, 1993; Dumont, 1996). More recently, however, it has been suggested that the megafan of Grande River did not follow a counter clockwise movement (Plotzki, 2013). The study of an interior delta/sedimentary lobe at the distal part of one of the Grande River palaeocourses, seems to indicate that the Grande River avulsive phase may have been triggered by an increase in river discharge or sedimentary load that could have been brought about by a mid- to late Holocene climate shift towards wetter conditions (Lombardo et al., 2012; Plotzki et al., 2013). Similarly, the avulsion of the Mamoré seems to be related to an increase in discharge probably caused by a change in climate (Plotzki et al., 2013). The interior delta formed by the Maniqui River (Fig. 3d) has been described by Hanagart and Sarmiento (1990); they observed that the formation of the Maniqui megafan was caused by a reduction in the slope, which induced frequent avulsions.

A situation similar to that of the Maniqui and Grande river deltas (Fig. 3c) is found in the distal part of a palaeocourse of the Beni River (Fig. 3b), where numerous crevasse splays formed. Although these fluvial forms have received little attention in the literature, their location on the southern border of the Cerrado Beniano (the Beni crevasse splays) and the transitional zone (the Maniqui River megafan), suggests that their formation could respond to a change in slope, as suggested by Hanagart and Sarmiento (1990) for the Maniqui, which was probably caused by a downstream uplift. The central and southern Llanos de Moxos are covered by many palaeolevees belonging to DFS. This is characteristic of actively subsiding basins (Hartley et al., 2010; Latrubesse et al., 2010; Weissmann et al., 2010, 2013; Rossetti et al., 2012). In order to study these DFS, 25 stratigraphic cores were retrieved along a 300 km line crossing the central LM (Fig. 4). All the profiles reveal intercalations of clays, loams, silts, fine sands and organic-rich palaeosols.

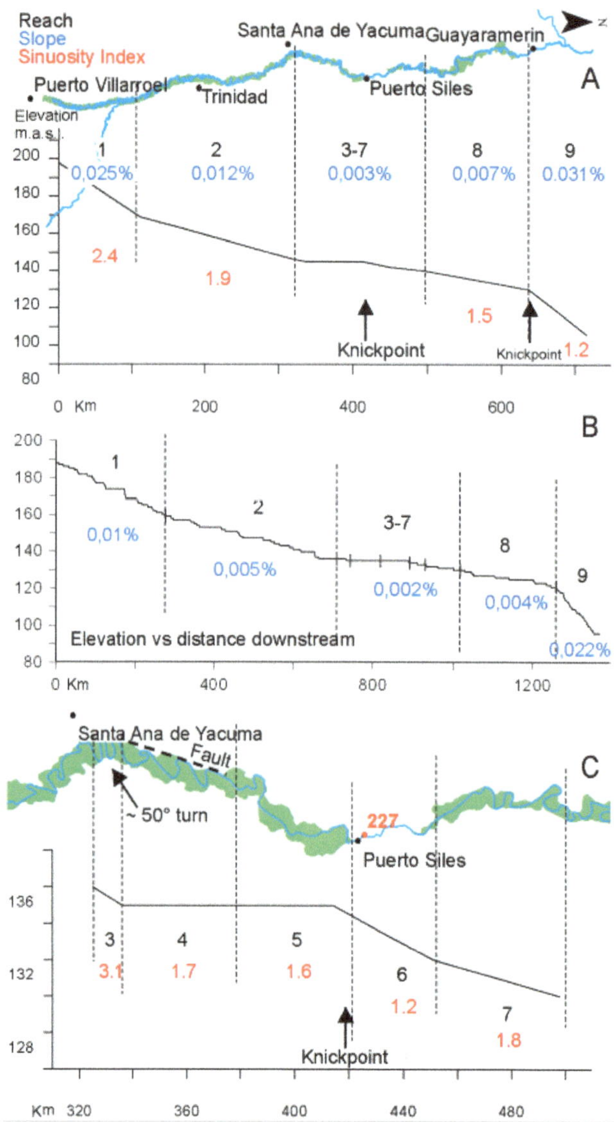

Figure 5. Geometric characteristics of the Mamoré River. (a) Valley river slope and sinuosity index (SI) of the Mamoré between Puerto Villarroel and Guayaramerín are plotted against each other. Here, the valley slope changes twice, indicating the presence of knickpoints: one in Puerto Siles and the other one close to Guayaramerín. (b) The slope is plotted against river length, showing the same pattern as in (a). In (c), a more detailed map of the section between Santa Ana de Yacuma and 100 km downstream of Puerto Siles shows a series of compressed meanders at Santa Ana, followed by a 50° turn and three asymmetric meanders. After Puerto Siles, the SI declines abruptly and the gallery forest disappears. Point 227 is where photos (c) and (b) in Fig. 7 were taken.

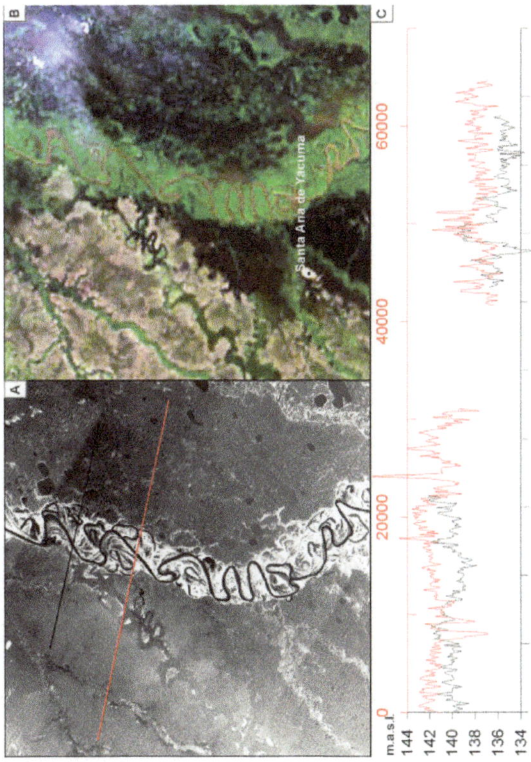

Figure 6. (b) Location of the topographic profiles shown in (c). (b) MODIS image of the flooded areas around Santa Ana de Yacuma in 2007 (black areas). (c) Topographic profiles across the asymmetric meanders of reach 4 (Fig. 5c), indicating that the western side of the river (the dry area shown in b) is about 5 m above the level of the eastern side of the Mamoré. This further supports the hypothesis of the presence of an active fault as indicated in Fig. 5c.

The association of levee sands and silts, splay sands, backwater loams and (organic) clays is characteristic of an avulsive fluvial setting, which is consistent with the formation of megafans in subsiding basins. Radiocarbon dating of palaeosols found below these alluvia indicate deposition during the mid- to late Holocene (Fig. 4). Data from drilling reported in Plafker (1964) reveal that this assemblage of Holocene fluvial sediments is similar to the rest of Quaternary and possibly Tertiary sediments constituting the 800 m of basinal infilling at the site of Perú. Although the evolution of specific rivers at any given time could have been influenced by different factors, the general landscape of the central and southern Llanos de Moxos is that of a typical subsiding basin. However, there are characteristic features in river patterns that can be associated with neotectonic activity more clearly than avulsive processes. The existence of underlying active tectonic processes can be noted from the analysis of features such as changes in the meandering pattern and the presence of knickpoints (Schumm et al., 2002; Burbank and Anderson, 2011).

Evidence of neotectonic activity in the LM can be inferred from the analysis of the Mamoré River. The sharp reduction in the sinuosity of the Mamoré River north of Puerto Siles (Fig. 5) has been considered direct evidence of uplifting along the Linea Bala–Rogaguado (Dumont and Fournier, 1994). The existence of this uplift is consistent with the presence of several rapids located downstream of Puerto Siles (Dumont and Fournier, 1994). In the lower part of the Mamoré and the Beni rivers, rapids are caused by the outcropping of pre-Cambrian rocks belonging to the Brazilian Shield (Hanagarth, 1993).

Figure 7. Changes in the geomorphic features of the Mamoré River at Puerto Siles. (a) Google Earth image showing the change in Mamoré's width and the outcropping of lateritic crusts; (b) The Mamoré River bank at point 227: a lateritic crust covered by old fluvial sediments; (c) Photograph of a well developed Luvisol at the top of the fluvial levee at location 227.

The presence of the Puerto Siles uplift is corroborated in this study by the analysis of the SRTM imagery (Fig. 5). The analysis of the topographic profile of the Mamoré River shows several changes along its course. In its first 400 km (reaches 1–5 in Fig. 5) the Mamoré River behaves as a free meandering river, with its slope and sinuosity decreasing downstream. This behaviour is interrupted at Puerto Siles, where the lowest sinuosity value is associated with a clear increase in slope and to the appearance of lateritic crusts within the river channel, causing the formation of rapids. The change experienced by the Mamoré River at Puerto Siles is also noted in the river's topography, which goes from concave to convex at this point. This change in Puerto Siles is preceded by three compressed meanders followed by three asymmetric meanders (reaches 3 and 4 in Fig. 5). A clear change in the direction of the river, with a clockwise turn of about 50° , is associated with the compressed meanders. Compressed meanders indicate a downstream increase of the river bank's resistance to erosion, often the result of the appearance of bedrock in the channel following localized uplift (Holbrook and Schumm, 1999). This situation is similar to what can be seen in the rest of the Amazon Basin, where discharge characteristics and bank stability play an important role in channel pattern variations (Baker, 1977). The topographic profiles across reach 4 of the Mamoré River show that the western side of the river is about 5 m higher than the eastern side (Fig. 6a). This difference in elevation between the two sides dramatically affects the flooding regime; the savannah in the north of Santa Ana de Yacuma (the transitional zone) is hardly ever flooded by the Mamoré outflow, whilst the savannah in the eastern side is severely flooded every year (Fig. 6b). Therefore, the asymmetric meanders and the 50° change in direction can be interpreted as the result

Figure 8. An erosional feature incising the savannahs of the transitional zone along the road from Santa Ana de Yacuma to lake Rogaguado (the dry area shown in Fig. 6b), indicating that these savannahs are established in uplands subject to erosion.

of a normal fault with its footwall on the western side of the river. In addition, the presence of underfit rivers in the transitional zone suggests that this subregion is uplifting. This uplift is evident in the formation of a new drainage network constituted by small creeks, which are a common element of the landscape easily seen in the field. Figure 8 shows one of these erosive features, which are formed when rain water travels from the waterlogged savannah to the underfit rivers.

The topographic profile of the Mamoré shows how, at Puerto Siles, the meander belt that borders the river for the first 400 km is replaced by two irregular forested levees. No oxbows are present here. A 40 % decrease in the channel's width is also observed here. The changes at Puerto Siles suggest that the Mamoré experiments acceleration and deepening of the channel. These changes increase the capacity of the river to transport sediment and, in turn, this changes its behaviour: from depositing sediments it now begins to erode its bed. A further confirmation of this change is that downstream from Puerto Siles the Mamoré's river banks are made by thick lateritic crusts covered by old fluvial levees, with no sign of recent fluvial deposition (Fig. 7b, c).

Here, welldeveloped Luvisols are found on top of the levees. About 200 km downstream, close to Guayaramerín, a second knickpoint appears. After this second knickpoint, the Mamoré reaches its steepest slope (reach 9 in Fig. 5). The analysis of the SRTM indicates that the general downslope topography of the LM is interrupted at Puerto Siles where the terrain becomes hillier. Despite the fact that these hills significantly change the slope of the LM, the river network seems to be unaffected by them. In fact, as we can see in the SRTM images shown in Fig. 9, all the rivers maintain their northward convergent path, cutting through these hills; no rivers flow south. This setting suggests that the courses of these rivers were established before the onset of the uplift, and that the rivers have overcome this obstacle by incising the uplifting rocks and sediments. This is consistent with the hypothesis that these hills constitute the southern margin of the Fitzcarrald Arch, which started uplifting about 4 M yr ago (Espurt et al., 2007). Not all the valleys crossing the uplifted area have been active at the same time, as shown by the palaeocourses of the Beni River: the Beni occupied at least two other valleys (now occupied by the Yata and Tapado rivers) before establishing its modern course (Dumont and Fournier, 1994).

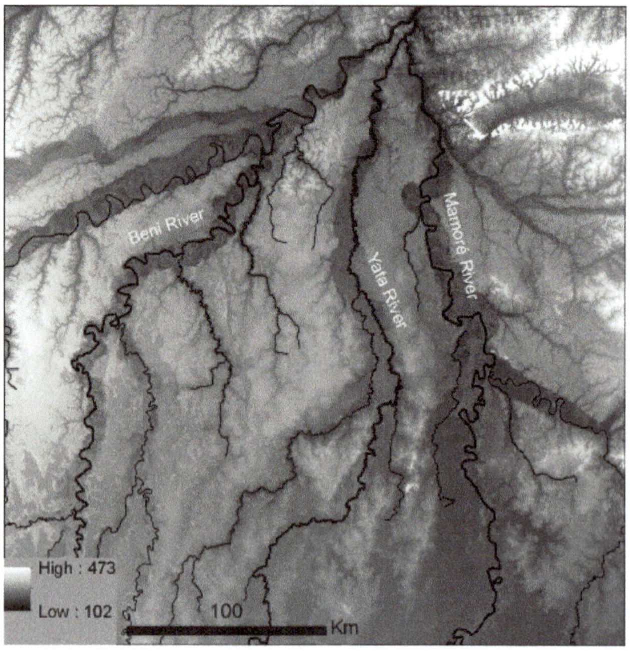

Figure 9. SRTM image showing an example of a river network in the Cerrado Beniano. All rivers flow north, against the slope of the general terrain, suggesting that the development of the drainage network preceded the uplift that crated the hilly landscape.

Lakes

There are more than 800 lakes in the LM, most of which are rectangular and oriented. Until recently, the most accepted hypothesis behind the formation of these rectangular and oriented lakes of the LM has been tectonic (Plafker, 1964; Allenby, 1988). According to Plafker, the lakes' rectangular shapes results from the propagation of bedrock fractures through unconsolidated sediments and the sinking of basement blocks. However, the coring of three of these lakes has shown that tectonics cannot explain their formation, as a horizontal palaeosol is found at the same depth both outside and inside the lakes (Lombardo and Veit, 2014). Given the available data, the most likely mechanism behind the formation of the rectangular and oriented lakes seems to be the erosive action of waves on the lakes' shores due to constant winds (Lombardo and Veit, 2014). Apart from the rectangular and oriented lakes, a second group of lakes, found mainly on the eastern and northern borders of the LM, are the ria lakes. Ria lakes are a valuable source of information to reconstruct past environments, sedimentary histories and tectonic processes (Behling et al., 2001; Irion et al., 2006, 2011).

They resemble water reservoirs resulting from a dam, where a preexisting valley is flooded and transformed into a lake. In the Bolivian Amazon, ria lakes have been reported in the southeastern and in the northern parts of the LM. In the southeastern side they have been formed by the rapid deposition of sediments by the Grande River, which dammed fluvial valleys previously eroded in the Brazilian Shield (Dumont and Fournier, 1994). Ria lakes that have been described in the northern LM are connected to the Linea Bala-Rogaguado (Hanagarth, 1993; Dumont and Fournier, 1994). Here there are two types of lakes. The first type, which includes Lago Largo and Agua Clara (Fig. 10), is the typical ria lake which results from a valley flooded by damming; hence, similar to those formed by the former Grande River. The second type, such as Rogaguado and Ginebra, is constituted by lakes that have formed within a valley that became flooded because of tilting (Dumont and Guyot, 1993; Dumont and Fournier, 1994).

Ria lakes of tectonic origin are not limited to the LM; they have also been described in the Marañon and Ucayali basins (Räsänen et al., 1987; Dumont, 1993). Many other ria lakes are found along the Negro, Xingú and Tapajó rivers in the Brazilian Amazonia. These seem to have formed because of the increased river incision caused by the drop of the ocean water level during the Last Glacial Maximum (Irion et al., 2011), but it has been suggested that tectonic movements could have also played a role. (Latrubesse and Franzinelli, 2002). The counter clockwise shift of the Beni River (Dumont, 1996) has left several palaeofloodplains, one of which is nowadays occupied by Lake Rogaguado. If we assume that, before the formation of the lake, when the Beni River flowed through the Rogaguado Valley its slope was relatively constant, the northern uplift or tilting that generated the lake, would have, therefore, also changed the slope of the original valley.

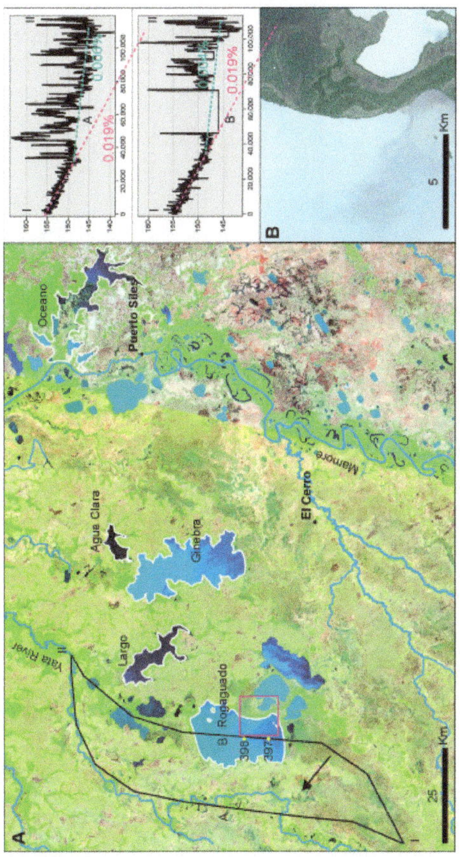

Figure 10. Landsat image of ria lakes in the Cerrado Beniano. (a) Yellow points in Lake Rogaguado indicate the location of the cores shown in Fig. 11. Topographic profiles from I to II show a reduction in the slope, coinciding with two 90° turns of the Yata River (profile A) and the southern shore of Lake Rogaguado (profile B). The topographic profile B also shows an inversion of slope north of Lake Rogaguado, which embanked the valley and formed the lake. (b) A palaeocourse of the Beni River which has been cut by the lake. Topographic profiles are based on the Hole-filled SRTM version 4 (Jarvis et al., 2008). The image of the palaeo-Beni River in (b) is from Google Earth.

The measurement of this change could provide constraints on the extent of the tectonic movement. Figure 10 shows the the original slope of Rogaguado's floodplain and the amount of vertical movement which created the lake. In order to confirm the mechanism at the origin of these lakes and constrain the timing of their formation, Lake Rogaguado was cored at several locations. Figure 11 shows the two cores retrieved from Lake Rogaguado that reached the bottom of the lacustrine sediments. The bottom part of core 398 includes the terrestrial phase that preceded the formation of the lake. This is a palaeosol characterized by abundant red oxides at its base, indicating recurrent wet and dry periods. The amount of organic matter increases towards the top of the terrestrial phase, which probably corresponds to a fossil Ah horizon. This terrestrial phase seems to correspond to the seasonally flooded backswamps of the ancient Beni River. The remains of the palaeochannel and levees of the ancient Beni River are still visible in the south-eastern margin of the lake (Fig. 10b). The stratigraphic units (as inferred from changes in colour and material) of core 398 correlate with those of core 397, which is located about 6.6 km further south. Figure 11 shows the two cores in their relative position. At the time of the coring, the depth of the water at 397 was 2.4 m. Assuming that the clay layer separating the terrestrial phase from the lacustrine one has, in this core, a similar thickness to that in core 398, the depth of the palaeosol can be estimated as being 3 m below the water level. The slope of the palaeosol below the lake, 0.0045 % (30 cm in 6.6 km), is consistent with the average slope of the valley northward from the lake. This suggests that the lake was formed by an uplift on its northern side, causing the tilting that dammed the valley. However, the depth of the palaeosol with respect to the savannah south of the lake indicates also the presence of a normal fault, with a southern footwall and a throw between 2 and 3 m, which probably de- fined the lake's southern border. The point where the valley's slope changes is about 8 km south of the lake's shore, coinciding with the sharp change in the direction of the Yata River (black arrow in Fig. 10). This data suggests that the formation of Lake Rogaguado, and probably of Lake Ginebra too, is the product of a combination of tilting and faulting, maybe resulting from an extensional setting related to an upward pulling north of the lake. In core 398, radiocarbon dating of a wood fragment from the bottom of the lacustrine phase yields the age of 5791 ± 155 cal yr BP, while the palaeosol that corresponds to the top of the terrestrial phase has been dated 10 910 ± 124 cal yr BP. The radiocarbon age of the palaeosol is based only on the soluble fraction, because no humins could be retrieved. Therefore, it should be interpreted as the average age of the palaeosol and not as the age of its burial.

The palaeosol is separated from the organic sediments of the lacustrine phase by a 6 cm thick layer of clays and silts. This layer of sediments contains no organic matter, suggesting it was deposited soon after formation of the lake in a quick event. The sediments could have come from the erosion of the recently formed lake shores. This preliminary data suggests that Lake Rogaguado was formed in the mid-Holocene. There is a group of ria lakes that have not yet been described. They are located on the eastern margin of the Mamoré River, north of Puerto Siles (Fig. 10). Lake Oceano is the largest. The damming of the valleys that formed these lakes was likely caused by the infilling of the valleys' bottom with sediments from the Mamoré river. Lake Oceano offers a modern analogue of the formation of ria lakes in the southeastern LM by Grande River.

DISCUSSION

In the present paper, geomorphological evidence of neotectonics in southern Amazonia is reviewed and corroborated with new data obtained from the analysis of remote sensing imagery and field work. A first attempt to establish the extent and timing of the neotectonic activity is made. Here, the three research questions formulated in the introduction are addressed in light of the new evidence presented.

What could be the general mechanism behind neotectonics in southern Amazonia?

None of the two mechanisms proposed in previous studies as being behind neotectonic activity in the LM, the reactivation of the Tapajó fault in central Brazil (Dumont and Fournier, 1994) and the uplift of the Andean forebulge (Aalto et al., 2003), seem to be able to explain all the geomorphological evidence of neotectonics reviewed here. The reactivation of a supposed south-western extension of the Tapajó fault in central Brazil could explain the existence of a fault system along the Linea Bala–Rogaguado. However, it cannot explain the meandering pattern of the Mamoré River, which responds to uplifts occurring about 100 kilometres south (Fig. 6). Nor can it explain the closeness of the Brazilian Shield to the surface, and the outcropping of it in several locations in the south and south-east of the Linea Bala–Rogaguado. Moreover, the neotectonic setting in the Tapajó region is constituted by two grabens trending NW–SE and NNE–SSW and a set of ENE– WSW strike-slip faults (Costa et al., 2001). It is unclear how these extensional structures in central Brazil could propagate into the normal faulting/tilting system associated with the Bala–Rogaguado region in northern Beni. Likewise, there is little evidence of uplift along the forebulge axis as identi- fied by Aalto et al. (2003) (Fig. 1b). The presence of the Andean forebulge can explain the uplift at Puerto Siles and the consequent change in the behaviour of the Mamoré River. However, it is not compatible with the region's overall fluvial pattern and the topography of the LM. The uplift of the forebulge is very small (a few hundred metres over a distance of several hundred kilometres) and an inherited (preflexural) topographic relief can be more important than the forebulge in determining the drainage divide (Garcia-Castellanos and Cloetingh, 2011). Nevertheless, in models of continental and exorheic basins, the effect of the forebulge on the drainage network is generally evident (Garcia-Castellanos, 2002). In an extremely flat landscape, such as that of the LM, an uplift along the forebulge axes which is able to create the Puerto Siles knickpoint would be visible in the subsequent formation of a divergent drainage network. Yet, along the alleged forebulge axes, rivers, and even recent small drainage creeks, seem to be unaffected by the forebulge. In addition (i) some of the lowest-lying areas of the LM are found right along the forebulge, which is very unusual for an uplifting region and (ii) the alleged forebulge would be orthogonal to the Linea Bala–Rogaguado, therefore unable to explain the formation of the ria lakes. This study suggests that the uplifting Fitzcarrald Arch is the most likely mechanism behind the observed uplift along the Linea Bala–Rogaguado, the general topography of the LM, the outcropping of the basement in the northern LM, the presence of underfit rivers crossing the savannahs in the transitional zone, the meandering pattern of the Mamoré River and the general shape of the fluvial network. The fault that caused the change in the direction of the

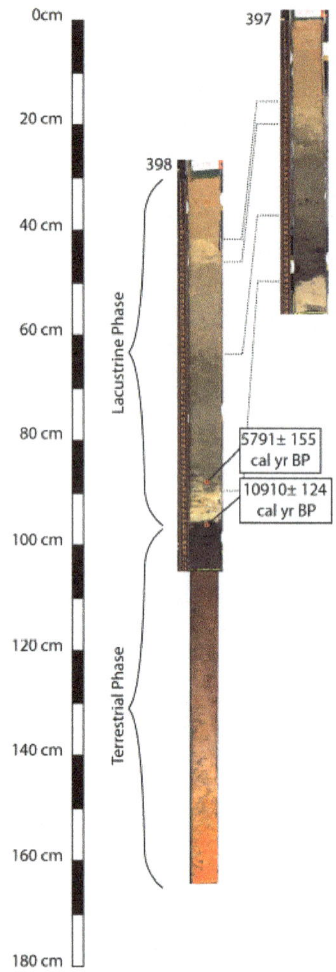

Figure 11. Cores from Lake Rogaguado. Core 398 includes a top part, which is the lacustrine phase that followed the formation of the lake, and a bottom part, which is the terrestrial phase which preceded the formation of the Lake. The red mottling at the base of the core indicates alternation of wet and dry periods, suggesting that savannah-like conditions prevailed in the backswamps of the palaeo-Beni River. The age of the beginning of the lacustrine phase is based on an AMS radiocarbon analysis of a wood sample performed at LARA AMS Laboratory in Bern (uncalibrated age 5104 ± 76 yr BP). The AMS radiocarbon age of the fossil Ah was measured by DirectAMS using the humate fraction (uncalibrated age 9685 ± 40 yr BP).

Mamoré River (Fig. 5, reach 4) and the direction of the Linea Bala–Rogaguado, is consistent with the orientation of the Nazca Ridge, suggesting that the Nazca Ridge subduction is the most likely mechanism behind the tilting/uplift along the Linea Bala–Rogaguado. The backward sedimentation induced by the rapid uplift of the Fitzcarrald Arch could have buried and hidden the Andean forebulge. In more general terms, the Bolivian Amazon can be divided into two areas under different depositional regimes: the uplifting north, crossed by incising rivers, which includes Pando and the northern LM; and the subsiding central and southern LM which are crossed by free meandering rivers. These two settings are the expression in the surface of the flat slab subduction in the north and normal subduction in the south. These two areas are separated by a transitional zone, which is flat, crossed by underfit rivers and also uplifting.

What is the extent and timing of the neotectonic activity?

It is unlikely that the different neotectonic elements of the landscape that have been reviewed here (the formation of ria lakes, the uplifted transitional zone, the avulsions of the Beni River, the formation of interior deltas and the changes in the meandering pattern of the Mamoré) River all happened at the same time and are the result of the same tectonic event. The data here presented suggest that the most likely scenario is that several tectonic events created and modified lakes and rivers at different times. The radiocarbon ages from the bottom of the lacustrine phase of Lake Rogaguado (which indicate that the lake formed about 6000 BP) and the radiocarbon ages of the palaeosols underlying fluvial deposits in central LM (which point to a mid- to late Holocene burial of the palaeosol) are, for now, the only fixed elements around which a relative chronology can be built. The formation of Lake Rogaguado is here interpreted as the last step in a series of correlated events.

Pleistocene

At the end of the Pleistocene, the Beni River flowed from east to west, up to a few kilometres north of Santa Ana de Yacuma, and then turned north in what is the modern Mamoré (Fig. 12a). The sediments that constitute the fluvial levees north of Puerto Siles (Fig. 7c) were probably deposited by the Beni River at this time. The Mamoré River used to flow further east, following the course now occupied by the far smaller Machupo River. Grande River was not a tributary of the Mamoré, but travelled north, discharging into the Iténez River, as well as the Mamoré (Fig. 12a).

Early Holocene

Probably during the late Pleistocene or at the beginning of the Holocene, and because of a fault along the Andean piedmont (Dumont, 1996), the River Beni shifted and occupied the valleys where, later, the lakes Rogaguado and Ginebra formed (Fig. 12b). Interestingly, the western border of the transitional zone coincides with the north–south oriented fault thought to have caused the migration of the deflection point of the Beni River, suggesting that the uplift of the transitional zone controlled the evolution of the Beni River. The same event also reduced the slope of the Maniqui River, inducing the formation of the Maniqui interior delta shown in Fig. 3d. During this time, the Beni River, and probably also the other rivers, was incising the northern LM in relatively deep valleys, also inducing erosion of the lateral valleys. Stratigraphic cores of these structures would allow us to date the underlying palaeosols and the time of the formation of both the Maniqui Delta and the Beni crevasse, allowing us to test whether or not their formation followed neotectonic uplifts. Palaeo-ecological records indicate that during the last glacial period (45.0–12.2 kyr BP) the climate in the LM was much drier than during the Holocene (Burbridge et al., 2004; Whitney et al., 2011). The first uplifting event that

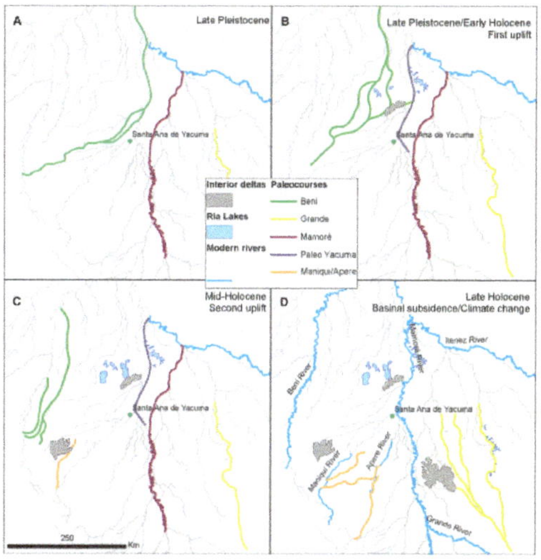

Figure 12. Reconstruction of the evolution of the landscape in the LM since the late Pleistocene. (a) shows the situation during the late Pleistocene with the Beni River flowing in what is now the Mamoré channel and the Mamoré River flowing towards the Iténez. Also the Grande River had a different course, flowing north into the Iténez. (b) shows the changes that followed the first uplift, with the Beni River shifting counterclockwise and the formation of a first group of ria lakes (Lake Oceano, Agua Clara and Largo). The crevasse splay of the Beni River in Fig. 3b probably formed during this time too. (c) shows how the Beni River shifted further west and how a second uplift formed the lakes Rogaguado and Ginebra. The DFS of the Maniqui River started forming at this time. (d) shows the modern setting, as it established during the late Holocene.

can be deduced from the geomorphological evidence happened north of 12.5° S and could be related to the formation of the knickpoint at Guayaramerín (Fig. 5). This uplift caused the backward sedimentation of the Beni River that, damming its lateral valleys, formed the lakes Agua Clara and Largo. This event could also be responsible for the formation of Lake Oceano and the other small ria lakes on the eastern margin of the Mamoré River (Fig. 12b). Further cores from lakes Largo, Agua Clara, Oceano and Ginebra are needed in order to date the beginning of the lacustrine sedimentation in these lakes and establish a more reliable chronology of the uplift events.

Mid-Holocene

A second neotectonic event, about 6 kyr BP, formed the knickpoint at Puerto Siles, caused the tilting along the Linea Bala–Rogaguado and formed the lakes Rogaguado and Ginebra (Fig. 12c). The further shift of the Beni River from the Rogaguado and Ginebra valleys towards its current position could have preceded or been synchronous to the Bala– Rogaguado tilting. Lake cores from oxbows located within ancient gallery forests of the Beni River palaeocourses could help determine the timing of the Beni River avulsions and whether the avulsions were synchronous to the formation of the ria lakes. After the first shift of the Beni and before the Mamoré moved to its modern position, the upper course of the modern Mamoré was occupied by an underfit river with a smaller catchment, here called the palaeo-Yacuma in Fig. 12b and c. The palaeo-Yacuma probably transported only fine sediments, because most of its catchment would have been within the LM. The tectonic events that created the lakes Agua Clara and Largo first and the lakes Rogaguado and Ginebra later, probably also dammed this small river, which, during the rainy seasons, became an ephemeral ria lake, resembling the lower sections of the modern Negro, Xingú and Tapajó rivers. During this period, clayish sediments could have deposited within the fluvial valley, forming the thick organic layers that outcrops today along the banks of the Mamoré. These layers have been interpreted as palaeosols formed in the floodplain during periods of reduced river sedimentation, (Plotzki, 2013). Along the bank of the Mamorè, about 50 km north of Santa Ana, the exposed organic layer is about 3 m thick, and radiocarbon data indicates it formed between 12 and 3 kyr BP (Plotzki, 2013).

However, the thickness of these palaeosols suggests that they are cumulative palaeosols, where deposition and pedogenesis coexisted (Kraus and Aslan, 1993). Their formation could be better explained by two cyclical processes working at different timescales. On a century to millennium scale, periods of more severe water ponding could have been cyclically alternated with periods of increased drainage. Clays would deposit during periods of water ponding due to downriver uplifts. These uplift events would create knickpoints that the rivers would start to erode immediately; the modern analogues for this situation are the knickpoints at Puerto Siles and Guyaramerín (Fig. 5). The erosion of the knickpoint would increase the drainage, lower the level of the river, reduce the sedimentation and allow for soil formation. On a seasonal scale, clays would have been deposited during the rainy season, when water was ponded within the river valleys; wetlands would have formed during the dry season, following the lowering of the water table.

These alternate phases of deposition of clays with soil formation could eventually form very thick layers of organic-rich clays as those found along the Mamoré banks. Evidence that uplift in the northern part of the LM could have induced water ponding, backward sedimentation and elevation of the water table several times during the Holocene, can also be seen in fossil wood (dated 5.575±105, 7.145±150, 9.995±150 and 10.085±150 yr BP) found embedded in clay layers along the rivers Acre, Beni and Madre de Dios (Campbell and Frailey, 1984; Campbell et al., 1985). According to Campbell, the presence of these clayey deposits over a very large area is indicative of a lacustrine depositional environment and is the result of the formation of a large lake, Lake Mojos, which would have had a southward drainage, belonging to the La Plata River basin (Campbell, 1989 cited in Hanagarth, 1993). It has been suggested that during the Holocene, the sedimentation of the Grande and Parapetí rivers closed the southern drainage and that Lake Mojos broke the northern enclosure, leading to the modern situation (Hanagart and Sarmiento, 1990).

However, the existence of such a lake is at odds with the general topography of the Cerrado Beniano, which is crossed by many fluvial valleys and not just one, as it would be in the case of Campbell's reconstruction. Moreover, if Lake Mojos drained towards the south, we should find in the Cerrado Beniano a palaeodrainage network which converges toward the south, but, as shown in Fig. 9, this is not the case. As for the palaeosols along the Mamoré, the presence of the layers of clay observed along many rivers in northern LM (Campbell and Frailey, 1984; Campbell et al., 1985), could be the result of water ponding and backwards sedimentations caused by downriver uplifts. If this hypothesis is correct, such deposits could serve as proxies to trace back neotectonic events.

Late Holocene

The situation of the Mamoré and Grande rivers did not change until the late Holocene, when the palaeo-Yacuma course was occupied by the Mamoré and the rest of the modern fluvial network was established (Fig. 12d). During the mid- to late Holocene, a large transformation occurred in the central and southern LM. Perhaps due to a climate shift towards wetter conditions (Lombardo et al., 2012; Plotzki et al., 2013), coupled with basinal subsidence (Hanagarth, 1993; Dumont and Fournier, 1994), the Mamoré established its modern position and the Grande River formed a large distributary system (Fig. 12d). It has also been suggested that, when the Grande River became a tributary of the Mamoré, the increased discharge triggered the Mamoré avulsion (Plotzki et al., 2013). This research shows that during the late Holocene, the western side of the LM was also subject to extensive fluvial depositions and the formation of distributary systems in the case of the rivers Maniqui and Apere, as indicated by the recurrent palaeosols shown in Fig. 4. As for the avulsion of the Grande River and for the fluvial deposition in the western LM, there is insufficient data yet to establish whether it was caused by an increase in precipitation, basinal subsidence or a combination of both.

How did neotectonics affect the LM hydrology during the Holocene?

Hydrology and landscape

The Fitzcarrald Arch's uplift played an important role in the evolution of the Amazon Basin and a major role in increasing its biodiversity (Espurt et al., 2010). However, in the case of the Bolivian Amazon, its effects have not yet been explored. Here, I suggest that the uplift of the Fitzcarrald Arch has determined the evolution of the landscape of the LM at both a regional and a local scale. Tectonics control the differences in soils and vegetation between the hilly landscape of the Cerrado Beniano, the flat and incised transitional zone, and the vast floodplain of the central and southern LM. In the Cerrado Beniano, the uplift causes the outcropping of lateritic crusts and the formation of a hilly topography, while, in the transitional zone, it prevents the deposition of fluvial sediments over the savannahs, with detrimental effects on soil fertility (Lombardo et al., 2013a).

The southern LM is a subsiding area, therefore subjected to the deposition of recent fluvial sediments, and hence more fertile. The transition between the Cerrado Beniano and the central/southern LM is also the transition between the uplifting north and the subsiding south. This exerts an important control over the river network of the western LM, determining the underfitting of rivers that flow north at approximately −14° S. When the uplift causes a decrease in river slope, interior deltas and wetlands form, like the large interior delta of the Maniqui River which could have been induced by the uplift of the transitional zone. The multiplying effect of precipitation on the severity of the floods, where an increase of less than 100 % in mean annual precipitation causes the size of the flooded area to increase by more than 300 %, is caused by the extremely flat landscape in the LM. Precipitation alone would not cause any floods if the LM had a better drainage. The savannah– forest ecotone in the central and southern LM is controlled by the duration and intensity of the seasonal floods (Mayle et al., 2007).

Hamilton et al. (2004) have pointed out that the outcrops of the Brazilian Shield in the northern LM constrain the river channels and contribute to a backwater effect at high water. Therefore, as floods determine the savannah– forest ecotone, drainage determines how floods respond to precipitation and neotectonics determines drainage, it can be concluded that uplifts in northern LM could have been a determining factor in the evolution of the savannah–forest ecotone in the central and southern LM during the Holocene. Uplift events in the northern LM reduced the drainage of the basin, but they were followed by stronger erosion by the rivers going through the uplifted rocks. This study suggests that uplifts were responsible for (1) the creation of interior deltas, (2) the elevation of the river beds with formation of ria lakes in some of the lateral valleys, and (3) the increase in sedimentation of fine silts and clays in the fluvial valleys, with a consequent flattening of the topography and creation of flat valley bottoms.

Conversely, periods between different uplifts were characterized by deeper river incision, improved drainage, lowering of the LM water table, reduction of the flooded areas and widespread changes in the process of soil formation and vegetation assemblages. This interplay between uplift and river incision could have been, together with climate, a major player in controlling inundation patterns in the LM and the most important factor determining the evolution of the LM landscape. Moreover, the changes in river slope due to downriver uplifts would have had an impact on the sedimentary load of the Madeira River and, consequently, of the Amazon River, as neotectonics not only affects the shape and behaviour of channels and flood plains but also determines the river's suspended-load transport capacity (Dunne et al., 1998).

Hydrology and people

Likewise, as people lived in the LM since the early Holocene (Lombardo et al., 2013b), it is to be expected that neotectonics had important effects on pre-Columbian peoples too. There is very little data yet about early Holocene human presence in southern Amazonia, the only study published so far (Lombardo et al., 2013b) suggests that hunter-gatherers settled in the southern LM from 10 500 till 4200 cal yr BP, approximately. It would seem that pre-Columbians were able to adapt to changes brought about by mid-Holocene neotectonic events; nevertheless, the formation of the Grande River megafan, approximately 4200 cal yr BP, which buried the hunter-gatherer sites, was followed by a period of about 2000 yr of archaeological hiatus.

Although the specific cause behind the Grande River avulsion is unclear, it is evident that it is basinal subsidence that, accommodating the sediment charge, created the preconditions for the formation of megafans. It is likely that the deposition of fertile sediments in southern LM created the necessary conditions for the subsequent development of complex societies in the southeastern LM, the so-called Earthmovers, about 1400 cal yr BP (Lombardo et al., 2012). It has been suggested that, late Holocene pre-Columbian settlement patterns, adaptive strategies and social complexity followed ecological gradients of soil fertility and flood intensity (Lombardo et al., 2011b). The data here presented shows that these ecological gradients are, to a great extent, the expression of the underlying geological control. 6

CONCLUSIONS

This paper has reviewed the published evidence of neotectonic activity in southern Amazonia and integrated it with new data from the analysis of remote sensing imagery and extensive field work in the LM. Based on the study of modern and palaeorivers, ria lakes, palaeosols and landscape topography in this region, this paper suggests that the uplifting Fitzcarrald Arch is the most likely mechanism behind the observed uplift along the Linea Bala–Rogaguado, the general topography of the LM, the outcropping of the basement in the northern LM, the presence of underfit rivers crossing the savannahs in the transitional zone, the meandering pattern of the Mamoré River and the general shape of the fluvial network. I suggest that the tectonic regime brought about by the Fitzcarrald Arch has determined the hydrology, flooding patterns, the forest-savannah ecotone and overall evolution of the landscape of the LM at both a regional and a local scale. These findings stress that, contrary to what many researchers have assumed, climate and human activities have not been the only factors influencing landscape evolution in the Bolivian Amazon during the Holocene. On the contrary, the interplay between the Fitzcarrald Arch uplift and river incision could have played a fundamental role in the evolution of the LM's landscape since the late Pleistocene. It is unlikely that the different neotectonic elements of the landscape that have been reviewed here were caused by a single uplift.

Instead, I propose a reconstruction of the evolution of the landscape of southern Amazonia in which at least two uplift events took place. More data is needed in order to confirm the hypothesis here presented and to better constrain the timing of the different events, such as the formation of ria lakes and interior deltas, river avulsions and deposition of crevasse spays. Nevertheless, the reconstruction here proposed represents a useful working model for future research and highlights the importance of taking into account neotectonics in palaeo-environmental reconstructions of southern Amazonia. In addition, this work highlights the need to disentangle the effects of human populations, climate change and neotectonics on the evolution of the LM, in order to better predict the impact of future climate change in the region and develop better adaptation strategies for local communities.

ACKNOWLEDGEMENTS

The present study has been funded by the Swiss National Science Foundation (SNSF), grants no. 200020 and 141277/1. I would like to thank H. Veit for his support and guidance. Lake Rogaguado has been cored in collaboration with the IRD project HydroGen; M. Grosjean and M-P. Ledru lent me the coring devices and taught me how to use them. L. Rodrigues and A. Giesche helped in several aspects of the fieldwork; J. Bocchetti helped me with several logistical problems in the field, including the fixing of the Waker motor. I am very grateful for the hospitality of the Community of Coquinal and, very specially, to Jonny Ibáñez, who pulled our car out of a crocodile-infested pond in the middle of the night. Further thanks are also due to E. Canal-Beeby, D. Garcia-Castellanos,

J. Carson, the editor, S. Mudd, and two anonymous reviewers who helped improve earlier versions of the manuscript.

REFERENCES

1. Aalto, R., Maurice-Bourgoin, L., Dunne, T., Montgomery, D. R., Nittrouer, C. A., and Guyot, J.-L.: Episodic sediment accumulation on Amazonian flood plains influenced by El Niño/Southern Oscillation, Nature, 425, 493–497, doi:10.1038/nature01990, 2003.
2. Allenby, R. J.: Origin of rectangular and aligned lakes in the Beni Basin of Bolivia, Tectonophysics, 145, 1–20, 1988. Angermeier, P. L.: The Natural Imperative for Biological Conservation, Conserv. Biol., 14, 373–381, doi:10.1046/j.1523- 1739.2000.98362.x, 2000.
3. Behling, H., Keim, G., Irion, G., Junk, W., and Nunes de Mello, J.: Holocene environmental changes in the Central Amazon Basin inferred from Lago Calado (Brazil), Palaeogeogr. Palaeocl., 173, 87–101, 2001.
4. Baker, V. R.: Adjustment of fluvial systems to climate and source terrain in tropical and subtropical environments, in: Fluvial Sedimentology, edited by: Miall, A. D., Memoir-Canadian Society of Petroleum Geologists, 211–230, 1977.
5. Boixadera, J., Poch, R. M., García-González, M. T., and Vizcayno, C.: Hydromorphic and clay-related processes in soils from the Llanos de Moxos (northern Bolivia), Catena, 54, 403– 424, doi:10.1016/s0341-8162(03)00134-6, 2003.
6. Bourrel, L. and Pouilly, M.: Hidrología y dinámica fluvial del Río Mamoré, in: Diversidad biológica en la llanura de inundación del Río Mamoré, edited by: Pouilly, M., Beck, S., Moraes, M., and Ibañez, C., Centro de Ecología Simón I. Patiño, Santa Cruz, Bolivia, 95–116, 2004.
7. Bourrel, L., Phillips, L., and Moreau, S.: The dynamics of floods in the Bolivian Amazon Basin, Hydrol. Process., 23, 3161–3167, doi:10.1002/hyp.7384, 2009.
8. Bridge, J. S.: Rivers and floodplains: forms, processes, and sedimentary record, John Wiley & Sons, Oxford, UK, 2009.
9. Buehler, H. A., Weissmann, G. S., Scuderi, L. A., and Hartley, A. J.: Spatial and Temporal Evolution of an Avulsion on the Taquari River Distributive Fluvial System from Satellite Image Analysis, Journal of Sediment. Res., 81, 630–640, doi:10.2110/jsr.2011.040, 2011.
10. Burbank, D. W. and Anderson, R. S.: Tectonic geomorphology, John Wiley & Sons, Chichester, UK, 2011. Burbridge, R. E., Mayle, F. E., and Killeen, T. J.: Fifty-thousandyear vegetation and climate history of Noel Kempff Mercado National Park, Bolivian Amazon, Quaternary Res., 61, 215–230, doi:10.1016/j.yqres.2003.12.004, 2004.
11. Bush, M. B. and Silman, M. R.: Amazonian exploitation revisited: ecological asymmetry and the policy pendulum, Front. Ecol. Environ., 5, 457–465, doi:10.1890/070018, 2007.
12. Campbell, K. E. and Frailey, C. D.: Holocene flooding and species diversity in southwestern Amazonia, Quaternary Res., 21, 369– 375, doi:10.1016/0033-5894(84)90075-9, 1984.
13. Campbell, K. E., Frailey, C. D., and Arellano, J. L.: The geology of the Rio Beni: Further evidence for Holocene flooding in Amazonia, Contributions in Science, Natural History Museum of Los Angeles County, 364, 1–18, 1985.
14. Carson, J. F., Whitney, B. S., Mayle, F. E., Iriarte, J., Prümers, H., Soto, J. D., and Watling, J.: Environmental impact of geometric earthwork construction in pre-Columbian Amazonia, P. Natl. Acad. Sci., 11, 10497–10502, doi:10.1073/pnas.1321770111, 2014.
15. Costa, J. B. S., Léa Bemerguy, R., Hasui, Y., and da Silva Borges, M. C.: Tectonics and paleogeography along the Amazon river, J. S. Am. Earth Sci., 14, 335–347, doi:10.1016/s0895- 9811(01)00025-6, 2001.
16. Dearing, J. A.: Landscape change and resilience theory: a palaeoenvironmental assessment from Yunnan, SW China, The Holocene, 18, 117–127, doi:10.1177/0959683607085601, 2008.
17. DeCelles, P. G. and Giles, K. A.: Foreland basin systems, Basin Research, 8, 105–123, doi:10.1046/j.1365-2117.1996.01491.x, 1996.
18. Dumont, J. F.: Type of lakes as related to neotectonics in western Amazonia, International symposium on the Quaternary of Amazonia, 8–13 November 1992, Manaus, Brazil, 1993.
19. Dumont, J. F.: Neotectonics of the Subandes-Brazilian craton boundary using geomorphological data: the Marañon and Beni basins, Tectonophysics, 257, 137–151, 1996.
20. Dumont, J. F. and Fournier, M.: Geodynamic environment of Quaternary morphostructures of the subandean foreland basins of Peru and Bolivia: Characteristics and study methods, Quatern. Int., 21, 129–142, doi:10.1016/1040-6182(94)90027-2, 1994.
21. Dumont, J. F. and Guyot, J. L.: Ria lac: ou, pourquoi?, Proceedings of the Third International Conference on Geomorphology, 23–29 August, 1993, Hamilton, Canada, 1993, Dunne, T., Mertes, L. A. K., Meade, R. H., Richey, J. E., and Forsberg, B. R.: Exchanges of sediment between the flood plain and channel of the Amazon River in Brazil, Geol. Soc. Am. Bull., 110, 450–467, doi:10.1130/0016-7606(1998)110< 0450:eosbtf> 2.3.co;2, 1998.
22. Erickson, C. L.: Amazonia: the historical ecology of a domesticated landscape, in: Handbook of South American archaeology, edited by: Silverman, H. and Isbell, W. H., Springer, Berlin, 157–183, 2008.
23. Espurt, N., Baby, P., Brusset, S., Roddaz, M., Hermoza, W., Regard, V., Antoine, P. O., Salas-Gismondi, R., and Bolaños, R.: How does the Nazca Ridge subduction influence the modern Amazonian foreland basin?, Geology, 35, 515–518, doi:10.1130/g23237a.1, 2007.
24. Espurt, N., Funiciello, F., Martinod, J., Guillaume, B., Regard, V., Faccenna, C., and Brusset, S.: Flat subduction dynamics and deformation of the South American plate: Insights from analog modeling, Tectonics, 27, TC3011, doi:10.1029/2007tc002175, 2008.

25. Espurt, N., Baby, P., Brusset, S., Roddaz, M., Hermoza, W., and Barbarand, J.: The Nazca Ridge and Uplift of the Fitzcarrald Arch: Implications for Regional Geology in Northern South America, in: Amazonia: Landscape and Species Evolution, edited by: Hoorn, C. and Wesselingh, F. P., Wiley-Blackwell Publishing Ltd., Oxford, UK, 89–100, 2010.

26. Fan, Y., Li, H., and Miguez-Macho, G.: Global Patterns of Groundwater Table Depth, Science, 339, 940–943, doi:10.1126/science.1229881, 2013.

27. Garcia-Castellanos, D.: Interplay between lithospheric flexure and river transport in foreland basins, Basin Research, 14, 89–104, doi:10.1046/j.1365-2117.2002.00174.x, 2002.

28. Garcia-Castellanos, D. and Cloetingh, S.: Modeling the Interaction between Lithospheric and Surface Processes in Foreland Basins, in: Tectonics of Sedimentary Basins, John Wiley & Sons Ltd., Chichester, UK, 152–181, 2011.

29. Gupta, A.: Tropical geomorphology, Cambridge University Press, New York, USA, 2011. Hajek, E. A. and Edmonds, D. A.: Is river avulsion style controlled by floodplain morphodynamics?, Geology, 42, 199–202, doi:10.1130/g35045.1, 2014.

30. Hamilton, S. K., Sippel, S. J., and Melack, J. M.: Seasonal inundation patterns in two large savanna floodplains of South America: the Llanos de Moxos(Bolivia) and the Llanos del Orinoco (Venezuela and Colombia), Hydrol. Process., 18, 2103–2116, doi:10.1002/hyp.5559, 2004.

31. Hanagart, W. and Sarmiento, J.: Reporte preliminar sobre la geoecología de la sabana de Espíritu y sus alrededores (Llanos de Moxos, departamento del Beni, Bolivia), Ecología en Bolivia, 16, 47–75, 1990.

32. Hanagarth, W.: Acerca de la geoecología de las sabanas del Beni en el noreste de Bolivia, Instituto de ecología, La Paz, Bolivia, 1993.

33. Hartley, A. J., Weissmann, G. S., Nichols, G. J., and Warwick, G. L.: Large Distributive Fluvial Systems: Characteristics, Distribution, and Controls on Development, J. Sediment. Res., 80, 167–183, doi:10.2110/jsr.2010.016, 2010.

34. Herzog, S., Maillard Z, O., Embert, D., Caballero, P., and Quiroga, D.: Range size estimates of Bolivian endemic bird species revisited: the importance of environmental data and national expert knowledge, J. Ornithol., 153, 1189–1202, doi:10.1007/s10336- 012-0850-2, 2012.

35. Hogg, A. G., Hua, Q., Blackwell, P. G., Niu, M., Buck, C. E., Guilderson, T. P., Heaton, T. J., Palmer, J. G., Reimer, P. J., Reimer, R. W., Turney, C. S. M., and Zimmerman, S. R. H.: SHCal13 Southern Hemisphere Calibration, 0–50 000 Years cal BP, Radiocarbon, 55, 1889–1903, 2013.

36. Holbrook, J. and Schumm, S. A.: Geomorphic and sedimentary response of rivers to tectonic deformation: a brief review and critique of a tool for recognizing subtle epeirogenic deformation in modern and ancient settings, Tectonophysics, 305, 287–306, 1999.

37. Irion, G., Bush, M. B., Nunes de Mello, J. A., Stüben, D., Neumann, T., Müller, G., Morais de, J. O., and Junk, J. W.: A multiproxy palaeoecological record of Holocene lake sediments from the Rio Tapajós, eastern Amazonia, Palaeogeogr. Palaeocl., 240, 523–535, doi:10.1016/j.palaeo.2006.03.005, 2006.

38. Irion, G., Mello, J. S. N., Morais, J., Piedade, M. F., Junk, W., and Garming, L.: Development of the Amazon Valley During the Middle to Late Quaternary: Sedimentological and Climatological Observations, in: Amazonian Floodplain Forests, edited by: Junk, W. J., Piedade, M. T. F., Wittmann, F., Schöngart, J and Parolin, P., Ecological Studies, Springer Netherlands, 27–42, 2011.

39. Jarvis, A., Reuter, H. I., Nelson, A., and Guevara, E.: Hole-filled SRTM for the globe Version 4, available atthe CGIARCSI SRTM 90m Database: http://srtm.csi.cgiar.org (last access: 15 December 2014), 2008.

40. Junk, W.: Current state of knowledge regarding South America wetlands and their future under global climate change, Aquatic Sciences – Research Across Boundaries, 75, 113–131, doi:10.1007/s00027-012-0253-8, 2013.

41. Junk, W. J., Bayley, P. B., and Sparks, R. E.: The flood pulse concept in river-floodplain systems, Canadian special publication of fisheries and aquatic sciences, 106, 110–127, 1989.

42. Kraus, M. J. and Aslan, A.: Eocene hydromorphic Paleosols; significance for interpreting ancient floodplain processes, J. Sediment. Res., 63, 453–463, doi:10.1306/d4267b22-2b26-11d7- 8648000102c1865d, 1993.

43. Langstroth, R. P.: Forest islands in an Amazonian savanna of northeastern Bolivia, University of Wisconsin-Madison, Unpublished PhD dissertation, 1996. Langstroth, R. P.: Biogeography of the Llanos de Moxos: natural and anthropogenic determinants, Geographica Helvetica, 66, 183–192, 2011.

44. Latrubesse, E. M. and Franzinelli, E.: The Holocene alluvial plain of the middle Amazon River, Brazil, Geomorphology, 44, 241– 257, doi:10.1016/S0169-555X(01)00177-5, 2002.

45. Latrubesse, E. M., Cozzuol, M., da Silva-Caminha, S. A. F., Rigsby, C. A., Absy, M. L., and Jaramillo, C.: The Late Miocene paleogeography of the Amazon Basin and the evolution of the Amazon River system, Earth-Sci. Rev., 99, 99–124, doi:10.1016/j.earscirev.2010.02.005, 2010.

46. Lombardo, U. and Prümers, H.: Pre-Columbian human occupation patterns in the eastern plains of the Llanos de Moxos, Bolivian Amazonia, J. Archaeol. Sci., 37, 1875–1885, doi:10.1016/j.jas.2010.02.011, 2010. Lombardo, U. and Veit, H.: The origin of oriented lakes: Evidence from the Bolivian Amazon, Geomorphology, 204, 502–509, doi:10.1016/j.geomorph.2013.08.029, 2014.

47. Lombardo, U., Canal-Beeby, E., Fehr, S., and Veit, H.: Raised fields in the Bolivian Amazonia: a prehistoric green revolution or a flood risk mitigation strategy?, J. Archaeol. Sci., 38, 502–512, doi:10.1016/j.jas.2010.09.022, 2011a.

48. Lombardo, U., Canal-Beeby, E., and Veit, H.: Eco-archaeological regions in the Bolivian Amazon: Linking pre-Columbian earthworks and environmental diversity, Geographica Helvetica, 66, 173–182, 2011b. Lombardo, U., May, J.-H., and Veit, H.: Mid- to late-Holocene

fluvial activity behind pre-Columbian social complexity in the southwestern Amazon basin, The Holocene, 22, 1035–1045, doi:10.1177/0959683612437872, 2012.

49. Lombardo, U., Denier, S., May, J.-H., Rodrigues, L., and Veit, H.: Human–environment interactions in pre-Columbian Amazonia: The case of the Llanos de Moxos, Bolivia, Quatern. Int., 312, 109–119, doi:10.1016/j.quaint.2013.01.007, 2013a.

50. Lombardo, U., Szabo, K., Capriles, J. M., May, J.-H., Amelung, W., Hutterer, R., Lehndorff, E., Plotzki, A., and Veit, H.: Early and Middle Holocene Hunter-Gatherer Occupations in Western Amazonia: The Hidden Shell Middens, PLoS ONE, 8, e72746, doi:10.1371/journal.pone.0072746, 2013b.

51. Lombardo, U., Denier, S., and Veit, H.: Soil properties and preColumbian settlement patterns in the Monumental Mounds Region of the Llanos de Moxos, Bolivian Amazon, SOIL Discuss., 1, 81–117, doi:10.5194/soild-1-81-2014, 2014. Mann, C. C.: Ancient earthmovers of the Amazon, Science, 321, 1148–1152, 2008.

52. Mayle, F. E., Burbridge, R., and Killeen, T. J.: Millennial-scale dynamics of southern Amazonian rain forests, Science, 290, 2291– 2294, doi:10.1126/science.290.5500.2291, 2000.

53. Mayle, F. E., Langstroth, R. P., Fisher, R. A., and Meir, P.: Longterm forest-savannah dynamics in the Bolivian Amazon: implications for conservation, Philos. T. Roy. Soc. B., 362, 291–307, doi:10.1098/rstb.2006.1987, 2007.

54. Melack, J. and Hess, L.: Remote Sensing of the Distribution and Extent of Wetlands in the Amazon Basin, in: Amazonian Floodplain Forests, edited by: Junk, W. J., Piedade, M. T. F., Wittmann, F., Schöngart, J., and Parolin, P., Ecological Studies, Springer Netherlands, 43–59, 2011.

55. Navarro, G.: Clasificación de la vegetación de Bolivia, Centro de ecología difusión fundación Simón I. Patiño, Santa Cruz, Bolivia, 2011.

56. Pitfield, P. E. J. and Power, G.: Geologic map MagdalenaSD 20-6 con parte de SD 20-2, Ordnance Survey 4/88/880413S, Southampton, UK, 1987. Plafker, G.: Oriented Lakes and Lineaments of Northeastern Bolivia, Geol. Soc. Am. Bull., 75, 503–522, doi:10.1130/0016- 7606(1964)75[503:olalon]2.0.co;2, 1964.

57. Plafker, G.: Tectonic implications of the oriented lakes and lineaments in northeastern Bolivia, First International Conference on the New Basement Tectonics, Salt Lake City, Utah, USA, 519– 527, 1974.

58. Plotzki, A.: Late Pleistocene to Holocene fluvial dynamics and environmental conditions in the Llanos de Moxos, Bolivian Amazon, PhD, Philosophisch naturwissenschaftlichen Fakultät, Universität Bern, Bern, Switzerland, 2013.

59. Plotzki, A., May, J. H., and Veit, H.: Past and recent fluvial dynamics in the Beni lowlands, NE Bolivia, Geographica Helvetica, 66, 164–172, 2011.

60. Plotzki, A., May, J. H., Preusser, F., and Veit, H.: Geomorphological and sedimentary evidence for late Pleistocene to Holocene hydrological change along the Río Mamoré, Bolivian Amazon, J. S. Am. Earth Sci., 47, 230–242, doi:10.1016/j.jsames.2013.08.003, 2013.

61. Räsänen, M. E., Salo, J. S., and Kalliola, R. J.: Fluvial Perturbance in the Western Amazon Basin: Regulation by Long-Term Sub-Andean Tectonics, Science, 238, 1398–1401, doi:10.1126/science.238.4832.1398, 1987.

62. Regard, V., Lagnous, R., Espurt, N., Darrozes, J., Baby, P., Roddaz, M., Calderon, Y., and Hermoza, W.: Geomorphic evidence for recent uplift of the Fitzcarrald Arch (Peru): A response to the Nazca Ridge subduction, Geomorphology, 107, 107–117, doi:10.1016/j.geomorph.2008.12.003, 2009.

63. Roddaz, M., Brusset, S., Baby, P., and Herail, G.: Miocene tidalinfluenced sedimentation to continental Pliocene sedimentation in the forebulge–backbulge depozones of the Beni–Mamore foreland Basin (northern Bolivia), J. S. Am. Earth Sci., 20, 351–368, doi:10.1016/j.jsames.2005.11.004, 2006.

64. Ronchail, J. and Gallaire, R.: ENSO and rainfall along the Zongo valley (Bolivia) from the Altiplano to the Amazon basin, Int. J. Climatol., 26, 1223–1236, doi:10.1002/joc.1296, 2006.

65. Ronchail, J., Bourrel, L., Cochonneau, G., Vauchel, P., Phillips, L., Castro, A., Guyot, J., and Deoliveira, E.: Inundations in the Mamoré basin (south-western Amazon—Bolivia) and seasurface temperature in the Pacific and Atlantic Oceans, J. Hydrol., 302, 223–238, doi:10.1016/j.jhydrol.2004.07.005, 2005.

66. Rossetti, D. F., Zani, H., Cohen, M. C. L., and Cremon, É. H.: A Late Pleistocene–Holocene wetland megafan in the Brazilian Amazonia, Sedimentary Geology, 282, 276–293, doi:10.1016/j.sedgeo.2012.09.015, 2012.

67. Schumm, S. A., Dumont, J. F., and Holbrook, J. M.: Active tectonics and alluvial rivers, Cambridge University Press, Cambridge, UK, 2002. Slingerland, R. and Smith, N. D.: River Avulsions and their deposits, Ann. Rev. Earth Planet. Sci., 32, 257–285, doi:10.1146/annurev.earth.32.101802.120201, 2004.

68. Tinner, W. and Ammann, B.: Long-term Responses of Mountain Ecosystems to Environmental Changes: Resilience, Adjustment, and Vulnerability, in: Global Change and Mountain Regions, edited by: Huber, U., Bugmann, H. M., and Reasoner, M., Advances in Global Change Research, Springer Netherlands, 133– 143, 2005.

69. UNASBVI-JICA: Estudio Hidrogeologico de Beni y Pando, Trinidad –Bolivia, 2009.

70. Urrego, D. H., Bush, M. B., Silman, M. R., Niccum, B. A., De La Rosa, P., McMichael, C. H., Hagen, S., and Palace, M.: Holocene fires, forest stability and human occupation in south-western Amazonia, J. Biogeogr., 40, 521–533, doi:10.1111/jbi.12016, 2013.

71. Vegas-Vilarrúbia, T., Rull, V., Montoya, E., and Safont, E.: Quaternary palaeoecology and nature conservation: a general review with examples from the neotropics, Quaternary Sci. Rev., 30, 2361–2388, doi:10.1016/j.quascirev.2011.05.006, 2011.

72. Walker, J.: Agricultural Change in the Bolivian Amazon,

Memoirs in Latin American Archaeology, University of Pittsburgh Latin American Archaeology Publications and Fundación Kenneth Lee, Trinidad, 2004.

73. Wallace, R., Martinez, J., Lopez-Strauss, H., Barreta, J., Reinaga, A., and Lopez, L.: Conservation Challenges Facing Two Threatened Endemic Titi Monkeys in a Naturally Fragmented Bolivian Forest, in: Primates in Fragments, edited by: Marsh, L. K. and Chapman, C. A., Developments in Primatology: Progress and Prospects, Springer New York, 493–501, 2013.

74. Weissmann, G. S., Hartley, A. J., Nichols, G. J., Scuderi, L. A., Olson, M., Buehler, H., and Banteah, R.: Fluvial form in modern continental sedimentary basins: Distributive fluvial systems, Geology, 38, 39–42, doi:10.1130/g30242.1, 2010.

75. Weissmann, G., Hartley, A., Scuderi, L., Nichols, G., Davidson, S., Owen, A., Atchley, S., Bhattacharya, J., Chakraborty, T., and Ghosh, P.: Prograding distributive fluvial systems: geomorphic models and ancient examples, New Frontiers in Paleopedology and Terrestrial Paleoclimatology: SEPM, Special Publication, 104, 131–147, 2013.

76. Whitney, B. S., Mayle, F. E., Punyasena, S. W., Fitzpatrick, K. A., Burn, M. J., Guillen, R., Chavez, E., Mann, D., Pennington, R. T., and Metcalfe, S. E.: A 45 kyr palaeoclimate record from the lowland interior of tropical South America, Palaeogeogr. Palaeocl., 307, 177–192, doi:10.1016/j.palaeo.2011.05.012, 2011.

77. Willis, K. J. and Birks, J.-B.: What is natural? The need for a long-term perspective in biodiversity conservation, Science, 314, 1261–1265, 2006.

CHAPTER 12

TECTONIC CONTROL ON DRAINAGE NETWORK

Girish Ch. Kothyari and B. K. Rastogi

INTRODUCTION

Cratonic part of central India has witnessed reactivation during Precambrian. Later during Cretaceous, a 1600 km long and 200 km wide ENE-WSW trending rift zone developed in central Indian peninsula [1]. Three pericontinental rift basins were formed in the western margin of the Indian shield. The Son-Narmada rift developed in parallel to the Satpura trend during late Cretaceous, while the Cambay graben formed in colinearity with the Dharwar trend during early Cretaceous [2]. The Son-Narmada fault (SNF) zone has a Precambrian ancestry which separates Vindhyan basin to its north and the Gondwana belt southwards [3]. In both northern and southern regions of the Son-Narmada lineament, vertical block movements have been reported [3, 4]. Due to these movements, the Bijawar group of rocks within the SNF zone got uplifted during the time of sedimentation in Vindhyan Supergroup [5]. Presence of series of ENE-WSW trending subsurface structures within the sediments of Cambay basin suggests extension of Satpura trend across the basin. The NE-SW Aravalli-Delhi trend is well reflected in northern part of Cambay basin [2, 6]. Faults and lineaments are mainly aligned parallel/subparallel to the major tectonic grains. Along the marginal faults of Cambay basin, the Deccan Traps have been downfaulted (3000–5000 m); throw being more in eastern marginal fault has resulted in asymmetric graben structure [7–10]. The Son-Narmada fault/lineament and Tapi north Fault have ENE-WSW trend and control the course of main drainage system [11]. The Son-Narmada-Tapi (SONATA) zone extends WNW up to the shore of the Arabian Sea. Represented by a fault pair, SONATA defines Narmada rift valley and Satpura horst. Tapi Fault marks the southern boundary of the Satpura range and northern boundary of Tapi alluvium. Several hot springs are reported in close proximity of this fault zone. Important lineaments present in this area include West Coast, Jaisalmer-Barwani, Rikhabdev, Chambal-Jamnagar, Kishangarh-Chipri (NE corner of Figure 1), and Pisangan-Vadnagar lineaments [2] (Figures 1(a) and 1(b)). The SNF zone divides the Indian plate into two halves and has a long tectonic history dating back to the Archaean time, trending ENE-WSW. Both Narmada and Tapi rivers all through their courses follow these tectonic trends (Figures 1(a)and 1(b)). Geophysical studies in the SNF zone reveal that this is a zone of intense deep seated faulting which reactivated around 2.5–2.2 and 1.5–0.9 Ga [12]. The faults dip steeply by 70°–80° [13]. The throw of the faults is of the order of 1500 to 1800 m [14]. The Narmada graben is displaced by 20–30 m along multiplicity of transverse tear faults. Pleistocene sedimentary terraces in SNF zone are tilted 20–25° in the tract Hardi-Jabalpur and are also displaced 10–15 m near Navagram [15]. The seismic structure and existence of sedimentary basin of SONATA zone is interpreted by Murty et al. [16] and Kumar et al. [17] using 2D crustal seismic velocity models and 2D gravity modeling by Mishra, [18], Verma and Banerjee [19], Sridhar and Tewari, [20] and Mishra et al. [21]. The high Bouguer gravity anomaly between Jabalpur and Mandla suggests localized uplift of SNSF and the presence of a magmatic body at different depths. Murty et al. [16] have concluded that the upper Vindhyans directly overlie the Archaean basement between Hirapur and Narsinghgarh. Between Narsinghgarh and Katangi, the Vindhyan basin consists of the upper and lower Vindhyans [22]. Narmada fault zone is divided into two parts, namely, Son-Narmada north fault (SNNF) and Son-Narmada south fault (SNSF). Using the basin morphology and drainage network, an attempt has been made to understand the role of geological structure in the evolution of drainage basin.

Citation: Girish Ch. Kothyari and B. K. Rastogi, "Tectonic Control on Drainage Network" Geography Journal, vol. 2013, Article ID 325808, 9 pages, 2013. doi:10.1155/2013/325808.

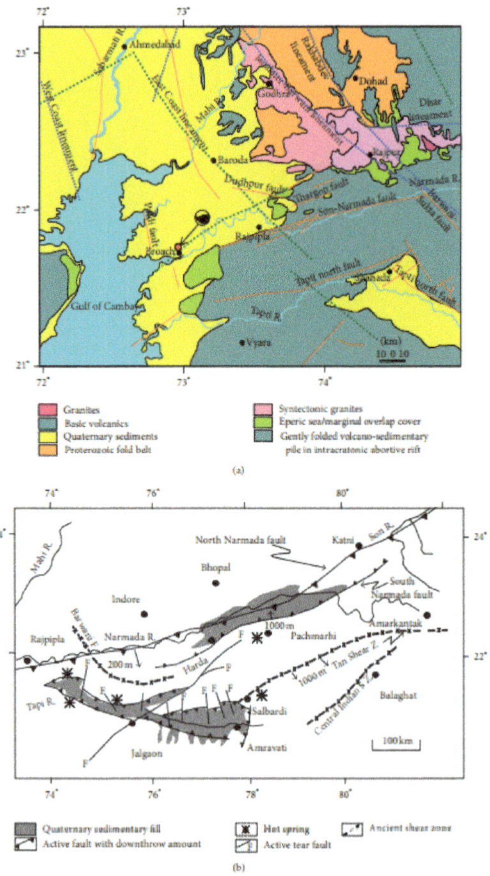

Figure 1: (a) Geological map of Narmada river valley shows major lineaments in the area [2]. (b) Map shows active grabens, the Narmada graben in the north, and Tapi graben in the south [15, 23]. Elevations of a few points are given.

TECTONIC SETTING AND STRATIGRAPHY OF (SONATA)

The Satpura range is bounded by Narmada graben in the north and the Tapi graben in the south. Seismic reflection and wide angle reflection data across the Son-Narmada lineament clearly bring out its graben character underneath the Deccan lavas in the west [24, 25]. These demonstrate presence of a horst like feature between the north Narmada fault and the south Narmada fault in the eastern part [22] (Figures 1(a) and 1(b)). The 1300 km long Narmada river in central India flows through three major bedrock gorges separated by alluvial basins. In the downstream direction, these are the Marble Canyon near Jabalpur, Punasa Gorge near Khandwa, and Dhadgaon Gorge west of Barwani [26]. These horst like features are associated with a midcrustal and Moho upwarp, implying its formation as a result of deep seated tectonics in the Precambrian time [27]. The grabens exhibit high heat flow of 290±50 mW per m^2 in the Tattapani-Jhor area in Narmada rift and 97–135 mW/m^2 at Salbardi in the Tapi valley [13]. The graben is continuing to subside along delimiting fault. The recent activity on the faults defining the graben is evident from abrupt eastward swing and reaches up to northerly flowing tributaries of the Narmada river along the foot of the alluvial scarp parallel to the graben fault. Various fault blocks and their alluvial cones are consequently sunken and tilted up to 300 m [15] in depth as seen in seismic profile [22] which shows that the basement-sediment contact is formed at 200–300 m depth. The Narmada-Tapi domain is thus a belt of tectonic resurgence where streams show deflection of channels as is evident from their longitudinal profiles (Figure 2). In the lower reaches of the Narmada, there was slow subsidence of the basin during the late Pleistocene time, followed by inversion of movement in the Holocene [11].

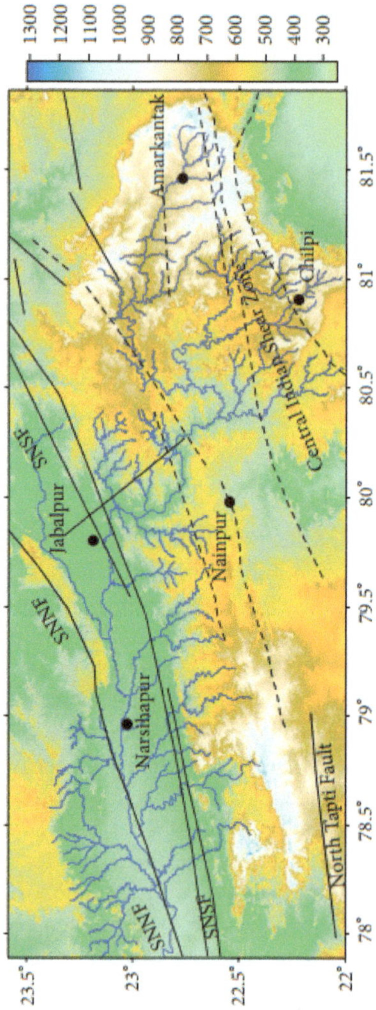

Figure 2: Digital elevation model (DEM) of upper Narmada valley, showing development of major landforms and position of major faults controlling the drainage basin. SNNF is Son-Narmada north fault, and SNSF is Son-Narmada south fault.

The SONATA zone in central Indian shield divides the Indian plate into two halves and has a long tectonic history dating back to the Archaean times [27] and trends in ENE-WSW direction and is laterally traceable for more than 1000 km [11]. It demarcates the Peninsular India into two geologically distinct provinces: the Vindhyan-Bundelkhand province to the north and the Deccan province to the south. The Narmada and Tapi rivers throughout their course follow these tectonic trends that are also known as central Indian tectonic zone (CITZ) [12, 28–30]. The zone witnessed large-scale tectonothermal events associated with large granitic intrusions around 2.5–2.2 and 1.5–0.9 Ga [12]. It was again reactivated during the Deccan volcanic eruption during Late Cretaceous-Palaeocene [11, 31]. The region is crisscrossed by E-W-trending dykes, which were probably formed due to main centre of eruptive activity [32], and the zone is thermomechanically and seismically vulnerable in the framework of contemporary tectonism [32]. Seismic reflection and wide angle reflection data of Dixit et al. [25] reveal three Tertiary sedimentary subbasins: the Mehmadabad-Jambusar basin, the Jambusar-Kosamba basin, and the Kosamba-Bilimora basin. These basins are vertically offset relative to each other along high-angle faults, and the maximum depth (~9.5 km) to the granitic/Proterozoic basement (5.8–6.0 km s^{-1}) in the region occurs in the Jambusar-Broach area. Near-surface sedimentary rocks within the subbasins are underlain by Deccan Trap volcanics (4.75–5.2 km s^{-1}), which have also been vertically offsetted [25]. The westward extension of this zone into the lower Narmada valley exhibits a less complex structural setting. In the lower Narmada basin, it is expressed as a single deep-seated fault (NSF) confirmed by the deep seismic sounding studies [25, 33]. Seismic reflection studies have firmly established that the NSF is a normal fault in the subsurface and becomes markedly reverse near the surface [34]. Reactivation of the fault in Late Cretaceous led to the formation of a depositional basin in which marine Bagh beds were deposited [35]. The NSF remained tectonically active since then with continuous subsidence of the northern block, designated as the Broach block, which accommodated 6-7 km thick Cenozoic sediments in Cambay basin [11, 35]. The Cenozoic sections within the SONATA zone shows offsetting of litho units [34]; however, the movements along this fault have not been unidirectional throughout. The general tendency of the basin to subside has been punctuated by phases of structural and tectonic inversion [34]. The N-S-directed compressive stresses during the early Quaternary folded the Tertiary sediments into a broad syncline, the Broach syncline, in the rapidly subsiding northern block [34].

GEOMORPHOMETRIC ANALYSIS

Methodology

It is well recognized that the commonly used geomorphometric indices of active tectonics are powerful tools to evaluate the relationship between tectonics and basin morphology to identify geologically recent deformations [36–38]. Calculation of the number of geomorphometric indices for a large region such as central peninsular India was made possible by the analysis of ASTER 1.5 arc second satellite data. The ASTER data have been found to be suitable for regional scale morphometric analysis [39, 40]. The Narmada river originates from Amarkantak at a height of 1068 m.a.s.l. (above sea level) and merges in gulf of Khambhat west of Bharuch. Longitudinal profile of Narmada river has been generated with the help of ASTER DEM to ascertain the role of major/minor structures in the evolution of valley landforms (Figure 2). For geomorphometry, four commonly used geomorphic indices of active tectonics such as stream length gradient index (SL), asymmetry factor (AF), transverse-topographic symmetry (T), and valley floor height Ratio (Vf) have been used which are calculated for the different segments. These indices are considered powerful tools to evaluate the relationship between tectonics and basin morphology on the regional or basin scale and to identify geologically recent deformation [36, 40].

Stream Length Gradient Index (SL)

The SL is used to infer stream power and rock erodibility [41] due to its sensitivity to disequilibrium state of channels due to tectonic and climatic perturbation in the channel slope [38]. The gradient index is expressed as

$$SL = \left(\frac{\Delta H}{\Delta L}\right) \times L \qquad (1)$$

where ΔH is the change in elevation of the reach, ΔL is the change in length of the reach, and L is the total length of the channel from the point where the SL index is being calculated upstream to the highest point on the channel [41]. Abrupt changes in slope (ΔH/ΔL) and the SL index have proved to be an excellent method to evaluate the ongoing processes of uplift of the underlying structures [42]. The SL index has been calculated all along the river profile which shows anomalous values in the vicinity of Shahpuri, Mandla, and Bhedaghat areas. The SL values range from 100 to 390 (between Shahpuri and Mandla) and from 100 to 1400 (between SNSF and Bhedaghat). The highest value (1400) is observed in vicinity of Bhedaghat area where the river passes through SNSF. In the SNSF zone, the river also shows high stream incision (Figure 3). The estimated incision depth in Amarkantak and downstream ranges between 10 and 120 m. In Shahpuri and Mandla area, the incision depth ranges between 5 and 50 m, whereas the incision depth of SNSF zone ranges between 5 and 40 m. Sudden increase of SL values implies differential uplift.

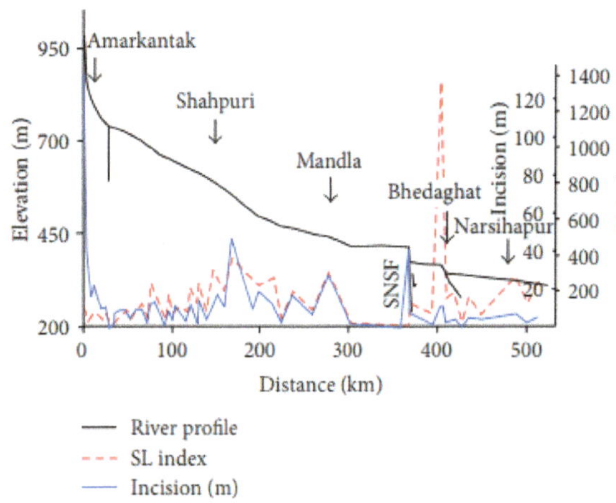

Figure 3: Longitudinal profile of Narmada river showing position of knick points at two places along the river. Changes in river gradient are marked by increase/decrease of SL index values as plotted by red brocken line. Solid blue line indicates bedrock depth incision along the longitudinal trace of river Narmada.

Basin Asymmetry (AF)

The outcome of stream length analysis is supported by evidence gathered from basin asymmetric factor. The asymmetry factor was developed to detect tectonic tilt by using the formula

$$AF = 100\left(\frac{Ar}{At}\right), \qquad (2)$$

where AF is the asymmetry factor, Ar is the area of the basin to the right, and At is total area of basin. The asymmetry factor of 40 points to regional tectonic tilt towards the NNW direction (Figure 4). Tilting of topography may be associated with activity of ENE-WSW trending SNF zone.

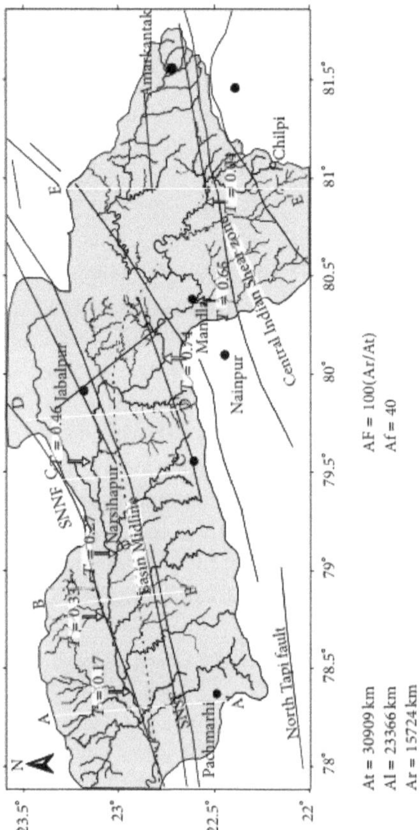

Figure 4: Drainage map of upper Narmada river basin showing calculated values of basin asymmetry (AF), topographic symmetric factor (*T*); solid white lines (A-A', B-B', C-C', D-D', and E-E') indicate cross-profile directions.

Transverse Topographic Symmetry (T)

Transverse topographic symmetry is another quantitative parameter to evaluate tectonic tilt of basin; it is expressed as

$$T = \frac{Da}{Dd} \tag{3}$$

where Da is the distance from midline of the drainage basin to the midline of the active meander belt. Dd is the distance from the basin midline to basin divide. In general, the value of T ranges from 0 to 1. T=0 implies a perfectly symmetrical basin and T=1 indicates a perfectly asymmetrical one [43]. Tectonic deformation impacts the drainage of that area through changes in the channel slope [44]. In present study, the observed value of for Narmada river ranges from 0.04 to 0.65 for (SNSF) implying ground tilting of river basins (Figure4).

Valley Floor Height and Width Ratio (Vf)

Valley floor height and width ratio (Vf) is defined as an index of the form shape of the valley; indicates whether the river is actively downcutting and incising [36] by

$$Vf = \frac{2Vfw}{((Eld - Esc) + (Erd - Esc))}, \tag{4}$$

where Vf is the valley floor width-to height ratio, Vfw is the width of the valley floor, Eld and Erd are elevations of the left and right valley divides, respectively, and Esc is the elevation of the valley floor. The downcutting of valley is commonly associated with the higher rate of uplift in tectonically active areas. The small values of Vf (<1) reflect deep narrow, V shaped valleys which are commonly associated with relative rapid uplift [40]. In contrast, high values (>1) are markers of wide open valleys in the regions of minimal uplift rates [37]. The index is used to evaluate tectonic activity along front impediment faults along the Son-Narmada south fault. We observed lower value of Vf towards the eastern limit of Jabalpur and comparatively higher value towards west which reflects that the eastern margin of Jabalpur is tectonically more active than the western limit. Near Jabalpur, Narsinghpur, and Hoshangabad, the Vf values are 24.31, 24.6 and 25.65 which are associated with broad open valleys indicating lateral erosion due to stability of base level and less tectonic activity. Near Amarkantak and Mandla, the obtained Vf ratios are 12.55 and 14.52 which are associated with intense tectonic perturbation in Quaternary time.

The elevation difference in N-S profiles shows that the southern block of basin is uplifted and the entire topography is tilted towards NNW direction (Figure 5). The southern block of basin is much elevated along the SNF and has rugged topography as compared to the northern block. The southern block is devoid of any Quaternary and indicates continuous uplift and erosion of the bed rocks. The presence of Quaternary deposits towards the north of basin suggests subsidence of the northern block relative to differential uplift of the southern block.

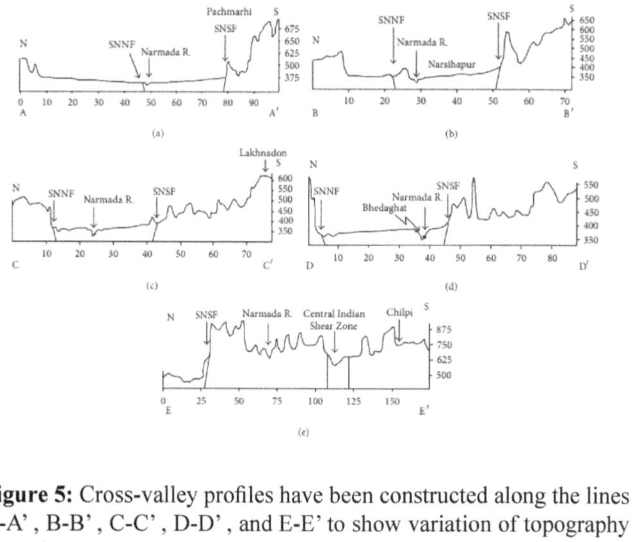

Figure 5: Cross-valley profiles have been constructed along the lines A-A', B-B', C-C', D-D', and E-E' to show variation of topography towards north and south directions. The elevation difference in N-S profiles shows that the southern block of basin is uplifted.

Neotectonic activity in Jabalpur area includes seismic activity and development of young geomorphic features. These features include development of strath terraces with huge Pleistocene (9–40 Ka) alluvial cover all along the river course and formation of gorges (~100 m deep) with steep vertical wall and development of 10 m high waterfall [26]. Incision and steepening of Narmada river floor with vertical wall probably formed due to vertical tectonic movement of Narmada faults during 60–40 ka [26].

Seismicity and GPS Observations

Historical and instrumental records indicate that the compressive stresses still continue to accumulate along the fault zone due to continued northward movement of the Indian plate. This is evidenced by the fault solution studies of the earthquakes at Broach (23 March 1970) and Jabalpur (22 May 1997) (Figures 1(a) and 6), which suggest a thrusting movement [45–47]. Results obtained from geomorphometric observations were validated and crosschecked with the results obtained from seismic and GPS observations of various workers. The ongoing movements are manifest in recurrent seismicity in upper Narmada catchment of central India. The enhanced earthquake activity in SONATA zone occurred during the period 1876 to 1993 which includes 23 March 1970 Bharuch earthquake of Mw 5.4, 1927 event of Mw 6.5 that occurred southwest of Rewa, 1938 earthquake of Mw 6.3 near Khandwa, 1945 earthquake of Mw 5.7 in Navagam, and the 1997 Jabalpur earthquake of Mw 6.0 (Figure 6). The 22 May 1997 Jabalpur earthquake occurred at a depth of 36±4 km along an ENE-WSE trending fault [48]. Fault plane solution indicates reverse faulting with a component of left lateral strike-slip motion [49]. This is the first significant strong earthquake that was recorded by the broadband stations of the national network, and the focal depth was well estimated using the depth phases [50]. It is observed that the earthquake was associated with the Narmada rift and was of deep crustal source. The seismic cross-section of, the main shock and aftershocks by Kayal [49], indicates that the south dipping Narmada south fault (NSF), the southern margin fault of the Narmada rift basin, was activated by reverse faulting and the earthquakes occurred at depth of 35 km and up. The seismic pattern in upper Narmada catchments is directly correlated with geomorphic indices suggesting that the eastern limit of upper Narmada river is more active than western limit. Occurrence of 1846, 1903, 1927, and 1997 tremors in and around the eastern limit of Jabalpur was correlated with lines of evidence gathered from geomorphic datasets. The observations by Kayal [49] further suggest that the rift basin boundary faults are mantle reaching; these faults are developed in tensional regime in the geological past and are now under compressive stress due to the NNE movement of the Indian plate [51].

Occurrence of seismicity in Son-Narmada fault zone was further allied with GPS observation of Banerjee et al. [52] and Apel et al. [51] (Figure 7). The velocity model of Apel et al. [51] shows northward motion in the south and southward motion in the north, across the SNF zone as indicated by black vectors in Figure 7, indicateing that the central part of India is under contraction [52]. The Son-Narmada lineament zone separates the Indian subcontinent into two blocks [51], for example, north Indian and south Indian blocks showing 4 mm/yr contraction in the east to 1 mm/yr in the west which suggests N-S contraction in central India along SNF (Figure 7). The Narmada-Son lineament through central India shows high heat flow and strain rate estimated from seismicity which is larger than that in many stable continental regions and suggests a concentration of intraplate deformation [53]. It is possible that the seismicity in the region is also enhanced by a thinned and weakened lithosphere due to passive-margin normal faulting in the Cretaceous [6, 25] and by heating from the plume head responsible for the late Cretaceous Deccan flood basalts [54, 55].

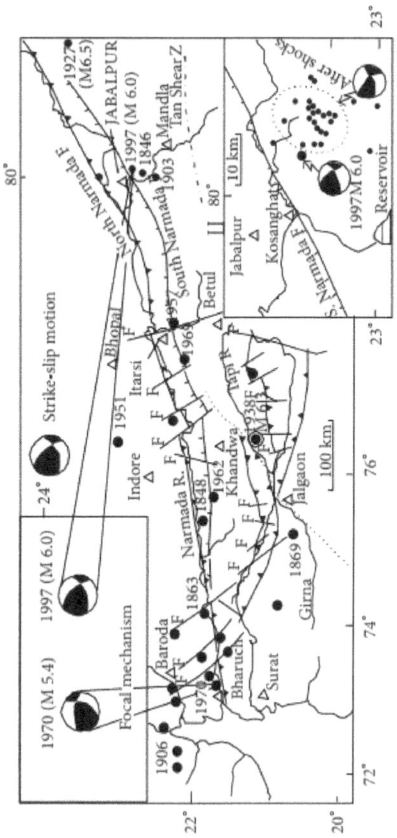

Figure 6: Tectonic map of SONATA shows epicentral distribution of earthquakes [15, 48,49]. The index beach ball shows thrust faulting with strike-slip component.

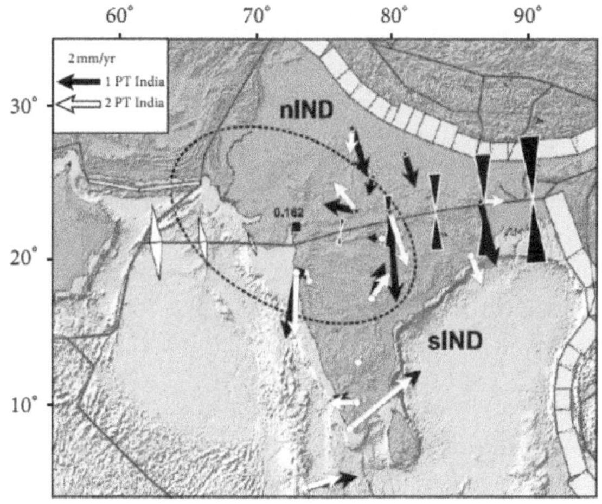

Figure 7: Residual velocities shown for the 1-block model (black vectors) and 2-block India model (white vectors). Separation of the In-

dian block (grey line) is along the Narmada-Son Lineament in Central India [51]. The relative pole of rotation for the 2-block India model is shown in the west with a 20 sigma error ellipse. The rotation rate of 0.162 deg/My predicts contraction along the Narmada-Son line from 5 mm/yr in the east to 0 mm/yr near the pole in the west (black triangles). Extensional rates in Arabian Sea are shown by White triangles [51].

DISCUSSION

Increase of river gradient index in two places in and around Jabalpur indicates vertical uplift due to N-S compression along SNF zone. Valley floor height and width ratio and basin asymmetric facture suggest that the entire basin is tilted towards NNW direction. The geomorphometric database and satellite data interpretation clearly suggest that the Son-Narmada south fault was activated during the Quaternary time. Various tectonically induced landforms were formed all along the Narmada river. Shifting of river channel (near Amarkantak) and waterfall; formation of gorge with nearly vertical wall in and around Bhedaghat, and development of strath terraces are common landform features which were formed due to movement along SNF. These observations suggest strain accumulation in the Son-Narmada fault zone.

The geomagnetic field variation and GPS observations also suggested strain accumulation in the region. A high geomagnetic anomaly field variation in Narmada river region ranging between ±0.06 and ±9.54 nT indicates accumulation of stress within the SNF zone [1, 51]. On the basis of aeromagnetic database, Anand and Rajaram [56] believe that the two ENE-WSW and NW-SE trending tectonic features were reactivated in Quaternary time. GPS observations of Apel et al. [51] suggest that the strain is accumulating beneath the Son-Narmada south fault. The accumulation of strain beneath this zone probably may cause strong earthquakes in future.

The morphometric indices and geomorphological observations in the Narmada river suggest the role of seismicity in the evolution of fluvial landform. Conventionally, tilting of the basin is attributed to tectonic control. The Narmada river is occupied by the rocks belonging to Vindhyan and Satpura towards upper reaches and Deccan basalt towards lower reaches. The river system is controlled by two parallel faults (SNNF and SNSF). The ongoing tectonic uplift along preexisting faults is conventionally examined through abrupt changes in river gradient. The observed SL values between Shahpuri and Mandla ((100–390) and (100–1400)) in and around Jabalpur suggest vertical tectonic processes along SNSF. The basin is tilting towards NNW due to tectonic movement along SNSF zone (Figure 3). Towards the upper reaches the river shows southwest migrations where T ranges between 0.04 and 0.65. The lower reaches show migration of river towards NNW which is supported by observed T (0.17–0.46). Migration of river can be assigned to low resistance bedrock or tectonic uplift along SNSF (Figure 4). The Vf value for upper reaches ranges between 12.55 and 14.52, and towards lower reaches, the Vf values are 24.31, 24.6, and 25.65. These abnormal changes in Vf towards upper and lower reaches could be associated with intense tectonic uplift. Further morphometric indices and geomorphic anomalies (water fall, gorges, etc.) together indicate that the terrain is active.

CONCLUSION

(i)Morphometric indices and the drainage pattern indicate that Shahpuri-Mandla and Jabalpur-Narsihapur segments in the SNF zone are active and showing enhanced uplift in Narmada river basin.(ii)Geodetic observations of Apel et al. [51] suggest that the SNSF zone is major stress accumulation zone in SONATA zone.(iii)Seismicity pattern and occurrence of moderate earthquakes in the region implie that the area has potential for another moderate earthquake in future.

ACKNOWLEDGMENTS

The study was supported by Gujarat State Disaster Management Authority. We thank the anonymous referees for their critical reviews and constructive comments which helped to improve the quality of the manuscript.

REFERENCES

1. S. Y. Waghmare and L. Carlo, “Geomagnetic Investivation in the seismoactive area of Narmada-Son lineament, Central India,” Journal of Indian Geophysical Union, vol. 12, no. 1, pp. 1–10, 2008.
2. Geological Survey of India, Projet Crumansonata: Geoscientific Studies of the Son-Narmada-Tapi Lineament Zone, 1995.
3. W. D. West, “The line of Narmada-Valley,” Current Science, vol. 31, pp. 143–144, 1962.
4. J. B. Auden, “Geological discussion of the Satpura hypothesis,” Proceedings of the National Institute of Sciences of India, vol. 15, pp. 315–340, 1949.
5. M. S. Krishnan and J. Swaminathan, “The great Vindhyan basin of northern India,” Journal of the Geological Society of India, vol. 1, pp. 10–30, 1959.
6. S. Biswas, I. Coutand, D. Grujic, C. Hager, D. Stöckli, and B. Grasemann, “Exhumation and uplift of the Shillong plateau and its influence on the eastern Himalayas: new constraints from apatite and zircon (uth-[sm])/he and apatite fission track analyses,” Tectonics, vol. 26, no. 6, 2007.
7. M. N. Qureshy, “Geophysical and Landsat lineament mapping—an approach illustrated from West-Central and South India,” Photogrammetria, vol. 37, no. 3–5, pp. 161–184, 1982.
8. A. K. Jain, N. Annup, and D. C. Singhal, “Crustal evolution of the Narmada-Son lineament and associated shear zones of the Indian lithosphere,” Indian Journal of Earth Science (CEISM Seminar), pp. 125–148, 1984.
9. D. M. Maurya, R. Rachna, and L. S. Chamyal, “Seismically induced deformational structures (seismites) from the mid-late Holocene terraces, lower mahi valley, Gujarat,” Journal of the Geological Society of India, vol. 51, no. 6, pp. 755–758, 1998.
10. D. M. Maurya, R. Rachna, and L. S. Chamyal, “History of tectonic evolution of Gujarat alluvial plains, western India during Quaternary: a review,” Journal of the Geological Society of India, vol. 55, no. 4, pp. 343–366, 2000.
11. L. S. Chamyal, D. M. Maurya, S. Bhandari, and R.

Raj, "Late Quaternary geomorphic evolution of the lower Narmada valley, Western India: implications for neotectonic activity along the Narmada-Son Fault," Geomorphology, vol. 46, no. 3-4, pp. 177–202, 2002. ·

12. S. K. Acharyya and A. Roy, "Tectonothermal history of the central Indian tectonic zone and reactivation of major faults/shear zones," Journal of the Geological Society of India, vol. 55, no. 3, pp. 239–256, 2000. ·
13. R. Shanker, Geothermal Atlas of India, Geological Survey of India, Kolkata, India, 1991.
14. D. C. Mishra and P. Kumar, "Characteristic of faults associated with Narmada-Son lineament and rock types of Jabalpur section," Current Science, vol. 75, pp. 308–310, 1998.
15. K. S. Valdiya, The Making of Indian Geodynamic Evolution, Macmillian Publication House, New Delhi, India, 2010.
16. A. S. N. Murty, D. M. Mall, P. R. K. Murty, and P. R. Reddy, "Two-dimensional crustal velocity structure along Hirapur-Mandla profile from seismic refraction and wide-angle reflection data," Pure and Applied Geophysics, vol. 152, no. 2, pp. 247–266, 1998.
17. P. Kumar, H. C. Tewari, and G. Khandekar, "An anomalous high-velocity layer at shallow crustal depths in the Narmada zone, India," Geophysical Journal International, vol. 142, no. 1, pp. 95–107, 2000.
18. D. C. Mishra, "Mid-continent gravity "high" of central India and the Gondwana tectonics,"Tectonophysics, vol. 212, no. 1-2, pp. 153–161, 1992.
19. R. K. Verma and P. Banerjee, "Nature of continental crust along the Narmada-Son lineament inferred from gravity and deep seismic sounding data," Tectonophysics, vol. 202, no. 2–4, pp. 375–397, 1992.
20. A. R. Sridhar and H. C. Tewari, "Existence of a sedimentary graben in the western part of Narmada Zone: seismic evidence," Journal of Geodynamics, vol. 31, no. 1, pp. 19–31, 2001.
21. D. C. Mishra, K. Arora, and V. M. Tiwari, "Gravity anomalies and associated tectonic features over the Indian Peninsular shield and adjoining ocean basins," Tectonophysics, vol. 379, no. 1–4, pp. 61–76, 2004. ·
22. H. C. Tewari, A. S. N. Murty, P. Kumar, and A. R. Sridhar, "A tectonic model of the Narmada region,"Current Science, vol. 80, no. 7, pp. 873–878, 2001.
23. V. S. Krishnaswamy and K. R. Raghunandan, "The Satpura uplift and the palaeoclimate of the Holocene and auxiliary evidence from the Valmiki Ramayana," Journal of the Geological Society of India, vol. 66, no. 2, pp. 161–170, 2005.
24. K. L. Kaila, P. R. K. Murty, D. M. Mall, M. M. Dixit, and D. Sarkar, "Deep seismic soundings along Hirapur-Mandla profile, central India," Geophysical Journal of the Royal Astronomical Society, vol. 89, no. 1, pp. 399–404, 1987.
25. M. M. Dixit, H. C. Tewari, and C. V. Rao, "Two-dimensional velocity model of the crust beneath the South Cambay Basin, India from refraction and wide-angle reflection data," Geophysical Journal International, vol. 181, no. 2, pp. 635–652, 2010.
26. A. Gupta, V. S. Kale, L. A. Owen, and A. K. Singhvi, "Late Quaternary bedrock incision in the Narmada river at Dardi Falls," Current Science, vol. 93, no. 4, pp. 564–567, 2007.
27. R. Shanker, "Thermal and crustal structure of "SONATA". A zone of mid continental rifting in Indian shield," Journal of the Geological Society of India, vol. 37, no. 3, pp. 211–220, 1991. ·
28. V. D. Choubey, "Narmada-Son lineament, India," Nature, vol. 232, pp. 38–40, 1971.
29. B. P. Radhakrishna and M. Ramakrishnan, "Archaean-Proterozoic boundary in India," Journal of the Geological Society of India, vol. 32, no. 4, pp. 263–278, 1988.
30. S. C. Jain, K. K. K. Nair, and D. B. Yedekar, "Geology of the Son-Narmada-Tapi lineament zone in Central India. Geoscientific studies of the Son-Narmada-Tapi lineament zone," in Project Crumansonata, vol. 10 of Special Publication, pp. 1–154, Geological Survey of India, 1995.
31. B. N. P. Agarwal, L. K. Das, K. Chakraborty, and C. H. Sivaji, "Analysis of the Bouguer anomaly over central India: a regional perspective," Journal of the Geological Society of India, vol. 31, pp. 469–493, 1995.
32. S. Bhattacharji, N. Chatterjee, and J. M. Wampler, "Zones of Narmada-Tapi rift reactivation and Deccan volcanism: geochronological and geochemical evidence," in Deccan Basalts, S. S. Deshmukh and K. K. K. Nair, Eds., pp. 329–340, Gondwana Geological Society, Nagpur, India, 1996.
33. K. L. Kaila, V. G. Krishna, and D. Mall, "Crustal structure along Mehmadabad-Billimora profile in the Cambay basin, India, from deep seismic soundings," Tectonophysics, vol. 76, no. 1-2, pp. 99–130, 1981. ·
34. T. K. Roy, "Structural styles in southern Cambay basin India and role of Narmada geofracture in the formation of giant hydrocarbon accumulation," Bulletin of the Oil and Natural Gas Commission, vol. 27, pp. 15–38, 1990.
35. S. K. Biswas, "Regional tectonic framework, structure and evolution of the western marginal basins of India," Tectonophysics, vol. 135, no. 4, pp. 307–327, 1987.
36. W. Bull and L. Mcfadden, "Tectonic geomorphology north and south of the Garlock Fault California," inGeomorphology in Arid Region, D. O. Doehring, Ed., pp. 115–138, State University of New York, Binghamton, NY, USA, 1977.
37. E. A. Keller and N. Pinter, Active Tectonics, Prentice Hall, Upper Seddle River, NJ, USA, 1996.
38. D. W. Burbank and R. S. Anderson, Tectonic Geomorphology, Blackwell Science, 2001.
39. D. Harbor and Y. Gunnell, "A long-strike escarpment heterogeneity of the Western Ghats: a synthesis of drainage and topography using digital morphometric tools," Journal of the Geological Society of India, vol.

70, no. 3, pp. 411–426, 2007.

40. V. S. Kale and N. Shejwalkar, "Uplift along the western margin of the Deccan Basalt Province: is there any geomorphometric evidence?" Journal of Earth System Science, vol. 117, no. 6, pp. 959–971, 2008.

41. J. T. Hack, "Stream profile analysis and stream gradient index," Journal of Research of the U.S. Geological Survey, vol. 1, pp. 421–429, 1973.

42. L. Seeber and V. Gornitz, "River profiles along the Himalayan arc as indicators of active tectonics,"Tectonophysics, vol. 92, no. 4, pp. 335–367, 1983.

43. P. W. Hare and T. W. Gardner, "Geomorphic indicators of vertical neotectonism along convergent plate margins, Nicoa Peninsula, Costa Rica," in Tectonic Geomorphology: Proceedings of the 15th Annual Binghamton Geomorphology Symposium, International Series, vol. 15, pp. 75–104, 1985.

44. V. Jain and R. Sinha, "Response of active tectonics on the Alluvial Baghmati river, Himalayan foreland basin, eastern India," Geomorphology, vol. 70, no. 3-4, pp. 339–356, 2005.

45. H. K. Gupta, I. Mohan, and H. Narain, "The Broach earthquake of March 23, 1970," Bulletin of the Seismological Society of America, vol. 62, pp. 47–61, 1972.

46. H. K. Gupta, R. K. Chadha, M. N. Rao et al., "The Jabalpur earthquake of May 22, 1997," Journal of the Geological Society of India, vol. 50, no. 1, pp. 85–91, 1997.

47. U. Chandra, "Earthquakes of peninsular India: a seismotectonics study," Bulletin of the Seismological Society of America, vol. 65, pp. 1387–1413, 1977.

48. K. Rajendran and C. P. Rajendran, "Characteristics of the 1997 Jabalpur earthquake and their bearing on its mechanism," Current Science, vol. 74, no. 2, pp. 168–174, 1998.

49. J. R. Kayal, "Seismotectonic study of the two recent SCR earthquakes in central India," Journal of the Geological Society of India, vol. 55, no. 2, pp. 123–138, 2000.

50. S. N. Bhattacharya, A. K. Ghose, G. Suresh, P. R. Baidya, and R. C. Saxena, "Source parameters of Jabalpur earthquake of May 22, 1997," Current Science, vol. 73, no. 10, pp. 855–863, 1997.

51. E. V. Apel, R. Burgmann, and P. Banerjee, "Indian plate motion, deformation, and plate boundary interactions," Geophysical Journal International. In press.

52. P. Banerjee, R. Bürgmann, B. Nagarajan, and E. Apel, "Intraplate deformation of the Indian subcontinent," Geophysical Research Letters, vol. 35, no. 18, 2008.

53. B. R. Rao, "Historical seismicity and deformation rates in the Indian peninsular shield," Journal of Seismology, vol. 4, no. 3, pp. 247–258, 2000.

54. B. L. N. Kennett and S. Widiyantoro, "A low seismic wavespeed anomaly beneath northwestern India: a seismic signature of the Deccan plume?" Earth and Planetary Science Letters, vol. 165, no. 1, pp. 145–155, 1999.

55. D. V. Chandrasekhar, R. Bürgmann, C. D. Reddy, P. S. Sunil, and D. A. Schmidt, "Weak mantle in NW India probed by geodetic measurements following the 2001 Bhuj earthquake," Earth and Planetary Science Letters, vol. 280, no. 1–4, pp. 229–235, 2009.

56. S. P. Anand and M. Rajaram, "Crustal structure of Narmada-Son Lineament: an aeromagnetic perspective," Earth, Planets and Space, vol. 56, no. 5, pp. 9–12, 2004.

CHAPTER
13 TECTONIC AND CLIMATIC CONTROLS

A. Margirier, L. Audin, J. Carcaillet, S. Schwartz, and C. Benavente

INTRODUCTION

In active mountain ranges, landslides are an important process in long-term erosion and thus contribute to the geomorphologic evolution of relief (Korup et al., 2007). Despite their importance in terms of hazards, landslide maps remain rare (Guzzetti et al., 2012) and information on the type, age or distribution of individual landslides is often lacking. Only a few publications deal with landslide triggering and/or evolution in arid contexts such as the western Andean flank, where several gigantic scarps disrupt the forearc piedmont (Audin and Bechir, 2006; Pinto et al., 2008; Strasser and Schlunegger, 2005; Wörner et al., 2002; Mather et al., 2014; Crosta et al., 2015). In contrast, because of the potential seismotectonic trigger (Keefer, 1984, 2002; McPhillips et al., 2014), landslide triggering along subduction active margins has been studied for a number of years, but most previous studies focused on humid climatic settings (Taiwan, New Zealand, Papua New Guinea, Japan; Meunier et al., 2008; Hovius et al., 2011). In southern Peru, the topographic gradient (average slope of 4 % between the coast and the Western Cordillera), the crustal seismotectonic activity and the aridity of the forearc region has been directly linked to Andean uplift and subduction of the Nazca Plate for the last 25 Myr at least (Devlin et al., 2012; Alpers and Brimhall, 1988; Dunai et al., 2005). However, Quaternary tectonic crustal activity and its role in the localization of landslides along the Western Andean Escarpment has never been explored in southern Peru and northern Chile. As a consequence, numerous questions remain concerning the importance of giant landslides in slope erosion relative to other nonseismic agents of erosion such as climatic forcing. The "Chuquibamba landslide" is a large complex zone of imbricated landslides (about 80 km^2) affecting the western Andean Cordillera in southern Peru. It belongs to the Andean arid piedmont, where geomorphic markers are minimally affected by erosion/transport processes (Hall et al., 2008). In this study, we aim to explore how seismotectonic activity and/or abrupt climate change may potentially have triggered the large Chuquibamba landslides. In this paper, we map out the area and characterize the tectonic and geomorphological settings, use terrestrial cosmogenic nuclides (TCN) to date pertinent markers of the last debris-flow event, and document the evolution of the landslide area.

Citation: Margirier, A., Audin, L., Carcaillet, J., Schwartz, S., and Benavente, C.: Tectonic and climatic controls, Earth Surf. Dynam., 3, 281-289, doi:10.5194/esurf-3-281-2015, 2015.

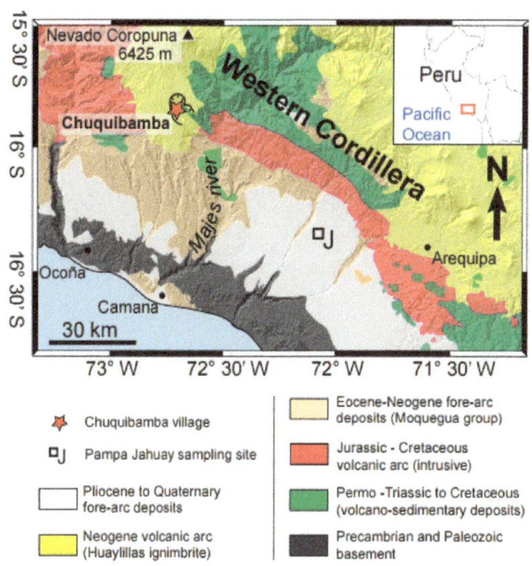

Figure 1. Simplified geological map of South Peru (modified from Roperch et al., 2006; INGEMMET, 2001), showing the village of Chuquibamba (red star), the destabilization zone and the Pampa Jahuay sampling site (J). Coordinates are given in WGS 84 longitude and latitude (degrees). Inset shows the study area location within Peru and a part of South America.

CONTEXT

Geologic and climatic setting of Chuquibamba region

The Andean Pacific arid front comprises different morphological units: the Coastal Cordillera 0–1000 m above sea level (a.s.l.), the Pacific piedmont 1000–1500 m a.s.l. and the Western

Cordillera 1500–6000 m a.s.l. In southern Peru, the Western Cordillera corresponds to a Jurassic to Cretaceous volcanic arc (Atherton et al., 1985) (Fig. 1). These magmatic and volcano-sedimentary rocks are emplaced within a Precambrian to Paleozoic basement. During the Eocene to Neogene, volcano-sedimentary deposits of the Moquegua group were overlain onto the Western Cordillera. This group is partly covered by the Huaylillas ignimbrite, produced during the Neogene volcanism and dated at 14.3–12.7 Ma (^{40}Ar / 39Ar; Thouret et al., 2007). The top of this ignimbrite forms a weathering surface. The Moquegua group and Huaylillas ignimbrite reach a maximum of 1500 m thick (Thouret et al., 2007; Gunnell et al., 2010). Some volcanosedimentary rocks were deposited on the coastal plains during the late Neogene. Starting at ~ 9–10 Ma, these units were incised by 1–3 km deep canyons that strike perpendicular to NW–SE-striking range (Schildgen et al., 2007, 2009; Thouret et al., 2007; Gunnell et al., 2010). The Western Cordillera and Pacific piedmont are affected by Quaternary tectonic deformation mainly expressed in the Western Cordillera by an active fault system striking parallel to the range (Sébrier et al., 1985; Hall et al., 2008, 2012) (Fig. 2). Along the Andes, the active subduction and induced crustal deformation largely control the geomorphologic evolution of the forearc (Keefer, 1994; Keefer and Moseley, 2004; Audin et al., 2006; Tavera et al., 2007; Perfettini et al., 2010). The piedmont region is exposed to extremely low denudation rates ranging from 0.1–1 mm kyr^{-1} in the coastal desert and Pacific piedmont to 1–46 mm kyr^{-1} in the Western Cordillera (values obtained for Quaternary timescale by TCN dating; Hall et al., 2008; Kober et al., 2007) due to the arid climatic conditions since at least the Neogene (Mortimer, 1980; Alpers and Brimhall, 1988; Dunai et al., 2005; Rech et al., 2006).

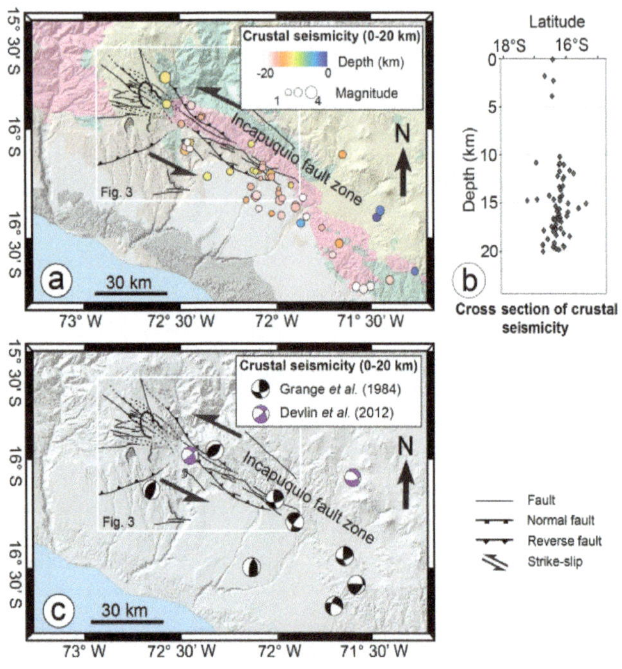

Figure 2. Regional crustal seismicity and focal mechanisms. (a) Crustal seismicity (depth < 20 km) and faults are represented on the geological map of South Peru (modified from Roperch et al., 2006; INGEMMET, 2001). (b) Cross section showing the vertical cluster of earthquakes. (c) Focal mechanisms from Grange et al. (1984) and Devlin et al. (2012); the kinematics of the Incapuquio fault system are indicated by black arrows.

Moreover, the Altiplano has been the target of many paleoclimatic investigations (Thompson et al., 1998, 2000; Cross et al., 2001; Baker et al., 2001; Placzek et al., 2006, 2013) that highlighted climate fluctuations for the last 130 kyr. Steffen et al. (2009, 2010), Carretier et al. (2013) and Bekaddour et al. (2014) highlighted the contribution of these wet climatic events to mass transport in major canyons from the Altiplano to the Pacific coast. In contrast, McPhillips et al. (2014) observed that the landslide frequency appears to have not changed during the Quaternary and thus questioned the impact of the climatic fluctuations on landsliding and erosion rate.

Fault geometry and kinematics

In southern Peru, the Incapuquio fault zone has a strong geomorphic imprint on the Andean range (Huaman, 1985; Sébrier et al., 1985). Based on microtectonic studies, Sébrier et al. (1985, 1988) and Schildgen et al. (2009) identified different kinematic episodes. Sébrier et al. (1985, 1988) de- fined major Tertiary and early Quaternary compressional phases and a minor late Quaternary episode of normal faulting. Schildgen et al. (2009) proposed that a Tertiary episode of strike-slip and normal faulting occurred between 14 and 2.2 Ma. However, relocated microseismicity (Grange et al., 1984) and teleseismic data (Devlin et al., 2012) demonstrate present-day reverse and strike-slip components on the Lluta fault segment and more regionally for the Incapuquio fault system in the Arequipa region (Fig. 2). The Neogene surface and Quaternary drainage network are also affected by these fault segments (Fig. 3). Channel orientations and captures evidence a sinistral strike-slip component along an extrado graben (Fig. 3c). These Quaternary offsets and inferred kinematics are consistent with the vertical geometry of the fault plane imaged at depth (20 km) by the distribution of crustal earthquakes (Fig. 2b) and focal mechanisms (Fig. 2c) (M_w < 5; < 20 km, Grange et al., 1984; Devlin et al., 2012). Local reverse and normal apparent surface movements are known to occur along major strike-slip faults in the forearcs of northern Chile (Victor et al., 2004) and southern Peru (Hall et al., 2012). The present-day main sinistral strike-slip kinematics of the Incapuquio fault system demonstrated by the seismicity is compatible with normal apparent displacements as observed by Schildgen et al. (2009) and Sébrier et al. (1985, 1988).

Geomorphological setting

The Chuquibamba landslide is located in the Majes River catchment, along the western flank of the Central Andes, between 1000 and 4000 m a.s.l. (Fig. 1). The Chuquibamba landslide comprises an imbricated set of rotational landslides, a debris-flow deposit and, in the lower area, a megafan and alluvial terraces (Fig. 3a). Upstream, the rotational landslides remobilize the thick Huaylillas ignimbrite formation fractured by the activity along the Incapuquio fault system.

We estimate the total volume mobilized for Chuquibamba landslide reaches ~ 40 km3 . These imbricated rotational landslides correspond to a succession of three head scarps (Fig. 3b). The scar of the rotational landslides, cutting the Huaylillas weathering surface, delimits the landslide area (Fig. 3b) and forms an elongated amphitheater trending in the NW–SE direction. This direction does not fit with the overall southwest-dipping topographic slope but rather corresponds to the structural trend of active faults (Fig. 3a).

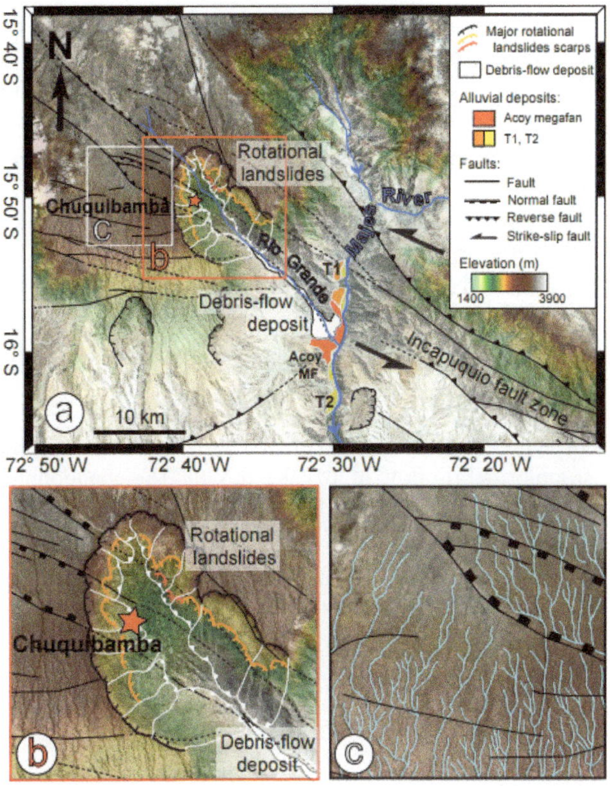

Figure 3. SRTM numerical elevation model overlay by Landsat image of the Chuquibamba destabilization zone. Coordinates are given in WGS 84 longitude and latitude (degrees). (a) Global view of the Chuquibamba area pointing major detachment scarps, debris-flow deposits, Acoy megafan, alluvial terraces and faults. (b) Close-up of the amphitheater-shaped scar of the rotational landslides showing the different head scarps. (c) Drainage network and faults.

Moreover, several fault planes appear to control the shape of the polylobed rotational landslides (Fig. 3b). The base of the cirque (2900 m a.s.l.) is formed by smooth, subhorizontal surfaces (Fig. 3b). These surfaces likely correspond to former lateral landslide deposits re-incised by the river after the initiation of the Chuquibamba landslide. The debris flow remobilizes the rotational landslide deposits. It consists of mixed angular clasts, breccias and numerous meter-size boulders embedded in a thin volcanic matrix (Fig. 4a), reworked from the Huaylillas ignimbrite. The debris-flow deposit displays a smooth and 100 m scale undulated surface (Fig. 4a). In its upper part, the debris flow rests directly on the basement (Fig. 4a), but 30 km downstream of the head scarp, in the vicinity of the Majes River (900 m a.s.l.), the debris flow overlays the Acoy megafan (Fig. 3a). The megafan, located downstream of the Chuquibamba village at the outlet of the Rio Grande, is about 100 m thick and 8 km long (Fig. 3a). Its stratigraphy indicates both a response to multiple phases of sediment production and surface erosion in the Rio Grande and fluvial incision in the Majes Canyon (Steffen et al., 2010). Steffen et al. (2010) described an alternation of sheet flood units and debris- flow deposits, and a typical clast- to matrix-supported fabric. Optically stimulated luminescence (OSL) dating places the Acoy megafan formation between 107.0 ± 5.0 ka (base) and 101.6 ± 4.9 ka (top) (Steffen et al., 2010). Two levels of alluvial terraces (T1, T2) have been identified in the Majes canyon in the vicinity of the megafan. The Acoy megafan and alluvial-terrace development has been linked to wetter episodes, identified in paleoclimate proxies from Altiplano paleolake sediments (Placzek et al., 2006, 2013), driving the fluctuations in the Colca/Majes hydrologic regime (Steffen et al., 2010)

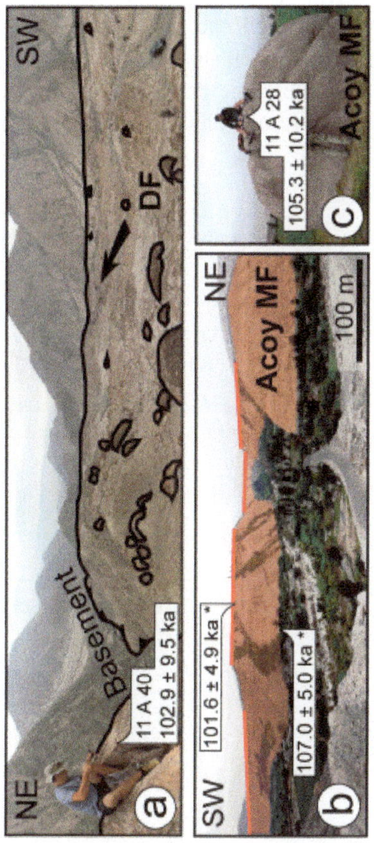

Figure 4. Field photographs showing the debris-flow deposit morphology and the Acoy megafan. (a) Characteristic block sampled on the debris-flow deposit. The arrow indicates flowing direction of the debris flow. (b) General view of the Acoy megafan (Acoy MF); red lines highlight the top of the terrace. The OSL ages obtained by Steffen et al. (2010) for the Acoy megafan are indicated with an asterisk. (c) Block sampled on the megafan

SAMPLING STRATEGY AND METHODS

The last debris-flow deposit is well preserved all along the Chuquibamba valley (Fig. 3a). Six meter-scale boulders of rhyolite entrapped in the Acoy megafan and debris-flow surfaces (Fig. 4a and c) have been sampled. The preserved surface of the boulders (evidenced by desert varnish) indicates minimal post-abandonment erosion; they are well anchored and sufficiently elevated on the debris-flow surface to minimize the possibility of post-depositional movement and potential covering by surficial material. In order to provide an additional constraint on the regional erosion rate and consequently improve the exposure age determination, we sampled a quartzite pebble from the preserved Pampa Jahuay surface in the pediplains 60 km southeast of Chuquibamba (Fig. 1, Table 1). Sample preparation was conducted at the Institut des Sciences de la Terre (ISTerre, Grenoble). As the rhyolitic samples contained enough quartz, we extracted in situ produced beryllium-10 (^{10}Be) using the chemical procedures developed by Brown et al. (1991) and Merchel and Herpers (1999). The

AMS 10Be measurements were performed at the ASTER AMS French national facility (CEREGE, Aix en Provence). Analytical uncertainties include uncertainties associated with AMS counting statistics, AMS external error (1 %), standard reproducibility and chemical blank measurements (^{10}Be / ^{9}Be blank = 1.60 ± 0.72 × 10^{-15}). External uncertainties include 6 % uncertainty in the production rate and 8 % uncertainty in the 10Be decay constant. Exposure ages were calculated using the online CRONUS calculator (Balco et al., 2008). Results are computed using the time-dependent scaling scheme of Lal (1991) modified by Stone (2000).

TCN DATING

In the debris-flow deposit and Acoy megafan samples, the ^{10}Be concentrations are relatively consistent and range from 6.67 ± 0.28 × 105 to 1.38 ± 0.08 × 10^{6} atoms per gram of quartz (at g^{-1} qtz) (Table 1). The high ^{10}Be content of the sample (1.33 ± 0.02 × 107 at g−1 qtz, 11A28) collected from the Pampa Jahuay suggests an extremely low erosion rate lasting at least for the last 2 Myr in the southern Peruvian forearc (1.9 ± 0.3 Myr). The computed erosion rate (0.21 ± 0.05 mm kyr^{-1}) agrees with rates published by Hall et al. (2012) for the South Peru and by Kober et al. (2007) for Chile (Oxaya formation). These results support the hypothesis of an insignificant erosion of the sampled boulders of the debris-flow deposit. TCN exposure ages deduced from debris-flow boulders (with an erosion rate of 0.21 mm kyr^{-1}) range from 96.1 ± 8.9 to 108.5 ± 10.2^{10}Beka (Table 1, Fig. 5). Considering the uncertainties, exposure ages are consistent and suggest a single remobilization event with a weighted average age of 101.9 ± 5.5 ka. The age of the large boulder sampled on the Acoy megafan (Fig. 4) is 105.3 ± 10.2^{10}Be-ka (Table 1).

DISCUSSION

Tectonic and climatic forcing on Chuquibamba landslide evolution

The weighted average age of debris-flow boulders indicates a last major debris flow at 101.9 ± 5.5 ka. The abandonment age of the megafan surface (105.3 ± 10.2^{10}Be-ka; Figs. 4 and 5) agrees with the OSL ages published by Steffen et al. (2010) (i.e., 107.0 ± 5.0 ka at the base of the megafan and 101.6 ± 4.9 ka near the top). The Ouki wet event has been evidenced from sediments collected in the eponym paleolake located in the higher Bolivian Altiplano (Placzek et al., 2006, 2013). The chronological framework deduced from U– Th dating on carbonates indicates the Ouki deep lake cycle between 120 and 98 ka (Placzek et al., 2006, 2013). Steffen et al. (2010) already suggested a correlation between wet time intervals on the Altiplano and sediment aggradation in the Majes Valley. According to Steffen et al. (2010), the Acoy megafan recorded two

Table 1. TCN results of the Chuquibamba Valley and the Pampa Jahuay

Sample	Latitude (° N)	Longitude (° W)	Elevation (m a.s.l.)	Production rate (atoms g^{-1} yr^{-1}) Spallation	Production rate Muons	Geomorphic scaling factor[a]	Sample thickness factor[b]	Quartz (g)	^{9}Be (mg)	$^{10}Be/^{9}Be$[c]	Uncertainty $^{10}Be/^{9}Be$ (%)
Blank											
BKGRE67	–	–	–	–	–	–	–	–	0.486	1.604×10^{-15}	44.74
Debris-flow deposit											
11A39	−15.9294	−72.5306	1315	7.15	0.27	0.99	0.98	9.54	0.291	3.998×10^{-13}	3.10
11A40	−15.5327	−72.9267	1356	7.43	0.28	0.99	0.99	6.02	0.254	3.257×10^{-13}	3.16
11A41	−15.5343	−72.9245	1391	7.39	0.28	0.98	0.97	5.10	0.290	2.566×10^{-13}	3.52
11A42	−15.9145	−72.5487	1543	8.40	0.29	0.99	0.98	10.98	0.292	5.468×10^{-13}	3.44
11A43	−15.8784	−72.5927	1899	10.54	0.33	0.98	0.98	10.44	0.291	6.981×10^{-13}	6.14
Acoy megafan											
11A28	−16.0091	−72.4892	817	5.13	0.24	1.00	0.99	9.77	0.294	3.196×10^{-13}	4.18
Pampa Jahuay											
11A62	−16.3408	−72.0840	1592	8.76	0.30	1.00	0.97	19.96	0.295	1.320×10^{-11}	1.71

Sample	^{10}Be concentration (× 10^3 atoms g^{-1} qtz)	^{10}Be concentration uncertainty (× 10^3 atoms g^{-1} qtz)	Erosion rate (m Myr^{-1})	Erosion rate external uncertainty (m Myr^{-1})	Age ^{10}Be (kyr) $\epsilon = 0$[d]	Age ^{10}Be internal uncertainty (kyr)[e]	Age ^{10}Be external uncertainty (kyr)[f]	Age ^{10}Be (ka) $\epsilon = 0.21$ m Myr^{-1}[d]	Age ^{10}Be internal uncertainty (kyr)[e]	Age ^{10}Be external uncertainty (kyr)[f]
Blank										
BKGRE67	31.374	14.036	–	–	–	–	–	–	–	–
Debris-flow deposit										
11A39	828.574	25.748	–	–	95.99	3.67	8.83	96.11	3.68	8.86
11A40	938.088	29.597	–	–	102.72	4.07	9.49	102.86	4.09	9.52
11A41	992.821	34.974	–	–	108.29	4.85	10.17	108.46	4.87	10.2
11A42	992.144	34.175	–	–	97.51	4.17	9.10	97.88	4.19	9.15
11A43	1375.656	84.507	–	–	106.7	8.30	11.42	106.87	8.34	11.46
Acoy megafan										
11A28	666.607	27.905	–	–	105.1	5.55	10.15	105.26	5.57	10.19
Pampa Jahuay										
11A62	13 318.936	227.531	0.21	0.05	1894.08	94.88	271.34	–	–	–

(a) The topographic scaling factor has been calculated following the method of Dunne et al. (1999). (b) The sample thickness correction has been calculated with a 2.7 density factor. (c) AMS (accelerator mass spectrometry) analyses were carried out at the French AMS facility ASTER. Calibration of ^{10}Be concentrations were done with NIST standard reference material 4325, using a $^{10}Be/^{9}Be$ ratio of 2.79×10^{-11} and a ^{10}Be half-life of $1.387 \pm 0.012 \times 10^{6}$ years (Chmeleff et al., 2010; Korschinek et al., 2010). Results have been corrected from the chemical blank ($^{10}Be/^{9}Be$ blank = $1.60 \pm 0.72 \times 10^{-15}$). Internal uncertainties consider the analytical uncertainties including counting statistics, the instrumental variability (1 %), the standard deviation and chemical blank. External uncertainties include 6 % uncertainty in the production rate and 8 % uncertainty in the ^{10}Be decay constant. (d) Ages have been computed with the CRONUS calculator (Balco et al., 2008) using the time-dependent production rate of Lal (1991) modified by Stone (2000). The production rate calibrate by Kelly et al. (2015) on recent timescale in Peruvian Andes is not relevant for this study. For our range of ages (100 kyr) geomagnetic variations have to be considered. Ages are presented with the internal (e) and the external uncertainties (f).

phases of Rio Grande catchment development characterized by landsliding during the Ouki wet climatic event. The data from Steffen et al. (2010) indicate that during the megafan emplacement, between 112.0 and 96.7 ka, the Rio Grande permits the sediment transport and aggradation in the Acoy megafan downstream. Similar to the Lluta catchment in Chile, landsliding might have been initiated by enhanced precipitation and reduced friction along a basal shear plane due to increasing hydrostatic pressure in the groundwater (Hoke et al., 2004; Strasser and Schlunegger, 2005). The weighted average age of the last debris-flow deposit (101.9 ± 5.5 ka) also correlates with the Ouki event (120–98 ka). In the western Andes, other landslides have been associated with wetter climatic conditions such as the older Lluta collapse (northern Chile, 18° S; Wörner et al., 2002), yielding a minimum age of 2.5 Ma (Strasser and Schlunegger, 2005), the Tomasiri landslide (southern Peru, 17°300 S), dated at 400 ka (Blard et al., 2009) and younger landslides in Argentina, which have been associated with the Minchin wet event (40–25 ka; Trauth et al., 2003; Hermanns and Schellenberger, 2008). We suggest that the Chuquibamba debris flow, which remobilized rotational landslide deposits, was triggered by an increase in pore-water pressure in the Huaylillas ignimbrite during the Ouki wet climatic event. After this event, none of the more recent wet climatic events identified on the Altiplano (e.g., Michin, 48–36 ka, and Tauca, 26–15 ka) triggered a large landslide or remobilized the Chuquibamba debris flow. Similarly, in Southern Peru, Keefer et al. (2003) identified only centimeter-scale debris flows for the last 38 kyr. These observations are consistent with the preservation of the Chuquibamba debris-flow morphology since its emplacement. We suggest that the Ouki event was the latest event whose duration and intensity was sufficient to trigger a large landslide in the arid western Andean flank (Wörner et al., 2002; Strasser and Schlunegger, 2005; Blard et al., 2009). Carretier et al. (2013) proposed that rare and strong erosive events contribute to more than 90 % of the long-term erosion of the arid western side of the Andean range. Accordingly, we suggest that climatic fluctuations that favor landsliding, such as the Ouki wet event, have a great impact on sediment transport and morphologic evolution of the western flank of the Andes.

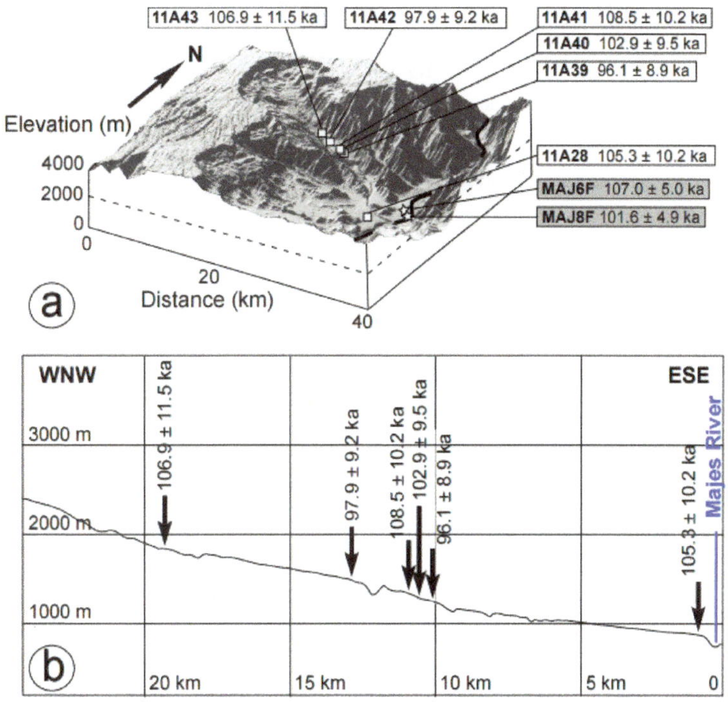

Figure 5. Samples location. (a) Digital elevation model (GeoMapApp, SRTM data, 90 m resolution) of the study area, showing samples location and TCN ages obtained for the debris- flow deposit and the Acoy megafan surface. OSL ages obtained by Steffen et al. (2010) for the Acoy megafan are noted in grey. (b) Elevation profile locating debris flow and the Acoy megafan ages.

The important role of the tectonic framework on the localization and flow direction of mega-landslides has already been suggested for the Andean forearc domain (Pinto et al., 2008; Antinao and Gosse, 2009; Mather et al., 2014). As the Chuquibamba landslide is elongated in a NW–SE trend guided by the Incapuquio fault system, we suggest that the localization and geometry of the landslide is mainly controlled by preferential fracture orientations (Fig. 3b). More broadly, in southern Peru and northern Chile, most of the large landslides are located in tectonically fractured regions (Audin and Bechir, 2006; Pinto et al., 2008; Strasser and Schlunegger, 2005; Wörner et al., 2002; Crosta et al., 2015). This preferential localization suggests that tectonic fracturing plays an important role in preconditioning the landscape for landsliding within this arid area.

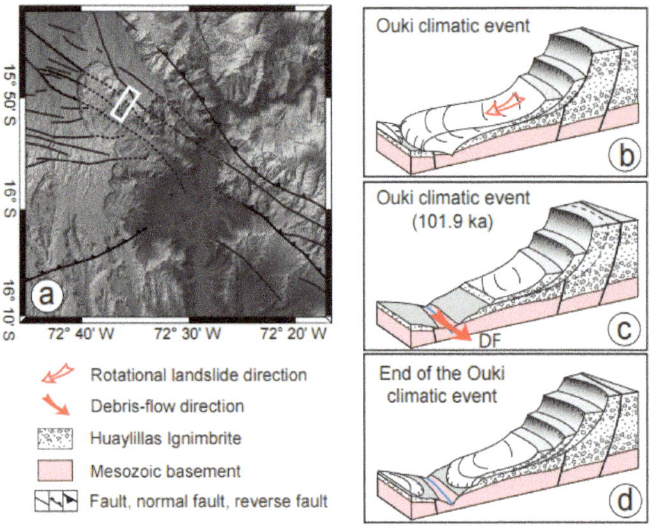

Figure 6. Map and block diagrams of the evolution of the Chuquibamba landslide. (a) SRTM numerical elevation model, the white rectangle localizes the block diagrams. (b) Block diagram showing rotational landslide initiation on the flank of the Chuquibamba Valley during the Ouki wet event. The landslide enlarges the valley and accumulates material at the bottom. (c) The accumulated materials are remobilized by debris flows during the wetter phases. (d) This remobilization allows new rotational landslides that enlarge the amphitheater-shaped valley.

No relationship between a M_w 8.0 subduction earthquake and a giant landslide has been previously documented in southern Peru (Lacroix et al., 2013). However, Keefer et al. (2003) point out that a close temporal succession of subduction earthquakes and El Niño events produces debris flows in the coastal region (southern Peru). In our study area, even if the Ouki wet phase would favor triggering, we cannot exclude a contemporaneous seismic triggering event (subduction or crustal earthquake) for the Chuquibamba debris flows. More generally, robust arguments indicating such a correlation for giant landslides are scarce along the South American margin; other triggering factors (increase in pore pressure, climate change, glacial debuttressing) are usually invoked in addition to a seismic triggering event (Keefer, 2002; Pinto et al., 2008).

Landscape evolution and tectono-climatic scenario

We propose a landscape evolution scenario based on geomorphic marker analysis and new TCN ages (Fig. 6). The initial drainage network flowed toward the coast (i.e., in a southerly direction) and incised the roof of the weathered Huaylillas ignimbrite (Fig. 6a). These parallel incisions are still preserved (Fig. 3b, c) and sometimes even abandoned. Afterwards, the development of a graben structured by the Incapuquio fault system captured the drainage network along a 110◦ N trending direction (Fig. 3c), creating a new tributary of the Majes River. Rotational landslides were initiated by slope instabiity or pore pressure increase at the base of the Huaylillas ignimbrite during wet climatic events, as proposed by Hoke et al. (2004) and Strasser and Schlunegger (2005) for northern Chile (Fig. 6b). Sliding surfaces of rotational landslides are localized on fault planes and progressively enlarge the valley, as evidenced by the successive rotational landslide scarps (Fig. 3). The rotational landslide deposits accumulated upstream in the valley (Fig. 6b) and were remobilized by debris flows during the Ouki event (120–98 ka; Fig. 6c). The accumulation of successive debris-flow and mudflow deposits at the outlet of the Rio Grande during the Ouki event formed the Acoy megafan (Steffen et al., 2010). Finally, at 101.9 ± 5.5 ka, the last major debris flow sealed the erosion/transport system (Fig. 6c). This pattern, with alternating upper slope destabilizations and sediment remobilization by debris flows, drove valley enlargement over time (Fig. 6d). At the present time, the arid climate on the western flank of the Cordillera preserves geomorphologic features from erosion; the water yield comes only from rivers.

CONCLUSIONS

In this study we address how local tectonic activity and climatic fluctuations may control landslide occurrence at the arid western Andean front. Indeed, for the Chuquibamba landslide region, the sequence of debris flows (identified in the Acoy megafan stratigraphy and overlaying its roof) seems to have been favored by the occurrence of a wet climatic event (Ouki event), even if a contemporaneous seismic triggering event cannot be excluded. Our results suggest sediment accumulation in the valley during wet periods and incision during dry periods. Wet climatic events then appear to control the growth of the drainage network, participating in regressive erosion and in the creation of new tributaries on the western Andean front. Specifically, our results show that, during a wet event in a tectonically fractured region, hillslope processes, rather than fluvial erosion, dictate the evolution of the landscape at the channel head in the arid and high-relief area. As rare and strong erosive events represent the most important contribution to the long-term erosion of the arid Western Cordillera (Carretier et al., 2013), we propose that events like the Chuquibamba landslide contribute signifi- cantly to the long-term erosion of the western escarpment of the Andes. We thus suggest a strong regional tectonic and climatic control of the long-term erosion of the arid western Andean front.

ACKNOWLEDGEMENTS

This project was supported by the Institut de Recherche pour le Développement (IRD). We acknowledge additional support from the Instituto Geologico Mineralogico y Metallurgico del Peru (INGEMMET, Peru). We thank all the staff from the ASTER AMS facility at CEREGE. We are grateful for detailed reviews from T. Schildgen and the anonymous reviewers, which helped to improve the manuscript. This work was carried out in the frame of the master's thesis of A. Margirier at ISTerre.

REFERENCES

1. Alpers, C. N. and Brimhall, G. H.: Middle Miocene climatic change in the Atacama Desert, northern Chile: Evidence from supergene mineralization at La Escondida, Geol. Soc. Am. Bull., 100, 1640–1656, 1988.
2. Antinao, J. L. and Gosse, J.: Large rockslides in the Southern Central Andes of Chile (32–34.5◦ S): tectonic control and signifi- cance for Quaternary landscape evolution, Geomorphology, 104, 117–133, 2009.
3. Atherton, M. P., Warden, V., and Sanderson, L. M.: The

Mesozoic marginal basin of Central Peru: a geochemical study of withinplate-edge volcanism, in: Magmatism at a Plate Edge: the Peruvian Andes, edited by: Pitcher, M. P., Cobbing, E. J., and Beckinsale, R. D., Blackie Halsted Press, London, 47–58, 1985.

4. Audin, L. and Bechir, A.: Active Tectonics as Determinant Factor in Landslides Along the western Cordillera?, Congreso Peruano de Geologia, Lima, Peru, 17–20 October, 237–239, 2006.

5. Audin, L., David, C., Hall, S., Farber, D., and Hérail, G.: Geomorphic evidences of recent tectonic activity in the forearc, southern Peru, Special Volume on Neotectonics in South America, Revista de la Asociación Geológica Argentina, 61, 545–554, 2006.

6. Baker, P. A., Seltzer, G. O., Fritz, S. C., Dunbar, R. B., Grove, M. J., Tapia, P. M., Cross, S. L., Rowe, H. D., and Broda, J. P.: The history of South American tropical precipitation for the past 25,000 years, Science, 291, 640–643, 2001.

7. Balco, G., Stone, J. O., Lifton, N. A., and Dunai, T. J.: A complete and easily accessible means of calculating surface exposure ages or erosion rates from 10Be and 26Al measurements, Quat. Geochronol., 3, 174–195, 2008.

8. Bekaddour, T., Schlunegger, F., Vogel, H., Delunel, R., Norton, K. P., Akçar, N., and Kubik, P.: Paleo erosion rates and climate shifts recorded by Quaternary cut-and-fill sequences in the Pisco valley, central Peru, Earth Planet. Sc. Lett., 390, 103–115, 2014.

9. Blard, P. H., Lavé, J., Farley, K. A., Fornari, M., Jiménez, N., and Ramirez, V.: Late local glacial maximum in the Central Altiplano triggered by cold and locally-wet conditions during the paleolake Tauca episode (17–15 ka, Heinrich 1), Quaternary Sci. Rev., 28, 3414–3427, 2009.

10. Brown, E. T., Edmond, J. M., Raisbeck, G. M., Yiou, F., Kurz, M. D., and Brook, E. J.: Examination of surface exposure ages of Antartic moraines using in situ produced 10Be and 26Al, Geochim. Cosmochim. Ac., 55, 2269–2283, 1991.

11. Carretier, S., Regard, V., Vassallo, R., Aguilar, G., Martinod, J., Riquelme, R., Pepin, E., Charrier, R., Hérail, G., Farias, M., Guyot, J. L., Vargas, G., and Lagane, C.: Slope and climate variability control of erosion in the Andes of central Chile, Geology, 41, 195–198, 2013.

12. Chmeleff, J., von Blanckenburg, F., Kossert, K., and Jacob, D.: Determination of the 10Be half-life by multicollector ICP-MS and liquid scintillation counting, Nucl. Instrum. Methods, 268, 192– 199, 2010.

13. Cross, S. L., Baker, P. A., Seltzer, G. O., Fritz, S. C., and Dunbar, R. B.: Late Quaternary Climate and Hydrology of Tropical South America Inferred from an Isotopic and Chemical Model of Lake Titicaca, Bolivia and Peru, Quat. Res., 56, 1–9, 2001.

14. Devlin, S., Isacks, B. L., Pritchard, M. E., Barnhart, W. D., and Lohman, R. B.: Depths and focal mechanisms of crustal earthquakes in the central Andes determined from teleseismic waveform analysis and InSAR, Tectonics, 31, TC2002, doi:10.1029/2011TC002914, 2012.

15. Crosta, G. B., Paolo, F., Elena, V., and Hermanns, R. L.: The Cerro Caquilluco–Cerrillos Negros Giant Rock Avalanches (Tacna, Peru), Engineering Geology for Society and Territory, 2, 921– 924, doi:10.1007/978-3-319-09057-3_159, 2015.

16. Dunai, T. J., González López, G. A., and Juez-Larré, J.: OligoceneMiocene age of aridity in the Atacama Desert revealed by exposure dating of erosion-sensitive landforms, Geology, 33, 321– 324, 2005.

17. Grange, F., Hatzfeld, D., Cunningham, P., Molnar, P., Roecker, S. W., Suarez, G., Rodrigues, A., and Ocola, L.: Tectonic implications of the microearthquake seismicity and fault plane solutions in southern Peru, J. Geophys. Res., 89, 6139–6152, 1984.

18. Gunnell, Y., Thouret, J. C., Brichau, S., Carter, A., and Gallagher, K.: Low-temperature thermochronology in the Peruvian Central Andes: implications for long-term continental denudation, timing of plateau uplift, canyon incision and lithosphere dynamics, J. Geol. Soc. London, 167, 803–815, 2010.

19. Guzzetti, F., Mondini, A. C., Cardinali, M., Fiorucci, F., Santangelo, M., and Chang, K.-T.: Landslide inventory maps: new tools for an old problem, Earth-Sci. Rev., 112, 1–25, 2012.

20. Hall, S. R., Farber, D. L., Audin, L., Finkel, R. C., and Mériaux, A. S.: Geochronology of pediment surfaces in southern Peru: implications for Quaternary deformation of the Andean forearc, Tectonophysics, 459, 186–205, 2008.

21. Hall, S. R., Farber, D. L., Audin, L., and Finkel, R. C.: Recently active contractile deformation in the forearc of southern Peru, Earth Planet. Sc. Lett., 337–338, 85–92, 2012.

22. Hermanns, R. L. and Schellenberger, A.: Quaternary tephrochronology helps define conditioning factors and triggering mechanisms of rock avalanches in NW Argentina, Quatern. Int., 178, 261– 275, 2008.

23. Hoke, G. D., Isacks, B. L., Jordan, T. E., and Yu, J. S.: Groundwatersapping origin for the giant quebradas of northern Chile, Geology, 32, 605–608, 2004.

24. Hovius, N., Meunier, P., Ching-Weei, L., Hongey, C., Yue-Gau, C., Dadson, S., Ming-Jame, H., and Lines, M.: Prolonged seismically induced erosion and the mass balance of a large earthquake, Earth Planet. Sc. Lett., 304, 347–355, 2011.

25. Huaman, R.: Evolution tectonique Cenozoïque et néotectonique du piémont Pacifique dans la région d'Arequipa (Andes du Sud Pérou), PhD thesis, Université Paris Sud, Centre d'Orsay, 220 pp., 1985.

26. INGEMMET: Mapa Geológico del cuadrángulo de Caravelí, Ocoña, La Yesera, y Chuquibamba, scale 1 : 100 000, Lima, 2001. Keefer, D. K.: Landslides caused by earthquakes, Geol. Soc. Am. Bull., 95, 406–421, 1984.

27. Keefer, D. K.: The importance of earthquake-induced landslides to long-term slope erosion and slope-failure hazards in seismically active regions, Geomorphology, 10, 265–284, 1994.

28. Keefer, D. K.: Investigating landslides caused by earthquakes – a historical review, Surv. Geophys., 23, 473–510, 2002.

29. Keefer, D. K. and Moseley, M. E.: Southern Peru desert shattered by the great 2001 earthquake: implications for paleoseismic and paleo-El Niño-Southern Oscillation records, P. Natl. Acad. Sci. USA, 101, 10878–10883, 2004.

30. Keefer, D. K., Moseley, M. E., and deFrance, S. D.: A 38 000 year record of floods and debris flows in the Ilo region of southern Peru and its relation to El Niño events and great earthquakes, Palaeogeogr. Palaeocl., 194, 41–77, 2003.

31. Kelly, M. A., Lowell, T. V., Applegate, P. J., Phillips, F. M., Schaefer, J. M., Smith, C. A., Kim, H., Leonard, K. C., and Hudson, A. M.: A locally calibrated, late glacial 10Be production rate from a low-latitude, high-altitude site in the Peruvian Andes, Quat. Geochronol., 26, 70–85, 2015.

32. Kober, F., Ivy-Ochs, S., Schlunegger, F., Baur, H., Kubik, P. W., and Wieler, R.: Denudation rates and a topography-driven rainfall threshold in northern Chile: multiple cosmogenic nuclide data and sediment yield budgets, Geomorphology, 83, 97–120, 2007.

33. Korschinek, G., Bergmaier, A., Faestermann, T., Gerstmann, U. C., Knie, K., Rugel, G., Wallner, A., Dillmann, I., Dollinger, G., Gostomski von C. L., Kossert, K., Maiti, M., Poutivtsev, M., and Remmert, A.: A new value for the half-life of 10Be by heavy-ion elastic recoil detection and liquid scintillation counting, Nuclear Instum. Methods, 268, 187–191, 2010.

34. Korup, O., Clague, J. J., Hermanns, R. L., Hewitt, K., Strom, A. L., and Weidinger, J. T.: Giant landslides, topography, and erosion, Earth Planet. Sc. Lett., 261, 578–589, 2007.

35. Lacroix, P., Zavala, B., Berthier, E., and Audin, L.: Supervised method of landslide inventory using panchromatic SPOT5 images and application to the earthquake-triggered landslides of Pisco (Peru, 2007, Mw8.0), Remote Sensing, 5, 2590–2616, 2013.

36. Lal, D.: Cosmic ray labeling of erosion surfaces: in situ nuclide production rates and erosion models, Earth Planet. Sc. Lett., 104, 24–439, 1991.

37. Mather, A. E., Hartley, A. J., and Griffiths, J. S.: The giant landslides of northern Chile: tectonic and climate interactions on a classic convergent plate margin, Earth Planet. Sc. Lett., 388, 249–256, 2014.

38. McPhillips, D., Bierman, P. R., and Rood, D. H.: Millennial-scale record of landslides in the Andes consistent with earthquake trigger, Nat. Geosci., 7, 925–930, 2014.

39. Meunier, P., Hovius, N., and Haines, J. A.: Topographic site effects on the location of earthquake induced landslides, Earth Planet. Sc. Lett., 275, 221–232, 2008.

40. Merchel, S. and Herpers, U.: An update on radiochemical separation techniques for the determination of long-lived radionuclides via accelerator mass spectrometry, Radiochim. Acta, 84, 215–219, 1999.

41. Mortimer, C.: Drainage evolution of the Atacama Desert of northernmost Chile, Rev. Geol. Chile, 11, 3–28, 1980.

42. Perfettini, H., Avouac, J.-P., Tavera, H., Kositsky, A., Nocquet, J.- M., Bondoux, F., Chlieh, M., Sladen, A., Audin, L., and Farber, D. L.: Seismic and aseismic slip on the central Peru megathrust, Nature, 465, 78–81, 2010

43. Pinto, L., Hérail, G., Sepúlveda, S. A., and Krop, P.: A neogene giant landslide in Tarapacá, northern Chile: a signal of instability of the westernmost Altiplano and palaeoseismicity effects, Geomorphology, 102, 532–541, 2008.

44. Placzek, C., Quade, J., and Patchett, P. J.: Geochronology and stratigraphy of late Pleistocene lake cycles on the southern Bolivian Altiplano: implications for causes of tropical climate change, Geol. Soc. Am. Bull., 118, 515–532, 2006.

45. Placzek, C. J., Quade, J., and Patchett, P. J.: A 130 ka reconstruction of rainfall on the Bolivian Altiplano: implications for the causes of tropical climate change, Earth Planet. Sc. Lett., 363, 97–108, 2013.

46. Rech, J. A., Currie, B. S., Michalski, G., and Cowan, A. M.: Neogene climate change and uplift in the Atacama Desert, Chile, Geology, 34, 761–764, 2006.

47. Roperch, P., Sempere, T., Macedo, O., Arriagada, C., Fornari, M., Tapia, C., Garcia, M., and Laj, C.: Counterclockwise rotation of late Eocene–Oligocene fore-arc deposits in southern Peru and its significance for oroclinal bending in the central Andes, Tectonics, 25, TC3010, doi:10.1029/2005TC001882, 2006.

48. Schildgen, T. F., Hodges, K. V., Whipple, K. X., Reiners, P. W., and Pringle, M. S.: Uplift of the western margin of the Andean plateau revealed from canyon incision history, southern Peru, Geology, 35, 523–526, 2007.

49. Schildgen, T. F., Hodges, K. V., Whipple, K. X., Pringle, M. S., van Soest, M., and Cornell, K.: Late Cenozoic structural and tectonic development of the western margin of the central Andean Plateau in southwest Peru, Tectonics, 28, TC4007, doi:10.1029/2008TC002403, 2009.

50. Sébrier, M., Mercier, J. L., Mégard, F., Laubacher, G., and CareyGailhardis, E.: Quaternary normal and reverse faulting and the state of stress in the central Andes of south Peru, Tectonics, 4, 739–780, 1985.

51. Sébrier M., Lavenu, A., Fornari, M., and Soulas, J. P.: Tectonics and uplift in Central Andes (Peru, Bolivia and Northern Chile) from Eocene to present, Géodynamique, 3, 85–106, 1988.

52. Steffen, D., Schlunegger, F., and Preusser, F.: Drainage basin response to climate change in the Pisco valley, Peru, Geology, 37, 491–494, 2009.

53. Steffen, D., Schlunegger, F., and Preusser, F.: Late Pleistocene fans and terraces in the Majes valley (southern Peru) and their relation to climatic variations, Int. J. Earth Sci., 99, 1975–1989, 2010.

54. Stone, J. O.: Air pressure and cosmogenic isotope production, J. Geophys. Res., 105, 23753–23759, 2000.

55. Strasser, M. and Schlunegger, F.: Erosional processes, topographic length-scales and geomorphic evolution

in arid climatic environments: the "Lluta collapse," northern Chile, Int. J. Earth Sci., 94, 433–446, 2005.

56. Tavera, H., Audin, L., and Bernal, I.: Source parameters of Sama (Tacna) earthquake (5.4 Mw) of 20 November 2006, Bol. Soc. geol. Perù, 102, 109–115, 2007.

57. Thompson, L. G., Davis, M. E., Mosley-Thompson, E., Sowers, T. A., Henderson, K. A., Zagorodnov, V. S., Lin P. N., Mikhalenko V. N., Campen R. K., and Bolzan J. F.: A 25,000-year tropical climate history from Bolivian ice cores, Science, 282, 1858–1864, 1998.

58. Thompson, L. G., Mosley-Thompson, E., and Henderson, K. A.: Ice-core palaeoclimate records in tropical South America since the Last Glacial Maximum, J. Quat. Sci., 15, 377–394, 2000.

59. Thouret, J. C., Wörner, G., Gunnell, Y., Singer, B., Zhang, X., and Souriot, T.: Geochronologic and stratigraphic constraints on canyon incision and Miocene uplift of the Central Andes in Peru, Earth Planet. Sc. Lett., 263, 151–166, 2007.

60. Trauth, M. H., Bookhagen, B., Marwan, N., and Strecker, M. R.: Multiple landslide clusters record Quaternary climate changes in the northwestern Argentine Andes, Palaeogeogr. Palaeocl., 194, 109–121, 2003.

61. Victor, P., Oncken, O., and Glodny, J.: Uplift of the western Altiplano plateau: evidence from the Precordillera between 20 and 21° S (northern Chile), Tectonics, 23, TC4004, doi:10.1029/2003TC001519, 2004.

62. Wörner, G., Uhlig, D., Kohler, I., and Seyfried, H.: Evolution of the West Andean Escarpment at 18° S (N. Chile) during the last 25 Ma: uplift, erosion and collapse through time, Tectonophysics, 345, 183–198, 2002.

CHAPTER 14

LANDSLIDE SUSCEPTIBILITY ASSESSMENT

G. De Guidi and S. Scudero

INTRODUCTION

Shallow landslides are very common in steep mountainous regions where they often constitute a major risk factor for human activities (Glade, 1998; Singh et al., 2005; Claessens et al., 2007; Salciarini et al., 2008; Gulla et al., 2008; Deb and ` El-Kadi, 2009; Harp et al., 2009). Their occurrence, often triggered by exceptional rainfall or earthquakes, can involve large regions with a huge numbers of events. The simultaneous occurrence of multiple landslides gives the exclusive opportunity to study their distribution in relation to several environmental factors. On 1 October 2009, more than one thousand shallow landslides occurred in the NE corner of Sicily (Italy) (Maugeri and Motta, 2010). The affected area extended over approximately 30 km2 and was located on the eastern sector side of the Peloritani mountain ridge. Slope instabilities were triggered by heavy rainfall and affected mainly the lower parts of 14 small catchments directly facing the Ionian Sea (Fig. 1). Landslides caused 35 fatalities, the total devastation of several villages in Messina province, and severe damage to the public heritage and the local economy (Fig. 2). Rainfall was very intense: 223 mm of rain fell in 7 h, with a peak of 10.6 mm in 5 min, on terrain already saturated by precipitation during previous weeks (~ 400 mm in 14 days) (Maugeri and Motta, 2010). Here we perform a susceptibility analysis for shallow landslides, by spatial combination of the inventory map of the events with some causative factors. We employ a statistical bivariate method that assigns numerical weight values to each class of causative factors. The susceptibility assessment yields two main results. Firstly, we are able to identify reliably which areas in this territory are more likely to be involved in future landslides. Secondly, we can investigate the connection between the most landslide-prone conditions and the active geodynamics of the area. The principal thesis is that geotectonic history is behind the factors controlling landsliding, playing a leading role in the slow but inexorable process of transformation of the morphological structure of this area. In fact eastern Sicily and southern Calabria are among the most tectonically active areas in the Mediterranean. Since the Middle Pleistocene, the entire Calabrian arc has undergone strong tectonic uplift (Montenat et al., 1991; Westaway, 1993; Tortorici et al., 1995; Rust and Kershaw, 2000), accompanied by activity of a normal fault belt since 700 ka BP (Fig. 1). Present-day activity is testified by the strong historical seismicity (Monaco et al., 1997; Monaco and Tortorici, 2000; Catalano et al., 2008). Several papers have highlighted the relationship between tectonics and the evolving landscape (Burbank and Pinter, 1999 and reference therein; Montgomery and Brandon, 2002; Snyder et al., 2002, 2003; Korup et al., 2007; Agliardi et al., 2009; Goswami et al., 2011). Similarly, we investigate the role of tectonics on surface processes with respect to geological, climatic and environmental factors.

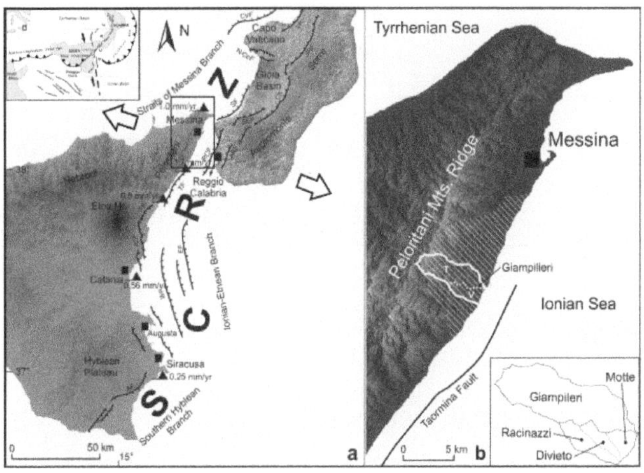

Figure 1. (a) Tectonic map of eastern Sicily and southern Calabria showing the Siculo–Calabrian Rift Zone (SCRZ), modified from Catalano et al. (2008). Arrows indicate the regional extension direction and triangles the uplift rates given by Catalano et al. (2003) and Catalano and De Guidi (2003); RIF: Rosolini–Ispica faults; AV: Avola fault; WF: western Ionian fault; EF: eastern Ionian fault; ASF: Acireale–S.Alfio faults; PF: Piedimonte fault; TF: Taormina fault; RCF: Reggio Calabria fault; AF: Armo fault; SF: Scilla fault; SEF: S.Eufemia fault; CF: Cittanova fault; SRF: Serre fault; VF: Vibo fault;

Citation: De Guidi, G. and Scudero, S.: Landslide susceptibility assessment, Nat. Hazards Earth Syst. Sci., 13, 949-963, doi:10.5194/nhess-13-949-2013, 2013.

CVF: Capo Vaticano fault; N-CoF: Nicotera–Coccorino faults. (b) the hatched zone represents the area affected by landslides on the 1st October 2009; the continuous white line indicates the study area; training area (T) and validation area (V) of the landslide susceptibility analysis are showed.

MORPHOSTRUCTURAL SETTING

The study area is part of the Peloritani mountain range belonging geologically to the Calabrian arc and representing the inner chain of the Apennine–Maghrebian mountain belt (Amodio Morelli et al., 1976; Bonardi et al., 1982). The mountain ridge extends longitudinally for about 50 km, with a SW–NE orientation (Fig. 1), resulting in peaks higher than 1200 m that shelve toward the Tyrrhenian and Ionian seas (NW and E respectively).

Tectonic outline

The Calabrian arc has a long and complex geological history. In fact, it has been involved in two orogenic cycles since late Palaeozoic times and many other Neogene deformation events. More recently (Middle Pleistocene), the entire Calabrian arc underwent strong tectonic uplift (Montenat et al., 1991; Westaway, 1993; Tortorici et al., 1995), with rates greater than 1.0 mm yr^{-1} , accompanied since 700 ka by the activity of a normal fault system called the Siculo– Calabrian Rift Zone (SCRZ) (Monaco et al., 1997; Monaco and Tortorici, 2000; Catalano et al., 2008) (Fig. 1). The corresponding regional tensional stress is oriented N ~ 115 E (D'Agostino and Selvaggi, 2004; Catalano et al., 2008) and is partially accommodated by deformation on the various fault segments. The study area is located at the footwall of one of these active normal faults (Taormina fault) (De Guidi et al., 2003). Taking into account the vertical deformation induced by its activity, the total tectonic uplift rate for the last 125 ka reaches a maximum value of 1.7 mm yr^{-1} (Catalano and De Guidi, 2003; Catalano et al., 2003) across all of eastern Sicily. Holocene uplift rates are even greater, and, in fact, this is one of the areas that incurred the highest uplift in this period for the entire Italian peninsula (Lambeck et al., 2004). Vertical rates measured in the area of the Peloritani Mts. peak at between 1.4 and 2.4 mm yr^{-1} for the last 5000–6000 yr (Stewart et al., 1997; Rust and Kershaw, 2000; Antonioli et al., 2003; De Guidi et al., 2003; Antonioli et al., 2009), although the true uplift with respect to base level (Ionian Sea level) is lower, because in the same period sea level rose at an average rate of about 1 mm yr^{-1} (Lambeck et al., 2004). However, the increase in the topographic gradient of the Peloritani Mts. is very notable.

Figure 2. Examples of some shallow landslides triggered by October 1st 2009 rainfall event in Giampilieri village: (a) front view; (b) downslope view.

Morphological features

The landscape of the Peloritani Mts. reflects the tectonic setting of the area, in particular the intense recent dynamics. Vertical movements of base level during the late Quaternary, caused by interaction between vertical tectonic deformation and eustatic sea level changes, controlled the shaping of the landforms. This resulted in a poly-cyclical evolution of the coastal slopes that shows imprinting of many sequential cycles of rejuvenation during Late Pleistocene and Holocene times (Montenat et al., 1991; Rust and Kershaw, 2000; Catalano and De Guidi, 2003; Catalano et al., 2003; De Guidi et al., 2003). One of the most prominent aspects is represented by a consequent drainage system, developed on both the northwestern and eastern slopes of the Peloritani mountain ridge. Other morphological features are represented by the remnants of marine terraces located at various elevations on the coastal slopes, and deep V-shaped valleys, typical of recent erosional stage, which cut pre-existing peneplaned surfaces. In particular, different fluvial landforms characterize the northwestern and eastern sectors of the Peloritani Mts. In the former, low-energy fluvial landforms mark the landscape; river valleys with smooth longitudinal profiles separate wide north–south elongated surfaces dipping toward the Tyrrhenian

Sea. In the latter the landscape exhibits more the results of its poly-cyclical evolution: deeply incised valleys separate interfluves constituted by remnants of marine terrace surfaces (APAT, 2008 and 2010). The longitudinalprofiles show several steady knickpoints due to stages of relative stability of the base level (Goswami et al., 2012). The most recent knickpoint is active and steadily migrates upslope while the steep flanks of the river valleys are affected by headward erosion. Several streams characterized by high gradient and short length, fed by intermittent and torrential rainfall, drain the Peloritani Mts., in particular the eastern (i.e. Ionian) sector. The local name for this kind of river is fiumara. The precipitation regime controls their activity, and therefore sudden, heavy floods usually interrupt longer period of inactivity. The morphostructural framework also influences flow regime, with active tectonic uplift being one of the main controlling features (Sabato and Tropeano, 2004). The overall topography is uneven with steep slopes ranging between 25◦ and 70◦ , but flat areas may also occur where remnants of Quaternary marine terraces are preserved; they originated through the interplay between tectonic uplift and the sea level high stands (Catalano and De Guidi, 2003; Antonioli et al., 2006).

Figure 3. Landslides scars: lateral (a) and frontal (b) view. Note the sharp contact between the colluvium (approximately 1.0 m thick) and the bedrock.

LANDSLIDE SUSCEPTIBILITY ASSESSMENT

The landslide susceptibility for a given landslide typology is the spatial probability that an event may occur in a given area considering a given selection of geo-environmental factors (van Westen at al., 2006; Fell et al., 2008). Susceptibility can be evaluated using many different approaches according to the aims of the susceptibility assessment itself and to the scale of the analysis. There are heuristic or empirical models, statistical analyses, deterministic models and probabilistic methods (Guzzetti et al., 1999; Brenning, 2005; van Westen et al., 2008; Cascini, 2008). Many of these have been developed in conjunction with geographic information systems (GIS) (Carrara et al., 1991; van Westen et al., 1997; Carrara et al., 1999). In particular, susceptibility assessment for shallow landslides recently has been evaluated using methods belonging to all four groups (Dai and Lee, 2003; Guzzetti et al., 2005; Singh et al, 2005; Deb and El-Kadi, 2009; Godt et al., 2008; Gulla et al., 2008; Salciarini et al., 2008; G ` unther and ¨ Thiel, 2009; Piacentini et al., 2012). All of the methods entail some basic assumptions (Guzzetti et al., 1999 and references therein): (i) each landslide leaves on the terrain some features that are recognizable and can be mapped in the field or remotely; (ii) slope failures obey mechanical laws, and the collection of instability factors leading to the failures allows assessment of the susceptibility models; (iii) future landslides will occur under the same condition that caused past failures; (iv) it is possible to infer the spatial and temporal occurrence of failures through the analysis of the geo-environmental factors. Thus, a given territory can be classified according to different landsliding probability.

We performed the susceptibility assessment with various different statistical models, but here we present only the most effective one. The model was performed in four small catchments of the eastern sector: Giampilieri, Divieto, Racinazzi and Motte (Fig. 1). These catchments were selected because they form the core of the area of landslides, where we have at our disposal the largest amount of data (Scudero and De Guidi, 2011) The basins extend over a ~ 13 km2 area within the eastern flank of the mountain range, directly facing the Ionian Sea. Moreover, the study area was subdivided into a “training area” where the analysis was performed, and a “validation area” where the effectiveness of the results was tested.

Landslide characteristics and inventory

Shallow landslides constitute the main geological hazard in the Peloritani Mts., even though some larger and more complex landslides are present. The latter are mainly induced by the downcutting of stream beds consequent to the overall rejuvenation process, but they were not considered in this study. All of the landslides took place within the patchy layer of loose, heterogeneous and often weathered material of variable thickness (1–3 m) covering the majority of the slopes (colluvium) (Fig. 3). This deposit is characterized by a high coarse fraction (gravel content more than 50 %) and a clay content that is often scarce. Its friction angle can reach more than 35°, allowing the material to remain, under normal conditions, on the steep slopes of the catchments. According to Lacerda (2007), the term colluvium refers to soil material that underwent little transport or derives from large landslides with sudden transport.

The landslides are usually triggered by the detachment of small boulders from natural scarps, rocky cliffs or ancient retaining walls used for terracing the slopes for agricultural purposes, over a saturated portion of colluvium (Agnesi et al., 2009). Another mechanism is the sudden fluidification of the colluvial material due to superficial or subsuperficial water currents or temporary springs rising from the bedrock where contrasts of hydraulic properties with the colluvial layer may occur (Agnesi et al., 2009). The volume of single events is relatively constant, usually comprising between 10^3 and 10^4 m^3 (Falconi et al., 2011). The flows usually developed downslope, even though they can also enlarge laterally and migrate upslope.

The material often canalized following the natural drainage, eroding and sometimes exhuming buried gullies. During their movement downwards, the flows eroded the base or the lateral edges of their channels, gaining further material and increasing their volume, until they stopped in the zone of accumulation, usually located at the foots of the slopes. The villages involved were located at the bases of large, steep slopes and were party swept away by the mobilized material. Even though different triggering mechanisms induced the failures, all of the events shared the same general geoenvironmental arrangement at the source points before the failure occurred. Therefore, we ascribe the failure mechanism to very local conditions and we will treat the instabilities as being a unique typology. The landslide inventory map of the Regional Civil Defence (Basile, 2010) was verified through field reconnaissance surveys and integrated with missing data to ensure accuracy in the presence and location of the landslides.

A total of 397 landslides have been mapped in the entire study area. The inventory map of Basile (2010) distinguishes, for each landslide, an area mainly characterized by erosion, and the fan, where the accumulation of transported material dominates. Within the areas experiencing loss of material, the source area and the channel have not been considered separately. For the susceptibility assessment of this type of landslide, we have to consider only the area where the first detachment originated. For this reason we considered as source areas for the shallow landslides a circle of 25 m in diameter comprising the initiation point and the upper part of each event (Salciarini et al., 2006). This value is congruent with the scale of our analysis (Hengl, 2006), and with the spatial variability of the causative factor observed in the field.

Spatial variables

Many factors can influence the occurrence of landslides, and understanding the determinant factors requires a good knowledge of the study area, of the phenomena and an assessment of the scale of analysis and the methodology (van Westen et al., 2006, 2008; Cascini, 2008). The experience gained during field activities (Scudero and De Guidi, 2011) allowed us to recognize six different controlling factors: lithology, geomechanical classification, geomorphology, distance to stream, slope angle, and aspect.

Many other geo-environmental factors have been recognized to have a role in instabilities (van Westen et al., 2008), but these have been excluded from our analysis for the following reasons: (i) if they act evenly, or almost evenly, over the entire study area; (ii) if we did not recognize their influence on landslides during field surveys; (iii) if they can be linked to other factors already taken into account. In particular we excluded any anthropogenic influences (e.g. proximity to transportation networks or buildings) since instabilities are widespread even on natural slopes, and they eventually interact with infrastructures or other anthropogenic features only during their development downslope. Factors such as drainage and lineament densities were ignored, because they are already incorporated in other indicators.

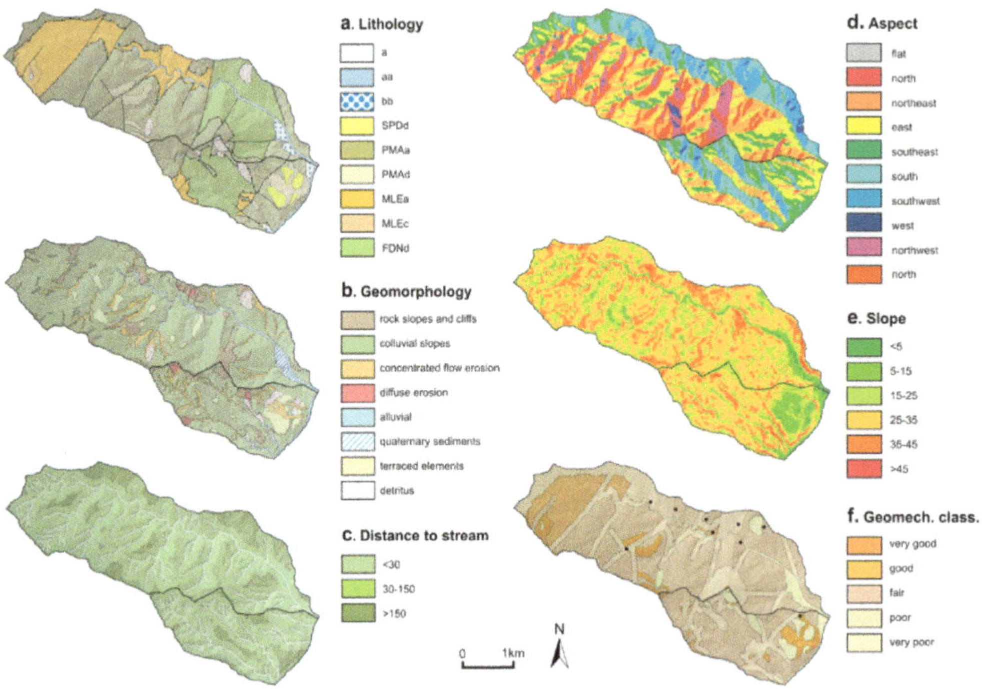

Figure 4. Spatial distribution of the six causative factors: (a) bedrock lithology and lineaments (abbreviations from Messina-Reggio C. Geological Map, Foglio 601 Messina-Reggio C. della Carta Geologica d'Italia, scala 1:50 000. APAT, Servizio Geologico d'Italia, Dipartimento Difesa del Suolo, Roma, 2008); (a) slope deposits; aa: present-day alluvial deposits; bb: recent alluvial deposits; SPDd: calcarenites; PMAa paragneiss of the Aspromonte unit; PMAd marble of the Aspromonte unit; MLEa: mica schists of the Mela unit; MLEc: marble of the Mela unit; FDNd: phyllite of the Mandanici unit); (b): geomorphological units; (c): distance to stream; (d): aspect of the slopes; (e): slope angle (degree); (f): geomechanical classification of rock types according to the RMR proposed by Bieniawski (1979); black dots indicate the measurement stations.

We mapped the spatial distribution of the causative factors (Fig. 4) on a 1:10 000 scale and acquired them as spatial layers in a G.I.S. environment with 10 m × 10 m pixel resolution. This definition is accurate enough considering the map scale and the minimum dimension of the observed objects (Hengl, 2006). The bedrock geology of the study area consists of three different types of metamorphic rock, including low-grade phyllites that are covered tectonically by higher-grade mica schist and paragneiss. These rocks represent the late Palaeozoic basement of the European plate that was thrust over the Apennine–Maghrebian chain (Giunta and Nigro, 1999; De Gregorio et al., 2003; Festa et al., 2004; Heymes et al., 2010), then deformed during the Tertiary (Monaco et al., 1996; Bonardi et al., 2002; Somma et al., 2005; Grande et al., 2009) and Quaternary (Tortorici et al., 1995; De Guidi et al., 2003; Catalano et al., 2008). Because of their geological history, the rocks are deeply fractured and weathered.

We performed a geomechanical characterization of the rock masses according to the rock mass rating (RMR) proposed by Bieniawski (1979) in order to assess their quality. For this purpose we surveyed random outcrops of bedrock, representing each lithotype, and estimated the rock quality designation index (RQD), the uniaxial compressive strength, the condition of the discontinuities (spacing, roughness, persistence, filling, alteration) and groundwater condition. In particular, the uniaxial compressive strength was estimated through the Schmidt hammer readings converted according to the empirical relationship proposed by Deere and Miller (1966). The collection of other parameters followed the scheme proposed by ISRM (1981). We noted a sharp decrease of mechanical characteristics approaching the tectonic lineaments at a distance of less than 50 m.

Therefore the localized RMR analyses were spatialized by combining rock type with distance to tectonic lineaments (Greco and Sorriso-Valvo, 2005; Borrelli et al., 2007; Pellegrino and Prestininzi, 2007). Such subdivision provides additional information concerning differences in the rock masses and represents a causative factor in describing landslide occurrence (Federici et al., 2007; van Westen et al., 2008). The geomorphological map distinguishes terrain units according to processes in action, the origin of the features, and the materials involved. The most relevant feature is the extensive presence of a colluvial layer covering the majority of the slopes. It is characterized by almost uniform thickness (~ 1 m) (Fig. 3), although thicker deposits, up to 3 m, may occur on some slopes. Rock slopes, concentrated flow erosion, and areas with diffuse erosion are other representative features. More than one geomorphological process could coexist; however we considered only the most prominent process in each part of the study area. We also generated a digital elevation model (DEM) from a 1:10 000 map with 10 m contour intervals and derived three topographic features: (i) slope; (ii) aspect; and (iii) distance from drainage.

Slope angle is one of the most relevant factors controlling landsliding. Steep slopes characterize the entire Peloritani ridge: the highest peaks can reach 1200 m a.s.l. within 5 km from the coast, and the mean slope angle in the study area is 32°. The slope aspect can influence hydrological processes, and therefore it also has an indirect bearing on weathering, vegetation and root development. But the aspect map can also suggest which slopes are more prone to landsliding considering the spatial organization of the drainage network. In fact, in active areas it is usually controlled by tectonic processes and provides information on the state of evolution of the fluvial basin (Guarnieri and Pirrotta, 2008). Finally, the distance from the drainage system has been considered because relief is intensely dissected by stream segments and erosion can influence the distribution of instabilities (van Westen et al., 2008).

Susceptibility analysis and mapping

The landslide susceptibility analysis was performed using a statistical bivariate method. Bivariate techniques have been used widely in the literature for susceptibility assessment (van Westen, 1997; Cevik and Topal, 2003; van Westen et al., 2003; Suzen and Doyuran, 2004; Yalcin, 2008, Nandi and Shakoor, 2009). Each of the spatial variables (layer) discussed in the previous paragraph was considered to be an independent variable causing landslides. In the bivariate approach each independent variable is correlated with the dependent one (landslide map), and their relationships with landslide occurrences were calculated through pairwise map crossing. Furthermore, the bivariate techniques assume that the independent variables are not inter-related and that they all have the same weight in landsliding. In particular, we performed landslide susceptibility analysis using the statistical index method proposed by van Westen (1997). In the statistical index method, the weights of the classes of each factor are calculated as the natural logarithm of the ratio between the density of landslides in the considered class and the density of landslides for the entire map; weights were calculated using the following equation:

$$w_{ij} = ln\left(f_{ij}/f\right) = ln\left[\left(A^*_{ij}/A_{ij}\right) * \left(A/A^*\right)\right] = ln\left[\left(A^*_{ij}/A^*\right) * \left(A/A_{ij}\right)\right] \quad (1)$$

where, w_{ij} is the weight of the class i of parameter j, f_{ij} is the landslide density of the class i of parameter j, f is the landslide density of the whole considered area, A^*_{ij} is the landslide area in the class i of parameter j, A_{ij} is the area of the class i of parameter j, A^* is the total landslide area, and A is the total surface of the considered area. Therefore the higher the weight of a class, the greater is its influence in the instabilities.

The method was first performed in the "training area" (T in Fig. 1) that represents about 75 % of the entire study area, and its robustness and prediction skill was then tested in an adjacent "test area". The numerical rating values for each class of the six causative factors are presented in Table 1. The calculated weights indicate that, for the geology factor, the highest value is given by Plio-Pleistocene calcarenites. They outcrop over a very small part of the basin, forming cliffs just above the initiation points of some of the landslides. For this reason, the weight is high but it is likely that there is no direct influence of this lithology on instability. Relatively high weights also occur where large bodies of detritus or loose deposits cover the slopes.

The Mandanici unit (phyllite) is by far the most favourable for landsliding compared to the other metamorphic units of the Peloritani Mountains. For the geomechanical characteristics of the rock types, a clear trend is evident. Good rocks (according to the RMR classification) have no influence on landsliding, while poor ones favour instabilities. For the geomorphological factor, the occurrence of large bodies of mobilized material and areas with concentrated flow erosion or strong erosional evidence have the highest influence. Rocky slopes and cliffs also have influence, because they can trigger shallow landslides through the fall of boulders. Landslides occur when the slope angle is greater than 15° . The class with the highest incidence of landslides corresponds to high slope angles (> 45°), while the frequency is lower for intermediate slope angles (25°–45°). Three classes of aspect factors influenced landsliding; in particular, south to southwest facing slopes were the most favourable.

Areas located farther from the drainage network system (more than 150 m) were the most unstable. Calculated weights have been spatially summed together in order to map the landslide susceptibility index (LSI) for the entire study area. The absolute value of LSI is meaningful only when it is classified into categories providing a qualitative estimate of susceptibility (Fell et al., 2008). For this reason the LSI has been classified into four different categories of susceptibility according to the most commonly used classification in the literature: low, medium, high and very high susceptibility. Several subdivision procedures have been published that classify the LSI (expert-based, binary, equal interval, natural breaks, standard deviation, percentage of correspondence with landslides). In this study subdivision within the classes follows the criterion of natural breaks according to the Jenks' algorithm (Ruff and Czurda, 2008; Ruff and Rohn, 2008; Falaschi et al., 2009; Nandi and Shakoor, 2009; Piacentini et al., 2012) and was performed by analysing the frequency distribution of LSI. The natural breaks subdivision was preferred to other classification criteria, because the frequency distribution of the LSI shows obvious jumps. The final susceptibility map is shown in Fig. 5.

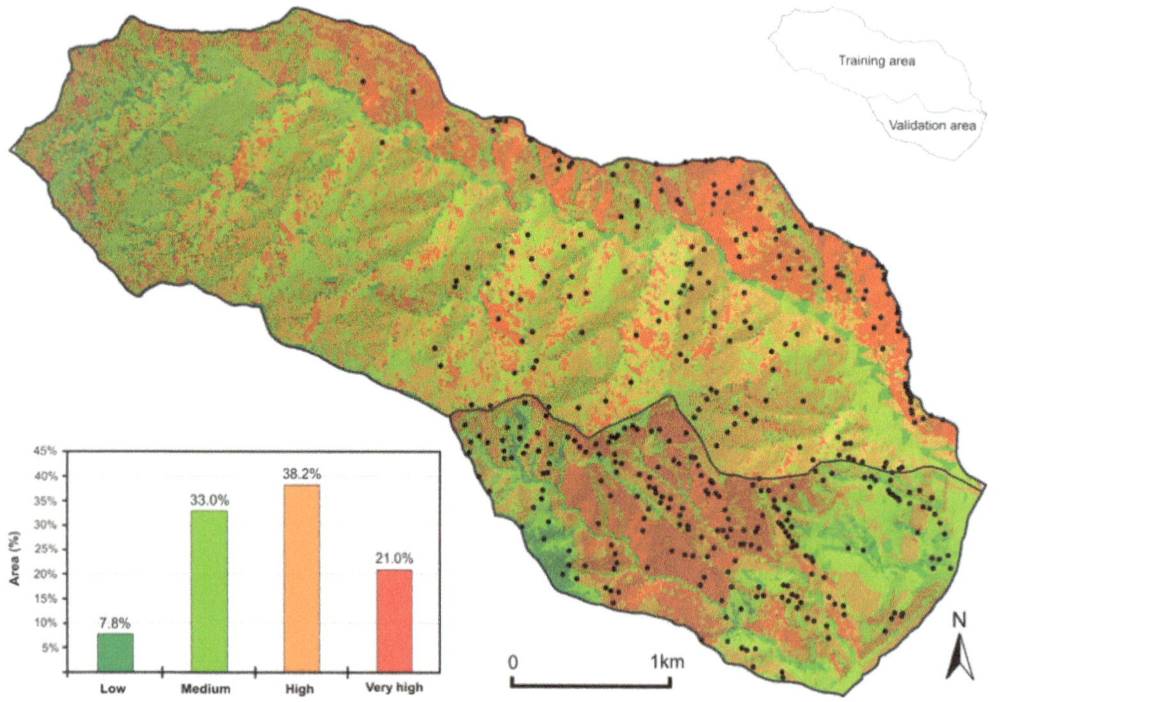

Figure 5. Landslide susceptibility map subdivided in four landslide susceptibility classes: green, low susceptibility; light green, medium susceptibility; orange, high susceptibility; red, very high susceptibility. The bars show the percentage area within the different landslide susceptibility classes; black dots indicate the location of the landslides.

Validation and prediction

Validation of the susceptibility model is an essential procedure that assesses of how well it matches the distribution of landslides. Moreover, the use of independent landslide information is necessary to evaluate its sensitivity in forecasting future instabilities (Chung ad Fabbri, 2003; Guzzetti et al., 2006). There are different criteria with increasing grades of accuracy in the validation procedure. Guzzetti et al. (2006) also proposed a classification of the quality level of the susceptibility models and associated maps. Low quality models do not meet any tests of reliability, or validate the model with the same landslide information used to obtain the models themselves. High quality models estimate the error associated with the predicted susceptibility index value and, in addition, validate the model using independent landslide information that was not used to produce the model. The model presented here has been tested following this last criterion. The effectiveness and the prediction skill of our model have been verified through a partition of the study area (Chung and Fabbri, 2003). We elaborated our model in the training area (T), and the validation was performed in the remaining part (V). We took into account the physiographical boundaries of the catchments, thus considering the watershed as boundaries of the training and validation areas (Fig. 5); 75 % of the study area, containing 182 landslides, was used as the training set, while the remaining 25 %, containing 215 landslides, was used for model validation. The validation area was excluded before assessment of the model; thus, the landslide susceptibility map was drawn up first and only subsequently combined with the landslide distribution data in the test area (V) that is therefore independent of the model. A quantitative assessment of the performance of the susceptibility model is provided by the success rate curve (Chung and Fabbri 2003; van Westen et al., 2003; Guzzetti et al., 2006; Poli and Sterlacchini, 2007; Blahut et al., 2010; Piacentini et al., 2012). This is drawn by plotting on the x-axis the cumulative percentage of the study area within each class of susceptibility and on the y-axis the cumulative percentage of landslides in each susceptibility class, ranked from the most to the least susceptible. The shape of the curve gives information about the degree of success of the model – the larger the area under the curve, the better is the performance in fitting the landslide distribution. For the susceptibility model in the study area, the curve rapidly departed from the x-axis and then approximated this axis in its final part, and the area under the curve (ACU) is 0.70, indicating a good success rate (Fig. 6). In detail half of the landslides (50 %) are found in the 26.1 % most susceptible zone, and approximately a quarter (24.8 %) of the landslides fall within the most susceptible 10 % of the area. The area evaluated and mapped as very highly or highly susceptible covers 84.4 % of the total mapped landslides, and only 1.8 % of the landslides are mapped in the lower susceptibility class. In the validation area we used independent landslide information in order to test the ability of the model to predict the locations of landslides. We plotted the cumulative percentage of landslides as a function of the spatial probability of occurrence (Fig. 7). Very few landslides (5.9 %) were within the lower class of susceptibility, while 70.9 % of landslides were correctly classified by the model as they occur in regions classified as being of high or very high susceptiblity.

Table 1. Calculated weights in the training areas for the classes of the six causative factors; * arbitrarily assigned

Instability factor/Classes	% Area covered by the class	Landslide density (%)	Weight
Lithology			
Slope deposit	1.33 %	2.7 %	0.69
Present-day alluvial deposits	2.06 %	0.0 %	0.00
Recent alluvial deposits	1.24 %	0.0 %	0.00
Calcarenites	0.09 %	2.2 %	*1.00
Aspromonte unit (gneiss)	48.35 %	41.9 %	−0.14
Mela unit (mica schist)	21.82 %	10.2 %	−0.76
Mandanici unit (phyllite)	20.61 %	42.9 %	0.73
Marble	4.49 %	0.0 %	0.00
Geomechanical classification			
Very good	3.62 %	0.0 %	0.00
Good	11.57 %	0.4 %	−3.49
Fair	49.41 %	67.8 %	0.32
Poor	30.67 %	26.9 %	−0.13
Very poor	4.73 %	4.9 %	0.04
Geomorphology			
Rock slopes and cliffs	8.70 %	13.9 %	0.47
Colluvial slopes	78.17 %	78.1 %	0.00
Concentrated flow erosion	5.09 %	2.5 %	−0.72
Areas with diffuse erosion	0.33 %	0.3 %	−0.07
Alluvial	2.06 %	0.0 %	0.00
Quaternary sediments	1.24 %	0.0 %	0.00
Terraced elements	3.20 %	2.6 %	−0.22
Detritus	1.22 %	2.6 %	0.77
Slope angle (°)			
< 5	5.11 %	2.5 %	−0.72
5–15	4.07 %	0.9 %	−1.55
15–25	19.03 %	14.9 %	−0.24
25–35	38.65 %	40.2 %	0.04
35–45	25.50 %	30.5 %	0.18
> 45	7.65 %	11.0 %	0.37
Aspect			
Flat	5.03 %	2.5 %	−0.70
North	8.67 %	6.4 %	−0.30
Northeast	18.68 %	18.6 %	−0.01
East	18.28 %	12.3 %	−0.39
Southeast	13.11 %	9.7 %	−0.30
South	11.64 %	17.2 %	0.39
Southwest	5.93 %	11.4 %	0.65
West	2.16 %	4.2 %	0.66
Northwest	7.40 %	4.2 %	−0.56
North	9.09 %	13.5 %	0.39
Distance to stream			
< 30	24.01 %	5.0 %	−1.58
30–150	60.05 %	69.0 %	0.14
> 150	15.94 %	26.0 %	0.49

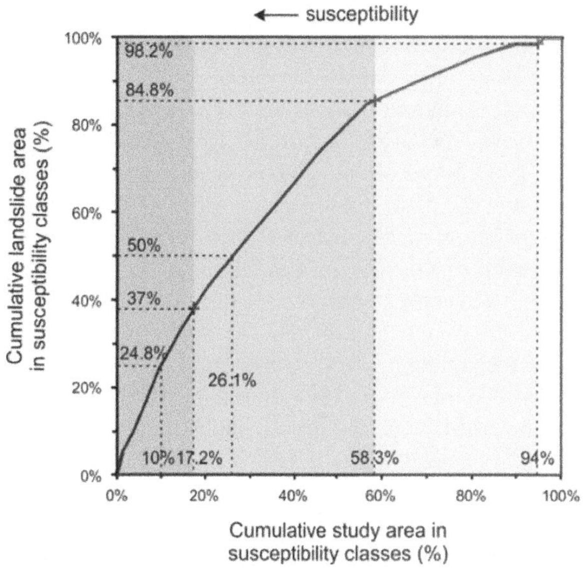

Figure 6. Analysis of the fitting degree of landslide susceptibility model.

TECTONIC IMPLICATIONS

The bivariate technique allowed us to perform and successively validate a susceptibility model for shallow landslides, proving the relevance of each geo-environmental factor in landsliding and the relative weight of each class. In this way we have been able to identify combinations of main factors leading to instability. These combinations generally consist of poor or very poor rock types, especially paragneiss and phyllite, steep (35°–45°) colluvial slopes or slopes covered by loose deposits, and at distance from the drainage network. The occurrence of such combinations can be considered in framework of the geological heritage of the region of the Peloritani Mts.

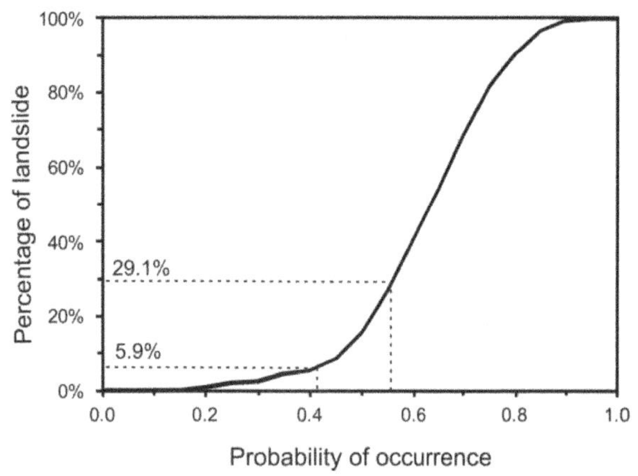

Figure 7. Analysis of the landslide susceptibility map in the validation area (V) in relation to the distribution of the landslides.

Vertical tectonic deformation, acting intensely especially during the last 5000–6000 yr, contributed to the topographic rejuvenation of the relief by inducing erosional processes (Cyr et al., 2010) and allowing the release of lithostatic load causing a shallow and dense net of discontinuities in the rock masses. The denudation process is also favoured by geological conditions of highly deformed and weatheringprone metamorphic rocks (Morgagni et al., 1993; Le Pera and Sorriso-Valvo, 2000; Snyder et al., 2003; Calcaterra and Parise, 2005; Marques et al., 2010; Goswami et al., 2011). Lithostatic release and erosion mutually enhance their effects by favouring degradation of the rock masses (Riebe et al., 2001; Hren et al., 2007) and consequently the acceleration of slope processes where the normal evolution of the landscape involves a slower cyclical accumulation and mobilization of loose superficial material (Dietrich et al., 1982; Burbank, 2002). In a mountain range bounded by active normal faults, relief responds very quickly to changes in the base level (Densmore et al., 1998; Goldsworthy and Jackson, 2000); even in a period of a century or less, river systems are able to adjust to the deformation (Keller and Pinter, 1996). Footwall uplift produces short, steep catchments in which the effects of tectonics are reflected both by the state of activity of the streams and the arrangement of the whole fluvial network. In the study area the activity of the Taormina fault controls coastline shape over a whole sector of the Peloritani mountain range. The long-term activity of the fault induced the formation of a consequent drainage network represented by the main streams of the catchments. In an attempt to reach equilibrium with the tectonic activity, the main stream incisions undergo intense linear erosion and the slopes become steeper, while the erosional processes become weaker moving upwards to the secondary network (Catalano and De Guidi, 2003; Goswami et al., 2012). This allows indirect coupling of factors such as aspect or distance from the network with recent tectonic activity. Slope aspect involved in instabilities is not only connected with hydrological or weathering processes, but also with the morphological evolution of the relief. In fact, the highest weights in landslide occurrence for S–SW and N-facing slopes (Table 1) are usually linked to the ~WNW–ESE directed consequent drainage (Fig. 8). Similarly, because the slopes of the consequent valleys are wider (transversely) and longer (longitudinally), the number of instabilities increases with the distance from the streams. These slopes are undisturbed by the continuous erosion along stream segments, and a greater amount of sediment can accumulate. In contrast, for areas that are very close to stream segments, the formation of colluvium is retarded, and a sufficient thickness of colluvial material necessary for the occurrence of the types of instabilities considered here does not accumulate. The relationship between recent tectonic activity and landslide distribution has also been highlighted by Goswami et al. (2011). These authors obtained the same result through frequency-area landslide statistics, showing that instabilities are mainly located where the effects, direct or indirect, of fault activity are the greatest.

CONCLUDING REMARKS

A landslide susceptibility analysis for a representative area of the Peloritani mountain range has been carried out. The results of the validation and prediction prove the effectiveness of the model; in fact, within the performance scheme of Guzzetti et al. (1999), the model shows an acceptable percentage of correctly classified landslides. Consequently it is possible to extend the model from a relatively small area to a larger region with similar geological and geomorphological conditions. The bivariate technique method separates each considered factor not providing their relative weights. For this reason the final map is a simple combination of factors which, although recognized as the most important ones, may not have equal incidence. For this reason, our results could be further improved by the application of more sophisticated methods for the analysis of susceptibility (i.e. multivariate techniques) that allow estimation of the relative influence of single causative factors. Nevertheless, the experience gained during field surveys allowed us to select the most relevant factors, enhancing the reliability of the results. Fieldwork is of fundamental importance for a reliable mapping of the factors contributing to landslides, such as lithology, lithotechnical parameters and colluvium coverage. In addition, we suggest that the influence of active tectonics, and its spatial variation over the entire Peloritani region, should be taken into account as causative factors in future small-scale susceptibility assessments.

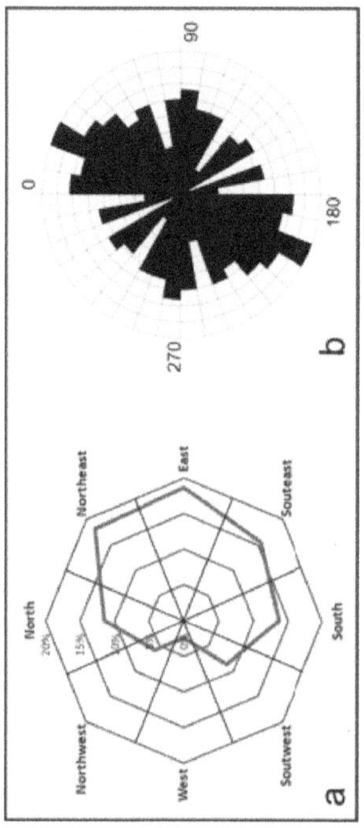

Figure 8. (a) frequency distribution of the slope aspect within the whole study area: the highest value is for east-dipping slope; relative

maxima are also for northeast- and south-facing slopes. (b) frequency distribution of the direction of the stream segments for the four considered catchments; note the peaks at SSW–NNE and at WNW–ESE directions.

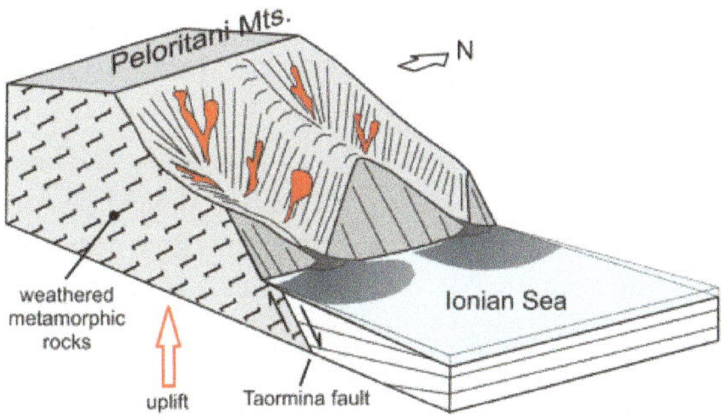

Figure 9. Sketch summarizing the morphological, geological and tectonic features of the Peloritani Mts. Modified from Goldsworthy and Jackson (2000).

The assessment of landslide susceptibility allowed us to speculate about the effect of geological dynamics on the geomorphological evolution of relief. Our investigations focused on four representative basins, and we were able to infer some direct relationships between the general setting, derived from the active endogenic dynamics, and the superficial phenomena of relief evolution for this sector. Erosion is favoured by geological conditions: metamorphic rocks in particular are very prone to chemical weathering. Tectonic uplift coupled with strong erosion allows the release of the lithostatic load originating as a shallow and dense net of discontinuities in the rock mass. These processes act together by favouring the weathering and consequently the formation of loose and heterogeneous material that covers most of the slopes and becomes cyclically unstable under critical conditions (e.g. extreme rainfall, earthquakes) (Fig. 9). It was not possible to calculate the recurrence interval for the triggering rainfall events as this is much longer than the available historical records (Vecchio, 2011).

Therefore, in a temporal perspective, these events are very rare and their in- fluence on relief processes is not decisive. The role played by fluvial bedrock incision, tectonic uplift and landslides in the evolution of relief is not universal, but rather changes from one mountain range to the other as well as with time (Korup et al., 2010). In different sectors of the Calabrian arc, several studies (Goswami et al., 2011, 2012; Santangelo et al., 2011) have recognized the direct influence of Quaternary tectonics (uplift and faulting) on mass movements and its greater relevance with respect the lithostructural and morphoclimatic factors. In conclusion, considering the almost homogeneous geological and environmental conditions in the eastern sector of the Peloritani Mts., the landslide susceptibility model developed here could be extended without any particular integration, to the entire ridge area. Of course our model contains some limitations for this extension (limited catalogue extent, single landslide type analysed), and also other landslide susceptibility assessment methods could be tested. The spatial distribution of landslide susceptibilities highlights how large portions of the study area (about 20 %), and likely of the whole Peloritani Mts. area, are characterized by very high susceptibility. Therefore triggering events similar to that of October 1st 2009 will certainly occur again in the future. Consequently the susceptibility analysis is important not only for an understanding of the geological dynamics, but also because it represents a fundamental step in landslide hazard mitigation in the Peloritani Mts.

ACKNOWLEDGEMENTS

We are grateful to Thomas Dewez (BRGM), Rajasmita Goswami (University of Manchester) and two anonymous reviewers for their constructive criticism on the originally submitted manuscript. We also thank Guido De Guidi (University of Catania) and Paul Taylor (Natural History Museum, London) for the English review. This work was supported by grants from the University of Catania (responsible C. Monaco and G. De Guidi).

REFERENCES

1. Agliardi, F., Zanchi, A., and Crosta, G. B.: Tectonic vs. gravitational morphostructures in the central Eastern Alps (Italy): Constraints on the recent evolution of the mountain range, Tectonophys., 474, 250–270, 2009.
2. Agnesi, V., Rasa, R., Puglisi, C., Gio ` e, C., Privitera, B., Cappado- ` nia, C., Conoscenti, C., Pino, P., and Rotigliano, E.: La franosita` diffusa dell'1 Ottobre 2009 nel territorio ionico-peloritano della Provincia di Messina: stato delle indagini e prime considerazioni sulle dinamiche geomorfiche attivate, Geologi di Sicilia, 23–30, 2009.
3. Amodio-Morelli, L., Bonardi, G., Colonna, V., Dietrich, D., Giunta, G., Ippolito, F., Liguori, V., Lorenzoni, S., Paglionico, A., Perrone, V., Piccarreta, G., Russo, M., Scandone, P., ZanettinLorenzoni, E., and Zuppetta, A.: L'Arco Calabro-peloritano nell'orogene appenninico-maghrebide, Mem. Soc. Geol. It., 17, 1–60, 1976.
4. Antonioli, F., Kershaw, S., Rust, D., and Verrubbi, V.: Holocene sealevel change in Sicily and its implications for tectonic models: new data from the Taormina area, northeast Sicily, Mar. Geol., 196, 53–71, 2003.
5. Antonioli, F., Kershaw, S., Renda, P., Rust, D., Belluomini, G., Cerasoli, M., Radtke, U., and Silenzi, S.: Elevation of the last interglacial highstand in Sicily (Italy): A benchmark of coastal tectonics, Q. Int., 145–146, 3–18, 2006.
6. Antonioli, F., Ferranti, L., Fontana, A., Amorosi, A., Bondesan, A., Braitenberg, C., Dutton, A., Fontolan, G., Furlani, S., Lambeck, K., Mastronuzzi, G., Monaco, C., Spada, G., and Stocchi, P.: Holocene relative sea-level changes and vertical movements along the Italian and Istrian coastlines, Q. Int., 206, 102–133, 2009.
7. APAT, Servizio Geologico d'Italia, Dipartimento Difesa del Suolo: Note illustrative della Carta Geologica d'Italia alla scala 1:50.000, Foglio 601 Messina. Reggio di Calabria, S.EL.CA, Firenze, 2008.
8. APAT, Servizio Geologico d'Italia, Dipartimento Difesa del Suolo: Note illustrative della Carta Geologica d'Italia alla scala 1:50.000, Foglio 613 Taormina, S.EL. CA, Firenze, 2010.
9. Basile, G.: Evento del 1 Ottobre 2009, Carta dei dissesti,

Dipartimento Regionale della Protezione Civile, Sicilia, 2010. Bieniawski, Z.T.: The geomechanics classification in rock engineering applications. Proceedings of 4th International Congress on Rock Mechanics, 2–8 September 1979. Montreux, Switzerland, 1979.

10. Blahut, J., van Westen, C. J., and Sterlacchini, S.: Analysis of landslide inventories for accurate prediction of debris-flow source areas, Geomorphology, 119, 36–51, 2010.

11. Bonardi, G., Cello, G., Perrone, V., Tortorici, L., Turco, E., and Zuppetta, A.: The evolution of the northern sector of the CalabriaPeloritani arc in a semiquantitative palynspastic restoration, Boll. Soc. Geol. It., 101, 259–284, 1982.

12. Bonardi, G., de Capoa, P., Di Staso, A., Martin-Martin, M., MartinRojas, I., Perrone, V., Tent-Manclus, J.E.: New constraints to the geodynamic evolution of the southern sector of the Calabria– Peloritani Arc (Italy), C. R. Geosci., 334, 423-430, 1982.

13. Borrelli, L., Greco, R., and Gulla, G.: Weathering grade of rock ` masses as a predisposing factor to slope instabilities: Reconnaissance and control procedures, Geomorphology, 84, 158–175, 2007.

14. Brenning, A.: Spatial prediction models for landslide hazards: review, comparison and evaluation, Nat. Hazards Earth Syst. Sci., 5, 853–862, doi:10.5194/nhess-5-853-2005, 2005.

15. Burbank, D. W.: Rates of erosion and their implications for exhumation, Mineral. Mag., 66/1, 25–52, 2002. Burbank, D. W. and Pinter, N.: Landscape evolution: the interactions of tectonics and surface processes, Basin Res., 11, 1–6, 1999.

16. Calcaterra, D. and Parise, M.: Landslide types and their relationships with weathering in a Calabrian basin, southern Italy, Bull. Eng. Geol. Environ., 64, 193–207, 2005.

17. Carrara, A., Cardinali, M., Detti, R., Guzzetti, F., Pasqui, V., and Reichenbach, P.: GIS techinques and statistical models in evalueting landslide hazard, Earth Process. Surf. Landforms, 16, 427–445, 1991.

18. Carrara, A., Guzetti, F., Cardinali, M., and Reichenbach, P.: Use of GIS Technology in the Prediction and Monitoring of Landslide Hazard, Nat. Hazards, 20, 117–135, 1999.

19. Cascini, L.: Applicability of landslide susceptibility and hazard zoning at different scales, Eng. Geol., 102, 164–177, 2008.

20. Catalano, S. and De Guidi, G.: Late Quaternary uplift of northeastern Sicily: relation with the active normal faulting deformation, J. Geodynam., 36, 445–467, 2003.

21. Catalano, S., De Guidi, G., Monaco, C., Tortorici, G., and Tortorici, L.: Long-term behaviour of the late Quaternary normal faults in the Straits of Messina area (Calabrian arc): structural and morphological constraints, Q. Int., 101/102, 81–91, 2003.

22. Catalano, S., De Guidi, G., Monaco, C., Tortorici, G., and Tortorici, L.: Active faulting and seismicity along the Siculo-Calabrian Rift Zone (southern Italy), Tectonophys., 453, 177–192, 2008.

23. Cevik, E. and Topal, T.: GIS-based landslide susceptibility mapping for a problematic segment of the natural gas pipeline, Hendek (Turkey), Environ. Geol., 44, 949–962, 2003.

24. Chung, C. F. and Fabbri, A. G.: Validation of Spatial Prediction Models for Landslide Hazard Mapping, Nat. Hazards, 30, 451– 472, 2003.

25. Claessens, L., Schoorl, J. M., and Veldkamp, A.: Modelling the location of shallow landslides and their effects on landscape dynamics in large watersheds: An application for Northern New Zealand, Geomorphology, 87, 16–27, 2007.

26. Cyr, A. J., Granger, D. E., Olivetti, V., and Molin, P.: Quantifying rock uplift rates using channel steepness and cosmogenic nuclide–determined erosion rates: Examples from northern and southern Italy, Lithosphere, 2/3, 188–198, 2010.

27. D'Agostino, N. and Selvaggi, G.: Crustal motion along the EurasiaNubia boundary in the Calabrian Arc and Sicily and active extension in the Messina Straits from GPS measurements, J. Geophys. Res., 109, B11402, doi:10.1029/2004JB002998, 2004.

28. Dai, F. C. and Lee, F.: A Spatiotemporal probabilistic modeling of storm-induced shallow landsliding using aerial photograps and logistic regression, Earth Surf. Proc. Land., 28, 527–545, 2003.

29. Deb, S. K. and El-Kadi, A. I.: Susceptibility assessment of shallow landslides on Oahu, Hawaii, under extreme-rainfall events, Geomorphology, 108, 129–233, 2009.

30. De Gregorio, S., Rotolo, S. G., and Villa, I. M.: Geochronology of the medium to high-grade metamorphic units of the Peloritani Mts., Sicily, Int. J. Earth Sci., 92, 852–872, 2003.

31. De Guidi, G., Catalano, S., Monaco, C., and Tortorici, L.: Morphological evidence of Holocene coseismic deformation in the Taormina region (NE Sicily), J. Geodynam., 36, 193–211, 2003.

32. Deere, D. U. and Miller, R. P.: Engineering classification and index properties for intact rock, Air Force Weapons Laboratory Technical Report, AFWL-TR-65-116, 277 pp. 1966.

33. Densmore, A. L., Ellis, M. A., and Anderson, R. S.: Landsliding and the evolution of normal-fault-bounded mountains, J. Geophy. Res., 103, 15203–15219, 1998.

34. Dietrich, W. E., Dunne, T., Humphrey, N. F., and Reid, L. M.: Construction of sediment budgets for drainage basins, Sediment Budgets and routing in forested drainage basins, Proceedings of the Symposium; 31 May – 1 June 1982; Corvallis, Oregon. Gen. Tech. Rep. PNW-141. Portland, Oregon: Pacific Northwest Forest and Range Experiment Station, Forest Service, U.S. Department of Agriculture, 5–23, 1982.

35. Falaschi, F., Giacomelli, F., Federici, P. R., Puccinelli, A., D'Amato Avanzi, G., Pochini, A., and Ribolini, A.: Logistic regression versus artificial neural networks: landslide susceptibility evaluation in a sample area of the Serchio River valley, Italy, Nat. Hazards, 50, 551–569,

2009.

36. Falconi, L., Campolo, D., Leoni, G., Lumaca, S., and Puglisi, C.: Geomorphology hazard assessment of Giampilieri and Briga river basins after the rainfall event on the October 1, 2009 (Sicily, Italy), The Second World Landslide Forum, "Putting science into practice". Global Promotion Committee of the International Programme on Landslides (IPL), Rome, Italy, 2011.

37. Federici, P. R., Puccinelli, A., Cantarelli, E., Casarosa, N., D'Amato Avanzi, G., Falaschi, F., Giannecchini, R., Pochini, A., Ribolini, A., Bottai, M., Salvati, N., and Testi, C.: Multidisciplinary investigations in evaluating landslide susceptibility. An example in the Serchio River valley (Italy), Q. Int., 171–172, 52–63, 2007.

38. Fell, R., Corominas, J., Bonnard, C., Cascini, L., Leroi, E., and Savage, W. Z.: On behalf of the JTC-1 Joint Technical Committee on Landslides and Engineered Slopes: Guidelines for landslide susceptibility, hazard and risk zoning for land use planning, Eng. Geol., 102, 85–98, 2008.

39. Festa, V., Messina, A., Paglionico, A., Piccarreta, G., and Rottura, A.: Pre-Triassic history recorded in the Calabria-Peloritani segmento f the Alpine chain, southern Italy. An overview, Per. Mineral., 73, 57–71, 2004.

40. Giunta, G. and Nigro, F.: Tectono-sedimentary constraints to the Oligocene-to-Miocene evolution of the Peloritani thrust belt (NE Sicily), Tectonophys, 315, 287–299, 1999.

41. Glade, T.: Establishing the frequency and magnitude of landslidetriggering rainstorm events in New Zealand, Environ. Geol., 35, 160–174, 1998.

42. Godt, J. W., Baum R. L., Savage, W. Z., Salciarini, D., Schultz, W. H., and Harp, E. L.: Transient deterministic shallow landslide modeling: Requirements for susceptibility and hazard assessments in a GIS framework, Eng. Geol,. 102, 214–236, 2008.

43. Goldsworthy, M. and Jackson, J.: Active normal fault evolution in Greece revealed by geomorphology and drainage patterns, J. Geol. Soc. London, 157, 967–981, 2000.

44. Goswami, R., Mitchell, N. C., and Brocklehurst, S. H.: Distribution and causes of landslides in eastern Peloritani of NE Sicily and western Aspromonte of SW Calabria, Italy, Geomorphology, 132, 111–122, 2011.

45. Goswami, R., Brocklehurst, S. H., and Mitchell, N. C.: Erosion of a tectonically uplifting coastal landscape, NE Sicily, Italy, Geomorphology, 171–172, 114–126, 2012.

46. Grande, A., Di Vincenzo, G., Prosser, G., and Caggianelli, A.: Direct evidence of Middle Oligocene extension in the Calabria– Peloritani terrane from co-seismic faulting: the pseudotachylytebearing shear zones of Palmi (southern Calabria, Italy), T. Nov., 21, 293–303, 2009.

47. Greco, R. and Sorriso-Valvo, M.: Relationships between joint apparent separation, Schmidt hammer rebound value, and distance to faults, in rocky outcrops, Calabria, Southern Italy, Eng. Geol., 78, 309–320, 2005.

48. Guarnieri, P. and Pirrotta, C.: The response of drainage basins to the late Quaternary tectonics in the Sicilian side of the Messina Strait (NE Sicily), Geomorphology, 95, 260–273, 2008.

49. Gulla, G., Antronico, L., Iaquinta, P., and Terranova, O.: Suscep- ` tibility and triggering scenarios at a regional scale for shallow landslides, Geomorphology, 99, 39–58, 2008.

50. Gunther, A. and Thiel, C.: Combined rock slope stability and shal- ¨ low landslide susceptibility assessment of the Jasmund cliff area (Rugen Island, Germany), Nat. Hazards Earth Syst. Sci., 9, 687– ¨ 698, doi:10.5194/nhess-9-687-2009, 2009.

51. Guzzetti, F., Carrara, A., Cardinali, M., and Reichenbach, P.: Landslide hazard evaluation: a review of current techniques and their application in a multi-scale study, Central Italy, Geomorphology, 31, 181–216, 1999.

52. Guzzetti, F., Reichenbach, P., Cardinali, M., Galli, M., Ardizzone, F.: Probabilistic landslide hazard assessment at the basin scale, Geomorphology, 72, 272–299, 2005.

53. Guzzetti, F., Reichenbach, P., Ardizzone, F., Cardinali, M., and Galli, M.: Estimating the quality of landslide susceptibility models, Geomorphology, 81, 166–184, 2006.

54. Harp, H. L., Reid, M. E., McKenna, J. P., and Michael, J. A.: Mapping of hazard from rainfall-triggered landslides in developing countries: Examples from Honduras and Micronesia, Eng. Geol., 104, 295–311, 2009.

55. Hengl, T.: Finding the right pixel size, Comput. Geosci., 32, 1283– 1298, 2006.

56. Heymes, T., Monie, P., Arnaud, N., P ´ echer, A., Bouillin, J. P., and ˆ Compagnoni, R.: Alpine tectonics in the Calabrian–Peloritan belt (southern Italy): New 40Ar/39Ar data in the Aspromonte Massif area, Lithos 114, 461–472, 2010.

57. Hren, M. T, Hilley, G. E., and Chamberlain, C. P.: The relationship between tectonic uplift and chemical weathering rates in the Washington Cascades: field measurements and model predictions, Am. J. Sci., 307, 1041–1063, 2007.

58. ISRM, International Society on Rock Mechanics: Basic geotechnical description of rock masses, Int. J. Rock Mech. Min. Sci. Geomech. Ab., 18, 85–110, 1981.

59. Keller, E. A. and Pinter, N.: Active Tectonics: Earthquakes, Uplift, and Landscape, Prentice Hall, Upper Saddle River, New Jersey, 338 pp., 1996.

60. Korup, O., Clague, J. J., Hermanns, R. L., Hewitt, K., Strom, A. L., and Weidinger, J. T.: Giant landslides, topography, and erosion, Earth Pl. Sci. Lett., 261, 578–589, 2007.

61. Korup, O., Densmore, A. L., and Schlunegger, G.: The role of landslides in mountain range evolution, Geomorphology, 102, 77–90, 2010.

62. Lacerda, W.: Landslide initiation in saprolite and colluvium in southern Brazil: Field and laboratory observations, Geomorphology, 87, 104–119, 2007.

63. Lambeck, K., Antonioli, F., Purcell, A., and Silenzi, S.: Sea-level change along the Italian coast for the past 10 000 yr, Quaternary Sci. Rev., 23 , 1567–1598, 2004.

64. Le Pera, E. and Sorriso-Valvo, M.: Weathering and

morphogenesis in a Mediterranean climate, Calabria, Italy, Geomorphology, 34, 251–270, 2000.

65. Marques, E. A. G., Barroso, E. V., Menez Filho, A. P., and Vargas, Jr. E. A.: Weathering zones on metamorphic rocks from Rio de Janeiro–Physical, mineralogical and geomechanical characterization, Eng. Geol., 111, 1–18, 2010.

66. Maugeri, M. and Motta, E.: Effects of Heavy Rainfalls on Slope Behaviour: The October 1, 2009 Disaster of Messina (Italy), Geotechnics/Earthquake Geotechnics towards Global Sustainability, Kyoto, Japan, 12–14 January, 2010.

67. Monaco, C. and Tortorici, L.: Active faulting in the Calabrian arc and eastern Sicily, J. Geodynam., 29, 407–424, 2000.

68. Monaco, C., Tortorici, L., Nicolich, R., Cernobori, L., and Costa, M.: From collisional to rifted basins: an example from the southern Calabrian arc (Italy), Tectonophys., 266, 233–249, 1996.

69. Monaco, C., Tapponier, P., Tortorici, L., and Gyllot, P. Y.: Late Quaternary slip rates on the Acireale-Piedimonte normal faults and tectonic origin of Mt. Etna (Sicily), Earth Planet. Sci. Lett., 147, 125–139, 1997. Montenat, C., Barrier, P., and d'Estevou, P.: Some aspects of the recent tectonics in the Strait of Messina, Italy, Tectonophys., 194, 203–215, 1991.

70. Montgomery, D. R. and Brandon, M., T.: Topographic controls on erosion rates in tectonically active mountain ranges, Earth Pl. Sci. Lett., 201, 481–489, 2002.

71. Morgagni, A., Marsigli, M., and Todini, E.: Assessment of soil erosion in a number of small ephemeral streams in Calabria and Sicily, Proc. Workshop on Soil Erosion in Semi-arid Mediterranean Areas, European Society for Soil Conservation, Centro Studi per l'Economia applicata all'Ingegneria, 137–148, 1993.

72. Nandi, A. and Shakoor, A.: A GIS-based landslide susceptibility evaluation using bivariate and multivariate statistical analyses, Eng. Geol., 110, 11–20, 2009.

73. Pellegrino, A. and Prestininzi, A.: Impact of weathering on the geomechanical properties of rocks along thermal–metamorphic contact belts and morpho-evolutionary processes: The deepseated gravitational slope deformations of Mt. Granieri– Salincriti (Calabria–Italy), Geomorphology, 87, 176–195, 2007.

74. Piacentini, D., Troiani, F., Soldati, M., Notarnicola, C., Savelli, D., Schneiderbauer, S., and Strada, C.: Statistical analysis for assessing shallow-landslide susceptibility in South Tyrol (south-eastern Alps, Italy), Geomorphology, 151/152, 196–206, 2012.

75. Poli, S. and Sterlacchini, S.: Landslide Representation Strategies in Susceptibility Studies using Weights-of-Evidence Modeling Technique, Nat. Resources Res., 16, 121–134, 2007.

76. Riebe, C. S., Kirchner, J. W., Granger, D. E., and Finkel, R. C.: Strong tectonic and weak climatic control of long-term chemical weathering rates, Geology, 29/ 6, 511–514, 2001.

77. Ruff, M. and Czurda, K.: Landslide susceptibility analysis with a heuristic approach in the Eastern Alps (Vorarlberg, Austria), Geomorphology, 94, 314–324, 2008.

78. Ruff, M. and Rohn, J.: Susceptibility analysis for slides and rockfall: an example from the Northern Calcareous Alps (Vorarlberg, Austria), Environ. Geol., 55, 441–452, 2008.

79. Rust, D. and Kershaw, S.: Holocene tectonic uplift patterns in northeastern Sicily: evidence from marine notches in coastal outcrops, Mar. Geol., 167, 105–126, 2000.

80. Sabato, L. and Tropeano, M.: Fiumara: a kind of high hazard river, Phys. Chem. Earth., 29, 717–715, 2004.

81. Salciarini, D., Godt, J. W., Savage, W. Z., Conversini, P., Baum, R. L., and Michael, J. A.: Modeling regional initiation of rainfallinduced shallow landslides in the eastern Umbria Region of central Italy, Landslides, 3, 181–194, 2006.

82. Salciarini, D., Godt, J. W., Savage, W. Z., Baum, R. L., and Conversini, P.: Modeling landslide recurrence in Seattle, Washington, USA, Eng. Geol., 102, 227–237, 2008.

83. Santangelo, M., Gioia, D., Fiorucci, F., Cardinali, M., Guzzetti, F., and Schiattarelli, M.: Relationships between mass movements and tectonic uplift in the upper Sinni River valley, Calabria-Lucania border, Southern Italy, Geophys. Res. Abstr., 13, EGU2011-11674, EGU General Assembly 2011, Vienna, Austria, 2011.

84. Singh, L. P., van Westen, C. J., Champati Ray, P. K., and Pasquall, P.: Accuracy assessment of InSAR derived inputmaps for landslide susceptibility analysis: a case study from the Swiss Alps, Landslides, 2, 221–228, 2005.

85. Scudero, S. and De Guidi, G.: Landslide processes and susceptibility mapping in NE Sicily, Italy. The Second World Landslide Forum, "Putting science into practice", Global Promotion Committee of the International Programme on Landslides (IPL), Rome, Italy, 2011.

86. Snyder, N. P., Whipple, K. X., Tucker, G. E., and Merritts, D. J.: Interaction between onshore bedrock-channel incision and nearshore wave-base erosion forced by eustasy and tectonics, Basin Res., 14, 105–127, 2002.

87. Snyder, N. P., Whipple, K. X., Tucker, G. E., and Merritts, D. J.: Channel response to tectonic forcing: field analysis of stream morphology and hydrology in the Mendocino triple junction region, northern California, Geomorphology, 53, 97–127, 2003.

88. Somma, R., Messina, A., and Mazzoli, S.: Syn-orogenic extension in the Peloritani Alpine Thrust Belt (NE Sicily, Italy): Evidence from the Al`ı Unit, C. R. Geosci., 337, 861–871, 2005.

89. Stewart, I. S., Cundy, A., Kershaw, S., and Firth, C.: Holocene coastal uplift in the Taormina area, northeastern Sicily: Implications for the southern prolongation of the Calabrian seismogenic belt, J. Geodynam., 24/14, 37–50, 1997.

90. Suzen, M. L. and Doyuran, V.: A comparison of the GIS based land- ¨ slide susceptibility assessment methods:

multivariate versus bivariate, Environ. Geol., 45, 665–679, 2004.

91. Tortorici, L., Monaco, C., Tansi, C., and Cocina, O.: Recent and active tectonics in the Calabrian arc (Southern Italy), Tectonophys., 243, 37–55, 1995.

92. van Westen, C. J.: Statistical landslide hazard analysis. In: Application guide, ILWIS 2.1 for Windows, ITC, Enschede, The Netherlands, 73–84, 1997.

93. van Westen, C. J., Rengers, N., Terlien, M. T. J., and Soeters, R.: Prediction of the occurrence of slope instability phenomena

CHAPTER 15

GEOCHRONOLOGY AND TECTONIC EVOLUTION

Hongyuan Zhang, Junlai Liu, and Wenbin Wu

INTRODUCTION

The Sanjiang-Indochina region is one key area to study the evolution of the Tethys [1–10] (Figure 1). The Changning-Menglian zone was considered as the major faunal break between Gondwanan assemblage to the west and Cathaysian to the east [11, 12]. The subduction-related magmatism occurred along the western edge of the Lanping-Simao-Indochina terrane throughout the Triassic and the closure of Palaeo-Tethys [12, 13]. The Lincang Batholith extends ~370 km from north to south, covering the Chiangrai-Chiang Mai region of Thailand, with an area of more than 10000 km^2 (Figure 2). It has been shown by many authors that there is great mineral potential [14].

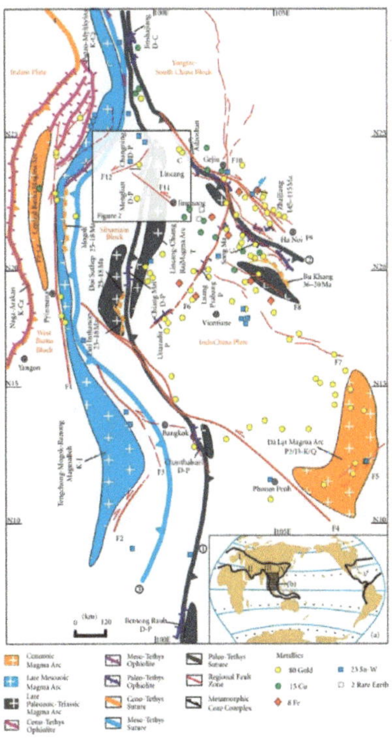

Figure 1: Map of plate tectonics and ore-forming metals in the southern Sanjiang and Indochina peninsula. (a) Scope of Tethys and zone divisions; (b) research area and distribution of major metals: (1) Changning-Menglian-Chiangmai Paleo-Tethys Suture (time and space after [1, 21, 22]); (2) Ailaoshan-Song Ma Paleo-Tethys Suture (time and space after [2, 6, 8]); (3) Bangong Co-Nujiang Meso-Tethys Suture (time and space after [5]); (4) Yanlung Tsangpo-Naga-Arakan New-Tethys Suture (time and space after [22, 23]); (F1) Nujiang-Saging Fault Zone; (F2) Ranong Fault Zone; (F3) Three Tower Fault Zone; (F4) Tonle Sap Fault Zone; (F5) Đà Lat-Bianhe Fault Zone; (F6) Điên Biên Phú Fault Zone; (F7) Changshan-Da Nang Fault Zone; (F8) Lanjiang Fault Zone; (F9) Ailaoshan-Red River Fault Zone; (F10) Xianshuihe-Xiaojiang Fault Zone; (F11) Lancang-Jinghong Fault Zone.

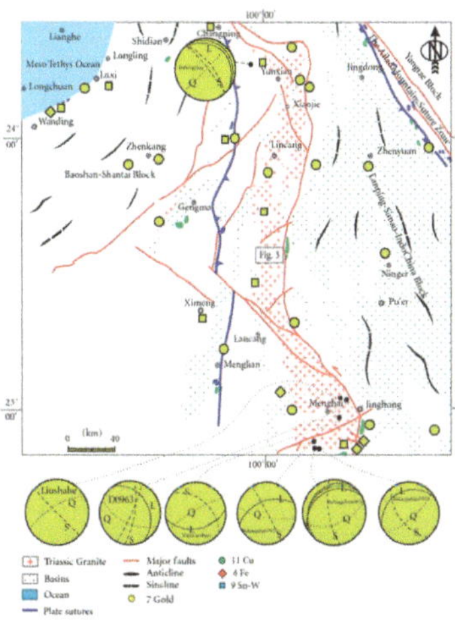

Figure 2: Regional map of Indian Period and research locality in western Yunnan. The distribution of metal deposits also indicates that there is an important boundary along the Lincang Granite Batholith. Stereographic maps of lower hemisphere projection indicating the initial occurrence of Lincang granite Batholith are also plotted in the regional map. L, S, Q refer to joints parallel to the initial foliation, parallel to the flow direction but upright to the initial foliation, upright to both the flow direction and initial foliation. Orientation of shear fractures can be generated by local stress system of the Batholith with emplacements [24].

Citation: Hongyuan Zhang, Junlai Liu, and Wenbin Wu, Geochronology and Tectonic Evolution, doi:10.1155/2012/287962.

The Lingcang Batholith was divided into three lithologic intervals by previous authors, the Xiaojie, Lincang, and Menghai intervals (Figure 2) [15–17]. The major intrusion is porphyritic monzonitic granite, which mainly contains quartz, zonal plagioclase, K-feldspar, and biotite. In addition, geochemical studies have shown that the granite mainly has source features of the crust mixing origin [16, 18].

The Lingcang Batholith was proposed to form in a passive continental margin according to geochemical research [19]. The intrusive timing spans from late Permian to late Triassic based on whole-rock Rb-Sr, mineral$^{40}Ar/^{39}Ar$, whole-rock $^{87}Sr/^{86}Sr$, Rb-Sr, Sm-Nd, and some other methods (Table 1).

Table 1: Isotopic age data for the Lincang Batholite in the earlier stage.

Lithologic section	Lithology	Measured objects and methods	Dating results, Ma	Sample positions	Testers and the time
Xiaojie	Granite porphyry	Whole-rock (Rb-Sr)		Laomaocun	Zhang et al., 2006 [25]
Xiaojie	Monzonitic granite	Whole-rock ($^{87}Sr/^{86}Sr$)	279	Yun County	Chen, 1991 [26]
Lincang	Biotite granite	Biotite ($^{40}Ar/^{39}Ar$)		Near the Lancangjiang River	Dai et al., 1986 [27]
Lincang	Monzonitic granite	Whole-rock (Rb-Sr)	279	Lincang	Chen, 1991 [26]
Lincang	Medium-grained equigranular granodiorite	Whole-rock (Rb-Sr)	263.8	Milestone along Lincang-Mengku road 327–387 km	Zhang et al., 1990 [28]
Lincang	Unequal-sized biotite granite	Whole-rock (Rb-Sr)	193	Shangyun-Xiaotang	Zhang et al., 1990 [28]
Lincang	Monzonitic granite	Whole-rock (Rb-Sr)		Lincang	Liu et al., 1989 [18]
Menghai	Monzonitic granite	Whole-rock ($^{87}Sr/^{86}Sr$)	279	Menghai	Chen, 1991 [26]
Menghai	Monzonitic granite	Biotite (Rb-Sr)	256	Menghai	Wang, 1984 [29]

But there are also some understanding differences on the crystallization period and evolution time of the Lincang Batholith. Many predecessors discussed mineral diagenesis and tectonic evolution through traditional data [16, 20], although, not from the viewpoint of the plate tectonics. On scientific problems of the Lincang Batholith, some solutions were put forward such as single period of mixed granite [16], quasi-situ metasomatic granite [20], or Neoproterozoic type I granite complex batholith with multiperiod transforming events [19].

Therefore, the formation and evolution of the Lincang Batholith is still the key to discuss the regional tectonic evolution of the Baoshan and Lanping-Simao-Indochina block. In this study, we conducted petrographic analysis of the Lincang Batholith and provide new LA-ICP-MS U-Pb data in order to constrain the tectonic evolution of the Batholith. We discuss these results of dating in the context of regional tectonic evolution.

DATING METHODS, SPECIMENS, AND RESULTS

Dating Methods

The U-Pb zircon age dating of monzonitic granite samples selected from the Mengku-Douge research area in Lincang segment (Figures 2 and 3) was designed and completed in the Institute of Geology and Geophysics, Chinese Academy of Sciences. Jobs of single zircon grain micro-area U-Pb geochronological analysis were taken through Laser Ablation Inductively Coupled Plasma Mass Spectrometry (LA-ICPMS) of 7500a-type producted by Agilent Co. Ltd., with which the laser beam diameter is 60 μm, the erosion frequency is 8 Hz, the energy density is 15~20 J·cm^{-2}, the erosion time is about 60 s. For detailed analysis procedures, please see Xu et al. [32]. The Zircon U-Pb isotope and the U, Th data processing are finished by software Glitter4. 0 [33], and computing of U-Pb concordia diagrams, weighted average ages and graphics were completed by software Isoplot3.0 [34].

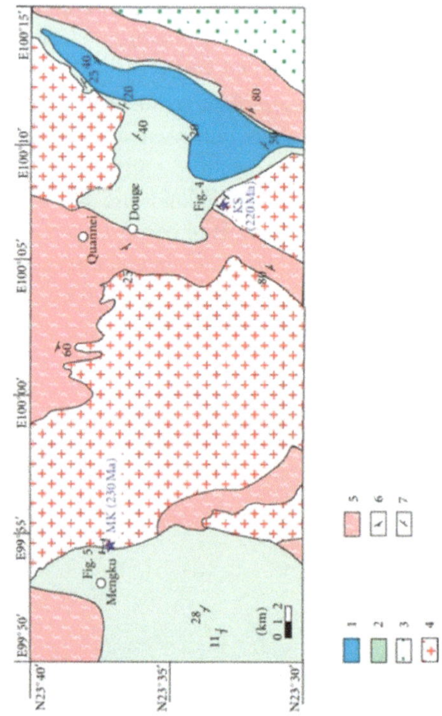

Figure 3: Geological sketch map in Mengku-Douge area (revised after [15–17, 30, 31]) (1) Area of ingression sea of the Meso-Tethys; (2) Jurassic terrigenous sediments; (3) Early Jurassic volcanic sedimentary rocks; (4) Lincang Triassic granite rock; (5) deep metamorphic wedge with local granodiorite quality migmatization Late Paleozoic; (6) schistosity occurrence; (7) bedding occurrence.

Specimens

Specimen from Kunsai Quarries

The biotite monzonitic granite samples were selected from the Kunsai quarries to the east of Quannei-Douge migmatite rock belt (Figures 3 and 4(a)). The north part of the rock unit is plunged into the Jurassic basin (Figure 3). Field study shows that both the rock unit and the Jurassic basin have experienced a period of ductile transformation events. The microstructure study shows brittle deformed feldspar (Figures 4(b) and 4(c)), dynamically recrystallized quartz (Figure 4(c)), microdeflected biotite (Figure 4(d)), zircon grains with relatively intact crystal, long column, some rounded output and a higher degree of porosity which is symbiotic to microcline grains (Figure 4(b)). The cathode luminescence (CL) images show that the typical magmatic zircon characteristics are colorless and transparent here with oscillatory zoning, the length of about 120 μm–420 μm and the aspect ratio of 1.2 to 4.1 (Figure 4(e)). Totally, for 22 points of 22 zircon grains of sample biotite granodiorite KS-2 from the Kunsai quarries we performed U-Pb isotope analysis in order to statistically gain a feasible dating result (Figure 4(f)). All points were selected on the edge of long-column euhedral grains where the magmatic zircon oscillatory zoning is clear.

(a)

(b)

(c)

(d)

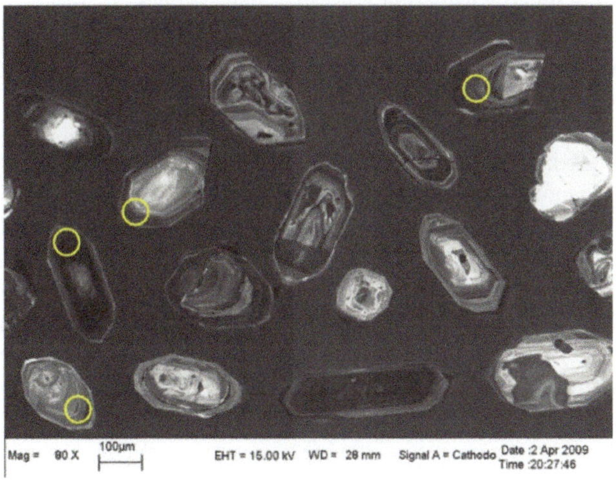

(e)

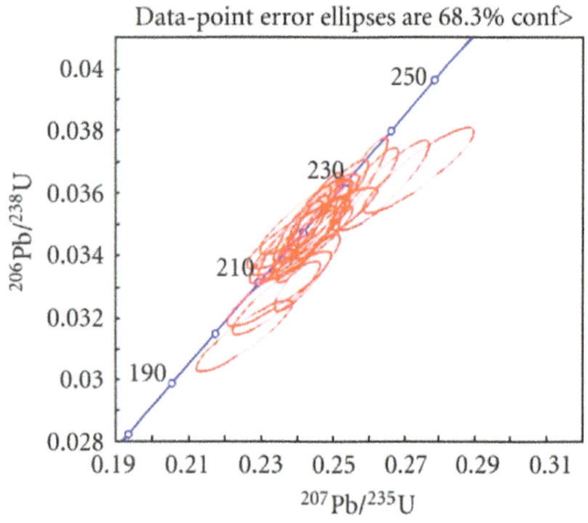

(f)

Figure 4: Geochronological research of granite sample KS-2. (a) Position photo of the biotite monzonitic granite sample in the field of Kunsai Quarry. (b) The zircon grows together with the microcline CPL. (c) The microstructure of complete recrystallized quartz indicates the rock once experienced high-greenschist facies of deformation, CPL. (d) One biotite crystal is folded, PPL. (e) Magmatic zircon is shown by the SEM cathodoluminescence (CL) image. The circled area is deduced to form during the crystallization of granite and is just the measured area by laser-ICPMS (f) Zircon La-ICPMs U-Pb dating figure, and the average age is about 220 Ma.

Specimens from Mengku Quarries

The monzonitic granite sample MK-4 is selected from the Mengku East Quarry at the eastern side of the Mengku Town in the Shuangjiang County, which separates Jurassic red clastic rocks with one bedding bottom granitic conglomerate by the paleo-weathering crust in between (Figures 5(a), 5(b), and 5(c)). Figure 5(a)shows the position photo of the biotite monzonitic granite sample in the Mengku East Quarry. From the field outcrop, both granitic intrusion (lower left) and its weathering top can be found to be cleavaged which had been covered by the upper Jurassic red detrital sediments with terrigenous origin, namely, the Huakaizuo Formation. The MK-4 zircon grains (Figure 5(d)) change greatly and have different shapes and sizes, such as long column, fan, and irregular granular. However, size of zircon grains is uniform with length between 90 μm and 150 μm and length-width ratio in 1.1 : 1–1.8 : 1. Zircon grains develop with quartz, plagioclase in rocks. The rock sample has been affected by weathering, showing retrograde metamorphism (Figure 5(e)). Characteristic oscillatory zoning of typical magmatic zircon is rendered on the cathode CL image. The idiomorphic degree of zircon is high, but some fragmented. On the sample of MK-4, 18 zircon grains and 18 points were analyzed by ICP-MS. Edge of the long-column idiomorphic or the chipped-hypidiomorphic zircon grains, and the clear zone of magmatic zircon oscillatory were always selected to perform the micro isotope analysis (Figure 5(f)).

(a)

(b)

(c)

(d)

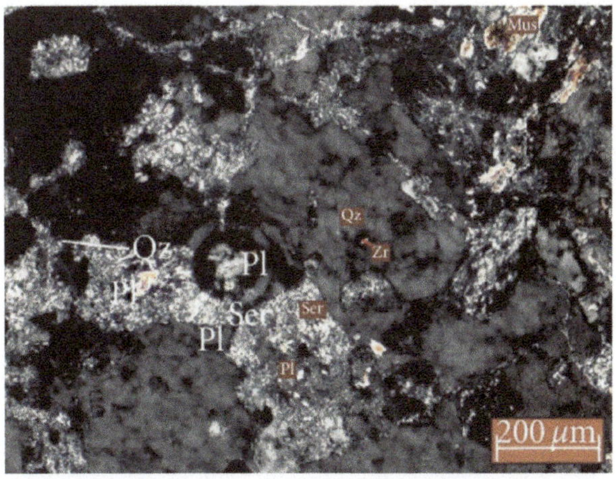

(e)

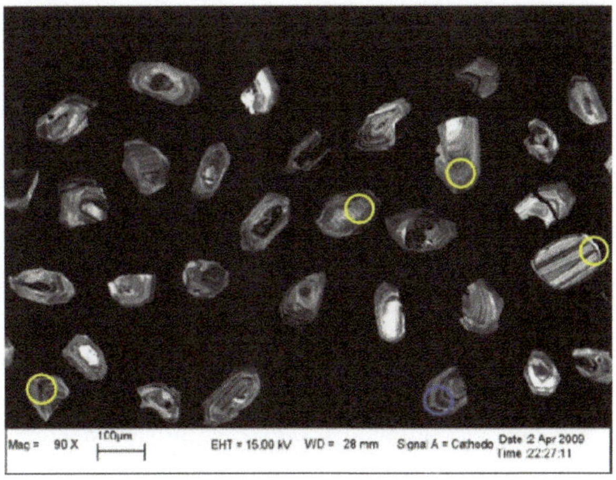

(f)

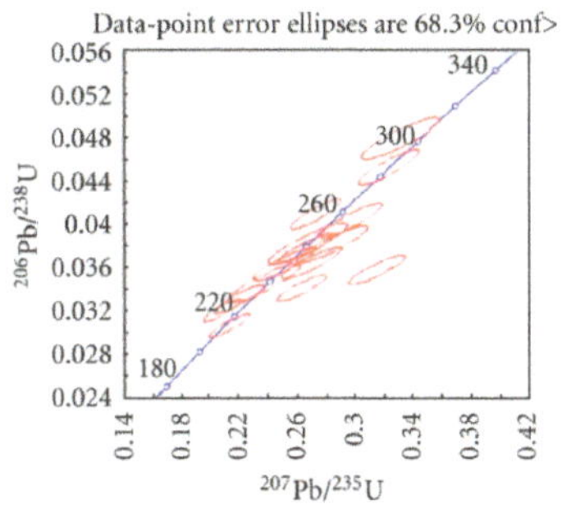

(g)

Figure 5: Geochronological research of the granite sample MK-4. (a) Position photo of the biotite monzonitic granite sample in a field Quarry of east Mengku. (b) Granitic intrusion (lower left) and its fossil weathering were cleavaged. (c) The red detrital material of terrigenous origin, Huakaizuo Formation in Jurassic, was normal gradational. (d) Granitic intrusions are selected as sample. (e) The micro characteristics of MK-4 and the zircon setting are shown. The zircon grains are associated with quartz and plagioclase and have been degraded by some weathering process, CPL, (f) Magmatic zircon is shown by the SEM cathodoluminescence (CL) image. The circled area is deduced to form during the crystallization of granite and is just the measured area by laser-ICPMS. (g) Zircon La-ICPMs U-Pb dating figure, and the average age is about 245 Ma.

Dating Results

As shown in Figure 4(f) and Table 2, the overall harmony values of KS-2 zircon age data range from 195 Ma to 245 Ma, but some of them are below the U-Pb harmonic line and deviated from the harmony line. This is probably because the micro region selected is too close to the boundary of the zircon grain. Some zircon grains (Figure 5(f)) also have some cracks in and this may also affect the dating result. From the harmonic visible figure, zircon dating values are focused to a range from 210 Ma to 230 Ma, and the average age is about 220 Ma. This means that the crystallization age of the Kunsai quarries biotite granite is about in the late Triassic Norian period.

Table 2: U-Pb isotopic dating for the single grain Zircon from granite of Kunsai.

No.	$^{207}Pb/^{206}Pb$		$^{207}Pb/^{235}U$		$^{206}Pb/^{238}U$		$^{208}Pb/^{232}Th$		$^{207}Pb/^{206}Pb$		$^{207}Pb/^{235}U$		$^{206}Pb/^{238}U$		$^{208}Pb/^{232}Th$	
	Ratio	Error (1)	Ratio	Error (1)	Ratio	Error (1)	Ratio	Error (1)	Ratio	Error (1)	Ratio	Error (1)	Ratio	Error (1)	Ratio	Error (1)
A77	0.05789	0.00231	0.25961	0.00949	0.03251	0.00089	0.01383	0.00076	526	38	234	8	206	6	278	15
A78	0.05188	0.00244	0.23562	0.00953	0.03294	0.00079	0.01035	0.00025	280	110	215	8	209	5	208	5
A79	0.05273	0.00141	0.24329	0.00604	0.03345	0.00081	0.01108	0.00055	317	26	221	5	212	5	223	11
A80	0.05213	0.00216	0.23471	0.00801	0.03265	0.00077	0.01025	0.00024	291	97	214	7	207	5	206	5
A81	0.05209	0.00131	0.25938	0.00608	0.0361	0.00086	0.01177	0.00034	289	25	234	5	229	5	237	7
A82	0.05014	0.00118	0.24192	0.00531	0.03498	0.00082	0.01113	0.00044	201	24	220	4	222	5	224	9
A83	0.05126	0.00122	0.24287	0.00541	0.03435	0.00081	0.01155	0.0006	253	24	221	4	218	5	232	12
A84	0.05027	0.00114	0.24534	0.00525	0.03539	0.00083	0.01176	0.00041	207	24	223	4	224	5	236	8
A89	0.0521	0.00237	0.22575	0.00875	0.03143	0.00075	0.00987	0.00023	290	107	207	7	199	5	199	5
A90	0.05305	0.00182	0.2649	0.0084	0.03621	0.00094	0.01376	0.00075	331	33	239	7	229	6	276	15
A91	0.05739	0.00113	0.23972	0.00441	0.03029	0.0007	0.00946	0.00032	507	23	218	4	192	4	190	6
A92	0.05076	0.00135	0.24604	0.00612	0.03515	0.00085	0.01173	0.00045	230	26	223	5	223	5	236	9
A93	0.05089	0.00279	0.24455	0.01197	0.03485	0.00086	0.01098	0.00025	236	127	222	10	221	5	221	5
A94	0.05114	0.00155	0.24377	0.00683	0.03457	0.00087	0.01167	0.00055	247	29	222	6	219	5	235	11
A95	0.05055	0.00109	0.24175	0.00488	0.03469	0.00081	0.01181	0.00045	220	24	220	4	220	5	237	9
A96	0.05018	0.00134	0.23857	0.00594	0.03448	0.00084	0.01149	0.00049	203	26	217	5	219	5	231	10
B02	0.05115	0.00119	0.24774	0.00541	0.03514	0.00084	0.01297	0.00054	248	25	225	4	223	5	260	11
B03	0.05067	0.00224	0.24165	0.00903	0.03459	0.00082	0.0109	0.00025	226	104	220	7	219	5	219	5
B04	0.05104	0.00235	0.24163	0.00953	0.03433	0.00082	0.01081	0.00025	243	109	220	8	218	5	217	5
B05	0.05079	0.00153	0.24968	0.00699	0.03567	0.0009	0.01274	0.00067	231	29	226	6	226	6	256	13
B06	0.05039	0.00095	0.24525	0.00437	0.03531	0.00082	0.01121	0.00038	213	25	223	4	224	5	225	8
B07	0.05429	0.00243	0.27463	0.01035	0.03669	0.00089	0.01147	0.00027	383	103	246	8	232	6	230	5
B08	0.05014	0.00223	0.24399	0.00913	0.0353	0.00085	0.01114	0.00026	201	104	222	7	224	5	224	5
B09	0.05141	0.00105	0.25807	0.00495	0.03643	0.00086	0.01414	0.00053	259	25	233	4	231	5	284	11

As shown in Figure 5(g) and Table 3, the overall harmony values of MK-4 zircon age data range between 190 Ma and 315 Ma. The values are somewhat scattered mainly because the granite of the Mengku quarries has been heavily weathered (Figure 5(d)). During the weathering process, chemical bonds of ZrO_2 in zircon grains of our MK-4 sample would be broken. Therefore, some zircon isotope escaped. The average age of 245 Ma might indicate that the late Permian to early-middle Triassic magmatic event happened in the Mengku Quarries area of the Lincang Batholith.

Table 3: U-Pb isotopic dating for the single grain Zircon from granite of Mengku.

Analysis	$^{207}Pb/^{206}Pb$		$^{207}Pb/^{235}U$		$^{206}Pb/^{238}U$		$^{208}Pb/^{232}Th$		$^{207}Pb/^{206}Pb$		$^{207}Pb/^{235}U$		$^{206}Pb/^{238}U$		$^{208}Pb/^{232}Th$	
no.	Ratio	Error (1)	Ratio	Error (1)	Ratio	Error (1)	Ratio	Error (1)	Ratio	Error (1)	Ratio	Error (1)	Ratio	Error (1)	Ratio	Error (1)
1	0.05174	0.00244	0.32303	0.0139	0.04528	0.00108	0.01152	0.00053	274	56	284	11	285	7	232	11
2	0.09115	0.00288	2.96352	0.08448	0.23585	0.0053	0.08061	0.00287	1450	25	1398	22	1365	28	1567	54
3	0.11509	0.00332	0.50952	0.01225	0.03212	0.00066	0.02102	0.00068	1881	19	418	8	204	4	420	13
4	0.04837	0.00265	0.22077	0.01108	0.03312	0.00085	0.01153	0.00063	117	69	203	9	210	5	232	13
5	0.04837	0.0022	0.27295	0.01145	0.04095	0.00092	0.01223	0.0006	117	57	245	9	259	6	246	12
6	0.05351	0.00279	0.29775	0.01415	0.04038	0.00104	0.01371	0.00078	350	62	265	11	255	6	275	16
7	0.0806	0.00218	2.27431	0.05585	0.2048	0.00394	0.05968	0.00155	1212	22	1204	17	1201	21	1172	30
8	0.0802	0.00341	2.36577	0.09085	0.2141	0.00577	0.05352	0.00277	1202	37	1232	27	1251	31	1054	53
9	0.05928	0.00248	0.29959	0.01132	0.03668	0.00082	0.00966	0.00046	577	45	266	9	232	5	194	9
10	0.04989	0.00375	0.23793	0.0163	0.03461	0.00119	0.00907	0.00072	190	96	217	13	219	7	182	14
11	0.07866	0.00333	0.34683	0.01279	0.032	0.00079	0.01233	0.00063	1164	37	302	10	203	5	248	13
12	0.0406	0.00217	0.22437	0.01119	0.04011	0.00095	0.01052	0.00055	−255	72	206	9	254	6	212	11
13	0.0639	0.00299	0.31538	0.01317	0.03582	0.00089	0.01126	0.00067	738	48	278	10	227	6	226	13
14	0.07481	0.00441	0.3517	0.01817	0.03412	0.00108	0.01178	0.00078	1063	56	306	14	216	7	237	16
15	0.06821	0.00263	0.97745	0.03376	0.10399	0.00236	0.03233	0.00106	875	37	692	17	638	14	643	21
16	0.05116	0.00255	0.26673	0.01227	0.03784	0.00088	0.01057	0.00052	248	64	240	10	239	5	213	10
17	0.05025	0.00304	0.25181	0.01395	0.03636	0.00102	0.01037	0.00064	207	77	228	11	230	6	209	13
18	0.05379	0.003	0.27018	0.01369	0.03645	0.00098	0.00965	0.0006	362	67	243	11	231	6	194	12
19	0.05083	0.00373	0.26778	0.01802	0.03823	0.00125	0.00987	0.00055	233	96	241	14	242	8	199	11
20	0.06022	0.00487	0.2998	0.0217	0.03612	0.00142	0.01137	0.00113	611	90	266	17	229	9	229	23
21	0.0473	0.00214	0.20932	0.0087	0.0321	0.0007	0.00892	0.00038	64	55	193	7	204	4	179	8
22	0.05405	0.00334	0.28442	0.01602	0.03817	0.00111	0.00858	0.0006	373	75	254	13	241	7	173	12
23	0.05907	0.0024	0.66212	0.02439	0.08128	0.00181	0.03438	0.00121	570	43	516	15	504	11	683	24
24	0.05064	0.00245	0.21459	0.00944	0.03072	0.00072	0.00797	0.00046	224	59	197	8	195	5	160	9
25	0.05594	0.00261	0.26352	0.01113	0.03415	0.0008	0.00964	0.00045	450	53	237	9	216	5	194	9
26	0.06769	0.00524	0.32217	0.02212	0.0345	0.00134	0.01049	0.0009	859	80	284	17	219	8	211	18
27	0.06354	0.00206	0.75553	0.02205	0.08618	0.00166	0.02056	0.00084	726	32	571	13	533	10	411	17
28	0.05014	0.00296	0.33191	0.01801	0.04797	0.00129	0.02068	0.0011	201	77	291	14	302	8	414	22
29	0.23782	0.00377	6.55914	0.08439	0.19984	0.00325	0.01824	0.00075	3105	12	2054	11	1174	17	365	15

TECTONIC IMPLICATIONS

Previous petrological and geochemical studies [18, 26, 28, 35] indicate that the Lincang granite mainly shows two types of granite with an initial strontium isotope ratio ranging from 0.71 to 0.78 and therefore it belongs to S-type granite and once may have been formed in an environment of tectonic collision. Combined with previous chronological data, we propose a three-stage model on the tectonic evolution between late Permian and Jurassic (Figure 6).

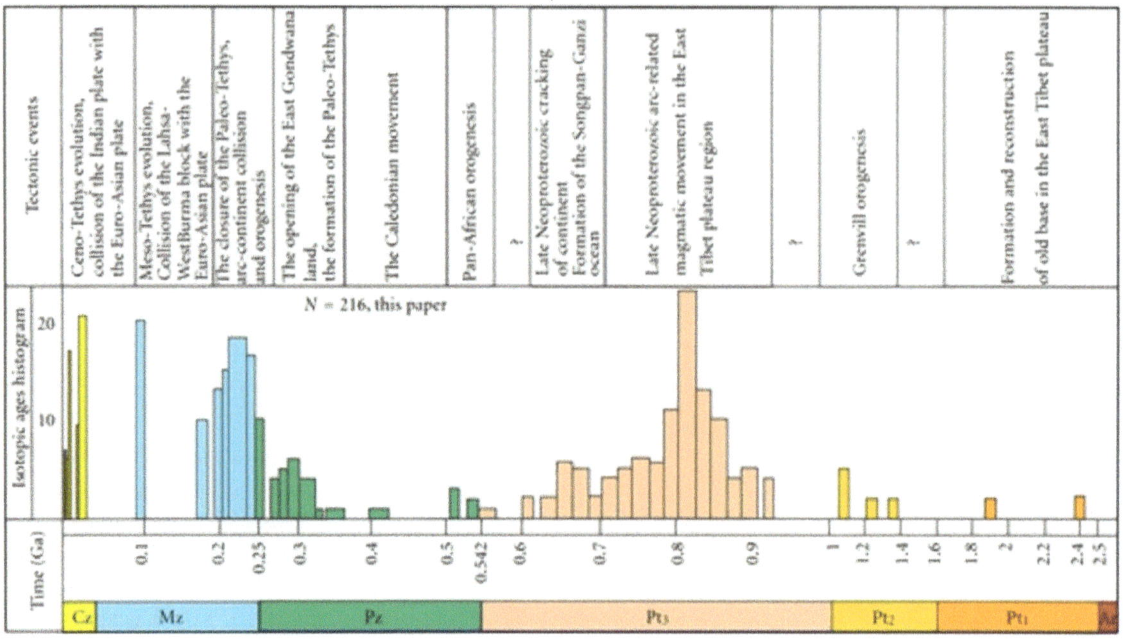

Figure 6: Isotopic age histogram showing the major tectonic event in the area east to the Tibet Peateau (after [8]).

(1) Regional Collision and Contraction before the Formation of the Lincang Batholith Before the formation of the Lincang Batholith, the Paleo-Tethys Ocean and many intraoceanic islands distributed between the Baoshan and the Lanping-Simao-Indosinian blocks [36]. In a regional contractional environment of subduction and collision, it is extremely common to find folding, thrusting, and uplifting phenomena (Figure 7(a)).

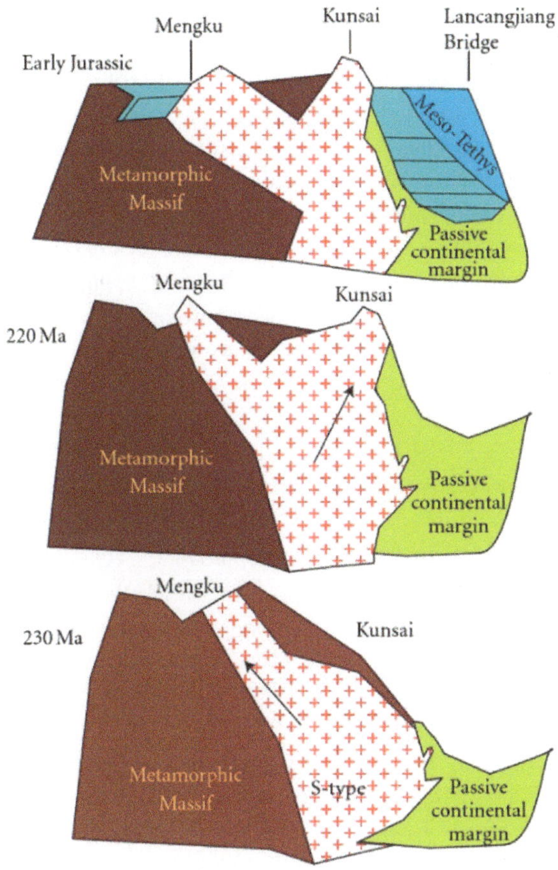

Figure 7: Formation and tectonic evolution model for the Lincang Batholith.

(2) Time Heterogeneity of In Situ Hybrid of the Lincang Batholith The Lincang Batholith was controlled by one deep-seated fault along the Lancang Jiang River and proposed in situ hybrid origin of the Batholith in the early to middle Triassic collision between the Baoshan-Shantai and Lanping-Simao-Indochina Blocks [16]. Our study shows an ununiform law result in the aspect of mixing time of the Batholith (Figure 7(b)). The intrusion time in the east Qunsai area was later than in the Mengku area, indicating that the regional contraction in Mengku area was weakened or stopped, and triggered magma poured at ~230 Ma, however, the contraction pattern may still remain in the Qunsai area to the east.

(3) Tectonic Denudation and Regional Exhumation Happened between Late Triassic and Middle Jurassic, Which Is Supported by ~201 Ma Biotite $^{40}Ar/^{39}Ar$ Age from the Lincang Batholith [27] This paper research shows that the upper part of the Lincang Batholith experienced long-term denudation, resulting in the development of the weathering crust (Figure 7(a)). The upper crust is the Jurassic Huakaizuo Formation, characterized by carbonate rocks of shallow sea facies, together with some volcano clastic rocks (Figure 5(b)). Unconformity relationship is shown between the Batholith and the Huakaizuo Formation (Figures 3, 5(a), 5(b), 5(c), and 7(c)). Apparently, the Meso-Tethys expansion also happened in this era and left some effect in the Lincang Area.

ACKNOWLEDGMENTS

The authors appreciate sincerely both the field research and inner discussion together with Professor Yong-Qing Chen from China University of Geosciences (Beijing), Research Fellow Ying-Xiang Lu from Yunnan Bureau of Geology and Mineral Resources, and Research Fellow Fang-Cheng Lin from Chengdu Institute of Geology and Mineral Resources. The authors thank Dr. Pengfei Li from University of Queensland for his perfect academic revision of the paper. The research was supported by the China Geology Survey (no. 200811008, no. 1212011121188), the Ministry of Science and Technology (no. 2006BAB01A03-3), the Chinese National Natural Foundation (no. 90814006) of the People's Republic, and the China University of Geosciences in Beijing (no. 2-9-2001-280).

REFERENCES

1. J. Q. Huang and B. W. Chen, The Evolution of the Tethys in China and Adjacent Regions, Geological Publishing House, Beijing, China, 1987.
2. P. Jian, D. Liu, A. Kröner et al., "Devonian to Permian plate tectonic cycle of the Paleo-Tethys Orogen in southwest China (II): insights from zircon ages of ophiolites, arc/back-arc assemblages and within-plate igneous rocks and generation of the Emeishan CFB province," Lithos, vol. 113, no. 3-4, pp. 767–784, 2009. · ·
3. H. D. Klemme and G. F. Ulmishe, "Effective petroleum source rocks of the world: stratigraphic distribution and controlling depositional factors," AAPG Bulletin, vol. 75, no. 12, pp. 1809–1851, 1991. ·
4. G. T. Pan, Z. L. Chen, and X. Z. Li, Geological Structure Formation and Evolution of East Tethys, Geological Publishing House, Beijing, China, 1997.
5. G. T. Pan, H. X. Zheng, Y. R. Xu, et al., "A preliminary study on Bangong Co-Nujiang Suture," in Editing Committee of Geological Memoirs of Qinghai-Xizang Plateau(12)-Geological Tectonics of "Sanjiang", pp. 229–242, Geological Publishing House, Beijing, China, 1983.
6. X. Wang, I. Metcalfe, P. Jian, L. He, and C. Wang, "The Jinshajiang-Ailaoshan Suture Zone, China: tectonostratigraphy, age and evolution," Journal of Asian Earth Sciences, vol. 18, no. 6, pp. 675–690, 2000. · ·
7. A. M. C. Sengor and U. J. Botum, "The Tethys orogenic system: an introduction," in Tectonic Evolution of the Tethyan Region, A. M. C. Sengor, Ed., pp. 1–22, Kluwer Academic, Boston, Mass, USA, 1989.
8. Q. Yan, Z. Wang, S. Liu et al., "Opening of the Tethys in southwest China and its significance to the breakup of East Gondwanaland in late Paleozoic: evidence from SHRIMP U-Pb zircon analyses for the Garze ophiolite block," Chinese Science Bulletin, vol. 50, no. 3, pp. 256–264, 2005. · ·
9. P. A. Ziegler, Evolution of Laurussia: A Study in Late Palaeozoic Plate Tectonics, Kluwer Academic, Dordrecht, The Netherlands, 1989.
10. B. P. Liu, Q. L. Feng, C. Chonglakmani, et al., "Framework of paleotethyan archipelago ocean of western Yunnan and its elongation towards north and south," Earth Science Frontiers, vol. 9, no. 3, pp. 161–171, 2002.

11. C. S. Hutchison, Geological Evolution of Southeast Asia, vol. 13 of Oxford Monographs on Geology and Geophysics, Clarendon Press, Oxford, UK, 1989.

12. H. R. Wu, C. A. Boulter, B. Ke, et al., "The Changning-Menglian suture zone: a segment of the major Cathaysian-Gondwana divide in Southeast Asia," Tectonophysics, vol. 242, no. 3, pp. 267–280, 1995. ·

13. H. Y. Zhang and J. L. Liu, "Plate tectonics and mineralization of the Tethyan ophiolite in the southern Sanjiang and Indo-China peninsular area," Earth Science, Journal of China University of Geosciences, vol. 36, no. 2, pp. 262–276, 2011 (Chinese). · ·

14. P. F. Fan, "Accreted terranes and mineral deposits of Indochina," Journal of Asian Earth Sciences, vol. 18, no. 3, pp. 343–350, 2000. · ·

15. M. J. Lin, "The volcanic rock sequence and its geological age along the bank of Lancang Jiang, West Yunnan" (Chinese), See: Geology Papers of Qinghai-Tibet Plateau (13)-Rocks in Sanjiang Area, pp. 151–158, 1983.

16. J. Y. Xiong, Y. M. Lin, and S. R. Qin, "A preliminary study on the genesis and the basic characteristics of the Lincang magmatic complex" (Chinese), See: Geology Papers of Qinghai-Tibet Plateau (13)-Rocks in Sanjiang Area, pp. 1–19, 1983.

17. S. Y. Yu, K. Q. Li, Y. P. Shi, and H. H. Zhang, "A study on the granodioritc in thc middle part of Lincang granite batholiths," Yunnan Geology, vol. 22, no. 4, pp. 426–442, 2003.

18. C. S. Liu, J. C. Zhu, X. S. Xu, X. J. Chu, D. K. Cai, and P. Yang, "Study on the characteristics of Lincang Composite Granite Batholith in West Yunnan," Yunnan Geology, vol. 8, no. 3-4, pp. 189–212, 1989 (Chinese).

19. X. L. Li, "Basic characteristics and formation structural environment of Lincang Composite granite batholiths," Yunnan Geology, vol. 15, no. 1, pp. 1–18, 1996 (Chinese).

20. J. C. Chen, "Discussion on the age division and the selects of isotopic age determination for granitic rock in Western Yunnan," Yunnan Geology, vol. 6, no. 2, pp. 101–113, 1987 (Chinese).

21. X. X. Mo, S. Y. Shen, Q. W. Zhu, et al., Volcanic-Ophiolite and Mineralization of Middle-Southern Part in Sanjiang Area of Southwestern China, Geological Publishing House, Beijing, China, 1998.

22. D. L. Zhong, Paleo-Tethys Orogenic Zone in Western Yunnan and Western Sichuan, Science Press, Beijing, China, 1998.

23. A. Socquet and M. Pubellier, "Cenozoic deformation in western Yunnan (China-Myanmar border),"Journal of Asian Earth Sciences, vol. 24, no. 4, pp. 495–515, 2005. · ·

24. R. G. Park, Foundations of Structural Geology, Chapman & Hall, New York, NY, USA, 2nd edition, 1989.

25. C. H. Zhang, J. S. Liu, D. L. Liu, and S. Yang, "Geological, geochemical characteristics, age and tectonic setting of Laomaocun small rockbody in South Lancangjiang Zone," ACTA Mineralogica Sinica, vol. 26, no. 3, pp. 317–324, 2006.

26. J. C. Chen , "Characteristics of Pb, Sr isotopic compositions in west Yunnan granites: discussion on the age and nature of the basement in west Yunnan," Scientia Geologica Sinica, vol. 2, pp. 174–183, 1991. ·

27. T. M. Dai, B. Q. Zhu, Y. Q. Zhang, Z. P. Pu, Q. F. Zhang, and A. S. Hong, "Collision and thermial history of Indian-Sandaland-Eurasian Plates as implicated by $^{40}Ar/^{39}Ar$ age spectra of granodiorites,"Geochimica, vol. 15, no. 2, pp. 97–107, 1986.

28. Y. Q. Zhang, Y. W. Xie, and J. W. Wang, "Rb and Sr isotopic studies of granitoids in Tri-river region,"Geochimica, vol. 19, no. 4, pp. 318–326, 1990 (Chinese).

29. W. Y. Wang, "Rb-Sr isotopic Data," Yunnan Geology, vol. 3, no. 1, pp. 79–96, 1984 (Chinese).

30. Yunnan Bureau of Geology and Mineral Resources (YBGM), Geoogical Map of the Jinggu Sheet at a Scale of 1:200, 000, Ministry of Geology and Mineral Resources, Beijing, China, 1983.

31. Yunnan Bureau of Geology and Mineral Resources (YBGM), Geological Map of the Gengma Sheet at a Scale of 1:200, 000, Ministry of Geology and Mineral Resources, Beijing, China, 1985.

32. P. Xu, F. Wu, L. Xie, and Y. Yang, "Hf isotopic compositions of the standard zircons for U-Pb dating,"Chinese Science Bulletin, vol. 49, no. 15, pp. 1642–1648, 2004. · ·

33. S. E. Jackson, N. J. Pearson, W. L. Griffin, and E. A. Belousova, "The application of laser ablation-inductively coupled plasma-mass spectrometry to in situ U-Pb zircon geochronology," Chemical Geology, vol. 211, no. 1-2, pp. 47–69, 2004. · ·

34. K. Ludwig, Isoplot/Ex version 2.0: A Geochronological Tool Kit for Microsoft Excel, Geochronology Center Berkeley, Special Publication, 1999.

35. B. W. Chappell and A. J. R. White, "I- and S-type granites in the lachlan fold belt, Southeast Australia," inGeology of Granites and their Metallogenetic Relations, X. Keqin and T. Guangchi, Eds., pp. 72–78, Science Press, Bejing, China, 1982.

36. B. P. Liu, Q. L. Feng, N. Q. Fang, et al., "Tectonic evolution of Paleo-Tethys Poly-island-ocean in Changning-Menglian and Lancangjiang Belts, south-western Yunnan, China," Earth Science, Journal of China University of Geosciences, vol. 18, no. 5, pp. 529–539, 1993 (Chinese).

CHAPTER 16

THE EFFECT OF A TECTONIC STRESS FIELD ON COAL AND GAS OUTBURSTS

Fenghua An and Yuanping Cheng

INTRODUCTION

A coal and gas outburst is a dynamic failure that ejects coal mass containing high-pressure gas in a short time. As one main failure of underground coal resource exploitation, it causes severe damage in the major global producing nations [1, 2].The knowledge available regarding the coal and gas outburst mechanism is still qualitative at present, and the ground stress, coal gas, and physical and mechanical properties of coal determine the formation and initiation of an outburst. Ground stress plays a primary role in coal mass that is broken before outburst initiation and provides potential energy for outburst initiation. Gas content of the coal seam is an essential element in an outburst [1] and is usually the main index for outburst danger. The coal gas aggravates the failure of coal mass before the outburst initiation. A portion of the coal gas would participate in outburst initiation. Afterwards, the enormous energy released by the coal gas would crush, strip, and carry the broken coal mass. The stress conditions can affect the coal mass failure, the potential energy accumulation for outburst initiation, and the outburst danger. A key factor for underground resource exploitation is the in situ ground stress. The crustal stress on the stratum is mainly caused by gravity stress and tectonic stress. The gravity stress has a regular distribution. The maximum stress is the vertical stress and increases with depth linearly, and the lateral stress can be calculated with the coefficient of lateral pressure and is related to Poisson's ratio. The tectonic stress is caused by geotectonic movement and is mainly in the horizontal direction similar to the crustal movement. The results measured in situ show that the horizontal stress is greater than the vertical stress universally in shallow strata. The ratio between the horizontal stress and the vertical stress ranged from approximately 0.5 to 2 [3–10]. Generally speaking, in shallow depths, the tectonic forces have a significant effect on the horizontal stress and lead to a stress type of $\sigma_H > \sigma_V > \sigma_h$ and to a stress type of $\sigma_H > \sigma_h > \sigma V$ in the intense tectonic zone. For greater depths, the impact of tectonics weakens, and the vertical stress increases and gradually changes the stress type to $\sigma_V > \sigma_H > \sigma_h$. To control coal and gas outbursts [1], the effect of ground stress (especially the tectonic stress) on outbursts has been confirmed. Due to the susceptibility of open cracks due to stress, the permeability is significantly affected by the stress [11, 12]. The fracture connectivity of sheared coal was destroyed by tectonic events [13] and reduced the coal seam permeability [14]. As a result, the migration and storage of coal gas is strongly affected by the tectonic stress field and its evolution [14–16]. Additionally, the high tectonic stress has a primary role in the stress conditions of outburst areas, and the gas and ground stress abnormity in tectonic areas and neighboring areas is frequent [17]. Moreover, tectonism reduces the coal mass strength [18]. From these effects, tectonics has a critical influence on outburst occurrence and distribution [1, 18]. In addition, the ground stress measurement in outburst areas shows the characteristics of high ground stress. A study of the relationship between coal and gas outbursts and the ground stress of the Xie'er Coal Mine in the Chinese Huainan Coalfield showed a positive correlation [19]. The in situ stress distribution measured by Han et al. showed that with tectonic stress, the horizontal stress in outburst areas was greater in comparison with other areas [20]. In addition to the high ground stress, the spatial differentiation may be induced and may influence the deformation and failure of the coal and rock mass near the roadway. In the meanwhile, the study of roadway stability has shown that the magnitude and direction of the horizontal stress could affect the roadway stability significantly. The roadway that is parallel to the maximum horizontal stress has the best stability, and the roadway that is perpendicular to the maximum horizontal stress has the worst stability [21–23]. The cut-through roadway is most stable for the direction parallel or perpendicular to the maximum horizontal stress [23]. Previous

Citation: Fenghua An and Yuanping Cheng, "The Effect of a Tectonic Stress Field on Coal and Gas Outbursts," The Scientific World Journal, vol. 2014, Article ID 813063, 10 pages, 2014. doi:10.1155/2014/813063

research on the effect of the magnitude and distribution of ground stress on an outburst considers the predictability of an outburst occurrence in engineering. However, more research is required regarding the specific mechanism that causes an outburst.This paper focuses on the effect of the magnitude and distribution of ground stress and the influence of tectonics on an outburst. A numerical model considering the effect of coal gas is established. The coal mass failure and the potential energy accumulation surrounding the roadway is analyzed for different ground stress fields, and then the effect on the outburst tendency is analyzed.

DESCRIPTION OF THE NUMERICAL MODEL

Governing Equation

As dual-porosity media, the stress state of the coal mass is affected by the gas pressure in pores and fractures. If compressive stress is positive, the effect of free gas on the coal stress state could be expressed by the effective stress equation of porous media [24]:

$$\sigma' = \sigma - \left(\gamma_m p_m + \gamma_f p_f\right)\delta_{ij}, \tag{1}$$

where p_m and p_f represent the gas pressure in the fracture and the coal matrix, respectively. The pore and fracture effective stress coefficients are represented by γ_p and γ_f, respectively, and can be expressed as [24]

$$\gamma_p = \frac{K}{K_m} - \frac{K}{K_s}$$

$$\gamma_f = 1 - \frac{K}{K_m}, \tag{2}$$

where K represents the bulk modulus of the fractured porous solid, that is, the coal mass; K_m represents the bulk modulus of the porous solid, that is, the coal matrix; and Ks represents the bulk modulus of the solid skeleton and is the coal skeleton without pores. Coal mass expands after gas adsorption and can be compared to thermal expansion [25]. Considering the effect of coal mass, the relationship between the effective stress and strain can be

$$\sigma'_{ij} = \lambda\varepsilon_v\delta_{ij} + 2G\varepsilon_{ij} - K\varepsilon^s_v\delta_{ij}, \tag{3}$$

where λ is the Lame constant, ′ G is the shear modulus, and ε^s_{LV} represents the maximum adsorption-induced volume strain. Research on the adsorption-induced deformation shows that the volumetric strain is approximately linear, is proportional to the adsorbed gas [26, 27], and can be approximated using the Langmuir-like equation [28]

$$\varepsilon^s_v = \frac{\varepsilon^s_{Lv} p_m}{p_m + p^s_L} \tag{4}$$

where p_s^L represents the Langmuir pressure of the adsorption induced strain. Except for the induced deformation, the gas adsorption reduces the coal strength [29, 30]. The strength reduction is related to the adsorbed gas, and the reduction factor can be expressed with a linear relationship that assumes [28]

$$\alpha = \frac{\alpha_0 p}{p + p_L}, \tag{5}$$

where α_0 represents the maximum reduction of the coal mechanical parameters. The Mohr-Coulomb matching DP yield criterion is chosen for the failure criterion of the coal mass [31]:

$$F = \alpha_{DP} I_1 + k_{DP} - \sqrt{J_2}, \tag{6}$$

where I_1 represents the first stress invariant, $_{J2}$ represents the second deviator stress invariant, and α_{DP} and k_{DP} are identified by the cohesion C and the friction angle $\varphi$with the inscribed circle way. Coal mass is an elastic-plastic material with a strainsoftening property. Hence, the strain-softening model is appropriate [32] and the mechanical parameters develop as the softening parameter [33]:

$$\omega = \begin{Bmatrix} \omega_0 - (\omega_0 - \omega_r)\dfrac{\gamma^p}{\gamma^{p^*}}, & 0 < \gamma^p < \gamma^{p^*} \\ \omega_r, & \gamma^p \geq \gamma^{p^*} \end{Bmatrix}, \tag{7}$$

where ω represents the mechanical parameter, ω_0 represents the original parameters before the stress peak, ω_r represents the residual parameters at the residual zone, γ^p represents the softening parameter, and γ^{p*} represents the transition value of the softening parameter from which the residual behavior starts. A common softening parameter is the equivalent plastic shear strain [34]:

$$\gamma^p = \sqrt{\frac{2}{3}\left(\varepsilon^p_1\varepsilon^p_1 + \varepsilon^p_2\varepsilon^p_2 + \varepsilon^p_3\varepsilon^p_3\right)}, \tag{8}$$

where ε^p_1 , ε^p_2 , and ε^p_3 are the principal plastic strains. The dual-porosity model proposed by Warren and Root [35] is used primarily for gas migration in the coal seam. The adsorbed gas is desorbed from the pore wall and diffuse into fractures from the coal matrix with various diffusion forms. The dominant diffusion form is gaseous phase diffusion, the driving force of which is the gas density difference [36]. The coal matrix gas pressure evolves with time [28]:

$$\frac{\partial p_m}{\partial t} = \left(-\frac{1}{\tau}\cdot\left(p_m - p_f\right)\right) \times \left(\left(\frac{V_L p_m}{p_L + p_m} - \frac{V_L p_L^2 p_m}{\left(p_L + p_m\right)^2} + \frac{\phi_m}{\rho p_0}\right)\frac{\rho RT}{V_M}\right)^{-1}, \tag{9}$$

where τ represents the adsorption time of the coal matrix; VL represents the maximum adsorption volume of coal; p_L represents the Langmuir pressure of coal; ϕ_m represents the porosity of the coal matrix; ρ represents the bulk density of coal; p_0 represents the atmospheric pressure; R represents the universal gas constant; T represents the gas temperature; and V_M represents the molar volume of gas. As the source term, the gas in the coal matrix trades with the fracture, and the gas mass conservation equation for the fracture is

$$\frac{\partial\left(\phi_f p_f\right)}{\partial t} - \frac{k}{\mu}\nabla\cdot\left(p_f\cdot\nabla p_f\right) - \frac{1}{\tau}\cdot\left(1-\phi_f\right)\left(p_m - p_f\right) = 0, \tag{10}$$

where ϕf represents the fracture porosity, k represents the coal seam permeability, and μ represents the gas viscosity coefficient (1.08E−5 Pa·s for methane).

Approach and Parameters of the Model

The governing equations were performed by COMSOL Multiphysics using the solid and PED module. The model includes the coal seam in the middle and roof and floor rocks as the geometry model as shown in Figure 1. The finite element mesh consists of 74,692 tetrahedrons. The xz face (y = 0) at the beginning of the roadway excavation is the symmetry boundary. The xy face (z = 21.5) at the top of the model is the load boundary. The xy face (z = −21.5) at the bottom of the model is fixed. The two yz faces(x = 0; x = 100) and the xz face (y = 150) at the sides are the slip boundaries or the stress load boundaries according to the ground stress conditions. The excavation direction is in the y direction, and the roadway is 10 m long in the model. Most outbursts occurred in the excavation advancement, and the coal mass in front of the working face is the most dangerous. The emphasis of this paper is on the analysis of the effect of ground stress conditions on coal mass deformation and failure, and the effect of roadway size on outburst danger was not analyzed. The mechanical properties of roof and floor rocks are simplified as hard elastoplastic material. The coal seam is a strain-softening elastoplastic material. The mechanical and gas migration parameters are obtained from previous studies and are shown in Table 1

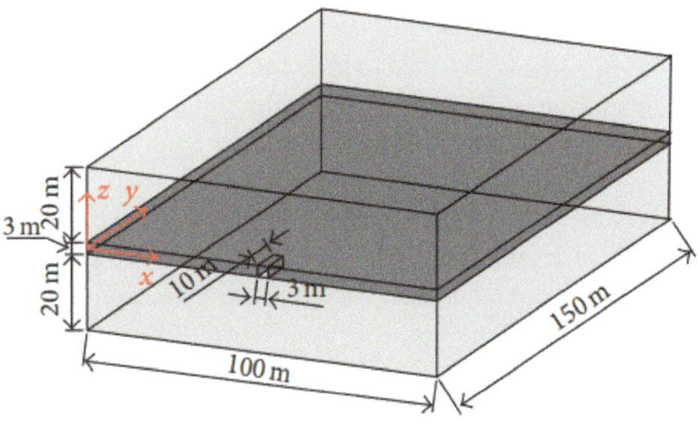

Figure 1: Geometric model.

Conditions of Tectonic Stress and Coal Gas

In the tectonic area, the horizontal stress of the gravity stress field increases with the effect of tectonics and forms stress field types of $\sigma_V > \sigma_H > \sigma_h$, $\sigma_H > \sigma_V > \sigma_h$, and $\sigma_H > \sigma_h > \sigma_V$. With an overlying strata load of 400 m deep and an average density of 2500 kg/m3 (10 MPa) as the reference stress, the ground stress of the stress field types above is loaded in the model as shown in Table 2. The ground stress studied includes the gravity stress field (Case 1), the stress field influenced by tectonics (Cases 2 and 3), and the stress field strongly influenced by tectonics (Cases 4 and 5). The slip boundary was set as the gravity stress field in Case 1. The difference between the coal seam roadway and the rock roadway is that the coal gas affects the stress state of the coal mass surrounding the roadway. With various stress and coal seam conditions, the gas pressure required for an outburst is different. 0.74 MPa (relative pressure) is recommended in China. The coal seams without coal gas and with gas pressure of 0.8 MPa and 1.5 MPa (9.3 m3 /t and 12.7 m3 /t correspondingly) are considered in this paper, and, with different stress conditions, the coal gas distributions surrounding the roadway are the same in the calculation as shown in Figure 2. Figure 2(a) shows the fracture pressure distribution surrounding the roadway with an initial gas pressure of 0.8 MPa, and Figure 2(b) shows the fracture pressure distribution with an initial gas pressure of 1.5 MPa. The numbers beside the frame are coordinate values. The values in the legends represent gas pressure in fractures and the unit is Pa. The gas distribution is obtained after 0.5 days of gas migration to the coal wall with a permeability of 0.0025 mD.

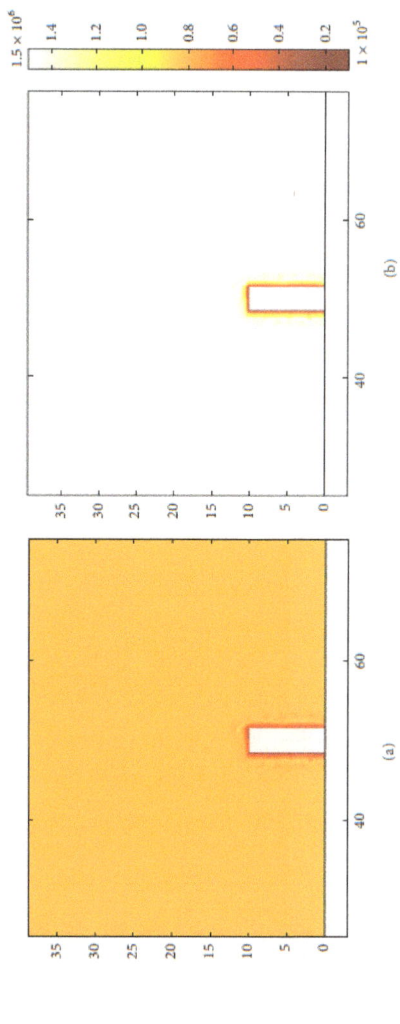

Figure 2: Distribution of coal gas pressure in fractures surrounding the roadway.

Table 1: Model parameters

Parameters	Value
Elastic modulus of coal (E_c, GPa)	2 [47]
Poisson ratio of coal (υ_c)	0.3 [47]
Bulk modulus of the coal matrix (K_m, GPa)	7.5 [48]
Bulk modulus of the coal skeleton (K_s, GPa)	8.9 [28]
Maximum adsorption-induced volume strain for CH_4 (ε^s_{Lv})	0.012
Langmuir pressure of adsorption-induced volume strain for CH_4 (p^s_L, MPa)	1
Initial cohesion of coal without gas (C_c, MPa)	0.923
Residual cohesion of coal without gas (C^*_c, MPa)	0.692
Friction angle of coal (φ_c, °)	30 [47]
Transition value of coal softening parameter (γ^{p*}, %)	2
Bulk density of coal (ρ_c, kg/m^3)	1300
Bulk density of rock (ρ_r, kg/m^3)	2500
Elastic modulus of rock (E_r, GPa)	20
Cohesion of rock (C_r, MPa)	20
Friction angle of rock (φ_r, °)	40
Poisson ratio of rock (υ_r)	0.3
Temperature of coal seam (T, K)	293
Initial permeability of coal seam (k_0, mD)	0.0025 [38]
Initial fracture porosity (ϕ_{f0}, %)	0.1 [49]
Matrix porosity (ϕ_m, %)	6 [38]
CH_4 Langmuir volume (V_L, m^3/t)	20
CH_4 Langmuir pressure (P_L, MPa)	1
Adsorption time (τ, day)	11.7 [47]

3. The Outburst Tendency Analysis Based on Its Energy Requirement The continuity condition required by the finite element method used above limits the analysis to broken and crush and is confined to the previous stage of dynamic failure of coal mass. Except for the strength failure as a precondition of the dynamic failure of the coal mass, the energy accumulated for an outburst initiation should be greater than the energy consumed. Based on this consideration, it can be deemed that the preparation of an outburst is an accumulation of potential energy for outburst initiation relative to the energy dissipated in coal mass crushing and stripping. The ratio of potential energy for outburst initiation and the energy consumed can be used to evaluate the outburst tendency. Gas pressure in the coal seam is much smaller than the stress, and the coal mass failure is mainly caused by the ground stress and the mining-induced stress. Gas pressure aggravates the failure and would crush and strip the broken coal mass in outbursts. The outbursts start in the failure zone of the coal mass. Therefore, the potential energy for outburst initiation in the failure zone is analyzed below and describes the outburst tendency using the ratio between the potential energy accumulated and the dissipation energy required in outburst initiation. The energy requirement for coal and gas outbursts is that the potential energy for outburst initiation Ei should be greater than the energy consumed in the coal mass crush, stripping, and movement Ew [37]. The energy released in outbursts includes the elastic energy of the coal mass and the internal energy of the free gas in the fractures and the desorbed gas from the coal matrix. The energy that participates in outburst initiation includes the elastic energy stored in coal mass Ee and the internal energy of partial gas Eg due to the short time for outburst initiation. The gas involved occupies a small fraction of the coal gas. The energy consumed in movement and in broken coal in the

Table 2: Conditions of ground stress and coal gas in the model.

Case	Ground stress in each direction*			Initial gas pressure in coal seam/MPa		
	X direction	Y direction	Z direction	I	II	III
1	Slip boundary		1			
2	0.5	1.5	1			
3	1.5	0.5	1	0	0.8	1.5
4	1.5	2	1			
5	2	1.5	1			

*The values of ground stress in the table are shown with 10 MPa as a reference.

follow-up process is provided by the sequentially desorbed gas from the coal matrix. Therefore, the potential energy for outburst initiation is defined as $E_i = E_g + E_e$. The energy consumed in outburst initiation is mainly for surface energy E_b and is consumed during the crushing and stripping of coal mass. $E_w = E_b$ is assumed. For the condition of three-dimensional stress with the generalized Hooke's law, the elastic energy per unit volume of coal mass is

$$e_e = \frac{1}{2E}\left[\sigma_1^2 + \sigma_2^2 + \sigma_3^2 - 2\upsilon\left(\sigma_1\sigma_2 + \sigma_2\sigma_3 + \sigma_3\sigma_1\right)\right] \tag{11}$$

The potential energy for outburst initiation by coal gas through expansion work can be calculated as

$$e_g = \frac{pV_0}{n-1}\left[\left(\frac{p}{p_0}\right)^{(n-1)/n} - 1\right] \quad (12)$$

where V_0 represents the volume of coal gas involved and n represents the polytropic exponent. The gas expansion in outburst is a polytropic process, and n is approximately 1.25 [38]. The coal gas amount involved in outburst initiation is difficult to measure. In addition to free gas, research by Valliappan and Zhang showed that desorbed gas from the coal matrix played an important role [39]. As dualporosity media, the gas in fractures is deemed to participate in the outburst initiation entirely, whereas the gas in the coal matrix participates partially. The potential energy for outburst initiation by coal gas is

$$e_g = \frac{p_f V_{f0}}{n-1}\left[\left(\frac{p_f}{p_0}\right)^{(n-1)/n} - 1\right]\rho + \frac{p_m V_{m0}}{n-1}\left[\left(\frac{p_m}{p_0}\right)^{(n-1)/n} - 1\right]\rho, \quad (13)$$

where V_{f0} represents the coal gas content in the fracture (mL/g) and V_{m0} represents the participant coal gas content from the coal matrix (mL/g). The participant coal gas from the coal matrix varies [40] and is related to the gas content, coal particle size, and diffusion coefficient.The desorbed gas from the coal matrix in outburst initiation can be estimated by the desorption ratio λ, and the participant coal gas from the coal matrix $Vm0$ can be expressed as

$$V_{m0} = \lambda V_{m\infty}, \quad (14)$$

where $V_{m\infty}$ represents the total desorbed gas of the coal matrix (mL/g). If the ambient pressure of the gas desorption is estimated according to atmospheric pressure, the total desorbed gas can be calculated based on the difference in the gas content in the coal matrix before outburst initiation and the gas content with the gas pressure as standard pressure. The participant internal energy would change from (17) to

$$E_g = \frac{p_f}{n-1}\frac{p_f\phi_f}{p_0}\left[\left(\frac{p_f}{p_0}\right)^{(n-1)/n} - 1\right] + \frac{p_m}{n-1} \times \lambda\left(\frac{p_m\phi_m}{p_0\rho} + \frac{V_L p_m}{p_L + p_m} - \frac{p_0\phi_m}{p_0\rho} - \frac{V_L p_0}{p_L + p_0}\right) \times \left[\left(\frac{p_m}{p_0}\right)^{(n-1)/n} - 1\right]\rho. \quad (15)$$

The desorption ratio is estimated by the following method. The outburst initiation is a transient process, and the desorption amount of coal gas in a short time is linearly related to the square root of the desorption time and is expressed as [41]

$$\frac{V_t}{V_\infty} = \frac{6}{\sqrt{\pi}}\sqrt{\frac{Dt}{r_p^2}}, \quad (16)$$

where V_∞ represents the total desorbed gas (mL/g); D represents the diffusion coefficient (cm^2 /s); and rp represents the diffusion radius (cm). The adsorption time τ is related to the diffusion coefficient and the matrix size. The cubic model of the coal matrix can be estimated as [42]

$$\tau = \frac{L^2}{3\pi^2 D}$$

where L represents the space between fractures (cm). If r_p is $L/2$, $\sqrt{D/r^2}\ p$ can be obtained from (15). From the observation record of a coal and gas outburst test site at the Zhongliangshan Mine in China in 1977, the first thud in sonic record occurred without a change to other parameters at 1.5 s after the inducing blast, and then three thuds occurred at 2.5 s, 3.5 s, and 4 s [43].The continuing time of outburst initiation is the average of the recorded occurrence time of the later three thuds. For the adsorption time in the model, the desorbed gas for outburst initiation is 0.677% of the total desorbed gas, and the desorption ratio is 0.677%.

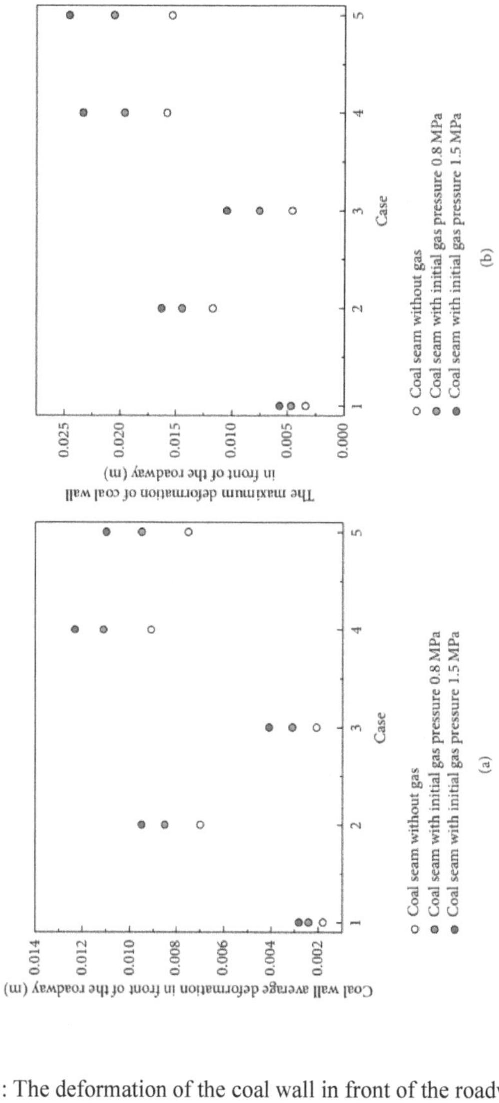

Figure 3: The deformation of the coal wall in front of the roadway.

The energy consumed in coal mass crushing is mainly the energy required by the increased surface and can be expressed by

$$e_b = s \cdot w \cdot \rho, \quad (18)$$

where s represents the increased surface area (cm^2/g) and w represents the specific energy (J/cm^2). The study of the specific energy of coal showed that soft coal is proportional to the firmness coefficient, has an outburst danger with less specific energy than hard coal, and is approximately 2×10^{-3} J/cm^2 [44, 45]. The additional specific surface area of the coal mass that is crushed by the hammer drop method was from 113 to 525 cm^2/g [44], and the specific surface area of the fragment from the rock burst test was 200 cm^2/g [46]. In this paper, the specific energy of 1×10^{-3} J/cm^2 and the additional specific surface area of 150 cm^2/g are used, and the surface energy consumed in outburst initiation is 0.15 J/g. The outburst initiation requires the accumulation of more potential energy than is consumed. Moreover, the outburst was initiated in the failure zone. Hence, the potential energy accumulated in the failure zone of the coal mass is related to the necessary energy in coal mass crushing and is analyzed to indicate the outburst tendency by the ratio

$$R_o = \frac{\int_{V_p} e_i dV}{\int_{V_p} e_b dV} = \frac{\int_{V_p} \left(e_e + e_g\right) dV}{\int_{V_p} e_b dV}, \tag{19}$$

where V_p is the plastic zone. When R_o is greater than 1, the potential energy for outburst initiation is sufficient and indicates the danger of an outburst.

RESULTS AND DISCUSSION

The Effect of Tectonic Stress on the Deformation and Failure of Coal Mass

The effect of horizontal stress on roadway stability is focused on the deformation and failure of coal walls and roofs [21–23]. Coal and gas outbursts occur in coal mass surrounding the roadway, and most outbursts occur in front of the heading face in excavation. The deformation and plastic zone surrounding the roadway with various ground stress conditions are analyzed, and the outburst tendency in the front is studied. The deformation of the coal wall is affected by the coal gas and stress conditions as shown in Figure 3. The tectonic stress increases the average and maximum deformation of the coal wall in front of the roadway, and the increase is affected by the distribution of tectonic stress. The coal mass deformation in front of the roadway is greater when the maximum horizontal stress is parallel to the direction of the roadway, but, as the tectonic stress increases, the effect of the stress distribution decreases. In addition, the coal gas has an important effect on the coal wall. For various stress conditions, the deformation increases as the coal gas increases. For various gas conditions, the size of the coal mass plastic zone in different stress conditions is shown in Table 3. The coal gas surrounding the roadway had an obvious effect on the plastic zone for pore pressure and strength reduction, causing a larger area with the same ground stress. The change in the ground stress condition affected the magnitude of the coal mass plastic zone and altered the distribution of the plastic zone. The plastic zone distribution for the middle cross-section of the coal seam with various ground stress fields is shown in Figure 4. The red zone is the plastic zone, and the blue zone is the elastic zone. With the alteration of coal mass failure, the outburst danger would be affected.

Table 3: The plastic zone of the coal mass in front of the roadway (m^3)

	Case 1	Case 2	Case 3	Case 4	Case 5
Coal seam without gas	3.40	4.44	3.70	3.41	4.57
Coal seam with initial gas pressure 0.8 MPa	5.44	5.55	5.13	4.93	5.67
Coal seam with initial gas pressure 1.5 MPa	6.32	6.85	6.39	5.86	6.62

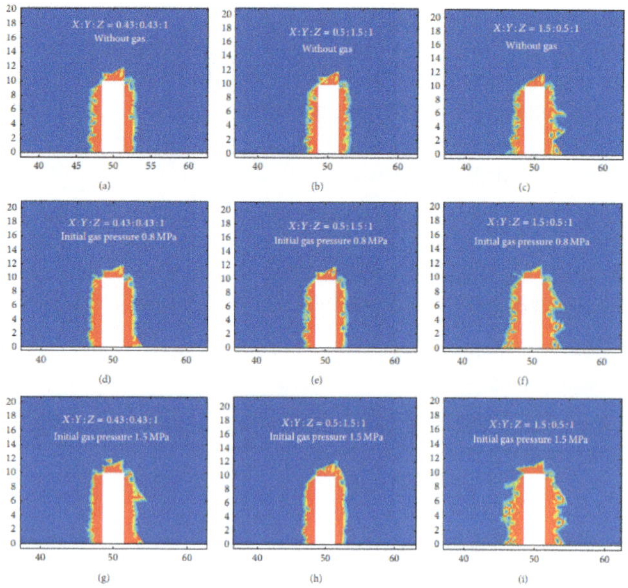

Figure 4: The plastic zone of the coal mass surrounding the roadway

The Effect of Tectonic Stress on Outburst Tendency

The energy accumulation for outburst initiation is affected by the coal gas and stress conditions as shown in Figure 5. The coal gas increased the potential energy for outburst initiation significantly, and, without coal gas, the accumulated energy had difficulty reaching the energy requirement for an outburst. The coal gas is the main influence factor for outburst initiation. Additionally, the tectonic stress has an important effect on the outburst tendency and is enhanced by tectonics as shown in Figure 5. For example, when the initial gas pressure is 0.8 MPa in the model, the tectonics could decide the occurrence of an outburst.This result verifies the previous ground stress measurement results that the horizontal stress was greater in the outburst area. Additionally, the stress distribution has an effect on the outburst. More potential energy is accumulated when the maximum horizontal stress is perpendicular to the roadway.

To further analyze the effect of tectonic stress, the energy composition for various tectonic stresses is studied as shown in Figure 6. The result showed that the proportion of elastic energy increases as the tectonic stress increase, and the proportion is greater when the maximum horizontal stress is perpendicular to the roadway. The elastic energy for outburst initiation that is increased by tectonic stress enhances the outburst tendency. Additionally, the distribution of the horizontal stress can alter the elastic energy. Additional elastic energy is held in excavation, when the maximum horizontal stress is perpendicular to the roadway and indicates a greater tendency for an outburst.

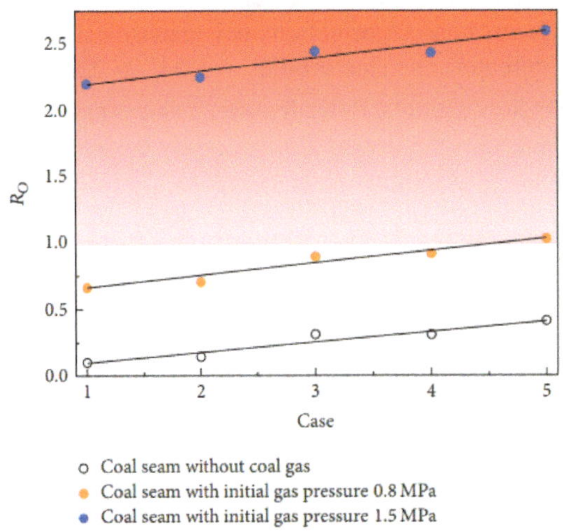

Figure 5: The effect of coal gas and tectonic stress on energy for outburst initiation in front of the roadway.

CONCLUSION

Affected by tectonics, the magnitude and spatial distribution of ground stress in mining areas are different. Previous research shows a close relationship between the outburst area and the tectonic stress. Tectonics and evolution cause the regional distribution of outbursts, and the outburst areas often have large tectonic stress. To analyze the effect of tectonic stress on outbursts, a numerical model of a roadway was established. In addition, a method for evaluating the outburst tendency was proposed based on the energy requirement for outburst initiation. The deformation and failure of a coal mass in front of a roadway was analyzed. The results showed that the deformation and failure zone increased with coal gas. The pore pressure of free gas and the strength reduction of the adsorbed gas aggravated the coal mass failure and had a larger plastic zone with more gas in the coal seam. The tectonic stress enhanced the deformation and had an effect on the distribution of the plastic zone. Based on the requirement of strength failure and energy for outburst initiation, the energy accumulation in the failure zone of coal mass is analyzed to show the outburst tendency using the ratio of potential energy for outburst initiation and the energy consumed. The results showed that coal gas was the dominant factor for energy accumulation in outburst initiation, and the increase in coal gas would enhance the outburst tendency significantly. Meanwhile, the ground stress change induced by tectonics has an important effect on an outburst. The outburst tendency increases as the tectonic stress increases and is verified in the previous measurement results in outburst areas. Moreover, the change in the composition of potential energy for outburst initiation showed that this result was due to the increase in elastic energy.

Conflict of Interests

The authors declare that there is no conflict of interests regarding the publication of this paper.

ACKNOWLEDGMENTS

This work is financially supported by the National Foundation of China (no. 51304204) and the National Basic Research Program of China (973 Program, no. 2011CB201204). The support of National Science Foundation of China (no. 51304188) and the Doctoral Program of the Ministry of Education (no. 20120095120001) are also acknowledged.

REFERENCES

1. R. D. Lama and J. Bodziony, "Management of outburst in underground coal mines," International Journal of Coal Geology, vol. 35, no. 1–4, pp. 83–115, 1998.
2. B. B. Beamish and P. J. Crosdale, "Instantaneous outbursts in underground coal mines: an overview and association with coal type," International Journal of Coal Geology, vol. 35, no. 1–4, pp. 27–55, 1998.
3. K. Hayashi, A. Sato, and T. Ito, "In situ stress measurements by hydraulic fracturing for a rock mass with many planes of weakness," International Journal of Rock Mechanics and Mining Sciences, vol. 34, no. 1, pp. 45–58, 1997.
4. H. Kang, X. Zhang, L. Si, Y. Wu, and F. Gao, "In-situ stress measurements and stress distribution characteristics in underground coal mines in China," Engineering Geology, vol. 116, no. 3-4, pp. 333–345, 2010.
5. B. C. Haimson, "The hydrofracturing stress measuring method and recent field results," International Journal of Rock Mechanics and Mining Sciences & Geomechanics Abstracts, vol. 15, no. 4, pp. 167–178, 1978.
6. G. Klee, F. Rummel, and A. Williams, "Hydraulic fracturing stress measurements in Hong Kong," International Journal of Rock Mechanics and Mining Sciences, vol. 36, no. 6, pp. 731–741, 1999.
7. H.-P. Kang, J. Lin, L.-X. Yan, X. Zhang, Y.-Z. Wu, and L.- P. Si, "Study on characteristics of underground in-situ stress distribution in Shanxi coal mining fields," Chinese Journal of Geophysics, vol. 52, no. 7, pp. 1782–1792, 2009 (Chinese).
8. H. Kang, T. Jiang, X. Zhang, and L. Yan, "Research on in-situ stress field in Jincheng mining area and its application," Chinese Journal of Rock Mechanics and Engineering, vol. 28, no. 1, pp. 1–8, 2009 (Chinese).
9. H. Kang, J. Lin, and X. Zhang, "Research and application of insitu stress measurement in deep mines," Chinese Journal of Rock Mechanics and Engineering, vol. 26, no. 5, pp. 929–933, 2007 (Chinese).
10. H. Kang, Z. Wu, F. Gao, and W. Ju, "Effect of geological structures on in-situ stress distribution in underground coal mines," Chinese Journal of Rock Mechanics and Engineering, vol. 31, no. 1, pp. 2674–2680, 2012 (Chinese).
11. S. Durucan and J. S. Edwards, "The effects of stress and fracturing on permeability of coal," Mining Science and Technology, vol. 3, no. 3, pp. 205–216, 1986.
12. D. Jasinge, P. G. Ranjith, and S.-K. Choi, "Effects of effective stress changes on permeability of latrobe valley brown coal," Fuel, vol. 90, no. 3, pp. 1292–1300, 2011.
13. H. Li, Y. Ogawa, and S. Shimada, "Mechanism of methane flow through sheared coals and its role on methane recovery," Fuel, vol. 82, no. 10, pp. 1271–1279,

2003.

14. H. Li, "Major and minor structural features of a bedding shear zone along a coal seam and related gas outburst, Pingdingshan coalfield, Northern China," International Journal of Coal Geology, vol. 47, no. 2, pp. 101–113, 2001.
15. X. Zhu and F. Xu, "The controlling effect of tectonic stress field and its evolution on coal and gas outburst," Journal of China Coal Society, vol. 19, no. 3, pp. 304–314, 1994 (Chinese).
16. J. Han, H.-W. Zhang, Z.-M. Zhu, and J.-C. Song, "Controlling of tectonic stress field evolution for coal and gas outburst in Fuxin basin," Journal of the China Coal Society, vol. 32, no. 9, pp. 934– 938, 2007 (Chinese).
17. J. Shepherd, L. K. Rixon, and L. Griffiths, "Outbursts and geological structures in coal mines: a review," International Journal of Rock Mechanics and Mining Sciences & Geomechanics Abstracts, vol. 18, no. 4, pp. 267–283, 1981.
18. Y. Cao, D. He, and D. C. Glick, "Coal and gas outburst in footwalls of reverse faults," International Journal of Coal Geology, vol. 48, no. 1-2, pp. 47–63, 2001.
19. X. Wang, "A study on the relationship between rock stress and outbursts of coal and gas," Journal of Geomechanics, vol. 3, no. 1, pp. 88–94, 1997 (Chinese).
20. J. Han, H. Zhang, W. Song, and Z. Wang, "In-situ stress field of coal and gas outburst mining area," Chinese Journal of Rock Mechanics and Engineering, vol. 27, no. 2, pp. 3852–3859, 2008 (Chinese).
21. W. J. Gale, "Strata control utilising rock reinforcement techniques and stress control methods, in Australian coal mines," Mining Engineer, vol. 150, no. 352, pp. 247–253, 1991.
22. R. N. Singh, I. Porter, and J. Hematian, "Finite element analysis of three-way roadway junctions in longwall mining," International Journal of Coal Geology, vol. 45, no. 2-3, pp. 115–125, 2001.
23. M. M. Gadde, Effect of In-Situ Stresses on the Stability of Coal Mine Development Workings, West Virginia University, 2003.
24. M. Chen and Z. Chen, "Effective stress laws for multi-porosity media," Applied Mathematics and Mechanics, vol. 20, no. 11, pp. 1207–1213, 1999.
25. I. Palmer and J. Mansoori, "How permeability depends on stress and pore pressure in coalbeds: a new model," in Proceedings of the SPE Annual Technical Conference and Exhibition, pp. 557– 564, October 1996.
26. E. P. Robertson and R. L. Christiansen, "Modeling permeability in coal using sorption-induced strain data," in Proceedings of the SPE Annual Technical Conference and Exhibition, pp. 4133–4142, October 2005.
27. X. Cui, R. M. Bustin, and L. Chikatamarla, "Adsorptioninduced coal swelling and stress: implications for methane production and acid gas sequestration into coal seams," Journal of Geophysical Research B: Solid Earth, vol. 112, no. 10, Article ID B10202, 2007.
28. F.-H. An, Y.-P. Cheng, L. Wang, and W. Li, "A numerical model for outburst including the effect of adsorbed gas on coal deformation and mechanical properties," Computers and Geotechnics, vol. 54, pp. 222–231, 2013.
29. D. R. Viete and P. G. Ranjith, "The effect of CO2 on the geomechanical and permeability behaviour of brown coal: implications for coal seam CO2 sequestration," International Journal of Coal Geology, vol. 66, no. 3, pp. 204–216, 2006.
30. M. S. A. Perera, P. G. Ranjith, and M. Peter, "Effects of saturation medium and pressure on strength parameters of Latrobe Valley brown coal: carbon dioxide, water and nitrogen saturations," Energy, vol. 36, no. 12, pp. 6941–6947, 2011.
31. D. C. Drucker and W. Prager, "Soil mechanics and plastic analysis or limit design," Quarterly of Applied Mathematics, vol. 10, no. 2, pp. 157–165, 2013.
32. M. E. Duncan Fama, R. Trueman, and M. S. Craig, "Two- and three-dimensional elasto-plastic analysis for coal pillar design and its application to highwall mining," International Journal of Rock Mechanics and Mining Sciences & Geomechanics Abstracts, vol. 32, no. 3, pp. 215–225, 1995.
33. E. Alonso, L. R. Alejano, F. Varas, G. Fdez-Manin, and C. ′ Carranza-Torres, "Ground response curves for rock masses exhibiting strain-softening behaviour," International Journal for Numerical and Analytical Methods in Geomechanics, vol. 27, no. 13, pp. 1153–1185, 2003.
34. V. Hajiabdolmajid and P. Kaiser, "Brittleness of rock and stability assessment in hard rock tunneling," Tunnelling and Underground Space Technology, vol. 18, no. 1, pp. 35–48, 2003.
35. J. Warren and P. J. Root, "The behavior of naturally fractured reservoirs," Old SPE Journal, vol. 3, no. 3, pp. 245–255, 1963.
36. X. Cui, R. M. Bustin, and G. Dipple, "Selective transport of CO2, CH4, and N2 in coals: insights from modeling of experimental gas adsorption data," Fuel, vol. 83, no. 3, pp. 293–303, 2004.
37. B. B. H.;.g, Coal and Gas Outburst, Translated by: S. Song, Y. Wang, China Industry Press, Beijing, China, 1966, (Chinese).
38. Q. Yu, Coal Mine Gas Prevention and Control, China University of Mining and Technology Press, Xuzhou, China, 1992, (Chinese).
39. S. Valliappan and W. Zhang, "Role of gas energy during coal outbursts," International Journal for Numerical Methods in Engineering, vol. 44, no. 7, pp. 875–895, 1999.
40. I. Gray, "Coal mine outburst mechanism, thresholds and prediction techniques," ACARP Report C, 2006.
41. D. M. Smith and F. L. Williams, "Diffusional effects in the recovery of methane from coalbeds," Society of Petroleum Engineers journal, vol. 24, no. 5, pp. 529–535, 1984.
42. C. A. Mora and R. A. Wattenbarger, "Analysis and verification of dual porosity and CBM shape factors," Journal of Canadian Petroleum Technology, vol. 48, no.

2, pp. 17–21, 2009.

43. Z. L. S. Mine, "In-situ measurement of an coal and gas outburst," Safety in Coal Mines, no. 1, pp. 45–50, 1979 (Chinese).

44. C. Cai and Y. Wang, "Experimental study of crushing energy of coal using mashing technology," Safety in Coal Mines, no. 7, pp. 13–18, 1988 (Chinese).

45. C.-G. Cai and Y.-X. Xiong, "Theoretical and experimental study on crushing energy of outburst-proneness coal," Journal of the China Coal Society, vol. 30, no. 1, pp. 63–66, 2005 (Chinese).

46. D. Li, X. Jia, J. Miao, M. He, and D. Li, "Analysis of fractal characteristics of fragment from rockburst test of granite," Chinese Journal of Rock Mechanics and Engineering, vol. 29, no. 1, pp. 3280–3289, 2010 (Chinese).

47. T. He and Y. Qin, Exploration, Exploitation and Utilization Technology of CBM, China University of Mining and Technology Press, Xuzhou, China, 2007, (Chinese).

48. S. Hol and C. J. Spiers, "Competition between adsorptioninduced swelling and elastic compression of coal at CO2 pressures up to 100 MPa," Journal of the Mechanics and Physics of Solids, vol. 60, no. 11, pp. 1862–1882, 2012.

49. C. D. Montemagno and L. J. Pyrak-Nolte, "Porosity of natural fracture networks," Geophysical Research Letters, vol. 22, no. 11, pp. 1397–1400, 1995.

CHAPTER 17

QUATERNARY ACTIVE TECTONIC STRUCTURES IN THE OFFSHORE BAJO SEGURA BASIN (SE IBERIAN PENINSULA – MEDITERRANEAN SEA)

H. Perea, E. Gracia, P. Alfaro, R. Bartolome, C. Lo Iacono, X. Moreno, E. Masana

INTRODUCTION

The present-day crustal deformation of the southeastern Iberian margin is driven mainly by the NW–SE convergence (4–5 mm yr^{-1}) between the African and Eurasian plates (Argus et al., 1989; DeMets et al., 1990; Serpelloni et al., 2007; Vernant et al., 2010; Koulali et al., 2011). This convergence is accommodated over a wide deformation zone with significant seismic activity south of the Iberian Peninsula (Buforn et al., 1995; Grimison and Cheng, 1986; Morel and Meghraoui, 1996). In the southeastern Iberian margin (Fig. 1), the Neogene and Quaternary faulting activity is dominated by a large left-lateral strike-slip fault system of sigmoid geometry known as the Eastern Betic Shear Zone (EBSZ) (De Larouziere et al., 1988; Silva et al., 1993). The ` EBSZ stretches more than 450 km from Alicante to Almer´ıa, and includes, from north to south, the Bajo Segura, Carrascoy, Alhama de Murcia, Palomares and Carboneras faults (e.g. Bousquet, 1979; Silva et al., 1993). Instrumental seismicity is mainly characterized by low to moderate magnitude events. Nevertheless, historically, major destructive events such as the 1522 Almer´ıa (IEMS98 IX), the 1829 Torrevieja (IEMS98 IX-X) and the 1910 Adra (IEMS98 VIII) earthquakes occurred in the area (IGN, 2011). In this paper, we focus on the northern termination of the EBSZ, the Bajo Segura fault zone (BSFZ). This area is marked by moderate instrumental seismic activity and low magnitude earthquakes (Figs. 1 and 2). Nevertheless, this

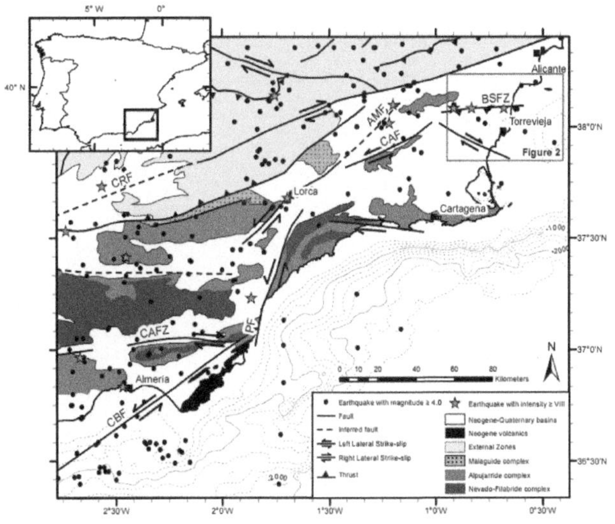

Figure 1. Regional bathymetric (contour interval 250 m) and simplified geological map (Sanz de Galdeano et al., 1995) of the southeastern Iberian margin with historical (gray stars) and instrumental (black dots) seismicity (IGN, 2011). Fault kinematics refers to Upper Neogene. BSFZ: Bajo Segura fault zone; CRF: Crevillente fault; AMF: Alhama de Murcia fault; PF: Palomares fault; CAFZ: Corredor de las Alpujarras fault zone; CBF: Carboneras fault; CAF: Carrascoy fault. The black outlined rectangle shows the study area corresponding to Fig. 2. Inset: Iberian Peninsula map, where the black outlined rectangle corresponds to the area depicted in this figure

zone has been affected by moderate to large historical earthquakes of which the largest is the 1829 Torrevieja earthquake (IEMS98 IX–X) (Lopez Marinas, 1976; Mezcua, 1982; ´ Rodr´ıguez de la Torre, 1984; Munoz et al., 1984; Albini and ˜ Rodr´ıguez de la Torre, 2001; Mart´ınez Solares and Mezcua, 2002). The onshore portion of the BSFZ comprises active faults and folds resulting from the transpressive setting established in the Plio-Pleistocene (Montenat, 1977; Taboada et al., 1993; Silva et al., 1993; Alfaro et al., 2002a, b, 2012).

Citation: Perea, H., Gràcia, E., Alfaro, P., Bartolomé, R., Lo Iacono, C., Moreno, X., Masana, E., Quaternary active tectonic structures in the offshore Bajo Segura basin (SE Iberian Peninsula - Mediterranean Sea), Nat. Hazards Earth Syst. Sci., 12, 3151-3168, doi:10.5194/nhess-12-3151-2012, 2012.

By contrast, the offshore area is almost devoid of tectonic data (Alfaro et al., 2002b). In September 2008, using high-resolution single channel seismic (SCS) sparker system, the marine geophysical EVENT-SHELF cruise sought to obtain images of the subseafloor structures in the offshore area of the EBSZ to be integrated with the onshore data to better understand the recent faulting history of the SE Iberian margin and characterize its seismic potential.

The specific objectives in the BSFZ were (a) to identify the main seismo-stratigraphic units of the area and localize the main tectonic structures in the offshore basin; (b) to correlate the offshore structures with the well-known onshore structures; (c) to determine which structures have been active during the Quaternary and are potential sources of future earthquakes; and (d) to consider the possibility that an offshore structure could be the source that produced the 1829 Torrevieja earthquake. The lack of offshore data is a common problem in coastal areas and particularly in the Mediterranean. This work shows how critical the integration of onshore and offshore data is to a better understanding and characterization of the seismic potential of the active faults located in these areas.

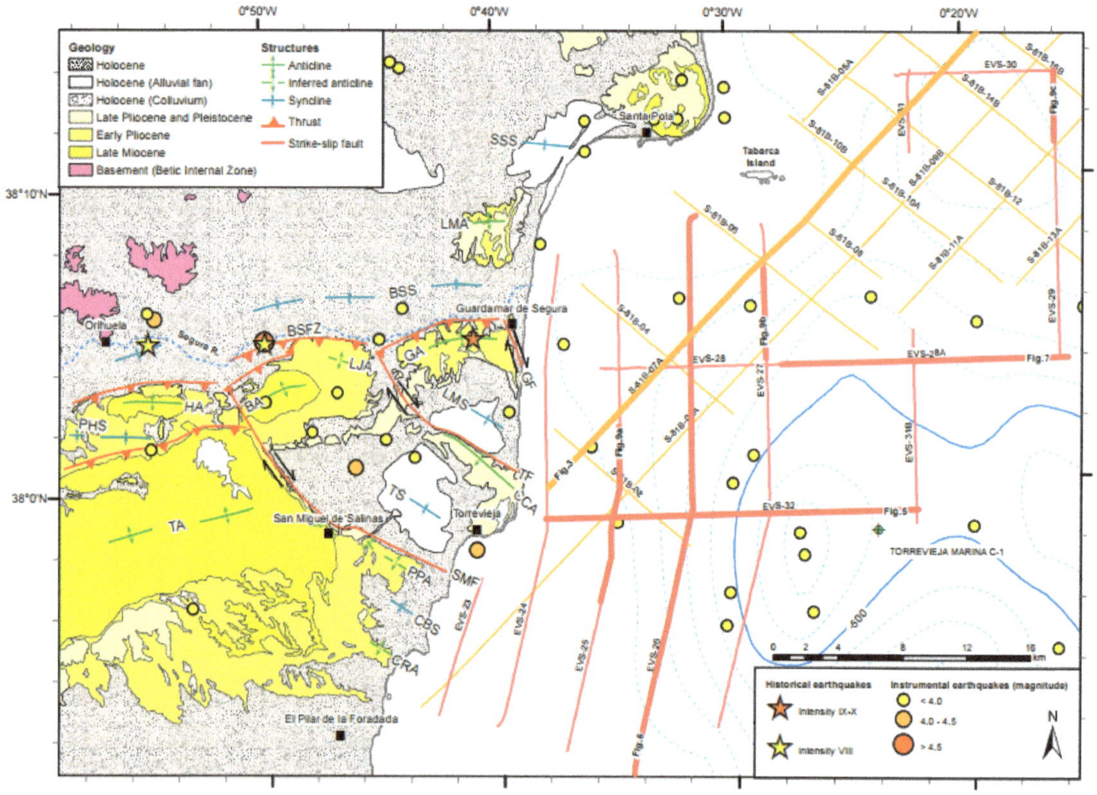

Figure 2. Bathymetric (dashed blue lines with contour interval 100 m) and geological (Alfaro et al., 2002a, b) map of the Bajo Segura basin. Stars and dots show the historical and instrumental epicenters, respectively (IGN, 2011). Orange lines correspond to commercial MCS profiles (survey S-81B 25/81; Archivo Tecnico de Hidrocarburos, 2012) and pink lines to SCS sparker profiles obtained during the EVENT- ′ SHELF cruise. Fault kinematics refers to Upper Neogene. Faults: BSFZ: Bajo Segura fault zone; GF: Guardamar fault; SMF: San Miguel de Salinas fault; TF: Torrevieja fault. Folds: BA: Benejuzar anticline; BSS: Bajo Segura syncline; CBS: Cala del Bosque syncline; CCA: ′ Cabo Cervera anticline; CRA: Cabo Roig anticline; GA: Guardamar anticline; HA: Hurchillo anticline; LJA: Lomas de la Juliana anticline; LMA: La Marina anticline; LMS: La Mata syncline; PHS: Pino Hermoso syncline; PPA: Punta Prima anticline; SSS: Salinas de Santa Pola syncline; TA: Torremendo anticline; TS: Torrevieja syncline.

GEODYNAMIC AND SEISMICITY SETTINGS OF THE BAJO SEGURA BASIN

The onshore Bajo Segura basin

The Bajo Segura basin is located in the Eastern Betic Cordillera (SE Spain), in the northern terminal splay of the EBSZ and extends eastwards into the Mediterranean Sea (Figs. 1 and 2). The basin is infilled by a succession of Upper Miocene to Quaternary sedimentary units that can be summarized (Montenat, 1977), from bottom to top, as the following: (a) Upper Miocene (Upper Tortonian) shallow marine (conglomerates and sandstones), slope and pelagic facies (marls and turbidites); (b) Upper Miocene (Late Tortonian–Messinian) slope and pelagic basin facies topped by shallow marine facies (evaporites, sandstones and coral reefs); (c) Pliocene unit comprising (Montenat, 1977), from bottom to top, the following: Hurchillo marls formation (pelagic marine facies), Rojales sandstone formation (coastal and shallow shelf facies), variegated lutite formation (river floodplain facies), and Segura conglomerates formation (fluvial channelized facies); and (d) Quaternary units, about 30 m thick, with the upper 20 m dated between 14 570 yr BP (14C age; Soria et al., 1999) and the present, and characterized by isolated outcrops of Pleistocene alluvial fans, marine terraces and a cover of unconsolidated sediments deposited along the present Segura River valley (Fig. 2). The Upper Miocene– Quaternary sedimentary cover overlies a basement forming part of the Internal Zone of the Betic Cordillera (Triassic calcareous metamorphic rocks of the Alpujarride complex). From the Upper Miocene to the present, the Bajo Segura basin area has been distinguished by a transpressive regime (Silva et al.,

1993; Alfaro et al., 2002a, b), with a NNW– SSE compressive stress field. In this geodynamic setting, the basement and the Upper Miocene to Quaternary sedimentary cover of the Bajo Segura basin have been folded and faulted (Fig. 2). This basin is bordered by two main faults: the Crevillente fault (Focault, 1974) to the north (Fig. 1), and the BSFZ (Montenat, 1977) to the south, both running approximately ENE–WSW. The BSFZ is characterized by several ENE–WSW blind thrusts that offset the Triassic basement of the Internal Zone and fold the Upper Miocene to Quaternary sedimentary cover (Montenat, 1977; Bousquet, 1979; Lopez Casado et al., 1987; Somoza, 1993; Taboada ´ et al., 1993; Alfaro, 1995; Alfaro et al., 2002a, 2012). The activity of these blind thrusts has produced an anticlinorium constituted by several north verging folds (Fig. 2), the most important being the Torremendo, Hurchillo, Benejuzar, Lo- ´ mas de la Juliana and Guardamar anticlines, and the Bajo Segura, Torrevieja and La Mata synclines. Other secondary folds, hectometric in scale, are also present in the basin. The folds are generally gentle or open with limbs usually dipping less than 30° , although strata of the Pliocene Segura conglomerate formation are sub-vertical in the northern limb of the Hurchillo anticline (Alfaro et al., 2002a). These active growth folds are interpreted as the result of several imbricated ENE–WSW blind thrusts limited by NW–SE San Miguel de Salinas (SMSF), Torrevieja (TF) and Guardamar dextral transfer faults (Alfaro et al., 2002a). The age of formation of the folds was established from the analysis of successive progressive unconformities in the fold limbs, and it is not synchronous along the entire basin. For instance, the E–W folds, such as La Marina and Santa Pola anticlines, initiated their activity during the Upper Miocene, whereas the NW–SE folds, such as the Cabo Cervera anticline and the Torrevieja syncline, became active during the Pliocene (Alfaro et al., 2002a; Gimenez et al., 2009). A vertical up- ´ lift rate of about 0.2 mm yr−1 has been calculated comparing high-precision leveling measurements obtained in 1976, 1997 and 2003 along a 30 km profile cutting across the BSFZ (Gimenez et al., 2009). This result confirms the recent activ- ´ ity of the faults and folds in the Bajo Segura basin.

The offshore study area

The offshore Bajo Segura basin is characterized by a wide shelf with an average width of 15 km and a smooth depression in its center (Fig. 2). The maximum depth in the area is around 700 m and is located to the east of the central depression. Tabarca island is located in the north of the basin where the shelf is wider. The integration of seismic reflection profiles, gravimetry, seismicity, cores and outcrop data showed that the compressive active structures described onshore propagate towards the east into the Mediterranean Sea (Alfaro et al., 2002b; Perea et al., 2010). The main feature already known and described in the offshore is the Tabarca anticline ridge (Fig. 3), which extends around 80 km from the coast in an ENE–WSW direction and is 20 to 30 km wide (Alfaro et al., 2002b). This ridge is bounded by thrust faults (GF4 and GF5, to the south, and SPF2, to the north, Fig. 3) verging in opposite directions and folding the Upper Miocene unit. The southern thrusts dip to the north and produce the uplift of the Guardamar high, whereas the northern thrusts dip to the south and raise the Santa Pola high. In the middle of the Tabarca anticline ridge is the BSFZ, which is composed of thrust faults that dip to the south and fold the Upper Miocene unit. To the north and south of this anticline ridge, there are the Alicante and Torrevieja growth synclines, respectively. These two synclines have operated as subsided zones since the start of the folding (Alfaro et al., 2002b). Along the Tabarca anticline ridge are secondary active anticlines and synclines. These synclines are filled with syntectonic sediments that reach the Quaternary.

Instrumental and historical seismicity

The Bajo Segura basin shows moderate seismic activity marked by earthquakes with magnitudes lower than 4.0 (Fig. 2). Nevertheless, there have been some earthquakes with higher magnitudes, such as the 1979 San Miguel de Salinas earthquake (mb 4.2), the composed 1919 Jacarilla earthquake characterized by two events (mb 5.2 and 5.1) and the 2003 east Torrevieja earthquake (mb 4.0). The seismicity observed in the onshore basin seems to be related to the main structures such as BSFZ, SMSF and TF faults. In addition, significant seismic activity has been recorded offshore. Given the uncertainty in the location of this seismicity, it is diffi- cult to associate the earthquakes to some specific structures. However, a group of earthquakes shows an E–W strike coinciding with the offshore prolongation of the BSFZ (Fig. 2).

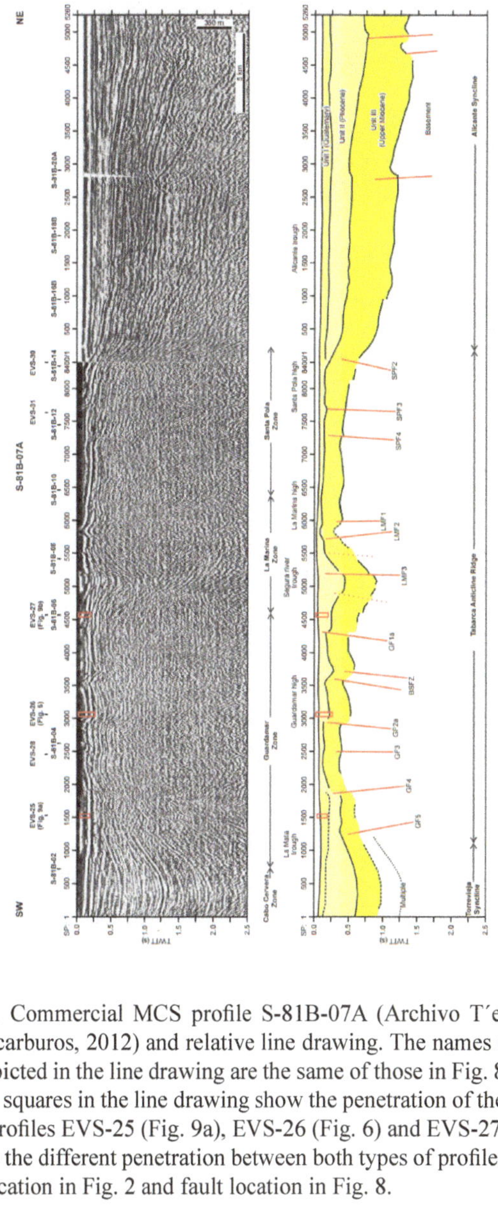

Figure 3. Commercial MCS profile S-81B-07A (Archivo T´ecnico de Hidrocarburos, 2012) and relative line drawing. The names of the faults depicted in the line drawing are the same of those in Fig. 8. The small red squares in the line drawing show the penetration of the SCS sparker profiles EVS-25 (Fig. 9a), EVS-26 (Fig. 6) and EVS-27 (Fig. 9b). Note the different penetration between both types of profiles. See profile location in Fig. 2 and fault location in Fig. 8.

The historical seismic catalog (IGN, 2011) shows that the area has been affected by some destructive earthquakes, such as the 1048 (IEMS98 = X), the 1482 (IEMS98 VIII) and the 1484 (IEMS98 IX) in Orihuela, the 1523 (IEMS98 VIII) in Guardamar del Segura, the 1746 (IEMS98 VII) in Rojales and the 1802, 1828, 1837, 1867 and 1909, all of them (IEMS98 VII) in Torrevieja (Bisbal, 1984; Giner et al., 2003; IGN, 2011). Nevertheless, the most destructive earthquake to have struck the Bajo Segura basin and the SE Iberian Peninsula was the 1829 Torrevieja earthquake (IEMS98 IX–X), which caused 389 fatalities and devastated Torrevieja and other small towns along the Segura River valley (Rodr´ıguez de la Torre, 1984; Munoz and Ud ˜ ´ıas, 1991; Albini and Rodr´ıguez de la Torre, 2001; Mart´ınez Solares and Mezcua, 2002).

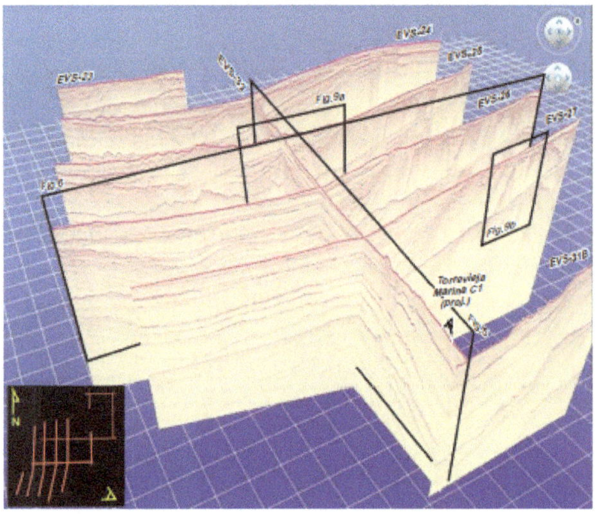

Figure 4. 3-D view from the SE of the SCS sparker profiles obtained during the EVENT-SHELF cruise on the Bajo Segura offshore basin. Note the shallowness of the Neogene basement (almost transparent seismic facies) to the north of the basin and how the Quaternary sedimentary basin (continuous reflector facies) thickens to the east and onlaps the basement to the north. Figures 5, 6, 9a and 9b are located in this figure. Lower-left inset; red lines: displayed profiles; dotted red lines: not displayed profiles; yellow symbol: point of view of the 3-D view. See location of profiles in Fig. 2.

DATA AND METHODS

In the offshore Bajo Segura basin during the EVENT-SHELF cruise, we sought to explore the prolongation towards the continental shelf of the onshore structures (faults and folds) active in the Quaternary and to provide evidence of their recent activity. To this end, a SCS sparker system was installed onboard the R/V Garc´ıa del Cid (CISC) in order to obtain high-resolution profiles to investigate seafloor ruptures and the shallow sub-surface structure. The source was a sparker GEO-SPARK 6 kJ manufactured by GEORESOURCES BV, specially designed to favor high frequencies. The triggering was every 2 s and the source used ranged between 4 and 6 kJ. The receiver consisted of a 9-m-long, 24- hydrophone single-channel streamer. The seismic signal was recorded by the Geotrace acquisition system with a sampling rate of 100 μs and a length of the record between 1.5 and 2.0 s TWTT, allowing up to 30 cm of vertical resolution and 400 m of penetration below the seafloor surface. Postprocessing consisted of change of polarity (the system uses a negative electric discharge instead of a positive pulse), debias, high-pass filter of 300 Hz and a low-pass band filter of 1400 Hz, AGC (10 ms window applied to the whole trace), gain, spherical divergence correction and a swell filter. Ten SCS sparker profiles (EVS-23 to EVS-32) were obtained, most of them being perpendicular to the onshore structures (Figs. 2 and 4). The signal was recorded in a SEG-Y format and later analyzed on SMT Kingdom Suite software. Commercial multichannel seismic (MCS) profiles were obtained and processed by the EXXON Company in 1982 during the S-81B (25/81) survey along the Alicante offshore basin (Archivo Tecnico de Hidrocarburos, 2012). The source ´ used comprised a tuned array of 6 clusters of 6 airguns, towed at a depth of 9.1 m (30 ft) and fired every 22.5 m (73.8 ft). Seismic data were recorded at 4 ms sampling interval, record length of 5 s TWTT, using a 203-channel Teledyne streamer towed at a depth of 10.6 m (35 ft). The seismic processing flow included exponential gain correction and pre-stack deconvolution (160 ms operator length). Later, CMP sorting, velocity analysis, NMO correction and post-stack deconvolution (256 ms operator length) were applied. Then, the pro- files were post-stack wave equation time migrated. Finally, MCS sections were time variant band-pass and time scaled filtered. The multibeam swath-bathymetry was obtained at the same time as the SCS sparker profiles. The used multibeamechosounder was an Elac Nautik SeaBeam 1050D, emitting a frequency of 180 kHz. Given that the objective of the cruise was the acquisition of SCS sparker profiles, the bathymetric coverage was limited to the areas crossed during the seismic survey. Thus, about 276 km with an average width of 300 m were recorded during the cruise. Nevertheless, small-scale morpho-structures such as sloping ramps and rocky reliefs up to 4 m high were found in the study area. However, a highresolution bathymetric map is necessary to understand the significance of these morpho-structures and the active processes that occur in the area.

SEISMO-STRATIGRAPHY AND STRUCTURES OF THE OFFSHORE BAJO SEGURA BASIN

Seismo-stratigraphic units

The commercial MCS and SCS sparker profiles enabled us to distinguish seismo-stratigraphic units at different resolutions and depth intervals. In the commercial MCS profiles three seismo-stratigraphic units were identified (Fig. 3), units I, II and III. Unit I, from top to bottom, is mainly preserved in the depressed areas with a maximum thickness of 0.2 s TWTT. It has sheet drape geometry and does not show internal reflectors at this resolution. Unit II has regional distribution and sheet drape geometry. Internal reflectors are roughly continuous and parallel with medium-to-high amplitude and low frequency. Divergent reflections can be observed close to the active zones. In the uplifted areas, reflectors from the top of unit II are eroded and onlap the lower unit III. Unit III is recognized in all the profiles. It shows sheet to sheet drape geometry with parallel and roughly continuous reflectors of high-to-medium amplitude and medium-to-low frequency. Comparison with the

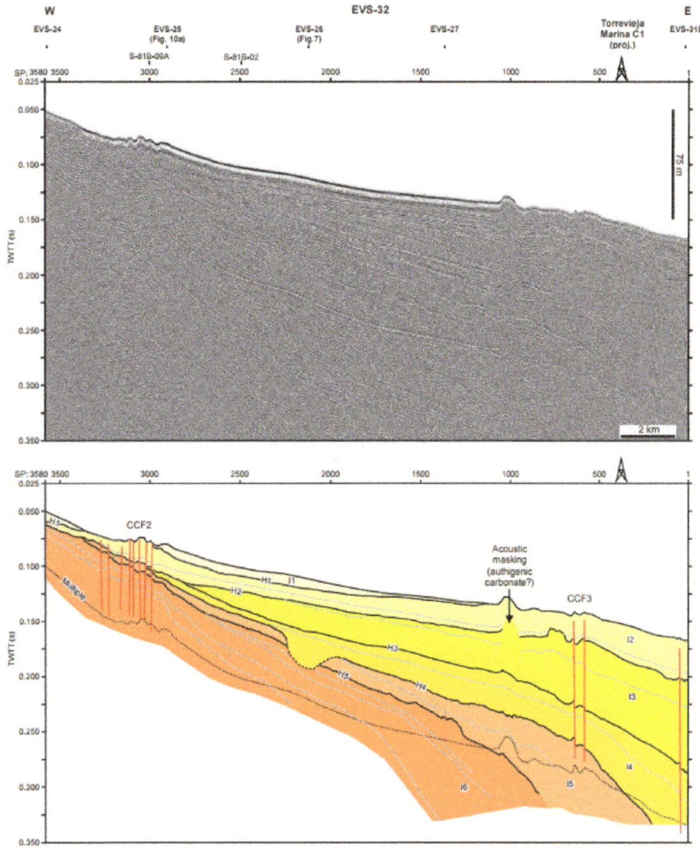

Figure 5. SCS sparker profile EVS-32 and relative line drawing showing the identified seismo-stratigraphic units and structures. Note that the CCF2 (Fig. 8) fault zone is offsetting the seafloor surface, thus demonstrating the recent activity on this fault zone. Around shot point (SP) 1000, there is an acoustic masking facies that could be related to authigenic carbonates (Lo Iacono et al., 2011). The position of the Torrevieja Marina C-1 well is projected. See profile and well location in Figs. 2 and 4.

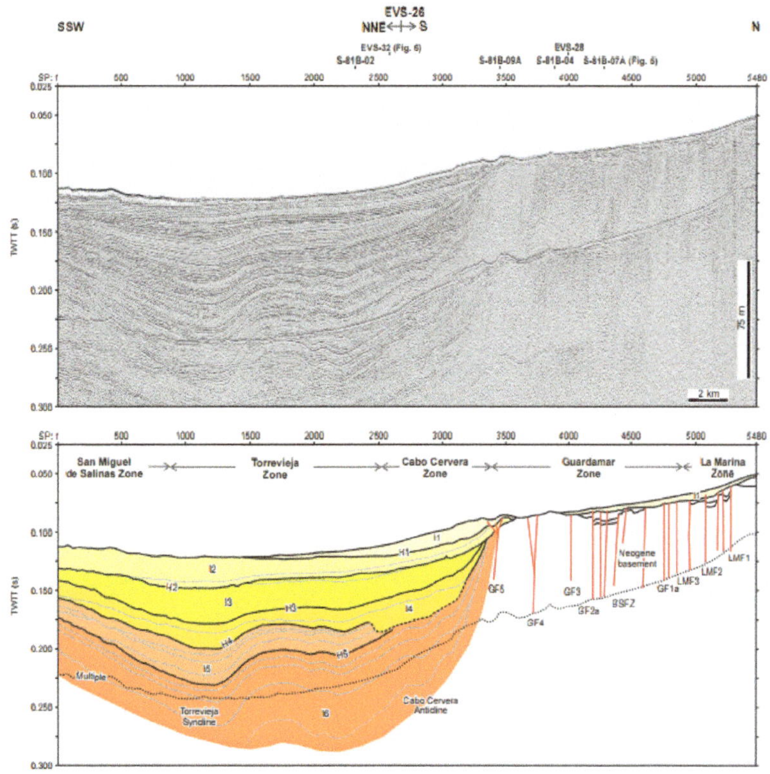

Figure 6. SCS sparker profile EVS-26 and relative line drawing showing the seismo-stratigraphic units and structures. Note how the folding affects unit I3 and some reflectors at the base of unit I2 in the Torrevieja zone and that some faults affect unit I1 in Guardamar and La Marina zones. See location of profile in Figs. 2 and 4.

work of Alfaro et al. (2002b) shows that units I and II are equivalent to their Pliocene–Quaternary unit, and unit III to their Upper Miocene unit. These authors established the age of these units based on a correlation with commercial wells located in the area (one of them the Torrevieja Marina C-1 described below and located in Fig. 2) and also with the geological units observed in the onshore basin and explained briefly in Sect. 2.1. On the SCS sparker profiles (Figs. 2 and 4), up to six units (I1 to I6) have been identified, bounded by five horizons (H1 to H5) that correspond to regional erosion surfaces (Figs. 5, 6 and 7). Units I1 to I6 are divided into several subunits limited by minor discordances (gray lines in Figs. 5, 6 and 7). A total of fifteen subunits were differentiated. Commonly, reflectors are continuous and parallel with a decreasing amplitude and frequency in depth. In general, the deeper zones of the basin show concordant geometry and the shallow areas reveal erosional truncation, although toplaps can occur (Figs. 4, 5, 6 and 7). The base of the units is generally distinguished by onlapping geometry in the uplifted zones. This geometry progressively becomes concordant and, finally, downlaps towards the center of the basin. The Torrevieja Marina C-1 commercial exploration well is located close to the EVS-32 SCS sparker profile (Figs. 2, 4 and 5). This well was operated by ENIEPSA in November 1978, and the recovered core was 578 m long (Lanaja et al., 1987). Three units were recognized in the well (Lanaja et al., 1987), from top to bottom: (a) 386 m of calcareous clay unit with fine sand layers, Quaternary in age; (b) 109 m of clay unit with fine layers of sand and limestone at the base, Upper Pliocene to Quaternary in age; and (c) 83 m of dolomite and dolomite limestone unit, Upper Pliocene in age. The sedimentary thickness imaged in the SCS sparker profile EVS-32 above the multiple (0.152 s TWTT) on the projection of Torrevieja Marina C-1 well (shot point 360) is around 120 m (velocity of seismic waves between 1500 and 1600 m s−1) and corresponds to the Quaternary unit in the well. Therefore, units I1 to I5, and most probably I6, must be Quaternary in age. Comparison of the SCS sparker pro- files with commercial MCS profiles shows that units I1 to I6 (Figs. 5, 6 and 7) coincide to the upper unit I (Fig. 3). Thus, unit I must be Quaternary in age. Unit I1 is the upper unit distinguished in the offshore basin. This unit becomes thinner and discontinuous from the coastline towards the basin, and its lower boundary corresponds to horizon H1, which is a regional erosional surface (Figs. 5, 6 and 7). Close to the coast, unit I1 attains a thickness of 15 m. Onshore, the last 20 m of the Quaternary unit were deposited during the last 14 570 yr BP. Thus, unit I1

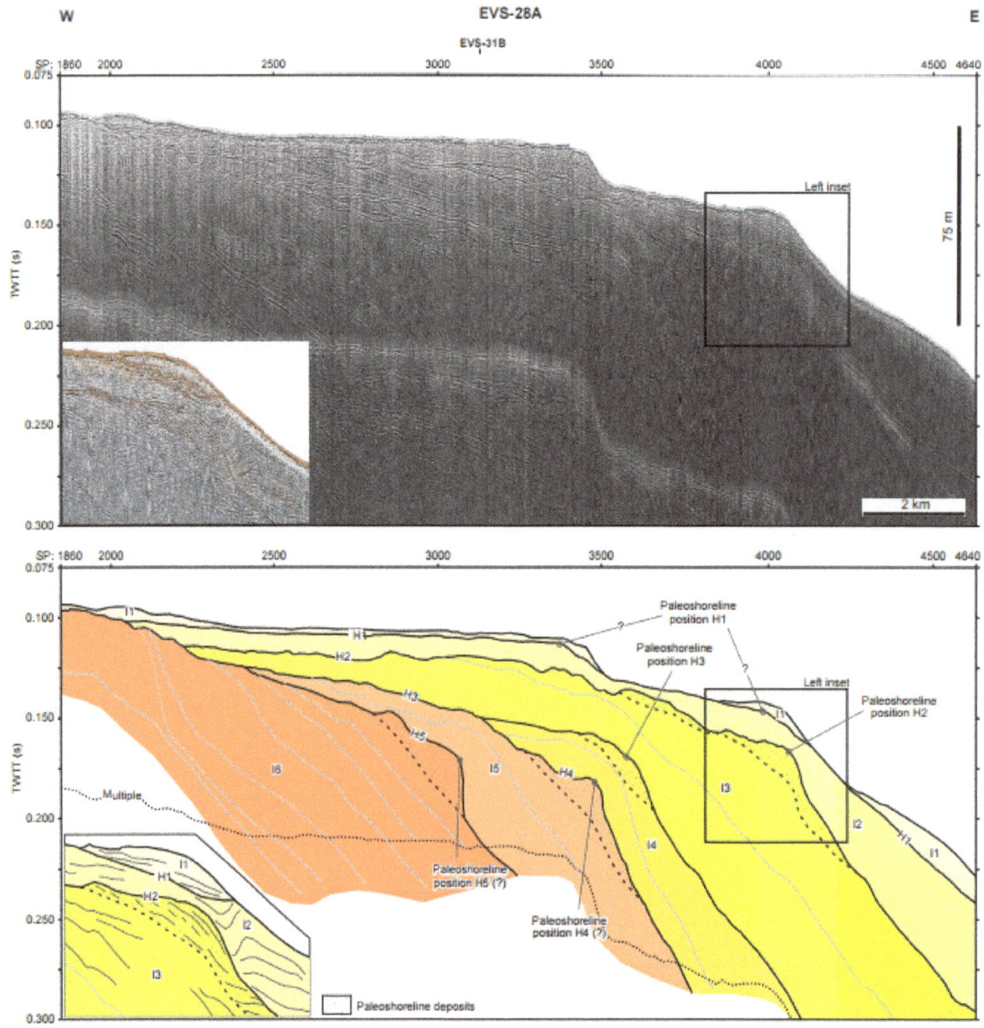

Figure 7. SCS sparker profile EVS-28A and relative line drawing showing the seismo-stratigraphic units. Note the presence of large erosional surfaces that in some cases end in ancient paleoshorelines indicated by the change in reflector slope and by the probable presence of shoreline deposits (detail in the left inset). These surfaces were used to determine the age of the seismic horizons H1 to H5. See profile location in Fig. 2

must be the continuation of the Quaternary onshore unit and it may be the offshore sedimentary unit deposited since the last glacial maximum (LGM). Accordingly, H1 may be related to the last global sea level lowstand, which is associated with Marine Isotopic Stage 2 (MIS2) and dates 20 ± 5 ka. In some SCS sparker profiles, this erosional surface ends in an abrupt change in slope (Fig. 8). This might correspond to the position of the LGM paleoshoreline, which at present is located approximately at 106 m below the present sea level (b.p.s.l.) (considering that the velocity of seismic waves propagation on sea water is 1500 m s^{-1} and in the first hundreds of meters of the sedimentary layer 1600 m s^{-1}). In the Gulf of Lions, the LGM paleoshoreline was placed between 10^2 and 10^7 m b.p.s.l. (Rabineau et al., 2006). These results are in agreement with the observations made in the Bajo Segura basin and support the hypothesis that H1 is the erosional surface developed during the LGM lowstand. Given that horizon H1 could be attributed to the LGM sea level lowstand, it is reasonable to assume that the other horizons, H2 to H5, correspond to other global sea level lowstands and could therefore be associated with different MIS (e.g. Rabineau et al., 2006; Ridente et al., 2008). Nevertheless, it should be noted that the present evolution of the Bajo Segura basin is controlled by Quaternary active structures. Comparison with other zones located in tectonically passive environments is therefore not straightforward. To avoid the effects of tectonic deformation, the following depth calculations of probable ancient paleoshorelines were made in the

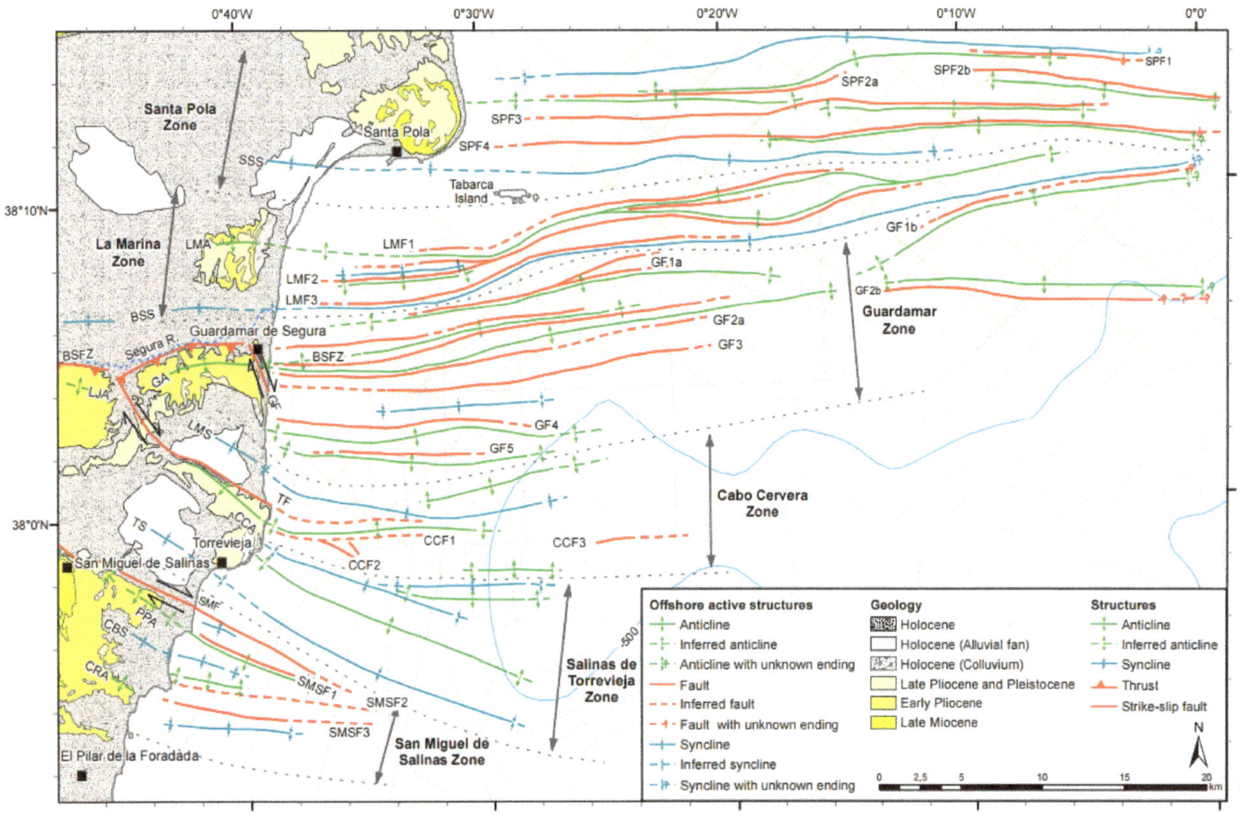

Figure 8. Geological (Alfaro et al., 2002a, b) and bathymetric map of the Bajo Segura basin showing the offshore geological structures (faults and folds). The deformation zones described La Mata syncline; PPA: Punta Prima anticline; SSS: Salinas de Santa Pola syncline; TS: Torrevieja syncline. Cala del Bosque syncline; CCA: Cabo Cervera anticline; CRA: Cabo Roig anticline; GA: Guardamar anticline; LJA: Lomas de la Juliana anticline; LMA: La Marina anticline; LMS: in the text are shown. Faults: BSFZ: Bajo Segura fault zone; GF: Guardamar fault; SMF: San Miguel de Salinas fault; TF: Torrevieja fault. Folds: BSS: Bajo Segura syncline; CBS:

E–W profiles that are parallel to the main active structures (Figs. 5 and 7). Horizon H2 is located at the base of unit I2 (Figs. 5, 6 and 7) and is interpreted as a regional erosional surface. In line with the assumptions made to date H1, H2 might also be ascribed to a global sea level lowstand. In profile EV- 28A, the erosional surface ends in an abrupt change in slope where some possible paleoshoreline deposits could be identified (Fig. 7). The depth of this ancient paleoshoreline is 121 m b.p.s.l., 15 m below H1. In the Gulf of Lions, the paleoshoreline associated with MIS6 is located 16 m below MIS2 (Rabineau et al., 2006). Because of the similar values obtained in both areas, it may be assumed that H2 developed during MIS6 (135 ± 5 ka). Minor discontinuities in unit I2 could correspond to (a) minor sea level oscillations similar to those in the Adriatic Sea that have been attributed to different stages of MIS5 (Ridente et al., 2008), or (b) tectonic activity. Horizon H3 is located at the base of unit I3 (Figs. 5, 6 and 7). As in the case of the aforementioned horizons, H3 is a regional erosional surface with an associated ancient paleoshoreline located at 125 m b.p.s.l., 4 m below H2 and closer to the present shoreline (Fig. 7). Given its regional character and after comparison with H1 and H2, H3 could also be correlated with an older global sea level lowstand. In the Gulf of Lions, the difference between the second and the third discontinuities ascribed to sea level drops is also 4 m, the latter discontinuity being assigned to MIS8.2 (Rabineau et al., 2006). Considering this and the similarities to the other horizons, H3 could correspond to MIS8.2 (247.6 ± 5 ka). The fourth horizon is H4 located at the base of unit I4 (Figs. 5, 6 and 7). This horizon also seems to be a regional erosional surface. The uncertainty is related to the fact that the zones where it can be best observed are uplifted and folded. Nevertheless, an ancient paleoshoreline related to H4 was identified approximately at 144 m b.p.s.l. (Fig. 7). Because of the similarities to H1, H2 and H3, horizon H4 could have been formed during a global sea level lowstand corresponding to MIS10 (341 ± 5 ka). Finally, the last regional discontinuity is horizon H5, which is located at the base of unit I5 (Figs. 5, 6 and 7). This is the last and the deepest regional erosional surface observed in the SCS sparker profiles. A probable ancient paleoshoreline was identified between 130 and 168 m b.p.s.l. (Fig. 7). This large range in the depth of the paleoshoreline is attributed to the difficulty of observing it in areas without deformation. Compared with the other horizons, H5 could also be related to a global sea level lowstand and assigned to MIS12 (434 ± 5 ka).

Offshore faults and fold systems

As stated above, the Tabarca anticline ridge is the main structure in the offshore Bajo Segura basin (Fig. 3). The analysis of the commercial MCS and SCS sparker profiles that cross the area enabled us to identify different structures (faults and folds) that produced the uplift of the Tabarca anticline ridge. Most of the structures trend WSW–ENE to W– E and have a compressive component in their kinematics, which is in agreement with the present regional stress field. The geological structures were grouped into six deformation zones coinciding with the main structures observed onshore and bounded by regional synclines. From north to south, these deformation zones are (a) Santa Pola zone, (b) La Marina zone, (c) Guardamar zone, (d) Cabo de Cervera zone, (e) Salinas de Torrevieja zone, and (f) San Miguel de Salinas zone. These structures were correlated between the different commercial MCS and SCS sparker profiles and with the onshore geology (Fig. 8). The Santa Pola zone is located in the north of the study area (Fig. 8). This deformation zone is bounded to the north and south by two synclines with an E–W direction that extends at least 50 km from the coastline into the sea. The northern syncline, which is a prominent regional fold that bounds the Tabarca ridge towards the north, marks the beginning of the Alicante syncline (Fig. 3). The southern syncline is a smooth fold that could be correlated with the Salinas de Santa Pola syncline observed onshore (Montenat, 1977; Somoza, 1993; Alfaro, 1995), although with some uncertainty. Between these two synclines is the Santa Pola high (Fig. 3), an E–W trending smooth ridge comprising narrowly spaced anticlines associated with faults with the same direction. The faults in the north dip slightly towards the south, whereas in the center of the zone these are sub-vertical (Figs. 3 and 9c). The commercial MCS profiles show that these faults affect unit III, but whether they disturb unit II is not clear. Nevertheless, the SCS sparker profiles show that the faults clearly offset reflectors under horizon H2 (fault SPF3 in Fig. 9c), indicating their Quaternary activity. South of the Santa Pola zone, La Marina zone (Fig. 8) has a WSW–ENE direction, extends at least 50 km and is located between two synclines: the Salinas de Santa Pola to the north, and the marine continuation of the onshore Bajo Segura syncline (Montenat, 1977; Taboada et al., 1993; Alfaro et al., 2002a, b; Gimenez et al., 2009) to the south (Fig. 3), which ′ continues along the offshore basin. The main faults are localized to the south and north of those synclines, respectively (Fig. 3). Associated with faults LMF1 and LMF2 is an anticline which in some areas has a double hinge, producing a small syncline in between. Towards the east, this anticline probably joins the southern anticline of the Santa Pola zone, leading to the disappearance of the Salinas the Santa Pola syncline. The faults dip slightly towards the central part of the zone. In the commercial MCS profiles, the faults affect unit III, and fault LMF3 affects unit II (Fig. 3). The higher resolution of the SCS sparker profiles shows that these faults are sealed by horizon H1, although in some places this horizon might be displaced (Fig. 6). The anticlines fold horizon H2 and probably fold H1.

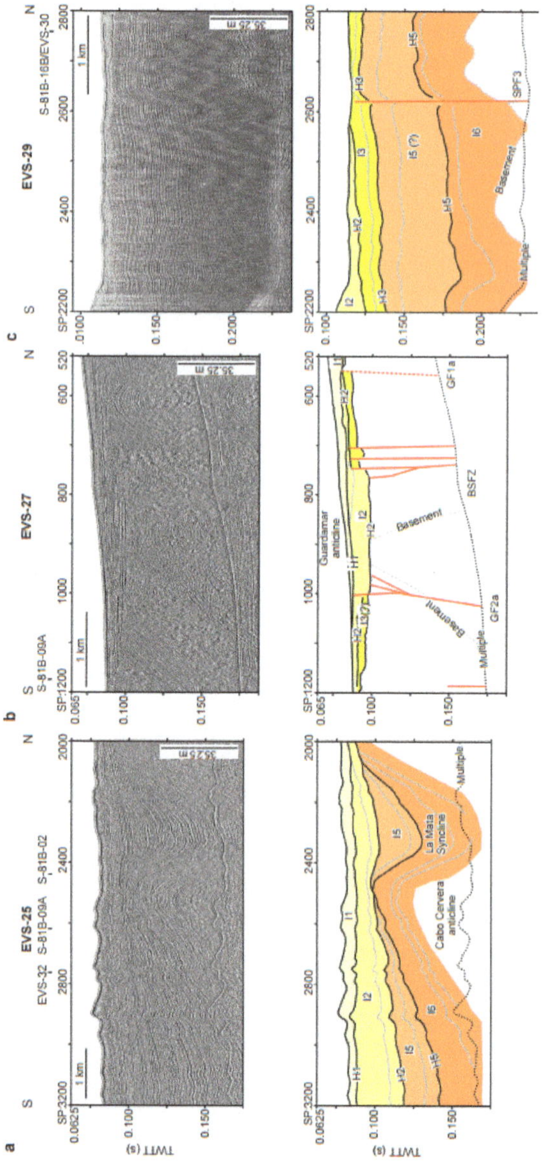

Figure 9. Details of SCS sparker profiles EVS-25 around the Cabo Cervera zone (a), EVS-27 around the Guardamar zone (b) and EVS-29 around the Santa Pola zone (c) and relative line drawings showing the seismo-stratigraphic units, faults and folds. Note how the deformation structures affect the most recent Quaternary units. See location of profiles in Figs. 2 and 4.

The Guardamar zone is a large deformation area with a WSW–ENE direction and a length of up to 50 km. It is located between the Bajo Segura and La Mata synclines (Gimenez et al., 2009), to the north and south, respec- ′ tively (Figs. 3 and 8). The La Mata syncline propagates around 20 km from the onshore towards the sea, progressively changing its direction from NW–SE to WSW–ENE. The faults located on both sides of the Guardamar zone produced the uplift of the area and a regional antiform. As a result, the Neogene basement is elevated and, in some areas, exposed on the seafloor (Figs. 6 and 9b). The structures located to the north of the antiform may be correlated with the BSFZ and the Guardamar anticline (Figs. 3, 6 and 9b). Both structures extend along the study area. The commercial MCS profiles show that the BSFZ is composed of thrust faults that dip towards the south (i.e. are folding units III and II).

Moreover, the SCS sparker profiles show that the BSFZ separates in some minor sub-vertical faults that offset horizon H1. The structures located to the south of the Guardamar high are not correlated with specific onshore structures and taper off towards the center of the shelf. In the commercial MCS profiles, the faults usually dip towards the center of the antiform and affect units III and II. The analysis of the SCS sparker profiles shows that most of the faults are sealed by horizon H1. However, some faults could displace reflectors above H1 (Figs. 6 and 9b). The folds clearly deform units III and II, which are very close to the seafloor surface. The horizons developed above these folds are apparently not folded in the SCS sparker profiles. The Cabo Cervera zone (Fig. 8) is located between the La Mata syncline to the north (Fig. 9a) and some small synclines to the south (Fig. 6).

The structures located along the Cabo Cervera zone have an E–W to WSW–ENE direction and continue up to 15 km offshore. In the commercial MCS pro- files, the faults are sub-vertical and offset unit III and probably unit II, which is folded (Alfaro et al., 2002b). The SCS sparker profiles show some faults reaching and deforming the seafloor (Fig. 5). The TF extends offshore with some uncertainty. Nevertheless, the CCF2 faults have been recently active (Fig. 5) and could be related to TF. The Cabo Cervera anticline also extends offshore. This anticline probably folds horizon H1, but decreases in amplitude before tapering off around 10 km from the coastline. The Salinas de Torrevieja zone is located south of the Cabo Cervera zone (Fig. 8). A number of synclines and anticlines continue offshore around 15 to 20 km from the coastline (Fig. 6). These folds have two different directions.

The northeastern ones have an E–W trend, a low amplitude and do not deform the reflectors above horizon H2. In contrast, the folds identified in the south have a WNW–ESE trend, a larger amplitude and slightly fold the reflectors above H2 and near H1 (Fig. 6). The Torrevieja syncline described onshore (Montenat, 1977; Somoza, 1993; Alfaro et al., 2002a; Gimenez et al., 2009) extends offshore in a WNW–ESE di- ´ rection for more than 20 km. The southernmost zone is the San Miguel de Salinas zone (Fig. 8). The structures identified in this zone change from WNW–ESE in the north to E–W in the south, and continue for 15 km offshore. Most of the faults are sub-vertical and offset horizon H2 and, in some places, also horizon H1. Both horizons are folded. As a result of the correlation with the structures described onshore (Montenat, 1977; Alfaro et al., 2002a; Gimenez et al., 2009), the SMSF, the Punta Prima ´ anticline and the Cala del Bosque syncline extend offshore.

DISCUSSION

Style of deformation

As stated above, the present regional stress field in the Bajo Segura basin is compressive and trends NW–SE to N–S. Accordingly, most of the described structures, faults and folds, have a compressive component in their kinematics and trend WSW–ENE to W–E (Figs. 3, 6 and 8). Taboada et al. (1993) considered three hypotheses in an attempt to explain the deformation undergone in the Bajo Segura basin and to account for the geometrical relationships between the main faults (BSFZ, TF and SMSF) and anticlines (the Guardamar, Cabo Cervera and Punta Prima). The first hypothesis considers that each anticline was created by an independent blind thrust. By contrast, the second and third hypotheses propose that all the faults branch into a common detachment layer at depth. Moreover, the Guardamar and Cabo Cervera anticlines would be associated with a thrust system with flatramp geometry. The difference between the two hypotheses is that one considers that both anticlines are related to blind thrusts, whereas the other assumes that only the Guardamar anticline is produced by a thrust. Analysis of the commercial MCS profiles S-81B-07A (Fig. 3) and S-81B-09 (Alfaro et al., 2002b), which have directions similar to those of the cross section provided by Taboada et al. (1993), shows the following: (a) there is no shallow (1 km depth) detachment layer that could be identified as the flat zone of flat-ramp thrust geometry; by contrast the BSFZ, TF and SMSF seem deeply rooted in the crust since they show high dip angles; (b) there is a predominance of thrust kinematics only in the BSFZ; (c) the TF and SMSF are more vertical and produce less uplift than the BSFZ, indicating that their thrust kinematics is much lower and that the strike-slip component is prevalent. In line with these observations, the second and third hypotheses of Taboada et al. (1993) could be discarded even though these hypotheses offer a better reproduction of the deformation at the base of the Pliocene. Furthermore, the first hypothesis should consider a main strike-slip component in the TF and SMSF. From the Bajo Segura fault to the north, there is a predominance of compressive structures (thrust faults and anticlines) that uplift the La Marina and Santa Pola highs (Fig. 3). There are few structures in this uplifted area the most significant being the anticline associated with fault LMF2, which has been correlated with La Marina anticline.

Timing of deformation

The activity of faults and folds identified in the Bajo Segura basin began in the Upper Miocene and continues during the Quaternary (Alfaro, 1995; Montenat et al., 1990). The commercial MCS profiles show that faults and folds deform units III and II, Upper Miocene and Pliocene in age, respectively, but it is not certain whether they affect unit I, which corresponds to the Quaternary (Fig. 3). However, the analysis of the SCS sparker profiles shows that most of the structures that were active during the Pliocene have been active during the Quaternary (Figs. 5, 6 and 9). As stated above, the different deformation zones show faults and folds that deform horizon H2. Some of these structures also affect horizon H1, and very few reach the seafloor surface (Figs. 5 and 9). In line with these observations and with the ages given to the different horizons, most of the structures were active until 135 ka and in some cases this activity postdates 20 ka. Considering the available slip rates of the main faults in the area, which vary between 0.12 and 0.40 mm yr−1 (Somoza, 1993; Garc´ıa-Mayordomo and Mart´ınez-D´ıaz, 2006; Gimenez et al.,

2009; Alfaro et al., 2012; Garc ́ ́ıa-Mayordomo et al., 2012), some authors have calculated the recurrence intervals of the large earthquakes, which range between 1700 and 30 000 yr (Alfaro et al., 2012; Garc ́ıa-Mayordomo et al., 2012). Thus, reflectors above horizon H1 were probably deformed by very few earthquakes. Moreover, the deformation produced by a single event could be distributed over a wide area given that the main structures of the zone are anticlines associated with blind faults. These two reasons would suggest that the accumulated offset in not necessarily visible in the resolution of the SCS sparker profiles. In conclusion, following this reasoning, the structures affecting the reflectors just below or above horizon H2 can be considered as active during the Quaternary and capable of causing large earthquakes.

Possible sources of the 1829 Torrevieja earthquake

On 21 March 1829, the province of Alicante was struck by a large earthquake known as the Torrevieja earthquake (IEMS98 IX–X). This is one of the largest historical earthquakes in Spain and the strongest shock of a seismic sequence that affected the province of Alicante during 1828 and 1829 (Lopez ́ Marinas, 1976; Mezcua, 1982; Rodr ́ıguez de la Torre, 1984; Munoz et al., 1984; Albini and Rodr ̃ ́ıguez de la Torre, 2001; Mart ́ınez Solares and Mezcua, 2002). According to historical records, the Torrevieja earthquake was felt over a large area, caused a great deal of damage to the epicentral area (e.g. the villages of Almorad ́ı, Benejuzar and Torrevieja were ́ completely destroyed) and about 389 casualties (Larramendi, 1829; Galbis, 1932). Considerable liquefaction occurred in the Bajo Segura Plain onshore (Larramendi, 1829; Rodr ́ıguez de la Torre, 1984; Alfaro, 1995; Alfaro et al., 1999; Soria et al., 1999). Earthquake magnitudes between 6.0 and 6.9 and focal depths between 3 and 7 km have been estimated for this event by different authors (Munoz and Ud ̃ ́ıas, 1991; Delgado et al., 1993; Giner, 1996; Mart ́ınez Solares and Mezcua, 2002; Garc ́ıa-Mayordomo and Mart ́ınez-D ́ıaz, 2006). The question still open is about which fault generated the 1829 Torrevieja earthquake. The most recent compilations and reevaluations of intensity data points include 67 localities where the 1829 Torrevieja earthquake caused damage to buildings and/or was felt by the population (Albini and Rodr ́ıguez de la Torre, 2001; Mart ́ınez Solares and Mezcua, 2002). The highest intensities (IEMS98 IX–X and IX) were felt between Benejuzar and Guardamar, from west ́ to east, and between Almorad ́ı and Torrevieja, from north to south. This geographical distribution of the highest intensities points to the BSFZ and TF as being the most likely sources of the earthquake (Fig. 10). Moreover, given the lack of information about intensity offshore and the location of some of the highest values close to the coastline, the earthquake could have resulted from the rupture of an offshore segment of one of these two faults. The analysis of the SCS sparker profiles obtained in this area could provide some new information about the identification of the earthquake source. The first candidate to be the 1829 earthquake source is the BSFZ (Figs. 2, 8 and 10). This is a WSW–ENE trending blind thrust with the hanging wall directed towards the NW. This is a blind fault that deforms the surface with several anticlines affecting the Plio-Quaternary units (Taboada et al., 1993; Alfaro, 1995). Using different geological markers and high-precision leveling data, an uplift rate of around 0.2 mm yr−1 has been estimated around the BSFZ (Somoza, 1993; Sanz de Galdeano et al., 1998; Garc ́ıa-Mayordomo and Mart ́ınez-D ́ıaz, 2006; Gimenez et al., 2009). Alfaro et ́ al. (2002a) interpreted this structure as the most likely source fault and located the focus between the Segura River and Torrevieja, considering the evidence of Quaternary activity and the southward dip of the BSFZ. This would explain the considerable damage and high intensities around Torrevieja as well as the liquefaction and local site effects observed in the basin villages. Analysis of the SCS sparker profiles shows that some minor faults, related to the BSFZ, offset horizon H2 and maybe H1 (Figs. 6 and 9b). Thus, the offshore part of the Bajo Segura fault has been active during the Upper Quaternary, with possible deformation events in the last 20 ka. The total length of the Bajo Segura fault onshore is 27 km, 10 km corresponding to the Hurchillo, 9 km to the Benejuzar and 8 km to the Guardamar segments (Alfaro et ́ al., 2012). On the basis of the correlation between the different seismic profiles, the BSFZ can be traced for at least 27 km offshore. Considering these different dimensions as the surface rupture length of an earthquake and using empirical relationships for reverse faults (Wells and Coppersmith, 1994), a moment magnitude (Mw) ranging between 6.1 and

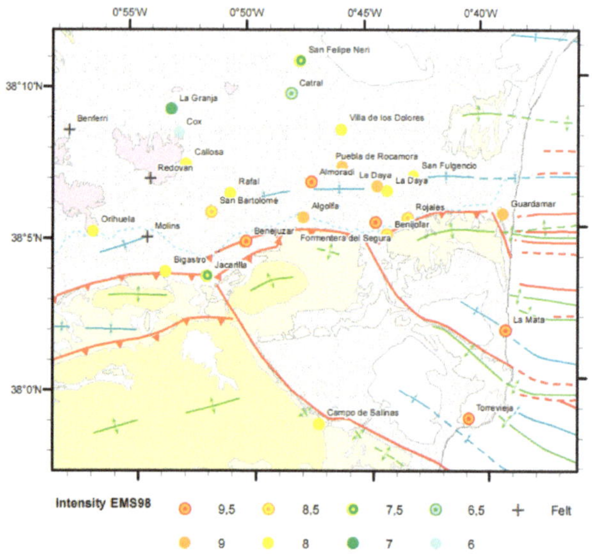

Figure 10. Geological map of the Bajo Segura basin (Alfaro et al., 2002a, b) showing the distribution of the intensity data points of the 21 March 1829 Torrevieja earthquake (Mart ́ınez Solares and Mezcua, 2002). Note that the area where the maximum intensities are located suggests that the BSFZ or the TF are the most likely sources of this earthquake.

6.7 is obtained. Thus, these ruptures would yield magnitudes in the range of magnitudes estimated by other authors for this earthquake. Nevertheless, the highest magnitude (Mw 6.9) is only obtained when the length of the Guardamar and the offshore segments are considered to be the earthquake rupture (35 km). This maximum magnitude is exceeded when assuming that the entire fault (onshore and offshore segments) ruptures together (54 km, Mw 7.1). The second candidate, the TF (Figs. 2 and 8), is a NW–SE trending right-lateral strike-slip fault with a reverse component. Although the fault is difficult to follow along the zone, its present activity is evidenced by the presence of the Cabo Cervera anticline where the Plio-Pleistocene sedimentary units are folded. Besides, the fault could have produced some lateral displacement of the BSFZ, which is active (Fig. 2). The estimated slip rate of the TF is 0.075 mm yr−1 because of the displacement of the Upper Pliocene structural surface (Garc´ıa-Mayordomo and Mart´ınez-D´ıaz, 2006). On the basis of a statistical study of the intensities assigned to some localities during the 1829 Torrevieja earthquake (Lopez Marinas, ´ 1976; Munoz et al., 1984; L ˜ opez Casado et al., 1992), and as- ´ suming that the source model is linear, Delgado et al. (1993) obtained a fault source that better reproduced these intensities. This fault trended NW–SE, extended offshore and may be correlated with the TF. Thus, as in the case of the BSFZ, an earthquake along the TF could also have accounted for the destruction in the area. Furthermore, rupture propagation from offshore towards the northwest could explain the considerable damage in the Bajo Segura basin. Nevertheless, a shallow earthquake with an estimated magnitude higher than 6.0 would produce surface rupture, which has not been described in the area. The existing reports concern a large fracture in a coastal and rocky outcrop between Torrevieja and La Mata (Larramendi, 1829), which could also be a landslide scar. As in the onshore, the offshore continuation of the TF is uncertain. However, the Cabo Cervera anticline can be identified and extends for around 10 km. The CCF2 faults are associated with both structures and reach and offset the seafloor surface (Fig. 5), indicating recent activity. Assuming that the Cabo Cervera anticline is directly related to the activity of the fault, the onshore and offshore lengths of the TF are similar, about 12 km each (Garc´ıa-Mayordomo and Mart´ınez-D´ıaz, 2006). Considering the length of these individual segments to be the maximum surface rupture and using empirical relationships for reverse and strike-slip faults (Wells and Coppersmith, 1994), a Mw 6.3 is obtained. Magnitude attains 6.7 when both segments are regarded as rupturing together. With this fault source, the maximum estimated magnitude for the earthquake (Mw 6.9) is not reached. As stated above, the BSFZ and the TF show Upper Quaternary activity. In some cases, this activity postdates horizon H1 or even implies the deformation of the seafloor surface. Thus, it may well be that the fault rupture or part of the rupture that caused the 1829 Torrevieja earthquake was located offshore. Nevertheless, no tsunami was reported to be associated with this event. The rupture of individual segments of these two faults or some combinations of them could produce an earthquake with magnitudes in the range estimated for this event. Much work has been done in the area substantially improving the imaging and characterization of the active/recent tectonic structures. However, there is no definite evidence yet about which fault caused the 1829 earthquake. Therefore, further studies (e.g. swath-microbathymetry, sub-bottom pro- filing or more SCS sparker profiles offshore) are warranted to improve our understanding of the recent Quaternary activity of both faults and of the seismic hazard in the northern termination of the EBSZ.

CONCLUSIONS

The analysis of the SCS sparker profiles obtained in the offshore Bajo Segura basin allowed us to identify six seismostratigraphic units (I1 to I6) limited by five horizons (H1 to H5). These horizons can correspond to regional erosional surfaces produced during the Quaternary global sea level lowstands and then can be correlated with marine isotopic stages (MIS). We propose that (a) H1 may be correlated to MIS2 (20 ka), (b) H2 to MIS6 (135 ka), (c) H3 to MIS8.2 (247.6 ka), (d) H4 to MIS10 (341 ka), and (e) H5 to MIS 12 (434 ka). A number of faults and folds affecting the Quaternary sediments were mapped for the first time in the Bajo Segura offshore basin. Most of these structures trend WSW–ENE to W–E and show a compressive component in their kinematics, which is in agreement with the present regional stress field. The main faults described onshore were correlated with those identified offshore. Six deformation zones bounded by regional synclines were proposed in line with the main onshore structures: (a) Santa Pola zone; (b) La Marina zone; (c) Guardamar zone; (d) Cabo Cervera zone; (e) Torrevieja zone; and (f) San Miguel de Salinas zone. Integrated analysis of the commercial MCS and SCS sparker profiles shows that (a) the main faults (e.g. BSFZ, TF and SMSF) have high dip angles and are deeply rooted in the crust; (b) the thrust kinematics predominates in the BSFZ and to the north, and is underlined by the formation of anticlines; and (c) the strike-slip kinematics prevails in the TF and SMSF, despite the fact that these faults formed the Cabo Cervera and Punta Prima anticlines, respectively. The relations between the seismo-stratigraphic discontinuities and the faults and folds that deform them provide evidence of recent Quaternary activity in the Bajo Segura basin. Most of the faults are sealed or offset horizon H2 (135 ka), and in some cases their activity postdates the formation of horizon H1 (20 ka). The structures that deform horizon H2 may therefore be considered active and capable of producing destructive earthquakes. Because of the distribution of the intensity data points and of the recent tectonic activity, both the BSFZ and TF could represent the most likely sources for the 1829 Torrevieja earthquake. Analysis of the length of both faults and fault segments onshore and offshore shows that almost all of the fault segments forming both BSFZ and TF are capable of producing a magnitude in the range of the magnitudes estimated for the 1829 earthquake (Mw between 6.0 and 6.9). Nevertheless, the maximum magnitude is attained or exceeded in the BSFZ when the rupture involves the Guardamar and offshore segments (Mw 6.9) or the entire fault (Mw 7.1). Further studies are warranted to provide more insight into the recent Quaternary activity of both faults.

ACKNOWLEDGEMENTS

The authors are indebted to Domenico Ridente, Juan Jose Mart ´ ´ınez D´ıaz and an anonymous reviewer whose helpful comments and suggestions have considerably improved the manuscript. Thanks are due to Daniela Pantosti for editing this manuscript. The authors acknowledge the support of the Spanish Ministry of Science and Innovation through National Projects IMPULS (REN2003-05996MAR), EVENT (CGL 2006-12861- C02-02) and SHAKE (CGL 2011-30005-C02-02); Acciones Complementarias EVENT-SHELF (CTM 2008-03346-E/MAR) and SPARKER (CTM 2008-03208-E/MAR); and ESF TopoEurope TOPOMED project (CGL 2008-03474-E/BTE). We are grateful to the captain, crew and science party onboard the RV Garcia del Cid. Hector Perea was a researcher at IDL-UL Associated Laboratory under contract no. 3/2010/ LAB IDL co-financed by the Portuguese FCT and FEDER and at UTM-CSIC under the Juan de la Cierva contract no. JCI-2010-07502 funded by the Spanish Ministry of Science and Innovation.

REFERENCES

1. Albini, P. and Rodr´ıguez de la Torre, F.: The 1828–1829 earthquake sequence in the provinces of Alicante and Murcia (S-E Spain): Historical sources and macroseismic intensity assessment, in: The use of historical data in natural hazard assessments, edited by: Glade, T., Albini, P., and Frances, F., 35–53, 2001.
2. Alfaro, P.: Neotectonica en la Cuenca del Bajo Segura (Cordillera ´ Betica Oriental), PhD Thesis, Universidad de Alicante, Alicante, ´ 217 pp., 1995.
3. Alfaro, P., Estevez, A., Moretti, M., and Soria, J. M.: Structures ´ sedimentaires de d ´ eformation interpr ´ et´ ees comme s ´ eismites dansle Quaternaire du bassin du Bas Segura (Cordillere b ` etique ori- ´ entale), C. R. Acad. Sci. Paris, 328, 17–22, 1999.
4. Alfaro, P., Andreu, J. M., Delgado, J., Estevez, A., Soria, J., and Teixido, T.: Quaternary deformation of the Bajo Segura ´ blind fault (eastern Betic Cordillera, Spain) revealed by highresolution reflection profiling, Geological Magazine, 139, 331– 341, 2002a.
5. Alfaro, P., Delgado, J., Estevez, A., Soria, J. M., and Y ´ ebenes, ´ A.: Onshore and offshore compressional tectonics in the eastern Betic Cordillera (SE Spain), Marine Geology, 186, 337–349, 2002b.
6. Alfaro, P., Bartolome, R., Borque, M.J., Est ´ evez, A., Garc ´ ´ıaMayordomo, J., Garc´ıa-Tortosa, F. J., Gil, A. J., Gracia, E., Lo ` Iacono, C., and Perea, H.: The Bajo Segura Fault Zone: Active blind thrusting in the Eastern Betic Cordillera (SE Spain), Journal of Iberian Geology, 38, 271–284, 2012.
7. Archivo Tecnico de Hidrocarburos: ´ http://hidrocarburos. mityc.es/ ath/, last access: 27 June 2012. Argus, D. F., Gordon, R. G., DeMets, C., and Stein, S.: Closure of the Africa-Eurasia-North America Plate Motion Circuit and Tectonics of the Gloria Fault, J. Geophys. Res., 94, 5585–5602, 1989.
8. Bisbal, L.: Estudio de la distribucion de intensidades s ´ ´ısmicas en el ambito valenciano. Sus incidencias en las obras p ´ ublicas. Tomo ´ V: Estudio monografico de la serie s ´ ´ısmica del 23-3-1748 al 8-4- 1748 con epicentro en Estubeny (tradicionalmente llamado terremoto de Montesa), Universitat Politecnica de Val ` encia, 1984. `
9. Bousquet, J. C.: Quaternary strike-slip faults in southeastern Spain, Tectonophysics, 52, 277–286, 1979.
10. Buforn, E., Sanz de Galdeano, C., and Udias, A.: Seismotectonics of the Ibero-Maghrebian region, Tectonophysics, 248, 247–261, 1995.
11. De Larouziere, F. D., Bolze, J., Bordet, P., Hernandez, J., Mon- ` tenat, Ch., and Ott d'Estevou, Ph.: The Betic segment of the lithospheric Trans-Alboran shear zone during the Late Miocene, ´ Tectonophysics, 152, 41–52, 1988.
12. Delgado, J., Giner, J. J., Lopez Casado, C., and Auernheimer, C.: ´ Analisis de la respuesta del suelo en intensidades. Aplicaci ´ on al ´ terremoto de Torrevieja, in: Problematica geoambiental y desar- ` rollo. V reunion nacional de geolog ´ ´ıa ambiental y ordenacion del ´ territorio II, edited by: Ortiz, R., 627–636, 1993.
13. DeMets, C., Gordon, R. G., Argus, D. F., and Stein, S.: Current plate motions, Geophys. J. Int., 101, 425–478, 1990. Focault, A.: Etude g ´ eologique des environs des sources du ´ Guadalquivir (provinces de Jaen et de Grenade, Espagne meridionale), PhD Thesis, University of Paris VI, 633 pp., 1974. ´
14. Galbis, J.: Catalogo s ´ ´ısmico de la zona comprendida entre los meridianos 5E y 20W de Greenwich y los paralelos 45N y 25N, Instituto Geografico Catastral y de Estad ´ ´ıstica, Madrid, 807 pp., 1932.
15. Garc´ıa-Mayordomo, J. and Mart´ınez-D´ıaz, J.J.: Caracterizacion´ s´ısmica del anticlinorio del Bajo Segura (Alicante): Fallas del Bajo Segura, Torrevieja y San Miguel de Salinas, Geogaceta, 40, 19–22, 2006.
16. Garc´ıa-Mayordomo, J., Insua-Arevalo, J. M., Mart ´ ´ınez-D´ıaz, J. J., Jimenez-D ´ ´ıaz, A., Mart´ın-Banda, R., Mart´ın-Alfageme, S., Alvarez-Gomez, J. A., Rodr ´ ´ıguez-Peces, M., Perez-L ´ opez, R., ´ Rodr´ıguez-Pascua, M. A., Masana, E., Perea, H., Mart´ınGonzalez, F., Giner-Robles, J., Nemser, E. S., Cabral, J., and the ´ QAFI Compilers Working Group: The Quaternary Active Faults DataBase of Iberia (QAFI v.2.0), Journal of Iberian Geology, 38, 285–302, 2012.
17. Gimenez, J., Borque, M., Gil, A., Alfaro, P., Est ´ evez, A., and ´ Surinach, E.: Comparison of long-term and short-term uplift ˜ rates along an active blind reverse fault zone (Bajo Segura, Se Spain), Studia Geophysica et Geodaetica, 53, 81–98, 2009.
18. Giner, J. J.: Sismicidad y peligrosidad s´ısmica en la comunidad autonoma valenciana. An ´ alisis de incertidumbres, PhD Thesis, ´ Universidad de Granada, 295 pp., 1996.
19. Giner, J. J., Molina, S., and Jauregui, P. J.: Sismicidad en la Comu- ´ nidad Valenciana (C.V.), F´ısica de la Tierra, 15, 163–187, 2003.
20. Grimison, N. and Cheng, W.: The Azores-Gibraltar plate boundary: focal mechanisms, depths of earthquakes and their tectonic implications, J. Geophys. Res., 91, 2029–2047, 1986.
21. IGN: Catalogo de terremotos en el ´ area de la Pen ´ ´ınsula Iberica e Islas Canarias (http://www.01.ign.es/ ign/layoutIn/ sismoFormularioCatalogo.do), last access October 2011.
22. Koulali, A., Ouazar, D., Tahayt, A., King, R. W., Vernant, P., Reilinger, R. E., McClusky, S., Mourabit, T., Davila, J. M., and Amraoui, N.: New GPS constraints on active deformation along the Africa–Iberia plate boundary, Earth Planet. Sci. Lett., 308, 211–217, 2011.
23. Lanaja, J. M., Querol, R., and Navarro, A.: Contribucion de la ex- ´ ploracion petrol ´ ´ıfera al conocimiento de la Geolog´ıa de Espana. ˜ IGME, Madrid. 465 pp., 1987.
24. Larramendi, J. A.: Memoria y relacion circunstanciada de los estra- ´ gos que la terrible catastrofe de los terremotos de 21 de Marzo ´ y siguientes, principalmente el Sabado Santo 18 de Abril hasta ´ el presente d´ıa, han causado en Torrevieja y demas pueblos de ´ la Gobernacion de Orihuela y sus inmediaciones, en la ciudad ´ de Murcia y algunos pueblos de este nombre, Imprenta Real, Madrid, 24 pp., 1829.

25. Lo Iacono, C., Sulli, A., Agate, M., Lo Presti, V., Pepe, F., and Catalano, R.: Submarine canyon morphologies in the Gulf of Palermo (Southern Tyrrhenian Sea) and possible implications for geo-hazard, Marine Geophys. Res., 32, 127–138, 2011.

26. Lopez Casado, C., Est ´ evez, A., Pina, J., and Sanz de Galdeano, C.: ´ Alineaciones sismotectonicas en el sudeste de Espa ´ na, Ensayo de ˜ delimitacion de fuentes s ´ ´ısmicas, Serie de Estudios Geologicos, ´ 6, 5–39, 1987.

27. Lopez Casado, C., Peinado, M. A., Delgado, J., Pel ´ aez, J. A., and ´ Fuentes, C.: La serie s´ısmica de Torrevieja de 1828 y 1829, I Congreso Iberoamericano sobre tecnicas aplicadas a la gesti ´ on de ´ emergencias para la reduccion de desastres naturales, Valencia, ´ 1992.

28. Lopez Marinas, J. M.: Sismo del 21 de Marzo de 1829, Madrid, ´ 1976. Mart´ınez Solares, J. M. and Mezcua, J.: Catalogo s ´ ´ısmico de la Pen´ınsula Iberica (880 a.C. – 1900), Instituto Geogr ´ afico Na- ´ cional – Monografia no. 18, 444 pp., 2002.

29. Mezcua, J.: Catalogo general de isosistas de la Pen ´ ´ınsula Iberica, ´ IGN, Madrid, 320 pp., 1982. Montenat, C.: Les bassins neog ´ enes et quaternaires du levant ` d'Alicante a Murcie (Cordill ` eres B ` etiques Orientales, Espagne), ` Stratigraphie, paleontologie et ´ evolution dynamique, PhD The- ´ sis, Laboratoire Geologique, University of Lyon, 345 pp., 1977. ´

30. Montenat, C., Ott D'Estevou, P., and Coppier, G.: Les bassins neog ´ enes entre Alicante et Cartagena, in: Les bassins n ` eog ´ enesdu domaine Betique orientale (Espagne), Doc. Trav. IGAL, 313– ´ 368, 1990.

31. Morel, J. L. and Meghraoui, M.: Goringe-Alboran-Tell tectonic zone: A transpression system along the Africa-Eurasia plate boundary, Geology, 24, 755–758, 1996.

32. Munoz, D. and Ud ˜ ´ıas, A.: Three large historical earthquakes in southern Spain, in: Seismicity, seismotectonics and seismic risk of the Ibero Maghrebian region, IGN, 175–182, 1991.

33. Munoz, D., Udias, A., and Moreno, E.: Reevaluaci ˜ on de los datos ´ del terremoto del 1829 (Torrevieja), in: Sismicidad historica de ´ la Pen´ınsula Iberica, Asociaci ´ on Espa ´ nola de Ingenier ˜ ´ıa S´ısmica, Madrid, 38–41, 1984.

34. Perea, H., Gracia, E., Bartolom ` e, R., Lo Iacono, C., and Masana, E.: ´ Structure and potential seismogenic sources of the offshore Bajo Segura Fault Zone, SE iberian Peninsula (Mediterranean sea). Looking for the source of the 1829 Torrevieja earthquake, in: Contribucion de la geolog ´ ´ıa al analisis de la peligrosidad s ´ ´ısmica, edited by: Insua, J. M. and Mart´ın-Gonzalez, F., 113–116, 2010. ´

35. Rabineau, M., Berne, S., Olivet, J.-L., Aslanian, D., Guillocheau, ´ F., and Joseph, P.: Paleo sea levels reconsidered from direct observation of paleoshoreline position during Glacial Maxima (for the last 500,000 yr), Earth Planet. Sci. Lett., 252, 119–137, 2006.

36. Ridente, D., Fracassi, U., Di Bucci, D., Trincardi, F., and Valensise, G.: Middle Pleistocene to Holocene activity of the Gondola Fault Zone (Southern Adriatic Foreland): Deformation of a regional shear zone and seismotectonic implications, Tectonophysics, 453, 110–121, 2008.

37. Rodr´ıguez de la Torre, F.: Los terremotos alicantinos de 1829, Publicaciones del instituto de estudios alicantinos, Alicante, 1984.

38. Sanz de Galdeano, C., El Hamdouni, R., Chacon, J., G ´ omez, A., ´ Salvador, F., Schulte, L., and Garc´ıa, A.: Neotectonica de la ´ fosa del Padul y del Valle de Lecr ´ ´ın, in: Itinerario geomorfologico por Andaluc ´ ´ıa Oriental, Publicaciones de la Universidad de Barcelona, 65–81, 1998.

39. Sanz de Galdeano, C., Lopez, C., Delgado, J., and Peinado, M. A.: ´ Shallow seismicity and active faults in the Betic Cordillera. A preliminary approach to seismic sources associated with specific faults, Tectonophysics, 248, 293–302, 1995.

40. Serpelloni, E., Vannucci, G., Pondrelli, S., Argnani, A., Casula, G., Anzidei, M., Baldi, P., and Gasperini, P.: Kinematics of the Western Africa-Eurasia plate boundary from focal mechanisms and GPS data, Geophys. J. Int., 169, 1180–1200, 2007.

41. Silva, P. G., Goy, J. L., Somoza, L., Zazo, C., and Bardaj´ı, T.: Landscape response to strike-slip faulting linked to collisional settings: Quaternary tectonics and basin formation in the Eastern Betics, southeastern Spain, Tectonophysics, 224, 289–303, 1993.

42. Somoza, L.: Estudio del Cuaternario litoral entre el Cabo de Palos y Guardamar (Murcia-Alicante). Las variaciones del nivel del mar en relacion con el contexto geodin ´ amico, Instituto Espa ´ nol de ˜ Oceanograf´ıa, 237 pp., 1993.

43. Soria, J. M., Alfaro, P., Estevez, A., Delgado, J., and Dur ´ an, J. ´ J.: The Holocene sedimentation rate in the Lower Segura Basin (eastern Betic Cordillera): eustatic implications, Bull. Soc. Geol. ´ France, 170, 349–354, 1999.

44. Taboada, A., Bousquet, J. C., and Philip, H.: Coseismic elastic models of folds above blind thrusts in the Betic Cordilleras (Spain) and evaluation of seismic hazard, Tectonophysics, 220, 223–241, 1993.

45. Vernant, P., Fadil, A., Mourabit, T., Ouazar, D., Koulali, A., Davila, J. M., Garate, J., McClusky, S., and Reilinger, R.: Geodetic constraints on active tectonics of the Western Mediterranean: Implications for the kinematics and dynamics of the Nubia-Eurasia plate boundary zone, J. Geodynamics, 49, 123–129, 2010.

46. Wells, D. L. and Coppersmith, K. J.: New empirical relationships among magnitude, rupture length, rupture width, rupture area, and surface displacement, Bull. Seismol. Soc. America, 84, 974– 1002, 1994.

CHAPTER 18 MESOZOIC TECTONIC DYNAMICS AND CHRONOLOGY

Quanlin Hou, Qing Liu,Hongyuan Zhang,Xiaohui Zhang,and Jun Li

INTRODUCTION

In recent years, Mesozoic tectonic regime inversion is becoming one of the focuses in the studies of tectonic evolution and geodynamics. The Mesozoic tectonic dynamics process in the eastern North China is one of the most typical examples. EW-trending structural frame was transformed to the NE-NNE-trending structural framework and a compressive tectonic system to extensional tectonic system during Mesozoic. This intracontinental geodynamic process has attracted sight of geologists from China and around the world. Several hypotheses, such as long-distance effect of Pacific Plate [1], comprehensive constraints of the adjacent blocks (including the subducted and extinct Kula Plate, [2]), adjustment of the stress after the deep subduction of the Dabie area [3], large-scale sinistral strike slip, mantle plume [4, 5], delamination of lithosphere, and/or continental root-plume tectonics, had been put forward during the last decade. These ideas, even though being of great benefit to the study on Mesozoic tectonic regime inversion, have neither given rise a complete theoretical framework nor concluded a uniform tectonic dynamics process. It should be noticed that China continent stands on the junction of three plates including the Tethyan oceans, the paleo-Asian ocean, and the paleo-Pacific ocean. The relative movement among the three plates, especially the Pacific plate which subducted NW warding beneath China continent, strongly affects China continent in the Mesozoic era. Many other important tectonic events, such as the continental deep subduction along the Dabie mountains, the lithosphere thinning of the eastern China, and the rapid uplifting of crust in western China, took place during a relative short period, although slightly different. With respect to the study of inversion process of tectonic regime in the eastern North China Block (NCB), the correlation and effect of the adjacent tectonic blocks should be considered, especially long-distance effect of the paleo-Pacific in the east and paleo-Tethyan in the west, and mantle upwelling would be considered as a reasonable tectonic model [6, 7]. It is recognized that the Mesozoic tectonic dynamics and chronology is one key problem to understand the Mesozoic tectonic regime inversion in the eastern NCB. Three areas, the Dabie area and the Jiaodong area at the south margin and the Yanshan tectonic belt at the north part of the eastern NCB, would be discussed here to recognize the Mesozoic tectonic dynamics process in the eastern NCB.

MESOZOIC TECTONIC DYNAMICS AT THE DABIE AREA

The Dabie area, located between NCB and Yangtze block in Early Mesozoic, is at the south margin of the eastern NCB. It is a key region to understand the Mesozoic tectonic regime inversion from compression to extension in the eastern NCB.

Collisional Time Constraints during the Early Mesozoic

The time range could be indicated by three aspects as follows. (1) The ultrahigh-pressure metamorphic assemblages started inversion at about 220–240 Ma and cooled down to 300°C at 180 Ma (time of the ^{39}Ar-^{40}Ar biotite clock start-up, [8, 9]). (2) Geomagnetism studies indicate that visual polar migration curve of Hehuai basin, located at the south margin of NCB, had not been consistent with that of the Yangtze block until 159 Ma [10], which indicated that the two blocks had jointed and the collision orogeny had completed by the time (159 Ma). (3) The sedimentation of the foreland molasse basin in the Badong-Echeng area in Hubei province began at Middle Triassic and finished in the Late Jurassic, which implies that the collisional orogeny developed from the Middle Triassic to the Late Jurassic. Therefore, the collisional orogeny of the Dabie area started at late stage of Early Triassic (~240 Ma) and ended at Middle Jurassic (~180 Ma).

Citation: Hou, Q., Liu, Q., Zhang, H., Zhang, X., & Li, J. (2012). Mesozoic Tectonic Dynamics and Chronology. Journal Of Geological Research, 2012, 1-11. http://dx.doi.org/10.1155/2012/291467

$^{40}Ar/^{39}Ar$ dating of biotite and hornblende from the Shuihou-Wuhe shear zone and the Taihu-Mamiao shear zone, at the north and south sides of ultrahigh-pressure metamorphic zone in the Dabie area, respectively, was carried out (Table 1 after [11]). The results indicated that the first stage deformation occurred at about 190 Ma, probably reflecting the inversion and cooling of the ultrahigh-pressure metamorphic rocks in the postorogeny. It was suggested that, with the weakening of the orogenic activity and increasing of potential energy of gravity of the orogenic belt in the postorogeny, the principal compressional stress direction changed from horizontal to vertical, leading to the orogenic laxity or collapse, and the rapid inversion of the ultrahigh-pressure metamorphic rocks; then the nearly completion of collision in the Dabie area is about 190 Ma. Perhaps this 190 Ma event extended to 160 Ma in the late Jurassic because of the subduction of the paleo-Pacific plate [12].

Table 1: Mineral ^{40}Ar-^{39}Ar data in the main shear zone of the eastern Dabie Orogenic Belt [11].

Shear zone	Mineral	Weighted mean plateau age (Ma)
Xiaotian-Mozitan shear zone	Biotite	124.17 ± 0.25
	Biotite	126.91 ± 0.30
Shuihou-Wuhe shear zone	Biotite	190.59 ± 0.42
Taihu-Mamiao shear zone	Hornblende	197.41 ± 0.46
	Biotite	189.42 ± 0.29
	Biotite	124.87 ± 0.21
Susong-Qingshuihe shear zone	White mica	194.01 ± 0.36
	Biotite	127.96 ± 0.30

The Late Mesozoic Extensional Detachment

The Northern Dabie metamorphic complex was identified as the crystalline metamorphic core of the Late Mesozoic extensional tectonics, which divided the Dabie area into two extensional tectonic systems, the north extensional tectonic system and the south extensional tectonic system (Figures 1 and 2).

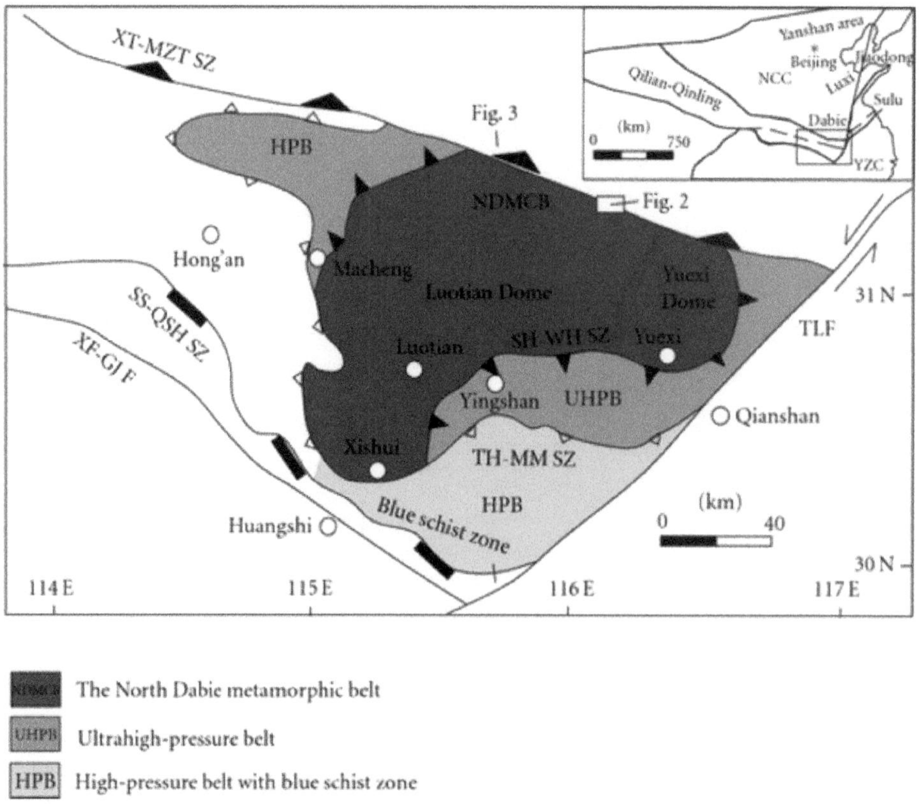

Figure 1: Sketch map of Late Mesozoic main shear zones in eastern Dabie Orogenic Belt, Central China (after [13]). NCP: Northern China Block; YZB: Yangzi Block; NHEB: North Huaiyang metamorphic belt; NDMCD: Northern Dabie metamorphic complex belt; UHPB: ultrahigh pressure metamorphic belt; HPB: high-pressure metamorphic belt; SH-WH SZ: Shuihou-Wuhe shear zone; TH-MM SZ: Taihu-Mamiao shear zone; SS-QSH SZ: Susong-Qingshuihe shear zone; XT-MZT SZ: Xiaotian-Mozitan shear zone; TLF: Tancheng-Lujiang fault (Tanlu Fault); XF-GJF: Xiangfan-Guangji fault.

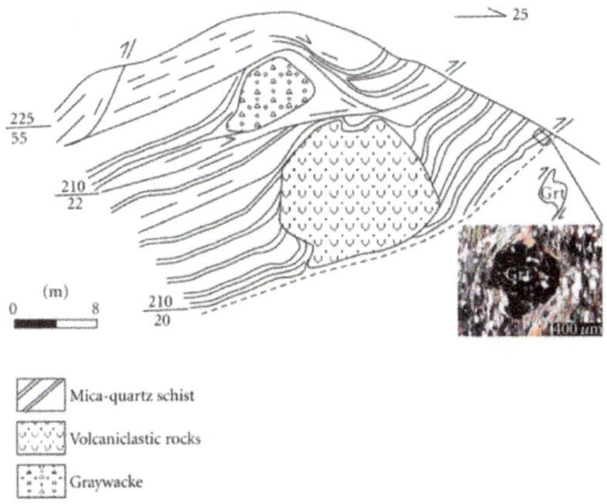

Figure 2: Late Jurassic volcaniclastic tectonite blocks were contained in mica-quartz schist (Pt_2n), north extensional detachment zone, implying the extension shearing after Late Jurassic. The microstructure picture of mica-quartz schist is shown in the right below the profile, which possibly indicates a dextral rotation of kinematics in the profile. The phenomenon is located near the Zhutang Village of Youdian Town in Jinzhai County of the Anhui Province.

The North Extensional Tectonic System

The identification of extensional crenulation cleavage (C′) in the north detachment zones indicated that it is a complete extensional tectonic system, extending NE or NNE along the Xiaotian-Mozitan faults where the principal extensional detachment zone was developed in the early Cretaceous (Figure 1). The shear strain (γ) in the central part of the north detachment zone is up to about 2.6 and gradually decreased outward. The measurements of rock finite strains in the detachment zone indicated that the shear displacement is at least 56 km.

South Extensional Tectonic System

The south extensional tectonic system is composed of, from north to south, the Shuihou-Wuhe shear zone, Taihu-Mamiao shear zone, and Susong-Qingshuihe shear zone, which could represent lower, middle, and upper detachment system, respectively, (Figure 1). Analysis of fabric and dislocation density of deformed quartz indicates that deformation temperature (>700°C, 650–700°C) and differential stress (92 Mpa, 70–84 Mpa) descended from north to south in the south extensional tectonic system. The shear displacement of the upper detachment zone is at least 12 km by measurements of deformed rock finite strains. The directions of the extensional detachment are SSW, SSE, and/or S. The Flinn parameters (K) of strain ellipsoids of the lower, middle and upper detachment systems are 0.01–0.1, 1, and >7 respectively, representing accordingly flattened strain (K = 0.01–0.1), plane strain ($K \approx 1$), and extensional strain ($K > 7$) from north to south, which implies that the extensional detachment could be caused mainly by magma intrusions, consisting with the timing indicates.

The Chronology Constraints on the Extensional Detachment

In north extensional detachment zone, the shear zone developing in the Xinyang Group (Pt_2n) mica-quartz schist contains some allochthonous blocks from less than one meter to several meters in size of Late Jurassic volcaniclastic rocks and tuff. In addition, there are some field-scale harmonic recumbent folds in the Late Jurassic volcaniclastic rocks and the Late Proterozoic mica schist. Both Hong'an Group (Pt) and the Yanshanian granite (J_3-K_1) experienced extensional shear deformation together in the south extensional detachment system, which implies that the extensional detachment of both the south and north detachment systems took place after the Late Jurassic. The biotite and hornblende $^{40}Ar/^{39}Ar$ data from the north extensional tectonic system and in the Susong-Qingshuihe shear zone in the south detachment system reveal that the age of latter deformation is about 125 Ma (Table 1, after [11] (Figure 1)), which should represent the time of extensional detachment after orogeny. As a whole, the extensional deformation in the Dabie area mainly took place about 120 Ma ago.

Genesis of Extensional Detachment

At about 130–120 Ma, a little earlier than the detachment deformation age, numerous extensive granite and ultrabasic-basic plutons intruded into the Northern Dabie complex core. The ^{40}Ar-^{39}Ar and Rb-Sr isotopic analyses of rocks in the North Dabie also recorded the time at about 110–130 Ma. These chronology data imply that the magma intrusion played a major rule during the extensional detachment, which were supported by the finite strain measure analysis of rocks in the south extensional tectonic system (Figure 3). In addition, the ductile deformation of the plagioclase and pyroxene of pyroxenite and gabbros (115–130 Ma) in the Northern Dabie metamorphic complex (mainly refer to the Luotian Dome) implies that they once reached granulite facies during deformation and then uplift to the surface. Therefore, it is concluded that voluminous magmatic emplacement and delamination of lithosphere in Early Cretaceous induced the rapid uplift of the Northern Dabie as the center area of UHPM rocks and extensional detachment on both sides of the Northern Dabie complex core. The intense denudation results in the voluminous sediments in the Hefei basin. The high-pressure and ultrahigh-pressure eclogites probably emplaced during this process on the basis of our data (Figure 3).

Figure 3: Extensional detachment model of Late Mesozoic, the Dabie Orogenic Belt.

Additionally, the distributions of the platinum group elements (PGEs) in the ultramafic and/or mafic rocks (~120 Ma) in the Northern Dabie complex core show that their source regional upper mantle in Later Mesozoic enriched PGE (Pd ≈ 6.2 ppb in mantle). The original mantle before Mesozoic, however, depleted the PGE (Pd ≈ 2.3 ppb), according to the PGE distribution in Bixiling ultramafic rocks (older than 240 Ma) in the Northern Dabie complex core [14, 15]. The

Later Mesozoic upper mantle with PGE enrichment in the Dabie area is supposed to be contaminated by about 8‰ Earth core materials, as the PGE contents in Earth core are much higher than in upper mantle [14, 15]. If thus, the enrichment mantle in the PGE must have been related to the Later Mesozoic extensional detachment in the Dabie area.

MESOZOIC TECTONIC DYNAMICS IN THE JIAODONG AREA

The Jiaodong area, especially the eastern Jiaodong peninsula is located in the eastern terminal of the Dabie-Sulu ultrahigh pressure metamorphic belt, the eastern part of the south margin of the eastern NCB. As part of the collision orogenic belt between NCB and Yangtze block in Early Mesozoic, the eastern Jiaodong peninsula is significant in regional tectonic dynamics study in the eastern NCB.

The eastern Jiaodong peninsula can be divided into four lithotectonic units by four shear zones, the Shidao, Rongcheng, Mishan, and Mouping faults from SE to NW, respectively (Figure 4).

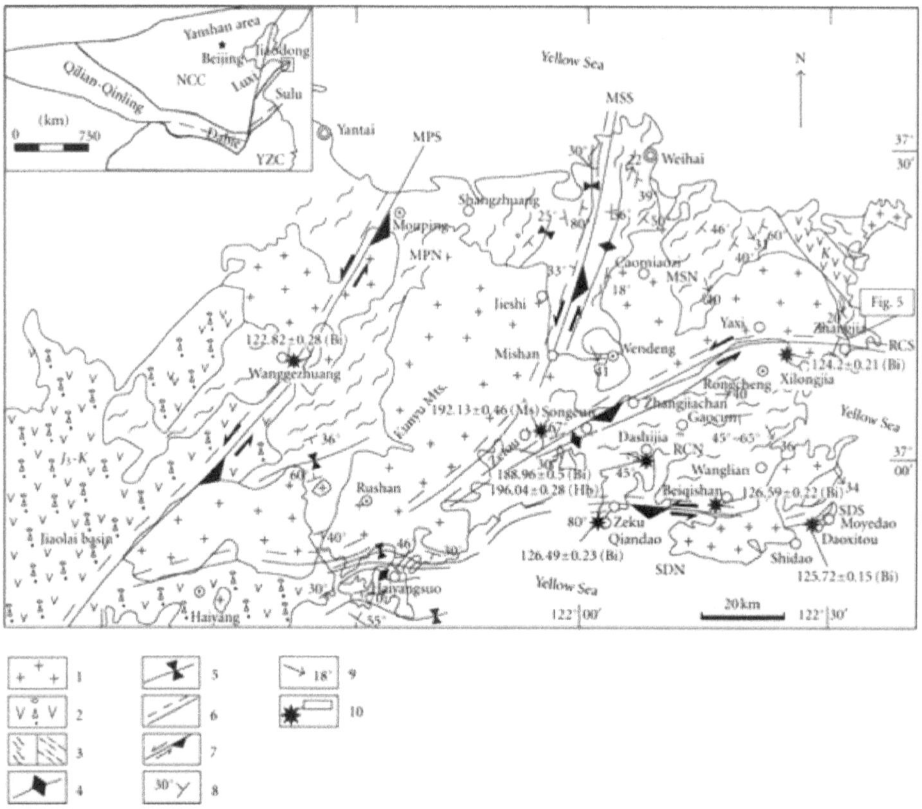

Figure 4: Geological map of the eastern Jiaodong peninsula (modified from [16, 17]). (Granite); (volcanic sedimentary rocks from Late Jurassic to Cretaceous); (gneiss rocks and mylonite zones); (syncline); (unticline); (brittle-ductile shear zones); (left lateral and reverse brittle shear sense); (layer occurrence); (lineation occurrence); (sample position and geochronological results). F_1, F_2, F_3, F_4: faults, namely, Shidao, Rongcheng, Mishan, and Mouping; I, II, II IV: nappes, namely, Shidao, Rongcheng, Mishan, and Mouping.

The Shidao shear zone is a broad ductile shear zone (>15 km in width) thrusting toward north and sinistral strike slipping in ENE-WSW direction. The total shear displacement is more than 90 km, in which the strike displacement is at least 19 km, the thrusting displacement >88 km, and the horizontal shrinkage >85 km, horizontal shrinkage ration about 68% [18]. If Rongcheng, Mishan, and Mouping nappes were taken as an entirety, it turned about 25° clockwise (Figure 4), similar to the Korea peninsula which turned 30.5° clockwise [19].

The Rongcheng shear zone is a sinistral strike slipping and thrusting ductile shear zone dipping S or SSE in early stage (Figure 4). Different rocks, such as porphyritic granite (γ3 5, ~120 Ma), granitic gneiss, and hornblende schists (Pt_1) of the Weideshan rock bodies, overrode one another as thrusting sheets (Figure 5), indicating a strong thrusting and sinistral strike slipping during late Yanshanian epoch. However, it could not be excluded that Rongcheng shear zone would experience extensional strike-slip nature at late stage.

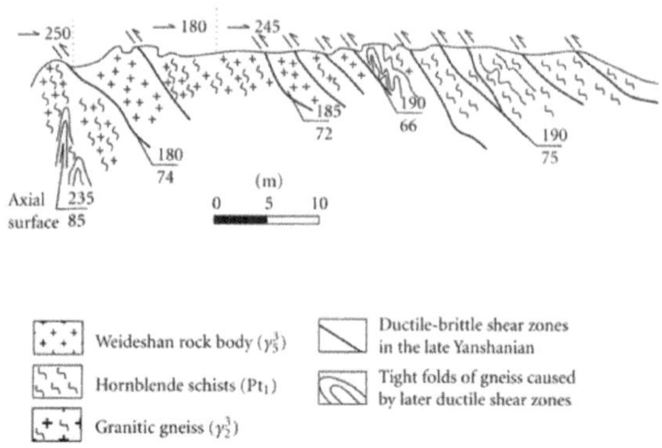

Figure 5: Structural outcrop sketch at Zhangjia, Rongcheng.

At outcrop scale the porphyritic granite ($\gamma 3$ 5), granitic gneiss, and hornblende schists (Pt_1) of the Weideshan rock body were found to be overriding together as the style of thrusting sheets.

The Mishan shear zone is an almost upright sinistral strike slipping fault zone striking NNE, in which a tectonoclastic rock belt of wide was formed. Rocks have a notable change across the fault. A few meters width of mylonite bands was observed in the fault zone, indicating that it could be a ductile shear zone in the early stages, and then was reconstructed later by brittle fractures. Therefore, it is difficult to identify the attitude and nature of the zone. However, it holds somewhat that the later stage of the shear zone was generally in agreement with the brittle shear zones in the Eastern Jiaodong peninsula.

The Mouping shear zone is a compression sinistral strike-slip fault striking NE-SW in steep dip. The shear zone expressed as a brittle-ductile tectonite belt (20–30 km in width) and had extensional characteristics in late stage. Mouping shear zone could be influenced by Tan-Lu faults, because similar tectonic behavior was proved in both the Mouping shear zone and the Tan-Lu fault zone, such as sinistral slip and long time of fault development.

The paleodifferential stress values of the Shidao, Rongcheng and Mouping shear zones are of 105 Ma, 99 Ma, and 85 MPa, respectively, showing a decrease tendency from south to north. The strain measurement and the quartz c-fabric analysis indicate that the deformation degree of those three shear zones decrease in order from south to north. The Flinn values (K) of the strain ellipsoid range between 0.6 and 0.75, implying flattening strain. Therefore, the thrusting in the Eastern Jiaodong peninsula should be from south to north in general.

$^{40}Ar/^{39}Ar$ chronological analyses on the biotite and hornblende from the Shidao, Rongcheng, and Mouping shear zones (Table 2, [20]) indicated two groups, ~190 Ma and ~120 Ma, amazingly similar to those of the Dabie area. The two groups might represent the ages of the reversion of ultrahigh pressure metamorphic rocks during the postorogenic structural relaxation and the slip compressional deformation after the orogeny, respectively.

As well known, in the Early Cretaceous, almost the whole North China area was tectonically in an extensional or extruding setting. However, the large-scale sinistral strike-slipping of the Tan-Lu fault zone resulted in the sinistral strike-slipping compressional settings at least at the shallow part of the crust in the east side front of the Tan-Lu faults, to form the sinistral strike-slipping thrust nappes and the clockwise rotation of the blocks bounded by the shear zones. The subduction of the west Pacific plate toward NW and/or NNW resulted in the strike slipping and the northward movement at the east side of the Tan-Lu fault zone. Meanwhile, the deep part of the crust in the Jiaodong peninsula might be under the extensional tectonic background, according to the relative data of basins and volcanic rocks.

Table 2: $^{40}Ar/^{39}Ar$ precise dating results of individual minerals in the eastern Jiaodong Peninsula [20].

Sample position and sample number	Lithology	Minerals selected	Serial number	$^{40}Ar/^{39}Ar$ plateau age, Ma	Tectonic setting of samples
Qiandao011120905	Granite gneiss	Biotite	R03020	126.49 ± 0.23	Shidao shear zone
Beiqishan011121002	Granite gneiss	Biotite	R03002	126.59 ± 0.22	Shidao shear zone
Daoxitou011121005	Granite gneiss	Biotite	R03026	125.72 ± 0.15	Shidao shear zone
Dashijia011120908B	Granite gneiss	Biotite	R03005	188.96 ± 0.50	Rongcheng nappe
Dashijia011120908H	Granite gneiss	Hornblende	R03019	196.04 ± 0.28	Rongcheng nappe
Zetou011120807	Granite mylonite	Muscovite	R03009	192.13 ± 0.46	Mishan nappe
Xilongjia011121101	Granite gneiss	Biotite	R03008	124.20 ± 0.21	Rongcheng shear zone
Wanggezhuang011121504	Quartzofeldspathic mylonite	Biotite	R03025	122.82 ± 0.28	Mouping shear zone

MESOZOIC TECTONIC DYNAMICS IN THE YANSHAN AREA

The Yanshan area, one famous intraplate orogen considered by some authors (e.g., [21, 22]), is located at the north part of the eastern NCB. The Yanshan area is the cradle of Chinese geologists. Since Wong [23] put forward "Yanshan Movement," lots of attentions have been paid by geologists in the aspect of the Mesozoic strong tectonic deformation and severe magma activities. In recent years, the Mesozoic complex intraplate deformation has become one major focus of continental dynamics research in China.

The eastern segment of the Yanshan area is characterized with two ductile shear zones trending EW and NNE respectively, which controlled the tectonic framework of the northern North China. Field observations and structural analyses reveal that the EW trending ductile shear zones are mainly contributed to dextral compressional deformation that resulted from top-to-the-southeast oblique thrust shearing, whereas the NNE trending ones are genetically related to sinistral strike slip and extensional faulting. The $^{40}Ar/^{39}Ar$ geochronology indicates that the top-to-the-southeast oblique thrusting event recorded by the EW-trending ductile shear zones mainly occurred in Late Triassic (Yanshanian, too) [24]. Recent studies have clarified a series of major Mesozoic contractional events in the Yanshan area [21, 25]. The Huairou thrust north of Beijing has overridden the Cambrian carbonates above the Middle Jurassic Tiaojishan formation (J_2) at Late Jurassic with a southern displacement direction. The Gubeikou-Pingquan fault features both south-vergent thrusting and dextral strike slipping with an age span from 148 to 132 Ma [21]. In summary, the activity period for contraction and thrusting at the Yanshan tectonic belt are mainly concentrated at Late Triassic and Late Jurassic.

Similar to most orogens, localized regions of large-magnitude extensional strain typified by metamorphic core complexes are also prominent geologic and physiographic features of the Yanshan tectonic belt. The examples include the Yunmengshan metamorphic core complex north of Beijing [26–28], the Kalaqin metamorphic core complex in the Inner Mongolia [29–31], and the Yiwulüshan metamorphic core complex in western Liaoning [31, 32].

As described by Han et al. [29] and Shao et al. [30], the Kalaqin metamorphic core complex is a major structural component of the Ma'anshan uplift with a length of about 100 km and a width of about 35 km (Figures 6(a) and 6(b)). The core of the uplift consists of Archean gneisses of epidote-amphibolite facies (>2.5 Ga), marbles and TTG series, Early Proterozoic schists and marbles of low green schist facies, Late Proterozoic Minganshan group slates, Cambrian carbonates, Early Permian-Middle Jurassic plutonic rocks, and Early Cretaceous granitic rocks. The eastern and western flanks of the uplift consist mainly of Jurassic-Cretaceous terrestrial volcanic and sedimentary strata. The Louzidian detachment fault, which consists of the Louzidian ductile shear zone and the related Louzidian normal fault, separates metamorphosed complexes in the footwall to the west from the unmetamorphosed Mesozoic and Cenozoic sedimentary covers in the hanging wall. Field observations and structural analyses reveal that the ductile shear zone was genetically related to sinistral strike slip and extensional faulting. $^{40}Ar/^{39}Ar$ geochronology established an Early Cretaceous age for its formation [31].

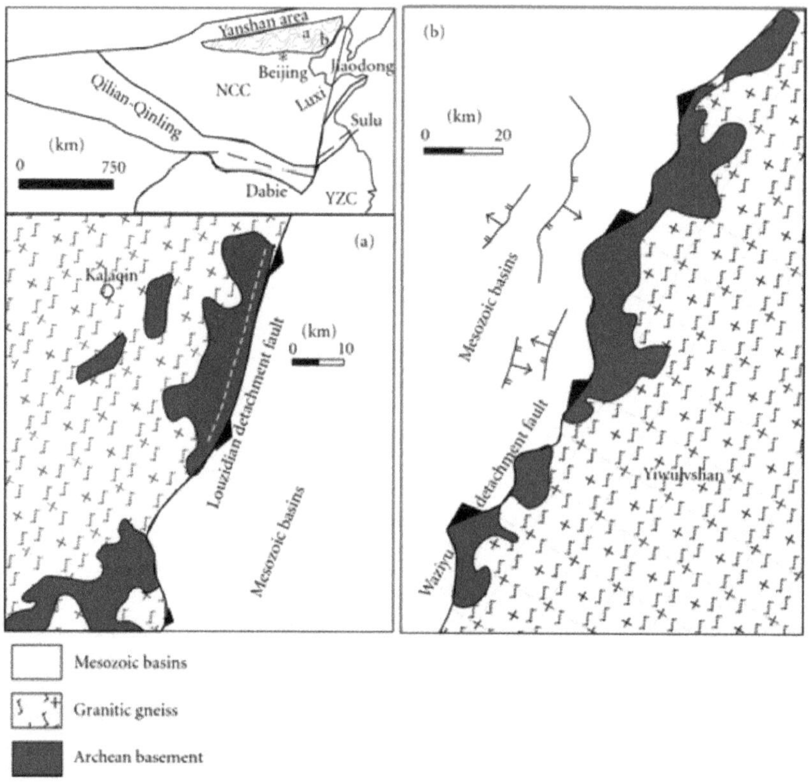

Figure 6: Geological map of the Eastern Yanshan Area. Simplified geological map showing major tectonic units of the Yanshan belt and its surrounding areas, with the locations of Kalaqin and Yiwulüshan metamorphic core complexes indicated by a rectangle. (a) Sketch geological map of the Kalaqin metamorphic core complex (modified after [29]); (b) Sketch geological map of the Yiwulüshan metamorphic core complex (adopted from [32]).

As first reported by Ma et al. [33], geochronologically constrained by Zhang et al. [31] and further defined structurally by Darby et al. [32], the Yiwulüshan of western Liaoning displays many of the structural, metamorphic, and igneous features that characterize Cordilleran metamorphic core complexes. Situated along the eastern segment of the Yanshan area, the Yiwulüshan metamorphic core complex has an east-west dimension of 25 km and a north-south dimension of 60 km (Figure 6(c)). The core consists predominantly of Mesozoic granitoid plutons and their country rocks, which include migmatite, epidote-amphibolite facies paragneiss, and orthogneiss of the Archean crystalline basement. Plutons are granodioritic to granitic and grade outward into augen gneiss or migmatite. At structurally higher levels, it is the well-developed NNE-trending Waziyu ductile shear which, together with the related normal fault above it along the eastern margin of the Fuxin basin, separates the core of the dome from the hanging wall of the unmetamorphosed to low-grade metamorphosed Middle-Late Proterozoic Changcheng System and the Mesozoic sedimentary and volcanic rocks. Three synkinematic biotites from the Waziyu ductile shear zone give a $^{40}Ar/^{39}Ar$ plateau age span of 130–116 Ma and thus establish its main activity period as Early Cretaceous [31, 32]. Zircon U-Pb geochronology indicates that three representative batholiths (Lüshan, Jianlazishan, and Hengshan) at the core were mainly emplaced during Middle-to-Late Jurassic times, with ages of 164, 153, and 153 Ma, respectively [34]. Four other mineral samples from the core and the eastern flank give $^{40}Ar/^{39}Ar$ ages of 133 Ma (muscovite), 140 Ma (biotite), 156 Ma (K-feldspar), and 165 Ma (hornblende), respectively, (unpublished data), indicting the extensional exhumation of the metamorphic core complex range from east to west.

As a case with the Basin and Range Province of the Western United States, the isolated metamorphic core complexes within the Yanshan area, such as the Yunmengshan, Kalaqin, and Yiwulüshan metamorphic core complexes, represent the high-strain extension regions within a broad region of more distributed extensional deformation during Early Cretaceous [25–27, 30, 33, 35–38]. Their formation are contributes to gravitational collapse of an orogenically thickened crust that was facilitated by thermal weakening due to localized deep-seated plutonism and paleo-Pacific plate boundary reorganization [25, 31, 32].

DISCUSSION AND CONCLUSION

The statistics indicate that the main Mesozoic magmatic, metamorphic, and tectonic events on the Dabie, Luxi, Jiaodong, and Yanshan areas of the eastern NCB from south to north are consistent in time and genesis (Table3, [39]). In the eastern NCB, the nearly S-N compression in Early Mesozoic dominated the formation of a series of thrusts trending almost E-W. In Cretaceous, the extension dominated the formation of a series of extensional detachment structures and metamorphic core complex. Many scholars had studied the tectonic setting of the tectonic regime inversion in eastern NCB and gotten different opinions [7, 15, 40–43]. On the base of our studies and former researches, this paper discusses this issue briefly.

Table 3: Time-range index of main Mesozoic structure events in the eastern North China Block [39].

Structure belt	Dabie			West Shandong	Yanshan	
Rock type	(Ultra) mafic rocks	Granite/shear zone	Metamorphic rock (migmatite)	Intermediate-mafic rock	Intrusive	Shear zone
Isotopic age of compression (Ma)	220–240	190	230–240	230–250	165–180	~150 ~219
Isotopic age of extension (Ma)	120–130	~120	117–130	115–130	111–130	116–140
Structural index of compression	The oldest deposit in foreland molasse basin is Mid-Triassic and the youngest one is upper Jurassic			γ_5^3 undeformed	J_2 or J_3 (Tiaojishan group) deformed	
Structural index of extension		J_3, γ_5^3 deformed	J_3 or K_1 deformed	K_1 deformed		

The Early Mesozoic Compression Setting

In the south of the NCB, the Mesozoic compressions occurred in Early-Middle Triassic-early Late Jurassic, such as the Dabie and Luxi areas, but, in the north, occurred a little later, such as the Yanshan area which experienced two periods: in the Late Triassic and the Late Jurassic.

The collision between the Yangtze block and the NCB began at Middle-Late Triassic, the Dabie-Sulu ultrahigh pressure metamorphic belt was formed at 242–224 Ma and cooled down to 300°C at 180 Ma [8, 9], which is approximately consistent with the early deformation age of ductile shear zone in Dabie-Sulu orogenic belt (Ar-Ar age of ~190 Ma, Tables 1 and 2, [11, 18]). Perhaps it was the geological record of orogenic relaxation at about 180–190 Ma, indicating the ending of collision orogenesis. The visual polar migration curve of NCB and Yangtze Block became consistency at 159 Ma [10]. In general, the collision between the NCB and the Yangtze Block might cause the nearly S-N strong compression thrusting in the southern NCB in Middle Triassic-Middle Jurassic (240 Ma–180 Ma) [20, 44].

The compressive shortening in the northern NCB was mainly affected by the collisions of the Mongolia block with the NCB and the Siberia block with the North China-Mongolia block. The collision time between the Mongolia block and the NCB was at about 250 Ma–200 Ma. This collision caused nearly S-N compression and formed a series of E-W striking thrust nappes of 250–200 Ma in Yanshan area, such as the thrust-nappes thrusting southward at 219 Ma in Yiwulűshan. The Mongolia-Okhotsk paleoocean gradually closed eastward in Early-Middle Jurassic (190–170 Ma) and caused collision between the Siberia block and the North China-Mongolia block. The collision might keep on until Late Jurassic (160–150 Ma) and even later [42, 45–48], which formed the Late Jurassic thrust-nappes thrusting southward in the Yanshan tectonic belt, such as the thrust nappes of the Cambrian overriding the Tiaojishan formation ($J_{2\text{-}3}$) in the Huaibei town of the Huairou area and the Gubeikou-Pingquan thrust fault in northern Beijing. During the relaxation period of compression, however, episodic extensional stress fields resulted in a series of Jurassic volcanic rocks.

Late Mesozoic Extensional Tectonic Setting

In the Late Mesozoic, the tectonic regime was inversed into extensional setting in the NCB. The extensions in the shallow crust were mainly displayed by extensional detachment faults and metamorphic core complex. The deformation age of extensional detachment shear zones in Yanshan tectonic belt is 140–100 Ma with a peak period of 120–110 Ma, while in the Dabie and the Luxi area it is 130–110 Ma. The intrusion ages of the magmatic rocks in the metamorphic core complex mostly are 150–110 Ma, only a few reached 170–160 Ma, a little earlier than the detachment shear deformation on the whole. Though some areas such as Yanshan tectonic belt still show compression in shallow crust where thrust napping was developing, underplating and delamination had developed deeper.

It usually takes about tens to hundreds million years for deep underplating and delamination to exhibit extensional detachment structure in the shallow crust. Chronology and deformed rock strain data (such as the Dabie extensional detachment belt (Figure 3)) showed that magma intrusion directly caused extensional detachment. Some geologists related the extension to the NNW or NW subduction of the west Pacific plate into the Eurasia plate. Many evidences indicate that East-Asia circum-Pacific active continental margin began at Middle Jurassic (180 Ma), peaking at Late Jurassic-Early Cretaceous, equivalent to the B episode of the movement in the Yanshan area [42].

In Jurassic, the NCB was acted by three plates, such as the Yangtze, Siberian, and west Pacific plates, and with the orogenesis going on, the crust gradually thickened, potential energy difference increased, and the main compression stress gradually changed from horizontal to vertical. Especially with the thickening of the NCB crust, the lithosphere became unbalanced and the asthenosphere upwelled. Lithosphere underplating and delamination could cause lithosphere rebound; then partial melting, magma intrusion and volcanism occurred, which resulted in the uplift extension detachment of the crust. But whether the underplating and delamination in the eastern North China, beginning at Late Jurassic-Cretaceous, was related to the west Pacific plate subduction still needs to be further studied.

Some researchers suggested that the Late Mesozoic extensional tectonic setting in the eastern North China were mainly resulted from mantle branch structures [3–5, 49, 50]. This is a good way to interpret the Late Mesozoic thermal dome-extensional tectonic setting of the eastern North China According to the study of the PGE distributions, it was suggested that the mantle was rich in PGE at 120 Ma, which is apparently different from the previous PGE-deplete mantle. Therefore, we speculate that the materials from the earth core was added to the Late Mesozoic mantle (at least 8‰, [15]), which supported the opinion of mantle branch structure to some degree.

In the Jiaodong peninsula, however, the sinistral strike-slip compressions dominated in Late Mesozoic, forming a series of thrust-nappes thrusting northward, and the deformation age of ~120 Ma (Table 2, [20]). The sinistral strike slipping of Jiaodong and the Tan-Lu faults could be related to the NE-ward subduction of the west Pacific plate. Therefore, sinistral strike-slip-compression stress field was generated in the east front of the Tan-Lu Faults zone, resulting in the sinistral strike-slip thrust faults and the clockwise rotation (at about 25°) of the blocks bounded by the thrust faults in the Jiaodong peninsula, similar to the Korea peninsula. Then deep structures could not be consistent with shallow ones and deep underplating and delamination; even magma intrusion and volcanism could have happened, but they were not immediately displayed in the shallow. It also suggested that the Tan-Lu Faults zone could be an important tectonic boundary, which caused significant difference in Mesozoic tectonic setting on the both sides of the Tan-Lu Faults.

ACKNOWLEDGMENTS

The authors would like to thank Professor Jiliang Li, Professor Wenjiao Xiao, Professor Mingguo Zhai, and Professor Zhihong Wang from Institute of Geology and Geophysics, Chinese Academy of Sciences, and Dr. Tianshan Gao from university of Science and Technology of China for their kind help in field work and discussion. This work has been supported by the following grants: the National Natural Science Foundation of China (Grants 41030422; 90714003).

REFERENCES

1. S. Hu, Y. Zao, Z. Hu, J. Guo, and B. Xu, "Evolution and development of tectonics and magmatism at the active continental margin of east China (E 106°) during Mesozoic and Cenozoic," Acta Petrologica Sinica, vol. 10, no. 4, pp. 370–381, 1994 (Chinese). . · .
2. J. S. Ren, B. G. Niu, and J. Zheng, "Tectonic frame and geodynamic evolution of eastern China,"Geological Research, vol. 29, no. 30, pp. 43–55, 1997 (Chinese). .
3. J. F. Deng, H. L. Zhao, and X. X. Mo, Continental Roots-Plume Tectonics of China, Beijing, China, Geological Publishing House, 1996.
4. J. Deng, S. Su, Y. Niu et al., "A possible model for the lithospheric thinning of North China Craton: evidence from the Yanshanian (Jura-Cretaceous) magmatism and tectonism," Lithos, vol. 96, no. 1-2, pp. 22–35, 2007. · . ·
5. S. V. Lysak, "Thermal history, geodynamics, and current thermal activity of lithosphere in China,"Russian Geology and Geophysics, vol. 50, no. 9, pp. 815–825, 2009. · . ·
6. M. Zhai, J. Yang, and W. Liu, "The large-scale metallogenesis and large-scale cluster of gold-deposits in eastern Shandong, China," Science in China D, vol. 44, no. 8, pp. 758–768, 2001. . ·
7. M. Zhai, R. Zhu, J. Liu et al., "Time range of Mesozoic tectonic regime inversion in eastern North China Block," Science in China D, vol. 47, no. 2, pp. 151–159, 2004. · . ·
8. B. R. Hacker, L. Ratschbacher, L. Webb, T. Ireland, D. Walker, and D. Shuwen, "U/Pb zircon ages constrain the architecture of the ultrahigh-pressure Qinling-Dabie Orogen, China," Earth and Planetary Science Letters, vol. 161, no. 1—4, pp. 215–230, 1998. · . ·
9. L. Ratschbacher, B. R. Hacker, A. Calvert et al., "Tectonics of the Qinling (Central China): tectonostratigraphy, geochronology, and deformation history," Tectonophysics, vol. 366, no. 1-2, pp. 1–53, 2003. · . ·
10. S. A. Gilder, P. H. Leloup, V. Courtillot et al., "Tectonic evolution of the Tancheng-Lujiang (Tan-Lu) fault via Middle Triassic to Early Cenozoic paleomagnetic data," Journal of Geophysical Research B, vol. 104, no. 7, pp. 15365–15390, 1999. . ·
11. Q. Hou, Q. Liu, J. Li, and H. Zhang, "Late Mesozoic shear zones and its chronology in the Dabie Mountains, Central China," Scientia Geologica Sinica, vol. 42, no. 1, pp. 114–123, 2007. . ·
12. S. W. Dong, J. M. Hu, S. Z. Li et al., "The Jurassic

deformation in the Dabie Mountains and its tectonic significances," Acta Petrologica Sinica, vol. 21, no. 4, pp. 1189–1194, 2005. .

13. S. Suo, Z. Zhong, and Z. You, "Extensional deformation of post ultrahigh-pressure metamorphism and exhumation process of ultrahigh-pressure metamorphic rocks in the Dabie massif, China," Science in China D, vol. 43, no. 3, pp. 225–236, 2000. . ·

14. Q. Liu, Study on the distribution of Platinum group elements in Dabie (ultra-) mafic rocks and Fuxin volcanic rocks [Ph.D. dissertation], Graduate University of the Chinese Academy of Sciences, Beijing, China, 2005.

15. Q. Liu, Q. L. Hou, X. H. Zhou, and L. W. Xie, "The distribution of platinum-group elements in gabbros from Zhujiapu, Dabie orogen," Acta Petrologica Sinica, vol. 21, no. 1, pp. 227–239, 2005. . ·

16. M. Zhai, J. Guo, Q. Wang, K. Ye, B. Cong, and W. Liu, "Division of petrological - tectonic units in the Northern Sulu ultra high pressure zone: an example of thick—skin thrust of crystalline units," Scientia Geologica Sinica, vol. 35, no. 1, pp. 16–26, 2000 (Chinese). . ·

17. M. Yang and G. Lü, The Geology-Geochemistry of Gold Deposits of the Greenstone Belt in Jiaodong District, China, Geological Publishing House, Beijing, China, 1996.

18. H. Y. Zhang, Q. L. Hou, and D. Y. Cao, "Study of thrust and nappe tectonics in the eastern Jiaodong Peninsula, China," Science in China D, vol. 50, no. 2, pp. 161–171, 2007. · . ·

19. X. X. Zhao, R. S. Coe, H. K. Chang et al., "Clockwise rotations recorded in Early Cretaceous rocks of South Korea: Implications for tectonic affinity between the Korea Peninsula and North China,"Geophysical Journal International, vol. 139, no. 2, pp. 447–463, 1999. .

20. H. Y. Zhang, Q. L. Hou, and D. Y. Cao, "Tectono-chronologic constraints on a Mesozoic slip and thrust belt in the eastern Jiaodong Peninsula," Science in China D, vol. 50, no. 1, pp. 25–32, 2007. · . ·

21. C. Zhang, G. Wu, D. Xu, G. Wang, and W. Sun, "Mesozoic tectonic framework and evolution in the central segment of the intraplate Yanshan orogenic belt," Geological Bulletin of China, vol. 23, no. 9-10, pp. 864–875, 2004 (Chinese). .

22. J. Hu, X. Liu, Y. Zhao, G. Xu, J. Liu, and S. Zhang, "On Yanshan intraplate orogene: an example from Taiyanggou area, Lingyuan, Western Liaoning Province, Northeast China," Earth Science Frontiers, vol. 11, no. 3, pp. 255–271, 2004 (Chinese). .

23. W. H. Wong, "Crustal movements and igneous activities in eastern China since Mesozoic time," Bulletin of the Geological Society of China, vol. 6, no. 1, pp. 9–37, 1927. · .

24. X. Zhang, T. Li, and Z. Pu, "A40r/A39r thermochronology of two ductile shear zones from Yiwulüshan, West Liaoning Region: age constraints on the Mesozoic tectonic events," Chinese Science Bulletin, vol. 47, no. 13, pp. 1113–1118, 2002. · . ·

25. G. A. Davis, Y. Zheng, C. Wang, B. J. Darby, C. Zhang, and G. E. Gehrels, "Mesozoic tectonic evolution of the Yanshan fold and thrust belt, with emphasis on Hebei and Liaoning provinces, northern China," inPaleozoic and Mesozoic Tectonic Evolution of Central and Eastern Asia: From Continental Assembly to Intracontinental Deformation, M. S. Hendrix and G. A. Davis, Eds., vol. 194 of Geological Society of America Memoir, pp. 171–197, 2001. .

26. G. A. Davis, X. Qian, Y. Zheng et al., "The Huairou (Shuiyu) ductile shear zone, Yunmengshan Mts.," inProceedings of the 30th international Geological Congress Field Trip Guide, Beijing Geological Publishing House, Beijing, China, 1996.

27. G. A. Davis, X. Qian, Y. Zheng et al., "Mesozoic deformation and plutonism in the Yunmengshan: a Chinese metamorphic core complex north of Beijing, China," in The Tectonic Evolution of Asia, A. Yin and T. M. Harrison, Eds., pp. 253–280, Cambridge University Press, Cambridge, UK, 1996. .

28. G. A. Davis, Y. Zheng, C. Wang, B. J. Darby, C. Zhang, and G. E. Gehrels, "Geometry and geochronology of Yanshan belt tectonics," in Proceedings of the 100th Anniversary Celebration of Peking University: Collected Works of International Symposium on Geological Science, pp. 275–292, Department of Geology, Peking University, Beijing, China, 1998.

29. B. F. Han, Y. Zheng, J. Gan, and Z. Chang, "The Louzidian normal fault near Chifeng, inner Mongolia: master fault of a quasi-metamorphic core complex," International Geology Review, vol. 43, no. 3, pp. 254–264, 2001. . ·

30. J. A. Shao, L. Q. Zhang, W. Jia, and P. Y. Wang, "Harkin metamorphic core complex in Inner Mongolia and its upwelling mechanism," Acta Petrologica Sinica, vol. 17, no. 2, pp. 283–290, 2001 (Chinese). . ·

31. X. Zhang, H. Wang, and Y. Ma, "A40r/A39r age constraints on two NNE-trending ductile shear zones, Yanshan intraplate Orogen, North China Craton," International Geology Review, vol. 45, no. 10, pp. 936–947, 2003. . ·

32. B. J. Darby, G. A. Davis, X. Zhang, F. Wu, S. Wild, and J. Yang, "The newly discovered Wuziyu metamorphic core complex, Yiwulü Shan, western Liaoning province, North China," Earth Science Frontiers, vol. 11, no. 3, pp. 145–155, 2004. .

33. Y. Ma, S. Cui, G. Wu et al., "The structural feature of metamorphic core complex in Yiwulüshan mountains, western Liaoning," Acta Geoscientia Sinica, vol. 20, pp. 385–391, 1999 (Chinese). .

34. X. H. Zhang, Q. Mao, H. F. Zhang, and S. A. Wilde, "A Jurassic peraluminous leucogranite from Yiwulüshan, western Liaoning, North China craton: age, origin and tectonic significance," Geological Magazine, vol. 145, no. 3, pp. 305–320, 2008. .

35. G. A. Davis, B. J. Darby, Y. Zheng, and T. L. Spell, "geometric and temporal evolution of an extensional detachment fault, Hohhot metamorphic core complex, Inner Mongolia, China," Geology, vol. 30, pp. 1003–1006, 2002. .

36. L. E. Webb, S. A. Graham, C. L. Johnson, G. Badarch,

and M. S. Hendrix, "Occurrence, age, and implications of the Yagan-Onch Hayrhan metamorphic core complex, southern Mongolia," Geology, vol. 27, no. 2, pp. 143–146, 1999. . ·

37. C. L. Johnson, L. E. Webb, S. A. Graham, M. S. Hendrix, and G. Badarch, "Sedimentary and structural records of Late Mesozoic high-strain extension and strain partitioning, east Gobi basin, Southern Mongolia," in Paleozoic and Mesozoic Tectonic Evolution of Central and Eastern Asia: From Continental Assembly to Intracontinental Deformation, M. S. Hendrix and G. A. Davis, Eds., vol. 194 of Geological Society of America Memoir, pp. 413–433, 2001. .

38. J. Ren, K. Tamaki, S. Li, and Z. Junxia, "Late Mesozoic and Cenozoic rifting and its dynamic setting in Eastern China and adjacent areas," Tectonophysics, vol. 344, no. 3-4, pp. 175–205, 2002. · . ·

39. Q. L. Hou, Y. D. Wu, F. Y. Wu, M. G. Zhai, J. H. Guo, and Z. Li, "Possible tectonic manifestations of the Dabie-Sulu orogenic belt on the Korean Peninsula," Geological Bulletin of China, vol. 27, no. 10, pp. 1659–1666, 2008. .

40. M. Zhai, R. Zhu, J. Liu, et al., "Time range of Mesozoic tectonic regime inversion in eastern North China Block," Science in China D, vol. 33, no. 10, pp. 913–920, 2003. .

41. Y. Zhao, Z. Yang, and X. Ma, "Geotectonic transition from Paleoasian system and Paleotethyan system to Paleopacific active continental margin in eastern Asia," Scientia Geologica Sinica, vol. 29, no. 2, pp. 105–119, 1994. . ·

42. Y. Zhao, G. Xu, S.-H. Zhang, et al., "Yanshanian movement and conversion oftectonic regimes in East Asia," Earth Sciences Frontiers, vol. 11, no. 3, pp. 319–328, 2004. .

43. Z. Yadong, G. A. Davis, W. Cong, B. J. Darby, and Z. Changhou, "Major Mesozoic tectonic events in the Yanshan belt and the plate tectonic setting," Acta Geologica Sinica, vol. 74, no. 4, pp. 289–302, 2000. . ·

44. R. Y. Zhang, J. G. Liou, and W. G. Ernst, "The Dabie-Sulu continental collision zone: a comprehensive review," Gondwana Research, vol. 16, no. 1, pp. 1–26, 2009. · . ·

45. A. M. Ziegler, P. M. Ree, D. B. Rowley, A. Bekker, Q. Li, and M. Hulver, "Mesozoic assembly of Asia: constraints from fossil floras. Tectonics, and paleomagnetism," in Tectonic Evolution of Asia, T. M. A. Harrison, Ed., pp. 371–400, Cambridge University Press, Cambridge, UK, 1996. .

46. A. Yin and S. Nie, "A Phanerozoic palinspastic reconstruction of China and its neighboring regions," inTectonic Evolution of Asia, A. Yin and T. M. Harrison, Eds., pp. 442–485, Cambridge University Press, Cambridge, UK, 1996. .

47. L. Zonenshain, M. Kuzmin, and L. Natapov, Geology of RSSR: A Plate Tectonic Synthesis, vol. 21 ofGeodynamic Series, American Geophysical Union, Washington, DC, USA, 1990.

48. Q. R. Meng, "What drove late Mesozoic extension of the northern China-Mongolia tract?"Tectonophysics, vol. 369, no. 3-4, pp. 155–174, 2003. · . ·

49. S. Niu, Z. Hou, and A. Sun, "The anti-gtavityu migration of metallogenic fluid from core and mantle,"Earth Science Frontiers, vol. 8, no. 3, pp. 95–101, 2001. .

50. J. Zheng, S. Y. O›Reilly, W. L. Griffin, F. Lu, M. Zhang, and N. J. Pearson, "Relict refractory mantle beneath the eastern North China block: significance for lithosphere evolution," Lithos, vol. 57, no. 1, pp. 43–66, 2001. · . ·

CHAPTER 19

ACTIVE TECTONICS ALONG THE SUBMARINE SLOPE OF SOUTH-EASTERN SICILY AND THE SOURCE OF THE 11 JANUARY 1693 EARTHQUAKE AND TSUNAMI

A. Argnani, A. Armigliato, G. Pagnoni, F. Zaniboni, S. Tinti, and C. Bonazzi

INTRODUCTION

The steep slope that bounds eastern Sicily and connect the shelf to the deep Ionian basin, has long been recognised to be part of a major tectonic feature of the central Mediterranean, the Malta Escarpment (Fig. 1; Scandone et al., 1981); its age, as well as the age of the adjacent Ionian Basin, remains controversial, ranging from Late Triassic to Early Cretaceous (e.g. Argnani and Bonazzi, 2005 and references therein). The south-eastern region of Sicily is currently the site of significant seismic activity that has been recorded in recent instrumental catalogues (Chiarabba et al., 2005; Musumeci et al., 2005) and historical chronicles (Boschi et al., 1995a and b). In particular, it has been affected by large historical earthquakes, one of which, the 11 January 1693 earthquake, has been considered as the largest earthquake in the history of Italy (Mw = 7.4; CPTI Working Group, 2004). This earthquake was accompanied by a large tsunami (estimated tsunami magnitude 2.3, Tinti et al., 2004), suggesting a source located in the near offshore. However, the identification of the tectonic structure responsible for the 1693 earthquake is still a matter of debate. Some authors proposed as source of the 1693 earthquake an active fault system identi- fied and mapped in detail off south-east Sicily (Argnani and Bonazzi, 2005). On the basis of macroseismic intensity field, some authors assumed an earthquake source located onshore (e.g. DISS Working Group, 2010). The suggested onshore faults, however, are still poorly documented, and have negligible tsunamigenic potential. For example, the two faults indicated by the DISS Working Group (2010), as possibly responsible for the 11 January 1693 earthquake (ITIS106 and ITIS074), are almost completely located inland, have magnitudes (6.0 and 6.6, respectively) significantly smaller than what reported in the CPTI catalogue and are then unable to produce any significant seafloor permanent deformation required for tsunami generation.

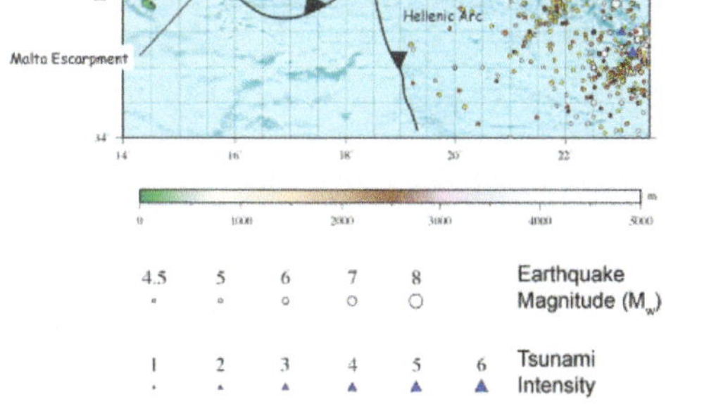

Figure 1. Seismicity and main geodynamic features of the Ionian region with the main geological features along the coast of Sicily and Calabria. The historical earthquakes have been compiled by suitably combining and homogenizing different national catalogues (see in particular CPTI Working Group, 2004; Papaioannou et al., 2000;

Citation: Argnani, A., Armigliato, A., Pagnoni, G., Zaniboni, F., Tinti, S., and Bonazzi, C.: Active tectonics along the submarine slope of south-eastern Sicily and the source of the 11 January 1693 earthquake and tsunami, Nat. Hazards Earth Syst. Sci., 12, 1311-1319, doi:10.5194/nhess-12-1311-2012, 2012.

Papazachos et al., 2007). Historical tsunamis are from Tinti et al. (2001b, 2004). The contour lines of maximum felt macroseismic intensity (over IX) for south-eastern Sicily are also shown (after Boschi et al., 1995a).

Making use of numerical modelling of tsunamigenic sources, this paper aims at confirming that (i) the active faults located off south-east Sicily, whose geometry was obtained as the result of seismic profiles interpretation, can account for the 1693 tsunami, as reported in historical chronicles, and (ii) a large submarine landslide, which might be related to the same fault activity, can be considered a possible additional factor for the generation of the tsunami.

GEOLOGICAL SETTING OF THE EASTERN SICILY SLOPE

The region encompassing south-western Calabria and eastern Sicily is known for being tectonically active (e.g. Argnani et al., 2009a and references therein). The disastrous historical earthquakes that struck eastern Sicily several times, and in particular Catania and Messina, as well as southern Calabria, were quite likely originated by marine tectonic structures (Argnani and Bonazzi, 2005). Moreover, the largest tsunamis that affected the Italian coasts originated from this marine region. Two examples are the 28 December 1908 tsunami, which hit north-eastern Sicily and southern Calabria with wave heights up to 13 m, and the 11 January 1693 tsunami that affected the whole coast of eastern Sicily and was felt as far as Malta, to the south, and the Aeolian Island, to the north (Tinti et al., 2004). The January 1693 event is a case of particular interest as the location of the source fault is still highly debated. Different onshore faults were proposed by various authors (D'Addezio and Valensise, 1991; Sirovich and Pettenati, 1999, 2001; DISS Working Group, 2010), most of which, however, have limited geological evidence. Further, the intensity map for this earthquake shows that the highest intensity values (X–XI and XI according to the EMS 98 scale) were found for towns located on the coast or very close to the coast (see Barbano and Rigano, 2001), suggesting that the earthquake source was located offshore. Moreover, the large tsunami that was associated with this earthquake cannot be explained by the proposed onshore faults without invoking a different source for the tsunami: an earthquake-triggered submarine landslide. On the other hand, recent geophysical surveys have shown that active faults do exist off eastern Sicily (Bianca et al., 1999; Argnani and Bonazzi, 2005) and furthermore that slope failures occurred offshore (Argnani and Bonazzi, 2005). The relevant characteristics of these active tectonic structures are described below, together with the description of a large submarine slide that has been identified off Augusta (hereafter named Augusta slide).

Faults

The segment of the slope extending north of Siracusa is characterised by the presence of NNW–SSE-trending, eastdipping extensional faults located along the morphological escarpment and a few km east of it (Fig. 2). Active faulting is well documented by seismic profiles that show half grabens filled by sediments with growth geometry and fault surfaces, producing morphological scarps at the sea floor (Argnani and Bonazzi, 2005). In addition, multibeam bathymetry data acquired in the western Ionian sea (Marani et al., 2004) show linear features correlating with seafloor topography, further supporting the presence of active faults cutting the sea floor. Profile MESC 08 shows a half-graben basin, filled with sediments up to 1 s (TWT) thick, and bounded to the west by two extensional faults, with the western one being an active splay of the master fault (Fig. 3). Wedging of reflections, identifying growth strata, can be observed within the graben; the age of the infilling sediments has been inferred to cover the whole Quaternary on the basis of seismo-stratigraphic relationship (Argnani and Bonazzi, 2005). On a depth migrated version of this profile the dip angle of the faults is about 30° . Although such dip is small for a typical normal fault, it is worth mentioning that the normal fault inferred to be responsible for the 1908 Messina earthquake has a similar small dip (e.g. Valensise and Pantosti, 1992; Pino et al. 2009).

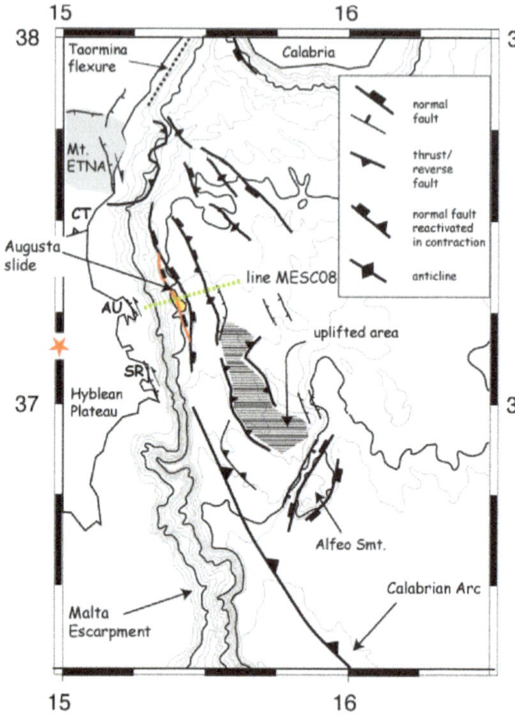

Figure 2. Bathymetry of the eastern Sicily offshore with the simplified structural map based on the MESC seismic survey (Argnani and Bonazzi, 2005). Onshore data are only schematic and are mainly taken from Lanzafame and Bousquet (1997) and Adam et al. (2000). The active fault is marked in red. The trace of seismic profile MESC 08 is indicated as a thick dashed green line, whereas the orange field represents the extent of the Augusta slide. The red star indicates the approximated location of the onshore 11 January 1693 earthquake epicenter suggested by CPTI Working Group (2004). CT, AU, and SR indicate the location of the cities of Catania, Augusta and Siracusa, respectively

The active tectonic features in the segment of the Malta Escarpment between Catania and the Alfeo Seamount present a remarkable complexity. Nevertheless, the extensional faults described above appear as the major feature. In particular, the normal fault splay that cuts the sea floor (Fig. 2) is long enough (about 30 km) to generate large earthquakes and tsunamis, and can be taken as a likely source for a 1693- type earthquake.

The map trace and the attitude of the active normal fault splay have been used as input for the tsunami numerical modelling.

Landslide

A large-scale submarine landslide has been imaged on seismic profiles at the foot of the eastern Sicily slope, along the Malta escarpment, at a water depth around 2000 m. This slide occurred at the base of the scarp of the active fault and affected the footwall sediments (Figs. 3 and 4). Seismic surveys allowed mapping of this slide, which covers an area of about 40 km^2 and can be as thick as 250 m, with an average thickness of about 100 m. From the extent of the slide scar a displacement of ca. 200 m has been estimated along the basal decollement (Fig. 4). The geometry at the toe of the slide is difficult to resolve, but it seems that ductile deformation is partly absorbing the downslope displacement. The estimated volume is about 4.8 km^3. Direct dating of the slide is difficult in principle and it had not been attempted during the marine survey. Nevertheless, the lack of visible sediments overlying the slide scar suggests that the slide may be a recent feature.

TSUNAMIGENIC POTENTIAL ASSESSMENT THROUGH NUMERICAL MODELLING

Numerical modelling was applied to determine the capability of both the tectonic and the gravitational slope failure sources described in the previous section to generate tsunami waves along the coasts of eastern Sicily, and to discuss whether they can be proposed as responsible for the 11 January 1693 historical tsunami

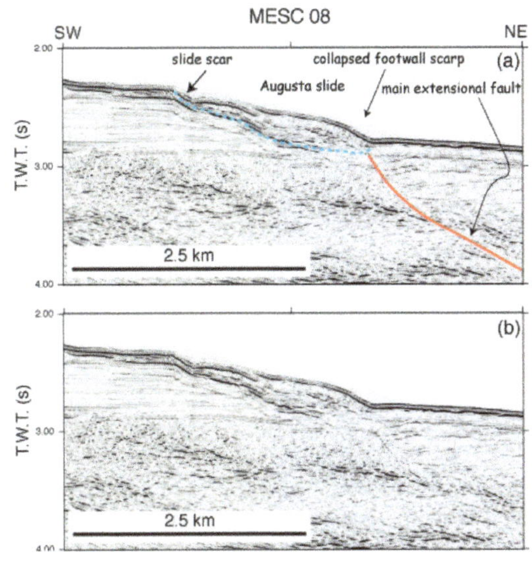

Figure 4. Detail of seismic profile MESC 08 (Fig. 3). The interpreted seismic line (a) shows the collapsed sediments in the footwall scarp of the active splay of the major normal fault. The body of the slide has a volume of ca. 5 km3 and has slid about 200 m. Vertical exaggeration ca. 1.7. (b) Non-interpreted seismic line.

From the numerical point of view, we treated separately the simulation of the generation and propagation of the tsunami waves.

Tsunami generation by the earthquake

In the case of the earthquake source, we used the approximation widely adopted in tsunami studies, consisting of taking the tsunami initial condition to coincide with the vertical component of the coseismic displacement of the sea-bottom caused by the earthquake. In turn, the coseismic deformation field is calculated by means of the Okada (1992) formulas, using an idealized figuration of a rectangular fault buried in a homogenous, isotropic and elastic half-space. A key step consists of determining the fault parameters to be used in the computation. The MESC seismic survey (Argnani and Bonazzi, 2005) puts constraints on the length, strike and dip of the fault. To represent approximately the irregular shape of the fault along strike, we adopted four different subfaults with different strike angles (varying in the range 330°– 360°) and different lengths (5.8 km to 10.2 km, for a total of 28.5 km), but sharing the same average dip angle (28°) and along-dip width (16.5 km). The rake angle of 270° was imposed on the basis of the tectonics of the Malta Escarpment, which is dominated by extension. Moreover, since the observed fault cuts the seafloor, the depth of the simulated fault's upper border was set to 0.5 km. The slip on the fault was estimated from the seismic moment M0 and the total fault area A through the law.

$$M_0 = \mu A \bar{u}$$

where $\bar{u}$ is the average slip and μ is the rigidity of the Earth's crust in the source zone. Given a reference rigidity value of 30 GPa, since the area of the fault is of limited extension, we need quite a high slip to obtain a seismic moment which approximates the one postulated for the 1693 earthquake: with $\bar{u}$ = 5 m we obtain M0 = 7×1019 Nm, corresponding approximately to a magnitude M = 7.2. The resulting vertical seafloor deformation is shown in Fig. 5. Due to the low-angle geometry of the fault, the deformation field is characterised by a pronounced subsidence facing the coastline and by a smaller uplift zone offshore. This pattern is adopted as the initial tsunami condition.

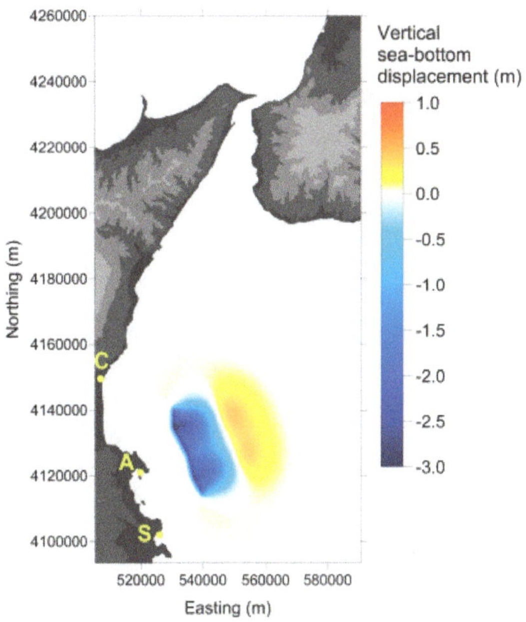

Figure 5. Vertical coseismic sea-bottom displacement, coinciding with the initial sea surface displacement, computed for the main active extensional fault. "C", "A" and "S" indicate the position of Catania, Augusta and Siracusa, respectively.

TSUNAMI GENERATION BY THE LANDSLIDE

The landslide simulation was carried out through the numerical code UBO-BLOCK2, developed by the Tsunami Research Team of the University of Bologna. The model is Lagrangian: the sliding mass is split into a 2-D matrix of interacting, deforming, constant-volume blocks. The code has been extensively applied to several cases of tsunamigenic landslides (e.g. Tinti et al., 2006). The simulation performed in this study takes advantage of the results already presented by Armigliato et al. (2007) and by Tonini et al. (2011). We will only recall here that the main parameters used in the landslide simulation involve a total volume of 4.8 km3 , a sliding mass of 12.8×109 kg, a density of 2000 kg m−3 , a maximum thickness of 250 m, and a minimum depth of the scar of 1850 m. The simulation performed with UBO-BLOCK2 reproduces the observation of a very limited landslide run-out distance (approximately 200 m).

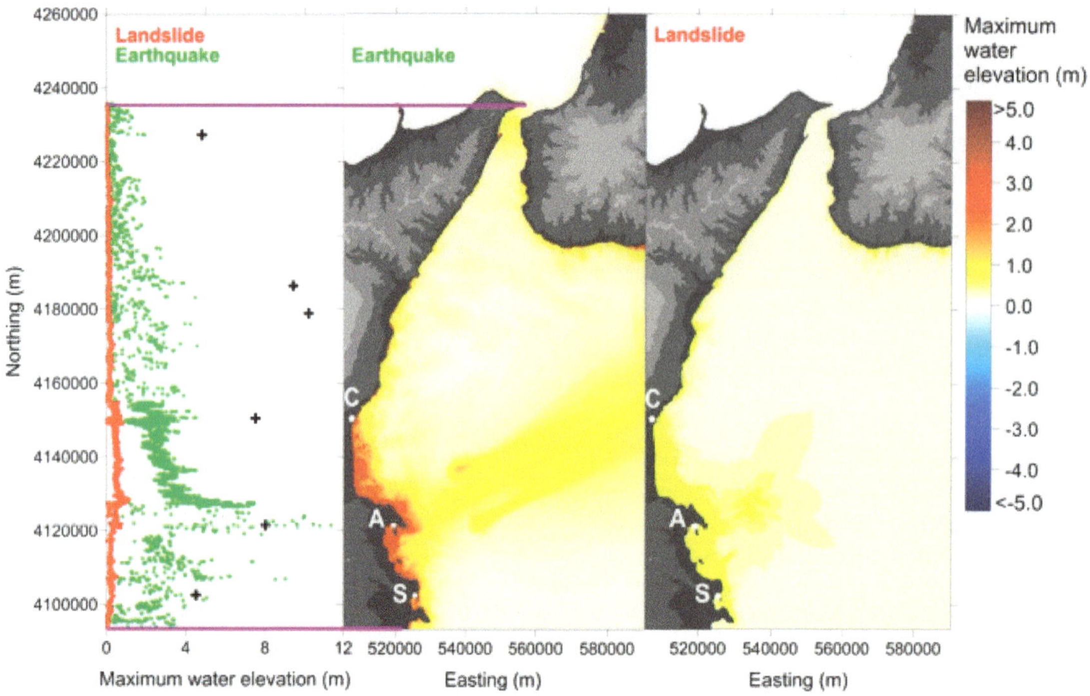

Figure 6. Maximum water elevations computed for the tsunami generated by the active fault (central panel) and by the landslide (right panel). The left panel shows the comparison of the along-shore maximum water elevation distributions. "C", "A" and "S" indicate the position of Catania, Augusta and Siracusa, respectively. The black crosses are the run-up values computed by Gerardi et al. (2008) on the basis of the Hills and Mader (1997) relations.

Tsunami propagation

The finite-differences numerical code UBO-TSUFD was used to simulate the tsunami wave propagation and impact on the coasts. The code, developed and maintained by the Tsunami Research Team of the University of Bologna, implements and solves the Navier-Stokes hydrodynamic equations in the approximation of shallow water, allowing to compute run-up and inundation. In the case of earthquakegenerated tsunamis, the coseismic deformation field is given as input to the code as a static initial condition. In the case of the landslide, the time derivative of the instantaneous sea surface elevation caused by the transit of the landslide on the sea-bottom represents the forcing term. In turn, this derivative is obtained from the time variation of the local thickness of the slide on the seafloor by means of a proper transfer function filtering out the high frequency signals. Further details are given in Tinti et al. (2006). UBO-TSUFD makes use of regularly-spaced grids that, depending on the application, can be nested to obtain fine space resolutions in selected target areas. The set of grids adopted in this study coincides with grids 3 (resolution of 200 m) and 4 (resolution of 40 m) used in the paper by Tonini et al. (2011). The topo-bathymetric data and the procedures used to obtain the computational grids are detailed in that paper.

Results

The historical information on the coastal effects produced by the 11 January 1693 tsunami mainly consists of the geographical distribution of the largest wave elevations along the eastern Sicily coastline and of the polarity of the first arrivals, which was negative in all the coastal places on the east coast of Sicily described in the coeval chronicles (see Piatanesi and Tinti, 1998; Tinti et al., 2004). Figure 6 summarizes and compares the maximum water wave elevation patterns computed for both the earthquakeand the landslide-induced tsunamis over the entire computational domain. The leftmost panel represents the along-coast distribution of the maximum elevations for both generating mechanisms. A first comment regards the relative efficiency in generating tsunami waves of the two sources: the elevation values obtained for the earthquake are much larger than for the landslide. Secondly, the pattern of the field related to the active fault is determined mainly by the position and directivity of the fault itself, and by the local bathymetry and coastal morphology. As a result, the zone between Catania and north of Siracusa is the one experiencing the largest tsunami impact.

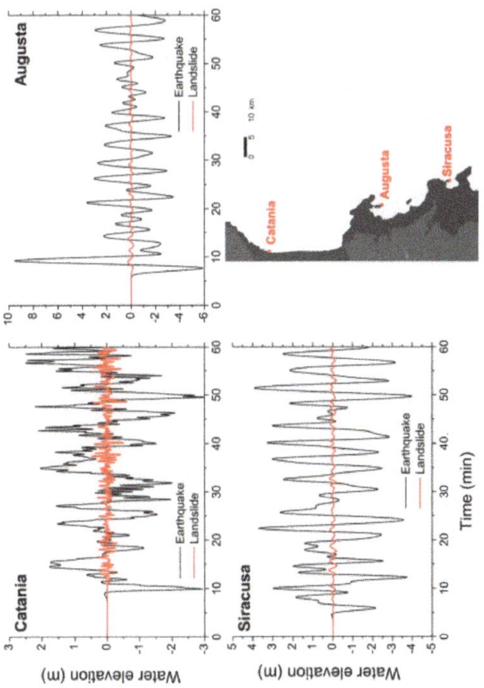

Figure 7. Synthetic marigrams in the cities of Catania, Augusta and Siracusa computed for the active fault and for the landslide tsunamis. The localities are indicated in the bottom, right panel.

In more detail, the leftmost panel of Fig. 6 indicates tsunami elevations larger than 4 m in Catania and the north of Siracusa, while the coast in the vicinity of Augusta is impacted by waves with elevations larger than 10 m. In contrast, the landslide is not able to produce waves larger than 1 m at any coastal location. Due to the landslide location the tsunami effects are the largest between Catania and just north of Augusta. Figure 7 shows the wave elevation time series computed at three different grid points in the areas of Catania, Augusta and Siracusa, respectively. The main result concerns the polarity of the first arrival: in all three virtual stations the first significant signal is undoubtedly negative, in agreement with the historical accounts. Figure 7 provides information on the first tsunami arrival times, which are in the range of 4– 8 min. Although this information is not provided on the coeval document, it is worth mentioning it from an early warning system perspective. Finally, the earthquake and the landslide produce signals with different frequency contents. The tsunami generated by the earthquake shows a typical period around 8 min, while the landslide induces shorter periods, in the order of few minutes. As a final observation, the fact that the virtual signal computed at Catania is more irregular than the ones obtained at Augusta and Siracusa should be attributed to the fact that Catania lies in the grid with the highest spatial resolution (40 m), while Augusta and Siracusa are found in the 200-m resolution grid.

DISCUSSION

The offshore fault that we have modelled, and that is part of the Malta Escarpment, is able to reproduce the tsunami waves, as described in the historical chronicles. On the other hand, the Augusta landslide that has been mapped at the base of the footwall scarp is not able to explain alone the observed tsunami, and in all grid points the computed maximum water elevations are always much smaller than those produced by the earthquake source. The problem judging the "degree of agreement" of the numerical simulations with the historical accounts for an event that occurred in 1693 is not trivial at all. The most important "physical" information we can retrieve from the sources, regards the polarity of the first arrivals, negative along the entire Sicily eastern coast. Assessing run-up values in any location is rather critical. Only two such "quantitative measurements" for the 1693 tsunami are available in Augusta, as reported in Gerardi et al. (2008). In that study, the authors computed run-up values in a number of other localities by adopting the Hills and Mader (1997) relations.

It should be noted that the relationships by Hills and Mader (1997) work well on highly idealised geometries and their application to realistic situations may introduce large uncertainties. One may be tempted to evaluate the matching between the computed values and our numerical results, as we do in the leftmost panel of Fig. 6; but we have to take these comparison with caution, because in Fig. 6 we are comparing two different tsunami metrics, the maximum water elevation (resulting from our modelling) and the run-up estimated by Gerardi et al. (2008). The numerical simulation of the run-up requires detailed information on the actual coastal topography in early 1693, which is not available to the authors. That is why our numerical model does not provide run-up values but elevations of waves that are normally expected to be smaller than run-ups. Nevertheless, it can be appreciated that our simulation reproduces well the one and only actual observation in Augusta.

The matching is satisfactory in Siracusa, while in Catania the run-up modelled by our simulation underestimates that from Gerardi et al. (2008), and the same happens, even in a more pronounced way, towards the north. Other faults have been proposed as responsible for the 11 January 1693 earthquake, either located onshore (e.g. D'Addezio and Valensise 1991; Sirovich and Pettenati, 1999, 2001; DISS Working Group, 2010; Di Bucci et al., 2010), or located offshore and related to the Calabrian Arc subduction fault plane (Gutscher et al., 2006). As already mentioned, the choice of onshore faults as the source for the 11 January 1693 earthquake is mainly based on the distribution of seismic intensities inland (Fig. 1), since the main argument used in support is that a coastal inland fault can explain satisfactorily the severe damage observed on the coast as well as the way damage decays inland.

It should be noted however that the estimates of the macroseismic intensity for an event that dates so far back in time as the January 1693 earthquake might suffer from severe uncertainties, especially because this event was preceded two days before by a strong disastrous foreshock, which contributed significantly to the total damage pattern. On the other hand, an onshore fault alone can hardly explain the tsunami generation and the main features of the observed impact (Tinti et al., 2001a).

One could try to solve the problem by invoking the occurrence of a suitable submarine landslide triggered by the earthquake; following this approach, Billi et al. (2010)

inferred the occurrence of a landslide to the north of Mount Etna that was triggered by a fault located in the Hyblean Plateau. Such a position, however, is probably too far from the inferred seismogenic fault for the sediments to be destabilised by the seismic load induced by the earthquake; moreover, a landslide in that position would not produce the expected tsunami response with maxima between Catania and Siracusa. Finally, and most important, no recent submarine landslides of suitable size have been found in the area north of Mount Etna (see the comments of Argnani et al., 2009b on a tsunamigenic slide supposedly located in the same position). It is also worth stressing that Gerardi et al. (2008), analysing the inundation pattern of the 1693 tsunami along the Sicily coast and using a discriminatory tool introduced by Okal and Synolakis (2004) for near-shore tsunami sources, came to the conclusion that the characteristics of this tsunami were more compatible with an earthquake rather than with a landslide source.

Our results support this conclusion, as we have demonstrated in this work that even a submarine landslide close to the coastal area, where the observed tsunami waves were larger, cannot reproduce the 11 January 1693 tsunami effects. As for offshore sources in relation with the Calabrian Arc, we have to mention that the Calabrian subduction fault surface may be capable of generating tsunamigenic earthquakes of the required magnitude (Gutscher et al., 2006). However, given the large area of the surface rupture and the expected predominant strike of the rupture, the tsunami waves modelled from such a source are impacting strongly against the Calabrian coastline, with heights actually larger than that of the waves along the coast of Sicily. This result is disproved by historical chronicles, which do not report any major tsunami event along the Calabrian coast in January 1693.

The Augusta landslide is located at the base of the fault scarp and straddles the fault plane. In onshore palaeoseismology landslides in the vicinity of an active fault are commonly considered seismically triggered (e.g. McCalpin, 2009). The close spatial relationship between fault and landslide, also in the Augusta case, suggests that this could be the case also in our study area, and the Augusta landslide can be considered as a process contributing to the degradation of an active fault scarp (e.g. Wallace, 1977). Although we cannot date the slide event, as the marine survey was not specifically planned for studying the landslide, no evidence arguing against a recent age has been found though. It is also worth considering that in marine area dating a landslide and correlating it to a seismogenic fault is a rather difficult task, even for recent events. For instance, both a seismogenic fault and a triggered landslide contributed to the July 1998 Papua-New Guinea tsunami (Matsumoto et al., 2003), making it one of the best documented examples of combined tsunami source. However, just from the data collected from the cruise purposely carried out after the July 1998 event (Matsumoto et al., 2003), it would be very difficult to date the landslide, if the earthquake and tsunami data were not pointing out the active source area. In spite of the limited displacement and the rather deep water, the large volume of the Augusta slide is enough to generate a tsunami, but not as large as the 1693 event. In our preferred scenario, this landslide was seismically triggered by the 1693 earthquake and contributed to the fault-generated tsunami.

CONCLUSIONS

The south-eastern region of Sicily has been affected by large historical earthquakes, one of which, occurring on the 11 January 1693, has been considered as the largest earthquake in the history of Italy. This earthquake has also been accompanied by a large tsunami, suggesting a source located in the near offshore. A realistic fault geometry, that has been derived from the interpretation of depth migrated seismic data, has been used for modelling the tsunamigenic source. Previous modelling, in fact, used only loosely constrained and idealized fault geometries (Tinti et al., 2001a; Tinti and Armigliato, 2003). The outcomes of the numerical simulations indicate that the coastal segment suffering the strongest tsunami impact is the one running from Catania to Siracusa, where the first tsunami arrival was negative, in agreement with the historical accounts of the 11 January 1693 tsunami. Recent offshore surveys have also identified a large size submarine slide (almost 5 km3) along the south-eastern Sicily slope. This slide, called Augusta slide, occurring just at the scarp of the active fault identified as the potential earthquake and tsunami source, and affecting the uplifted footwall sediments, has proven to have tsunamigenic potential. The tsunami of the 11 January 1693 can be explained by means of the offshore faulting alone, but a combination of earthquake faulting triggering the Augusta submarine slide is very likely.

ACKNOWLEDGEMENTS

The beginning of this work was partly funded by a GNDT project 2000–2002 “A Study to Evaluate the Hazard Related to Submarine Geological Processes: Earthquake, Tsunamis and Landslides”. It is also acknowledged that part of this study was possible thanks to funds of the European project TRANSFER (Contract no. 037058). The comments of the guest editor D. Pantosti and of the two reviewers, S. Barbano and L. McNeill, greatly improved the paper. The language help given by L. McNeill has been particularly appreciated.

REFERENCES

1. Adam, J., Reuther, C.-D., Grasso, M., and Torelli, L.: Active fault kinematics and crustal stresses along the Ionian margin of the southeastern Sicily, Tectonophysics, 326, 217–239, 2000.
2. Argnani, A.: Evolution of the southern Tyrrhenian slab tear and active tectonics along the western edge of the Tyrrhenian subducted slab, in: Collision and Collapse at the Africa-ArabiaEurasia Subduction Zone, edited by: Van Hinsbergen, D. J. J., Edwards, M. A., and Govers, R., Geol. Soc., Spec. Publ., 311, 193–212, 2009.
3. Argnani, A. and Bonazzi, C.: Malta Escarpment fault zone offshore eastern Sicily: Pliocene-Quaternary tectonic evolution based on new multichannel seismic data, Tectonics, 24, TC4009, doi:10.1029/2004TC001656, 2005.
4. Argnani, A., Brancolini, G., Bonazzi, C., Rovere, M., Accaino, F., Zgur, F., and Lodolo, E.: The results of the Taormina 2006 seismic survey: Possible implications for active tectonics in the Messina Straits, Tectonophysics, 476, 159–169, 2009a.
5. Argnani, A., Chiocci, F. L., Tinti, S., Bosman, A., Lodi,

M. V., Pagnoni, G., and Zaniboni, F.: Comment on "On the cause of the 1908 Messina tsunami, southern Italy" by Andrea Billi et al., Geophys. Res. Lett., 36, L13307, doi:10.1029/2009GL037332, 2009b.

6. Armigliato, A., Tinti, S., Zaniboni, F., Pagnoni, G., and Argnani, A.: New contributions to the debate on the cause of the January 11th, 1693 tsunami in eastern Sicily (Italy): Earthquake or offshore landslide source (or maybe both)?, Eos, Transactions of the American Geophysical Union 88 (52), Fall Meeting Supplement, Abstract S53A–1019, 2007.

7. Barbano, S. and Rigano, R.: Earthquake sources and seismic hazard in southeastern Sicily, Ann. Geofis.-Rome, 44, 723–738, 2001.

8. Bianca, M., Monaco, C., Tortorici, L., and Cernobori, L.: Quaternary normal faulting in southeastern Sicily (Italy): a seismic source for the 1693 large earthquake, Geophys. J. Int., 139, 370– 394, 1999.

9. Billi, A., Minelli, L., Orecchio, B., and Presti, D.: Constraints to the Cause of Three Historical Tsunamis (1908, 1783, and 1693) in the Messina Straits Region, Sicily, Southern Italy, Seismol. Res. Lett., 81, 907–915, doi:10.1785/gssrl.81.6.907, 2010.

10. Boschi, E., Favali, P., Frugoni, F., Scalera, G., and Smriglio, G.: Mappa della massima intensita macrosismica risentita in Italia, ` Stabilimento L. Salomone, Roma, 1995a.

11. Boschi, E., Guidoboni, E., and Mariotti, D.: Seismic effect of the strongest historical earthquakes in the Syracuse area, Ann. Geofis.-Rome, 38, 223–252, 1995b.

12. Chiarabba, C., Jovane, L., and Di Stefano, R.: A new view of Italian seismicity using 20 years of instrumental recordings, Tectonophysics, 395, 251–268, 2005.

13. CPTI Working Group: Catalogo Parametrico dei Terremoti Italiani, versione 2004 (CPTI04), INGV, Bologna, 2004.

14. D'Addezio, G. and Valensise, G: Methodology for the characterization of the responsible seismogenetic structure of the earthquake of December 13, 1990, in: Contributi allo studio del terremoto della Sicilia orientale del 13 Dicembre 1990, Contributions to the earthquake studies of eastern Sicily, December 13, 1990, edited by: Boschi, E. and Basili, A., 115–125, 1991.

15. Di Bucci, D., Burrato, P., Vannoli, P., and Valensise, G.: Tectonic evidence for the ongoing Africa-Eurasia convegence in central Mediterranean foreland areas: A journey among long-lived shear zones, large earthquakes, and elusive fault motions, J. Geophys. Res., 115, B12404, doi:10.1029/2009JB006480, 2010.

16. DISS Working Group: Database of Individual Seismogenic Sources (DISS), Version 3.1.1: A compilation of potential sources for earthquakes larger than M 5.5 in Italy and surrounding areas, available at: http:/diss.rm.ingv.it/diss/, Istituto Nazionale di Geofisica e Vulcanologia, 2010.

17. Gerardi, F., Barbano, M. S., De Martini, P. M., and Pantosti, D.: Discrimination of tsunami sources (earthquake vs. landslide) on the basis of historical data in eastern Sicily and southern Calabria, B. Seismol. Soc. Am., 98, 2795–2805, 2008.

18. Gutscher, M. A., Roger, J., Baptista, M. A., Miranda, J. M., and Tinti, S.: Source of the 1693 Catania earthquake and tsunami (southern Italy): New evidence from tsunami modelling of a locked subduction fault plane, Geophys. Res. Lett., 33, L08309, doi:10.1029/2005GL025442, 2006.

19. Hills, J. G. and Mader, C. L.: Tsunami produced by the impact of small asteroids, Ann. NY Acad. Sci., 882, 381–394, 1997. Lanzafame, G. and Bousquet, J. C.: The Maltese escarpment and its extension from Mt. Etna to the Aeolian Islands (Sicily): Importance and evolution of a lithosphere discontinuity, Acta Vulcanol., 9, 113–120, 1997.

20. Marani, M. P., Gamberi, F., Bortoluzzi, G., Carrara, G., Ligi, M., and Penitenti, D.: Seafloor bathymetry of the Ionian Sea, in: From seafloor to deep mantle: architecture of the Tyrrhenian backarc basin, edited by: Marani, M. P., Gamberi, F., and Bonatti, E., Memorie Descrittive della Carta Geologica D'Italia, Vol. 44, Plate 3, 2004.

21. Matsumoto, T., Tappin, D. R. and SOS Onboard Scientific Party: Possible Coseismic Large-scale Landslide off the Northern Coast of Papua New Guinea in July 1998:

22. Geophysical and Geological Results from SOS Cruises, Pure Appl. Geophys., 160, 1923– 1943, 2003.

23. McCalpin, J. P.: Paleoseismology, International Geophysics Series, Academic Press, 2nd Edn., 613 pp., 2009

24. Musumeci, C., Patane, D., Scarfi', L., and Gresta, S.: Stress directions and shear-wave anisotropy: observations from local earthquakes in southeastern Sicily, Italy, B. Seismol. Soc. Am., 95, 1359–1374, doi:10.1785/0120040108, 2005.

25. Okada, Y.: Internal deformation due to shear and tensile faults in a half-space, B. Seismol. Soc. Am., 82, 1018–1040, 1992.

26. Okal, E. A. and Synolakis, C. E.: Source discriminants for near- field tsunamis, Geophys. J. Int., 158, 899–912, 2004.

27. Papaioannou, Ch. A., Papazachos, C. B., and Scordilis, E. M.: A catalogue of earthquakes in Greece and surrounding area for the period 550BC-1999, Publ. Geophys. Laboratory, University of Thessaloniki, 1, 333 pp., 2000.

28. Papazachos, B. C., Comninakis, P. E., Scordilis, E. M., Karakaisis, G. F., and Papazachos, C. B.: A catalogue of earthquakes in the Mediterranean and surrounding area for the period 1901–2007, Publ. Geophys. Laboratory, University of Thessaloniki, 2007.

29. Piatanesi, A. and Tinti, S.: A revision of the 1693 eastern Sicily earthquake and tsunami, J. Geophys. Res., 103, 2749–2758, 1998.

30. Pino, N. A., Piatanesi, A., Valensise, G., and Boschi, E.: The 28 December 1908 Messina Straits Earthquake (Mw 7.1): A Great Earthquake throughout a Century of Seismology, Seismol. Res. Lett., 80, 243–259, 2009.

31. Scandone, P., Patacca, E., Radoicic, R., Ryan, W. B., Cita, M. B., Rawson, M., Chezar, H., Miller, E., Mackenzie,

J., and Rossi, S.: Mesozoic and Cenozoic rocks from Malta Escarpment (central Mediterranean), AAPG Bull., 65, 1299–1319, 1981.

32. Sirovich, L. and Pettenati, F.: Seismotectonic outline of SouthEastern Sicily: an evaluation of available options for the scenario earthquake fault rupture, J. Seismol., 3, 213–233, 1999.

33. Sirovich, L. and Pettenati, F.: Test of source-parameter inversion of the intensities of a 54,000-death shock of the seventeenth century in southeast Sicily, B. Seismol. Soc. Am., 91, 792–811, 2001.

34. Tinti, S. and Armigliato, A.: The use of scenarios to evaluate the tsunami impact in southern Italy, Mar. Geol., 199, 221–243, 2003.

35. Tinti, S., Armigliato, A., and Bortolucci, E.: Contribution of tsunami data analysis to constrain the seismic source: the case of the 1693 eastern Sicily earthquake, J. Seismol., 5, 41–61, 2001a.

36. Tinti, S., Maramai, A., and Graziani, L.: A new version of the European tsunami catalogue: updating and revision, Nat. Hazards Earth Syst. Sci., 1, 255–262, doi:10.5194/nhess-1-255-2001, 2001b.

37. Tinti, S., Maramai, A., and Graziani L.: The new catalogue of Italian tsunamis, Nat. Hazards, 33, 439–465, 2004. Tinti, S., Pagnoni, G., and Zaniboni, F.: The landslides and tsunamis of the 30th of December 2002 in Stromboli analysed through numerical simulations, B. Volcanol., 68, 462–479, 2006.

38. Tonini, R., Armigliato, A., Pagnoni, G., Zaniboni, F., and Tinti, S.: Tsunami hazard for the city of Catania, eastern Sicily, Italy, assessed by means of Worst-case Credible Tsunami Scenario Analysis (WCTSA), Nat. Hazards Earth Syst. Sci., 11, 1217–1232, doi:10.5194/nhess-11-1217-2011, 2011.

39. Valensise, G. and Pantosti, D.: A 125 Kyr-long geological record of seismic source repeatability: the Messina Straits (southern Italy) and the 1908 earthquake (Ms 7 1/2), Terra Nova, 4, 472–483, 1992.

40. Wallace, R. E.: Profiles and ages of young fault scarps, north-central Nevada, Bull. Geol. Soc. Am., 88, 1267–1281, 1977.

INDEX